1622302-1
10-22-04

SPRINGER HANDBOOK OF
AUDITORY RESEARCH

Series Editors: Richard R. Fay and Arthur N. Popper

Springer
New York
Berlin
Heidelberg
Hong Kong
London
Milan
Paris
Tokyo

SPRINGER HANDBOOK OF AUDITORY RESEARCH

Volume 1: The Mammalian Auditory Pathway: Neuroanatomy
Edited by Douglas B. Webster, Arthur N. Popper, and Richard R. Fay

Volume 2: The Mammalian Auditory Pathway: Neurophysiology
Edited by Arthur N. Popper and Richard R. Fay

Volume 3: Human Psychophysics
Edited by William Yost, Arthur N. Popper, and Richard R. Fay

Volume 4: Comparative Hearing: Mammals
Edited by Richard R. Fay and Arthur N. Popper

Volume 5: Hearing by Bats
Edited by Arthur N. Popper and Richard R. Fay

Volume 6: Auditory Computation
Edited by Harold L. Hawkins, Teresa A. McMullen, Arthur N. Popper, and Richard R. Fay

Volume 7: Clinical Aspects of Hearing
Edited by Thomas R. Van De Water, Arthur N. Popper, and Richard R. Fay

Volume 8: The Cochlea
Edited by Peter Dallos, Arthur N. Popper, and Richard R. Fay

Volume 9: Development of the Auditory System
Edited by Edwin W Rubel, Arthur N. Popper, and Richard R. Fay

Volume 10: Comparative Hearing: Insects
Edited by Ronald Hoy, Arthur N. Popper, and Richard R. Fay

Volume 11: Comparative Hearing: Fish and Amphibians
Edited by Richard R. Fay and Arthur N. Popper

Volume 12: Hearing by Whales and Dolphins
Edited by Whitlow W.L. Au, Arthur N. Popper, and Richard R. Fay

Volume 13: Comparative Hearing: Birds and Reptiles
Edited by Robert Dooling, Arthur N. Popper, and Richard R. Fay

Volume 14: Genetics and Auditory Disorders
Edited by Bronya J.B. Keats, Arthur N. Popper, and Richard R. Fay

Volume 15: Integrative Functions in the Mammalian Auditory Pathway
Edited by Donata Oertel, Richard R. Fay, and Arthur N. Popper

Volume 16: Acoustic Communication
Edited by Andrea Simmons, Arthur N. Popper, and Richard R. Fay

Volume 17: Compression: From Cochlea to Cochlear Implants
Edited by Sid P. Bacon, Richard R. Fay, and Arthur N. Popper

Volume 18: Speech Processing in the Auditory System
Edited by Steven Greenberg, William Ainsworth, Arthur N. Popper, and Richard R. Fay

Volume 19: The Vestibular System
Edited by Stephen M. Highstein, Richard R. Fay, and Arthur N. Popper

Volume 20: Cochlear Implants: Auditory Prostheses and Electric Hearing
Edited by Fan-Gang Zeng, Arthur N. Popper, and Richard R. Fay

Volume 21: Plasticity of the Auditory System
Edited by Thomas N. Parks, Edwin W Rubel, and Richard N. Popper

Continued after index

Geoffrey A. Manley
Arthur N. Popper
Richard R. Fay
Editors

Evolution of the Vertebrate Auditory System

With 101 illustrations

Springer

Geoffrey A. Manley
Lehrstuhl für Zoologie
Technische Universitaet Muenchen
85747 Garching, Germany
geoffrey.manley@bio.tum.de

Arthur N. Popper
Department of Biology
University of Maryland
College Park, MD 20742-4415, USA
apopper@umd.edu

Richard R. Fay
Department of Psychology and Parmly
　Hearing Institute
Loyola University of Chicago
Chicago, IL 60626, USA
rfay@wpo.it.luc.edu

Series Editors: Richard R. Fay and Arthur N. Popper

Cover illustration: Modified after Manley 2000c (see Chapter 1, Fig. 5), © 2000 Wiley-VCH Verlag GmbH and used with permission.

QL
948
.E96x
2004

ISBN 0-387-21089-X (HA)　　ISBN 0-387-21093-8 (SA)　　　　Printed on acid-free paper.

© 2004 Springer-Verlag New York, LLC
All rights reserved. This work may not be translated or copied in whole or in part without the written permission of the publisher (Springer-Verlag New York, LLC, 175 Fifth Avenue, New York, NY 10010, USA), except for brief excerpts in connection with reviews or scholarly analysis. Use in connection with any form of information storage and retrieval, electronic adaptation, computer software, or by similar or dissimlar methodology now known or hereafter developed is forbidden.
The use in this publication of trade names, trademarks, service marks, and similar terms, even if they are not identified as such, is not to be taken as an expression of opinion as to whether or not they are subject to proprietary rights.

Printed in the United States of America.　　(MVY)

9 8 7 6 5 4 3 2 1　　　SPIN 10940518 (HA)　　SPIN 10940525 (SA)

www.springer-ny.com

Springer-Verlag is a part of *Springer Science+Business Media*

springeronline.com

This book is dedicated to our families.

GM: Many thanks to Christine for putting up with all the work involved in preparing this book for publication.

ANP: This volume is dedicated to the ladies in my life—my wife Helen, my daughters Michelle and Melissa, and my late mother Evelyn. Helen and our girls have provided infinite love, support, guidance, tolerance, and understanding during my career. My mother, a consummate teacher, was the ultimate role model for an aspiring professor.

RRF: I dedicate this volume to my wife Catherine who has imagined and made possible our life, and the lives of our two children, Christian and Amanda, in spite of my academic career.

Series Preface

The *Springer Handbook of Auditory Research* presents a series of comprehensive and synthetic reviews of the fundamental topics in modern auditory research. The volumes are aimed at all individuals with interests in hearing research including advanced graduate students, postdoctoral researchers, and clinical investigators. The volumes are intended to introduce new investigators to important aspects of hearing science and to help established investigators to better understand the fundamental theories and data in fields of hearing that they may not normally follow closely.

Each volume presents a particular topic comprehensively, and each serves as a synthetic overview and guide to the literature. As such, the chapters present neither exhaustive data reviews nor original research that has not yet appeared in peer-reviewed journals. The volumes focus on topics that have developed a solid data and conceptual foundation rather than on those for which a literature is only beginning to develop. New research areas will be covered on a timely basis in the series as they begin to mature.

Each volume in the series consists of a few substantial chapters on a particular topic. In some cases, the topics will be ones of traditional interest for which there is a substantial body of data and theory, such as auditory neuroanatomy (Vol. 1) and neurophysiology (Vol. 2). Other volumes in the series deal with topics that have begun to mature more recently, such as development, plasticity, and computational models of neural processing. In many cases, the series editors are joined by a co-editor having special expertise in the topic of the volume.

RICHARD R. FAY, Chicago, Illinois
ARTHUR N. POPPER, College Park, Maryland

Volume Preface

One of the defining characteristics of vertebrate organisms is that they carry three paired, major sense organs in the head, the nasal epithelia, the eyes, and the ears. These sense organs of the head region are protected by being partially or almost completely enclosed within elements of the skull. This enclosure means that even in fossil vertebrates, some information concerning the size and structure of the major sense organs can be obtained, even though soft tissues are very rarely preserved during fossilization. During the evolution of vertebrates, these sense organs underwent changes associated with changes in lifestyle (for example, the transition from water to land living) but also with developments that led to improvements in the performance of the organs as interfaces to the outside world. This book is concerned with the evolution of one of these organs, the ear, and, to a lesser extent, the brain pathways associated with the processing of the auditory information supplied by the ears.

In Chapter 1, Manley and Clack provide a framework for understanding the evolutionary relationships between the various lineages of vertebrates, without which it is impossible to understand the changes that have occurred over evolutionary time. In spite of current disagreements regarding the exact status and relationships and even the nomenclature of some vertebrate groups, this systematic framework is adhered to throughout the book, to avoid confusion. Manley and Clack also provide an overview of the subsequent chapters, to help the reader better integrate the information provided for each of the major groups of vertebrates. Here, the emphasis is on the amniotes, the reptiles, birds, and mammals, since these lineages show not only the greatest variety of structural arrangements but also the largest and most complex inner ears.

When biologists speak of the "design" of a structure in an organism, they are referring to the result of selection pressures, sometimes over very long periods of time, that have shaped the morphology and the function of that structure. These selection pressures are, in the special case of sense organs, hugely influenced by the physical and chemical nature of the stimuli to be perceived. In the present case, the stimuli are sound-pressure waves in the media water and/or air. As is generally the case for sensory organs, selection pressures existed for attaining very high sensitivities, in other words for the detection of extremely small signal strengths even in the presence of noise and competing signals from various sources.

In Chapter 2, Lewis and Fay discuss some of the important ecological and physical factors that would have been omnipresent throughout the history of vertebrate evolution and must have had an impact on the evolution of vertebrate hearing. They conclude that in addition to having acoustic transducers, the sine qua non for any auditory system to function in such a world is the ability to isolate signals from their individual sources. Furthermore, this quintessential task must be achieved largely by neural computation, and this requirement establishes essential constraints on the acoustic transducers, on the signal-processing structures associated with them, and on the brain itself. In the appendix at the end of the book, Lewis provides a physical-mathematical framework for understanding the nature of the stimulus and the resulting constraints that operated during ear evolution. These constraints were responsible for molding the form and function of vertebrate inner ears to those we see today.

The sensory cells that subserve the function of hearing are among the oldest of cell types, with possible forerunners among chordate ancestors and perhaps even other invertebrate organisms. Thus "hair" cells not only are found in the inner ear, but also are the typical sensory cell of fish and amphibian lateral-line organs. In fact, they are the primary sensory cells used by vertebrates for the detection of mechanical signals from external sources. Since mechanoreceptive organs can have their stimuli filtered by accessory structures and be required to provide information on different aspects of the stimuli (acceleration, velocity, displacement), it is not surprising to find that selection pressures have greatly influenced the characteristics of such accessory structures. In parallel, however, we find modifications to hair cell structure and function that are large enough to justify the recognition of different types of hair cells. In Chapter 3, Coffin et al. describe the variety of hair cell types in different vertebrates and provide us with a framework for classifying them and understanding the functional background for hair cell specialization.

It is clear that the sense of hearing in vertebrates arose well before the origin of sensory organs especially dedicated to that sense. In fact, the rich acoustic world of fishes is testimony to the fact that fishes, which form by far the largest group among vertebrates (more than 50% of species), have created a very useful hearing system by utilizing preexisting sensory epithelia of the vestibular part of the inner ear. Since these animals not only live in water but also have a body largely made up of water, acoustic signals from the environment have no difficulty in penetrating their tissues and are thus not attenuated at the water-to-body interface. The only requirement for them to be able to detect acoustic stimuli is that relative motion at the sensory epithelium be produced, and this is only possible at an impedance transition such as from water to a dense inner-ear otolith or at the boundary between tissue and the fish's air bladder. It is thus evident that in fishes, sound detection is based on the use of sensory epithelia that are also used to detect linear acceleration, and there must be some kind of compromise in the organs' sensitivities to these different stimuli. In Chapter 4,

Ladich and Popper describe how fishes achieved this compromise, which organs are involved, and the constraints imposed by the requirement to retain vestibular responses.

Together with comparative studies of form and function in a variety of modern vertebrate groups, the study of fossil organisms is the most direct source of information as to which changes occurred when over evolutionary time. The transition to land during the development of the first land vertebrates (the labyrinthodont amphibians) greatly changed the selection pressures that operated on the various major sense organs. The nasal epithelium was moistened and could continue to operate with few changes. The eye required little more than a change in the optical properties of the lens to accommodate the fact that the air–corneal interface provided much more refraction of light than a water–corneal interface. Similarly, the largest acoustic problem for the inner ear was to overcome the huge impedance mismatch between air and body tissues. Or at least, that was the thinking that influenced the interpretation of the fossil story for many years.

In Chapter 5, Clack and Allin have gathered together a wealth of data from vertebrate fossils, providing a detailed interpretation as to what can be read out concerning the status of the inner ear and especially any special bony connections between it and the outside world. These authors indicate that tympanic middle ears as we know them were not characteristic of the earliest land vertebrates, but in fact arose quite late in their evolution, many millions of years after the origin of the amphibians. This implies not only that important initiating impulses were missing, but also that the early amphibians and, indeed, early reptiles were well served by the inner ear without a tympanic middle ear. Certainly the early land vertebrates had large heads, short necks, and short limbs, implying that in fact the head was in close or often direct contact with the ground and that sound and vibrational stimuli from the substrate thus had no problem influencing the inner ear. Clack and Allin demonstrate that tympanic middle ears have arisen a number of times independently in different lineages, a crucial finding that makes it possible to understand the great variety of inner-ear structures and the parallel developments that they show.

Before tympanic middle ears arose, the land vertebrates had already diversified into a number of different lineages. Existing alongside the amphibians, several lineages had arisen from the stem reptiles, the lepidosaurs, the archosaurs, and the mammal-like reptiles that later gave rise to the mammals. The lack of a tympanic middle ear as we now know it probably meant that during those periods of time, selection pressures for the development of larger and more complex inner-ear hearing epithelia were missing or weak. It is apparent that, while the amniote descendants of the stem reptiles all inherited a rather simple hearing papilla largely supported by the basilar membrane, the ancestors of modern amphibians developed two separate epithelia. As Smotherman and Narins describe in Chapter 6, it is possible, perhaps even probable, that strictly speaking, neither of these epithelia is homologous with the basilar papilla of amniotes (in spite of one of them bearing the same name). Nonetheless, as these authors also point out, all of these epithelia originate during ontogeny from one common epithelium with the vestibular sensory patches. The long history of the amphibians since the development of a tympanic middle ear was obviously accompanied by the evolution of different papillar specializations in different amphibian groups. In the anuran amphibians, these papillae are well enough developed, though small, to support sophisticated acoustic communication behaviors.

In the lepidosaurs, especially the lizards, the hearing papilla remained relatively simple and evolution led only in one group—the geckos—to acoustic communication. In the apparent absence of strong selection pressures for the maintenance of a particular size and development of the inner-ear hearing epithelium, the various lineages of the lizards demonstrate a remarkable degree of structural variety. In Chapter 7, Manley describes the lizard hearing organ and shows that in spite of this great structural variety, the function of the ear as seen at the level of the auditory nerve is much less variable. This can be attributed to a situation of neutral evolution, where in the absence of strong selection pressures, structural configurations can drift, provided that a basic degree of functional integrity is maintained. The huge variety of structural configurations seen in the lizard basilar papilla are thus not accompanied by any documented differences in behavior. Nonetheless, the structural variety does result in some functional differences that provide a window into the ear, helping to understand the function of variable components such as that of the enigmatic tectorial membrane.

Those who know the structure of the archosaur (birds and *Crocodilia*) inner ears will not find it surprising that crocodiles and their relatives are the closest living relatives of the birds. These organisms elongated and elaborated on the ancestral stem-reptile inner ear and have large numbers of hair cells in their auditory epithelia. These have been strongly specialized into two groups, where the extreme forms differ greatly in morphology and innervation pattern. As Gleich et al. describe in Chapter 8, these specializations parallel to a remarkable degree those observed in mammalian inner ears. This suggests that in this lineage, and completely independently from the mammals, two hair cell populations that are placed neurally and abneurally on the papilla developed that are, respectively, specialized for signal detection and for signal amplification. There is no doubt that this specialization is responsible for the exquisite hearing ability of these animals and for their extensive use of acoustic signals in behavior.

The treatment of the mammals has been placed at the end of the animal-group series only out of tradition. It should not be assumed that this is a series of continuously increasing quality of hearing. Indeed, in some respects, as for example in frequency selectivity, the mammalian ear has a poorer performance than many nonmammalian ears. What sets mammalian ears apart is something different. During the evolution of mammals, but of no other group, the independent development of a tympanic middle ear occurred at the same time as the improvement of the jaw joints. Mammals have a secondary jaw joint. This fortuitous coincidence meant that the mammals were able to incorporate redundant jaw-joint bones into their middle-ear ossicular chain. Much later in evolution, when mammals were able to move toward acquiring the ability to perceive higher frequencies of sounds (above, say, 10 kHz), it turned out that, quite unplanned, as it were, this kind of middle ear was much better at transmitting high frequencies than was the second-order lever system used by all the nonmammalian groups. Thus one of the features characterizing the mammalian hearing organ—the perception of very high frequencies—was not primarily a result of any specialization of the hearing epithelium, but a result of the first-order lever system of the middle ear. Hair cell specializations, such as specialized active processes, almost certainly developed later, together with a sometimes dramatic papillar elongation. As noted above, the mammalian inner ear shows hair cell specializations that are in parallel to those of birds. In mammals, the huge amount of data available on structure and function make it clear that the inner hair cells on the neural side of the papilla or organ of Corti are the signal detectors, and the outer hair cells are the signal manipulators and amplifiers. The evolutionary story behind the ears of the various mammals, some highly specialized, is told by Vater et al. in Chapter 9.

Most of the information transduced by the inner ear is transformed into neural signals that travel along afferent nerve fibers to the first neural processing station in the brain stem, the cochlear nucleus. From this nuclear complex, the neural information is passed to higher and higher centers of the auditory pathway that are often specialized to process certain aspects of the sound stimulus. During vertebrate evolution, the brain enlarged, often to a very great extent and independently in different lineages. During this enlargement, however, the basic structure of the auditory pathway remained very similar and recognizable in all lineages while specific neural centers sometimes became very large. Group-specific differences are seen, and differences in the pattern of the processing of neural information are sometimes quite large even within one group of vertebrates. In Chapter 10, Grothe et al. present an overview of the evolution of this pathway in the brain and examine the question as to whether we can detect clear evolutionary trends in the organization and function of the afferent auditory pathway.

Thanks to a huge research effort over the last 50 years or so, we are now in a position to much better understand what happened to middle- and inner-ear structure during evolutionary processes that lasted more than 400 million years. This book attempts to organize these facts into a coherent pattern and make this area of scientific knowledge accessible to those who are interested. As Manley outlines in the summary in Chapter 11, we have made huge strides in integrating large amounts of data. Nonetheless, many exciting questions, some of them very fundamental to understanding the hearing processes, are still open for the devoted investigator. The editors hope that this book will be an inspiration for young colleagues to tackle these problems in the near future.

As with most other volumes in the *Springer Handbook of Auditory Research* (SHAR) series, the readers of many chapters in this volume will benefit from reading chapters in earlier volumes. Thus, sensory hair cells were also discussed in several chapters in *The Cochlea* (volume 8) and chapters by Hoy and others in *Comparative Hearing: Insects* (volume 10). Hearing by fishes and amphibians was discussed in detail in volume 11 (*Comparative Hearing: Fish and Amphibians*), while hearing in reptiles and birds was the topic of *Comparative Hearing: Birds and Reptiles* (volume 13). Mammalian hearing has been the subject of chapters in many volumes in the SHAR series, but most notably, for the context of the current volume, in *Comparative Hearing in Mammals* (volume 4) and *Hearing by Bats* (volume 5). Finally, the structure and function of the central nervous system was first dealt with in volume 1 of this series, *The Mammalian Auditory Pathway: Neuroanatomy*, and volume 2, *The Mammalian Auditory Pathway: Neurophysiology*, and in the more recent overview in volume 15, *Integrative Functions in the Mammalian Auditory Pathway*.

> September, 2003
> GEOFFREY A. MANLEY, Garching, Germany
> ARTHUR N. POPPER, College Park, Maryland
> RICHARD R. FAY, Chicago, Illinois

Contents

Series Preface vii
Volume Preface ix
Contributors xvii

Chapter 1   An Outline of the Evolution of Vertebrate Hearing
            Organs 1
            GEOFFREY A. MANLEY AND JENNIFER A. CLACK

Chapter 2   Environmental Variables and the Fundamental
            Nature of Hearing 27
            EDWIN R. LEWIS AND RICHARD R. FAY

Chapter 3   Evolution of Sensory Hair Cells 55
            ALLISON COFFIN, MATTHEW KELLEY,
            GEOFFREY A. MANLEY, AND ARTHUR N. POPPER

Chapter 4   Parallel Evolution in Fish Hearing Organs 95
            FRIEDRICH LADICH AND ARTHUR N. POPPER

Chapter 5   The Evolution of Single- and Multiple-Ossicle Ears
            in Fishes and Tetrapods 128
            JENNIFER A. CLACK AND EDGAR ALLIN

Chapter 6   Evolution of the Amphibian Ear 164
            MICHAEL SMOTHERMAN AND PETER NARINS

Chapter 7   The Lizard Basilar Papilla and Its Evolution ... 200
            GEOFFREY A. MANLEY

Chapter 8   Hearing Organ Evolution and Specialization:
            Archosaurs 224
            OTTO GLEICH, FRANZ PETER FISCHER,
            CHRISTINE KÖPPL, AND GEOFFREY A. MANLEY

Chapter 9	Hearing Organ Evolution and Specialization: Early and Later Mammals..............................	256
	MARIANNE VATER, JIN MENG, AND RICHARD C. FOX	

Chapter 10	The Evolution of Central Pathways and Their Neural Processing Patterns........................	289
	BENEDIKT GROTHE, CATHERINE E. CARR, JOHN H. CASSEDAY, BERND FRITZSCH, AND CHRISTINE KÖPPL	

Chapter 11	Advances and Perspectives in the Study of the Evolution of the Vertebrate Auditory System	360
	GEOFFREY A. MANLEY	

Appendix: Useful Concepts from Circuit Theory 369
     EDWIN R. LEWIS
Index 401

Contributors

EDGAR ALLIN
Chicago College of Osteopathic Medicine, Midwestern University, Downers Grove, IL 60515, USA

CATHERINE E. CARR
Department of Biology, University of Maryland, College Park, MD 20742, USA

JOHN H. CASSEDAY
Department of Psychology, University of Washington, Seattle, WA 98195, USA

JENNIFER A. CLACK
University Museum of Zoology, Cambridge, CB2 3EJ, England

ALLISON COFFIN
Department of Biology, University of Maryland, College Park, MD 20742, USA

RICHARD R. FAY
Department of Psychology and Parmly Hearing Institute, Loyola University of Chicago, Chicago, IL 60626, USA

FRANZ PETER FISCHER
Lehrstuhl für Zoologie, Technische Universitaet Muenchen, 85747 Garching, Germany

RICHARD C. FOX
Laboratory for Vertebrate Paleontology, Department of Biological Sciences, University of Alberta, Edmonton, Alberta T6G 2E9, Canada

BERND FRITZSCH
Creighton University, Department of Biomedical Sciences, Omaha, Nebraska 68178, USA

OTTO GLEICH
ENT-Department, Universität Regensburg, 93042 Regensburg, Germany

BENEDIKT GROTHE
Max-Planck-Institut für Neurobiologie, D-82152 Martinsried, Germany

MATTHEW KELLEY
National Institute on Deafness and Other Communication Disorders, National Institutes of Health, Rockville, MD, 20850, USA

CHRISTINE KÖPPL
Lehrstuhl für Zoologie, Technische Universitaet Muenchen, 85747 Garching, Germany

FRIEDRICH LADICH
Institute of Zoology, University of Vienna, A-1090 Vienna, Austria

EDWIN R. LEWIS
Department of Electrical Engineering and Computer Science, University of California–Berkeley, Berkeley, CA 94720, USA

GEOFFREY A. MANLEY
Lehrstuhl für Zoologie, Technische Universitaet Muenchen, 85747 Garching, Germany

JIN MENG
Division of Paleontology, American Museum of Natural History, New York, NY 10024, USA

PETER NARINS
Department of Physiological Science, University of California–Los Angeles, Los Angeles, CA 90095, USA

ARTHUR N. POPPER
Department of Biology and Neuroscience and Cognitive Science Program, University of Maryland, College Park, MD 20742, USA

MICHAEL SMOTHERMAN
Department of Physiological Science, University of California–Los Angeles, Los Angeles, CA 90095, USA

MARIANNE VATER
Department of Biochemistry and Biology, University of Potsdam, 14471 Potsdam, Germany

1
An Outline of the Evolution of Vertebrate Hearing Organs

GEOFFREY A. MANLEY AND JENNIFER A. CLACK

The aim of this introductory chapter is twofold. First, we provide a phylogenetic framework to enable the reader to place each animal group being discussed in the correct historical context. Second, some major background themes concerning the evolution of the structure and the function of vertebrate hearing organs are briefly introduced. With this background information, the reader should be better able to fit the themes being discussed into their relevant place in the context of vertebrates as a whole.

1. The Phylogenetic Framework

No evolutionary analysis can take place without the underlying foundation of a well-supported phylogeny. To reconstruct the pathways of evolution of any suite of structures or behaviors, such as the evolution of the ear and hearing, we need to assess the phylogenetic or evolutionary relationships of the animals before we draw inferences about evolutionary sequences. Otherwise, our conclusions about how, when, or why characters were acquired or events occurred must remain purely speculative scenarios. If possible, the phylogeny that is used should be constructed independently of the characters that are being studied, to avoid the risk of circular reasoning.

Recently, cladistics has been the method favored for reconstructing phylogeny. Briefly, this method groups organisms according to characters that they share uniquely, called derived characters or apomorphies, rather than by primitive and more generally distributed ones, called plesiomorphies (for more details of cladistic methodology, see Kitching et al. 1998). One of the consequences of using cladistic methods that some people find counterintuitive is the way the resulting groups are defined and named. Evolutionary phylogeny, used almost universally until about 20 years ago, allowed groups to be defined by primitive characters, and the result was that the groups so defined often reflected "lay" or "folk" classifications. The cladistic or phylogenetic method, by contrast, aims to have its resulting classification reflect relationships rather than simple physical appearance. Groups that evolved from a common ancestor (monophyletic or "natural" groups) are placed into a series of more and more inclusive nested sets. Monophyletic groups include all the descendants of a single common ancestor with no group arbitrarily excluded. (If a group does not include all the descendants of a common ancestor, but omits some, the resulting group is known as "paraphyletic." An example of a paraphyletic group would be "reptiles" if birds—descendants of dinosaurs—are omitted from it, as discussed below.) Mammals, along with birds, crocodiles, lizards, snakes, and turtles, are all amniotes, defined by characters of the amniotic egg, and the living groups constitute the Crown Group. The results of the use of such methods can be seen in Figure 1.1 and Table 1.1, which give alternative ways of expressing hierarchical relationships among vertebrates. Figure 1.1 sets out a consensus scheme of relationships of vertebrates, and Figure 1.2 sets out a timetable showing the times of appearance of major groups.

To give another example of cladistic analysis, in phylogenetic classification, tetrapods belong within the Sarcopterygia (lobe-fins). They belong with other forms, such as the lungfishes and coelacanths as well as a series of extinct forms, within the lobe-fins. Because "sarcopterygian" is usually translated as "lobe-finned fish," it is sometimes a problem to appreciate that tetrapods really belong here. "Sarcopterygian" is better understood as "fleshy-limbed vertebrate." Similarly, Sarcopterygia, along with their sister group the Actinopterygia, comprise the Osteichthyes. Even though the word *osteichthyan* is sometimes translated as "bony fish," it is actually a more inclusive term that includes all bony vertebrates. Thus humans, as well as being mammals, tetrapods, and sarcopterygians, are also osteichthyans. "Fish" is another term that is meaningless phylogenetically, as it emphasizes appearance at the expense of relationship. More recently, the term *osteognathostome* has been coined to encompass the bony jawed vertebrates.

The Sarcopterygia includes the tetrapods as their main extant representatives, the other survivors being the three genera of lungfish and the two species of coelacanth, of which *Latimeria chalumnae* is the better known. Numerous fossil sarcopterygians are known, mainly from the late Paleozoic era, including a stem-tetrapod lineage known as "osteolepiforms." Older literature refers to "osteolepiforms" as members of the "Rhipidistia" or "Crossopterygia." However, both of these groups are now recognized to be paraphyletic in their original senses, and the names are no longer used in the same way. "Crossopterygia" has fallen into disuse. The extant tetrapods form the crown group of a lineage of tetrapod-like fish and fish-like tetrapods, collectively known as Tetrapodomorpha. (Confusingly, "Rhipidistia" has recently been used to mean tetrapodomorphs plus lungfishes and their immediate relatives, originally specifically excluded from "Rhipidistia.") All of the tetrapodomorphs, excluding tetrapods themselves, are extinct. Tetrapods radiated into many lineages during the Paleozoic era, of which only two survive, the amphibians and amniotes. These and their immediate relatives constitute Crown Group Tetrapoda.

It is now believed that the amphibian and amniote lineages have been separate since at least the early Carboniferous, with the first true amniotes found in the

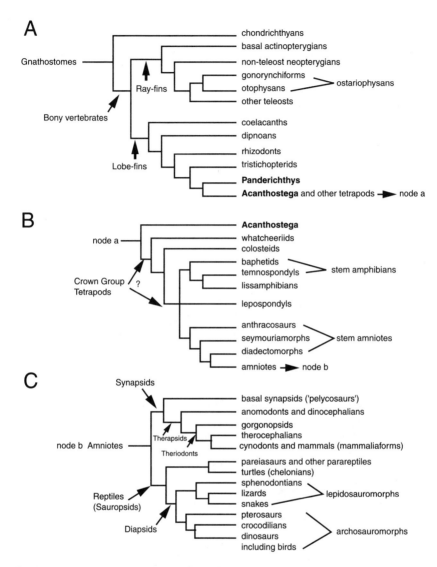

FIGURE 1.1. Consensus cladogram of vertebrates. The group names of the major nodes are in bold face.

late Carboniferous. Modern amphibians (Lissamphibia) are not known before the late Triassic, when the earliest anurans, urodeles, and caecilians appear. The assemblage formerly known as "labyrinthodonts" encompasses early members of both amphibian and amniote lineages, artificially uniting early members of these two disparate groups. The term and the concept it represents are no longer considered to be valid. There is a reasonable consensus that amniotes fall into

TABLE 1.1. Hierarchical arrangement of vertebrate taxa.

- Vertebrates
  - Hagfishes (these are sometimes excluded from vertebrates, sometimes included within vertebrates)
  - Lampreys
  - Other jawless vertebrates from the fossil record, e.g., cephalaspids
  - **Gnathostomes (jawed vertebrates)**
    - Placoderms
    - Acanthodians
    - **Chondrichthyans (cartilaginous vertebrates)**
      - Elasmobranchs
        - fossil forms
        - Selachians
          - Squaloids (sharks)
          - Batoids (skates and rays)
      - Holocephalans (rat-fishes, rabbit-fishes, chimaeroids)
    - **Osteichthyans (bony vertebrates)**
      - **Actinopterygians (ray-finned fishes)**
        - *Polypterus*
        - *Mimia* and relatives, early actinopts
        - Chondrosteans (e.g., sturgeon, paddlefish)
        - *Amia, Lepisosteus* and relatives
        - Teleosts (salmon, herring, sea horses, and almost all other fish)
      - **Sarcopterygians (lobe-finned vertebrates)**
        - Actinistians (coelacanths) ⎫ sometimes you will see these lumped together as
        - Dipnoans (lungfishes)   ⎭ "crossopterygians"
        - Tetrapodomorphs
          - Osteolepiforms   ⎫
          - *Eusthenopteron*   ⎬ sometimes you will see these called "rhipidistians," when lumped
          - *Panderichthys*   ⎭ with some extinct lungfish relatives
          - **Tetrapods**
            - stem tetrapods such as *Acanthostega* and *Ichthyostega*
            - **Crown-group tetrapods**
              - **Amphibians (and stem amphibians)**
                - Anurans (frogs, toads)
                - Urodeles (newts, salamanders)
                - Caecilians (apodans)
              - **Amniotes (and stem amniotes)**
                - stem amniotes
                - mammals (synapsids)
                - turtles (parareptiles, sometimes called "anapsids")
                - diapsids
                  - lepidosaurs
                    - *Sphenodon*
                    - lizards
                    - snakes
                  - archosaurs
                    - crocodiles
                    - pterosaurs
                    - dinosaurs
                    - birds

In the table, bold face names are higher (more inclusive) taxa.

1. Evolution of Vertebrate Hearing Organs 5

Millions of years ago			Event
Recent	1		
Cainozoic	65		● major radiation of mammals
			Major extinction event
Mesozoic	144	CRETACEOUS	● last dinosaurs
	205	JURASSIC	● earliest birds, lizards
			● earliest frogs, salamanders
	248	TRIASSIC	● earliest crocodile ancestors, dinosaurs, turtles, mammals
			Major extinction event
Paleozoic	295	PERMIAN	● first major diversification of amniotes
	325	LATE (Pennsylvanian) CARBONIFEROUS	● earliest true amniotes
			● major diversification of tetrapods
	355	EARLY (Mississippian)	● first terrestrial tetrapods
	418	DEVONIAN	● **Acanthostega**
			● **Panderichthys, Eusthenopteron**
			● earliest tetrapodomorphs
	443	SILURIAN	earliest sarcopterygians and actinopterygians
	495	ORDOVICIAN	
	545	CAMBRIAN	

FIGURE 1.2. Geological timetable showing time of origin of major groups. Names of genera are in bold face.

two divergent groups, with the Synapsida including mammals representing a separate radiation from that including all other amniotes. Diapsida, including crocodiles and birds (Archosauria), lizards and snakes (Lepidosauria), plus turtles (Chelonia), form a major radiation known either as Sauropsida or by the more familiar term *Reptilia*. However, the latter traditionally has excluded birds. Now that birds are recognized as descendants of dinosaurs, they should rightly be included here, and the term *Reptilia* is preferred by some authorities because it is the older. Within the mammalian lineage, there is some debate about the relationships of the three extant groups—monotremes, marsupials, and placentals—to each other and to fossil groups, with resulting differences of nomenclature. Readers will notice differences, for example, between the mammalian cladogram in Chapter 5 (Clack and Allin) and that in Chapter 9 (Vater et al.). However, these debates are representative of those within systematic biology, and so long as explicit sources or reasons are given, they should cause no undue concern.

The Actinopterygia includes a large number of extinct forms, some of which have a very few living representatives, while the vast majority belong within the Teleostei. The actinopterygians comprise the most numerous of vertebrate groups and includes most fishes alive today. Fossil actinopterygians first appeared in the Devonian, but diversified from the late Carboniferous onward, leaving an extensive fossil record. A few nonteleost fishes survive today, all fresh-water and mostly air-breathing forms. These include the chondrosteans (sturgeon and paddlefish), bowfin, and gar. The relationships of the latter two to teleosts and the interrelationships of actinopterygians generally are hotly debated, but one thing is agreed. The old terms *palaeoniscid* and *holostean* are paraphyletic grade groups with no phylogenetic coherence. Teleosts originated in the late Triassic, followed by an explosive radiation in the late Jurassic and Cretaceous. Among teleosts, the hearing specialists include a large group of mainly freshwater teleosts called the Ostariophysi, characterized by a unique complex of specializations of the inner ear, swim bladder, and vertebral column. These are not found in the most derived group of teleosts, the Neoteleostei, though some neoteleosts have developed other specialized features for hearing. For more details of fish and tetrapod anatomy and relationships, see Carroll (1987), Janvier (1996), or a more general vertebrate biology text book such as Pough et al. (2002).

2. The Early History of Hair Cells and Ears

In common with other sensory organs, the evolution of vertebrate hair cells was strongly influenced by the environment in which the cells are placed. Their ability to detect signals, separate them from other stimuli, and provide information that allows the brain to identify the signal developed in the context of a sensory cell surrounded by fluids and tissues and to a good degree isolated from direct contact with the outside world. The theoretical background to the signal-processing requirements and the interaction between environment and signal-detecting cells is discussed by Lewis and Fay in Chapter 2. Lewis's appendix at the end of the book discusses the physics of sound and the biophysics of acoustic sensors in animals.

The earliest intensive studies of hair cell structure and physiology were carried out on the lateral line of fish and the vestibular apparatus of amniotes (lepidosaurs, archosaurs, and mammals) (for a discussion see Coffin et al., Chapter 3). These early studies were concerned not with audition, but with the detection of disturbances in water and the organs responding to angular and linear acceleration of the head.

Whereas in early descriptions of fish and amphibian lateral line organs only one basic hair cell type was described, with little variation, vestibular hair cells of the land vertebrates were described as showing morphological and physiological heterogeneity. Tetrapod vestibular epithelia (canal cristae, saccule, utricle, and lagena in all but therian mammals) possess two distinct hair cell types, called type I and type II. Type II hair cells, considered to be phylogenetically older, have cylindrical cell bodies that are contacted by small afferent and efferent terminals (see Coffin et al., Chapter 3). Type I cells have flask-shaped somata and a thicker hair bundle and their basal region is surrounded by an afferent nerve calyx to which efferent fibers make contact.

A diversity in fish hair cells was established when cell-biological studies identified molecular markers that differentiate between two distinct populations of hair cells (for a discussion see Coffin et al., Chapter 3; Ladich and Popper, Chapter 4). Fish vestibular hair cells were then also found to differ in their innervation patterns. The distribution of these different types of hair cells does not seem to be related in any way to the fact that in fishes the hair cells of a few vestibular organs respond quite sensitively to sound. There are, however, too few studies to conclude that there are no obvious hair cell specializations in fish that relate to the evolutionary development of hearing.

But what about the transition to land? Was that transition accompanied by specializations in hair cells for terrestrial hearing? One of the closest living organisms to the ancestors of terrestrial vertebrates is *Latimeria*, the coelacanth. Fritzsch (1987) described a sensory papilla in *Latimeria* that resembles the amniote basilar papilla in structure, position, and innervation pattern. Whether or not this papilla is homologous to either the amphibian or the amniote basilar papilla is uncertain. It is possible that this papilla and those of both major lineages of terrestrial vertebrates were all developed independently.

The ear of modern Amphibia, the Lissamphibia, generally contains two sensory organs subserving audition—the basilar and amphibian papillae. Neither of these organs is clearly homologous to the basilar papilla of amniotes, but this is still a matter for discussion (see Smotherman and Narins, Chapter 6), and at present the relationships between the two papillae of amphibians and the single papilla of amniotes are uncertain. At present it can be argued that modern amphibian and amniote papillae are different because of separate origins, i.e., each may have evolved from an aquatic or even a semiaquatic early tetrapod.

Whereas the basilar papilla of amphibians appears to be a simple organ with little or no frequency differentiation between hair cells, the amphibian papilla is larger and, in contrast to probable forerunner papillae of the vestibular systems of fish, is tonotopically organized (see Smotherman and Narins, Chapter 6). In the evolution of the different lineages of the amphibia, different epithelial components were manipulated independently within each sense organ, while the organs themselves were moved about within the inner ear. The amphibian ear appears to have been a very pliable neural substrate that was capable of adapting in concert with each species' unique behavioral traits.

The inner ear of amniotes contains a highly sensitive and frequency selective mechanoreceptive sense organ that is connected by afferent and efferent nerve fibers to specialized brain centers. Like all large sense organs of vertebrates, the hearing organs of amniotes are the result of long evolutionary processes that in fact began rather late in evolution. Indeed, the hearing portion of the inner ear of amniotes was the last of the paired sense organs to arise and to undergo major formative evolution. This chapter outlines the issues involved and the major evolutionary events involving the ears of different amniote lineages. Detailed information for each lineage is provided in subsequent chapters.

3. The Origin of the Basilar Papilla of Amniotes

The origin of the auditory organ of amniotes, the Papilla basilaris, in mammals also called the organ of Corti, is not yet clear. This new epithelium developed between the saccular and lagenar maculae, two sensory epithelia that arise in ontogeny from the same group of cells. In the course of the separation of the lagenar hair cell precursor epithelium during ontogeny, a third, initially small, group of cells separated off between the saccular and lagenar maculae (Fig. 1.3). Like the lagenar macula, an otolith-covered epithelium, it formed part of the wall of a finger-shaped extension of the endolymphatic space called the cochlear duct. This third sensory epithelium is called the Papilla basilaris or basilar papilla. It was the first hair cell epithelium that lay wholly or partly over a freely moving membrane, the basilar membrane, that separated the endolymphatic and perilymphatic spaces of the inner ear. During the evolution of the descendants of the Paleozoic stem reptiles, the modern amniotes, this epithelium evolved over a period of more than 300 million years into the different kinds of basilar papillae of archosaurs and lepidosaurs and into the organs of Corti of mammals (Manley and Köppl 1998).

The amniotes show a fascinating structural variation in their hearing organs. Since the soft parts of the sense organs do not fossilize, however, an investigation of the origins of this structural variation is virtually restricted to using comparative studies of modern species. The questions to be posed of such comparative studies are, for example: What was the probable structure of the "original" hearing organ of the stem reptiles? Was the evolution of the hearing organ different in different groups of amniotes? When did the unique structural configurations in the various amniote groups arise?

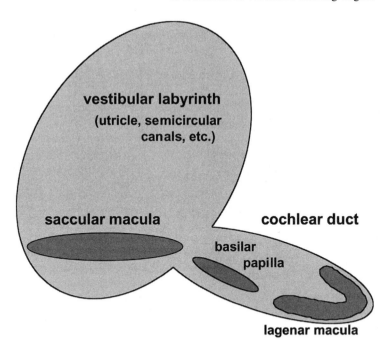

FIGURE 1.3. Highly schematic diagram of the saccular area of the vestibular labyrinth in a stem reptile. During the evolution of the amniotes, the lagenar recess becomes longer and incorporates on its wall a new hair cell epithelium, the basilar papilla. This papilla lies between the maculae of the sacculus and lagena and derives from the same group of hair cells during ontogeny. The basilar papilla lies on a thin membrane that forms part of the wall of the cochlear duct, separating it from the perilymphatic spaces.

4. Structural Evolution of the Basilar Papilla in the Main Amniote Lineages

The evolution of the amniotes can be traced back to common ancestors in middle Paleozoic times, specifically during the Devonian. From these early stem reptiles, many lines of development arose, some of which still exist today (see Fig. 1.4, from which those groups that went extinct, such as the plesiosaurs, pterosaurs, and "dinosaurs," have been left off).

4.1 The Primitive Condition

In some lineages (Fig. 1.4), specifically in the turtles and the Tuatara, the evidence from comparative studies of extant organisms suggests that the hearing organ has not changed a great deal since Paleozoic times. The sensory epithelium shown by turtles and the Tuatara can thus be regarded as being of a basal type (Wever 1978; Miller 1980) and thus reflecting the condition present in the

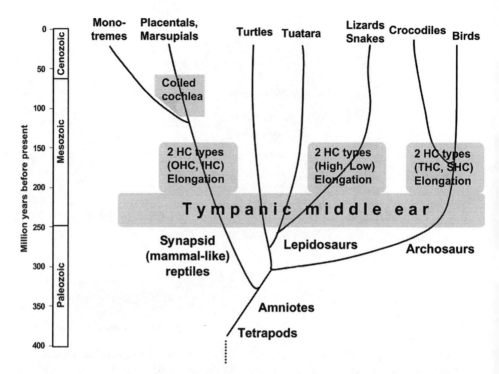

FIGURE 1.4. A family tree of the amniote vertebrates emphasizing the major events that occurred in the evolution of amniote hearing organs. Primitive papillae, probably not too different from that of the stem reptiles, are found in turtles and the Tuatara. During the Jurassic period, both the mammals and the archosaurs developed two specialized groups of hair cells arranged across the basilar papilla. These are called the inner (IHC) and outer hair cells (OHC) of mammals, and the tall (THC) and short hair cells (SHC) of birds. Following the divergence of the monotremes from the other mammals during the Cretaceous period, the latter reduced the number of IHC and OHC rows and also coiled the papilla into a cochlea. During the Cretaceous period also, the lizard papilla diverged in the evolving families, evolving two hair cell types arranged in this case along the length of the papilla and differentiated according to their frequency responses. (Modified from Manley 2000b.)

stem reptiles. In both turtles and the Tuatara, the tympanic middle ear (which is likely to have been an independent development, see Clack and Allin, Chapter 5) is less efficient. Turtles, for example, are substantially less sensitive than are almost all lepidosaurs, and this may have reduced selection pressures for alteration of the basal-type basilar papilla. An attempt can thus be made to circumscribe plesiomorphic and (syn)apomorphic features on the basis of this papillar type. The characteristics of such primitive papillae can be summarized as follows (see also Miller 1992). The papilla is small to medium in size (i.e., smaller than about 1 mm) and contains between a few hundred and about one thousand loosely packed hair cells that are contacted by both afferent and efferent nerve fibers. All the hair cells are covered by a well-developed tectorial membrane, and their stereovillar bundles are all abneurally oriented. Extension of the hair cell region beyond the free basilar membrane and on to the surface of the limbic support at both ends is typical for turtles and the Tuatara *Sphenodon* (Wever 1978).

From this plesiomorphic condition, each of the main lines of amniotes, shown in Figure 1.4 as lepidosaurs, archosaurs, and mammal-like reptiles, independently specialized their auditory papillae, each showing unique apomorphies. One of the main questions underlying this development is: What evolutionary event, or what selective pressures, led to this? The clue seems to lie in the fact that approximately during the Triassic (220 to 250 million years ago, Fig. 1.4), each of the groups independently developed a tympanic ear (Clack and Allin, Chapter 5). The term *tympanic* emphasizes developments in the region of the middle ear (which can be observed in fossil remains) that entailed the development of an eardrum or tympanum. These changes led to the development of middle-ear systems that were substantially more sensitive to airborne sounds, forming a powerful selective pressure for the expansion and specialization of the hearing organs of the inner ear.

4.2 The Diapsid Archosaur Descendants— Crocodilia and Aves (Birds)

Except for the mammals and perhaps the turtles, the living derivatives of the stem reptiles can be regarded as descendants of diapsids. Modern diapsids almost all belong to one of two great radiations that led on the one hand to the archosaurs and on the other to the lepidosaurs (Fig. 1.4). While many important groups of archosaurs (such as those included under the term *dinosaurs* and not shown on Fig. 1.4) died out at the end of the Mesozoic era, many modern species also belong to this group, a small number to the *Crocodilia* (crocodiles, alligators, etc.), but most to the birds, *Aves*. Recent archosaurs show a number of characteristic synapomorphic features, such as a basilar papilla that is wider and more elongated compared to that of stem reptiles. These basilar papillae are populated by thousands of hair cells that fall morphologically into a continuum across and along the basilar papilla, with the extreme forms arranged similarly to the situation in mammals (Gleich et al., Chapter 8). Archosaur hair cells are arranged in an irregular mosaic, rather than in clear rows, and are covered by a thick tectorial membrane. The number of hair cells in any given transect of the archosaur papilla can be large (up to about 50 at the apical end, Fig. 1.5), and they have comparatively large numbers of stereovilli in their bundles (up to 300 at the basal end in some species).

On the neural side of the archosaur papilla, and covering more of the papilla's width in the apex than in the base, are so-called tall hair cells. In *Crocodilia* there is a clear step transition from these tall hair cells to the short hair

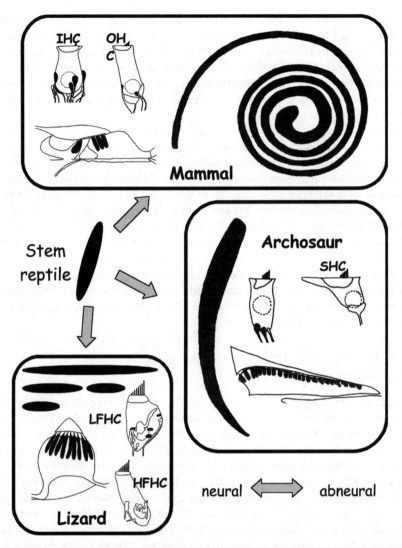

FIGURE 1.5. The evolutionary trends in hearing-organ shape and hair cell types in the three major lines that derived from the stem reptiles. The stem reptile papilla was most likely relatively short and contained one uniform kind of hair cell that responded to low frequencies. Papillae elongation and hair cell differentiation occurred independently in mammals, archosaurs, and lizard lines. In each line, the time course and the resulting structural patterns are unique to that group (see text). The black areas represent typical hearing organs of the respective group, as seen from above. In all three groups, a typical cross section of the hearing organ is shown (approximately to scale, neural side to the left), together with sketches of the respective two hair cell types. Afferent nerve endings are filled black, efferent endings are drawn open. INC, OHC, inner, outer hair cells; THC, SHC, tall, short hair cells; LFHC, HFHC, low-frequency, high-frequency hair cells. (After Manley 2000c, © 2000 Wiley-VCH Verlag GmbH, with permission.)

cells lying abneurally, but in birds the tall hair cells simply form one extreme of a size continuum along and across the papilla (for a review see Manley 1990). Tall hair cells (THCs) are cylindrical in shape and, as their name implies, taller than they are broad (Fig. 1.5). They receive both afferent and efferent innervation (Fischer 1994). The rest of the papilla's surface area (between supporting cells, of course) is occupied by so-called short hair cells (SHCs). In the group *Aves*, there is a unique defining feature of these cells. Whereas previously, SHC were arbitrarily characterized as being those cells whose width exceeded their height, they are now defined as those hair cells that, amazingly, receive no afferent innervation (Fischer 1994). They are contacted only by large efferent endings (Fig. 1.5). This remarkable situation, in which a large group of sensory cells (up to 40% of the hair cells in some bird species) is devoid of an afferent contact to the brain, is unique. Whether the same applies to the closely related *Crocodilia* is not known. In any case, it is clear that birds and *Crocodilia* share most features of the organization of their unique auditory papilla, and thus this basic structural pattern almost certainly arose before they diverged later in the Jurassic period (Fig. 1.4). The origin of THC and SHC has thus been placed in the Triassic in Figure 1.4, roughly covering the same time period as the origin of the (independently derived) hair cell groups of mammals.

4.3 The Diapsid Lepidosaur Descendants— Mainly Snakes and Lizards

It will be assumed for the purposes of this chapter that the turtles, which, according to some analyses (see, e.g., Hedges and Poling 1999) are closely related to some diapsids, are a plesiomorphic sister group of lepidosaurs plus archosaurs. An early divergence within the lepidosaurs gave rise to the Tuataras or *Rhynchocephalia*, remarkably primitive lizard-like animals that survive only on a few islands off the coast of New Zealand. The primitive structure of their auditory papilla has been noted above.

The lizards and snakes form the largest groups that arose within the diapsid lepidosaurs, and these two groups diverged in the early Cretaceous period (Fig. 1.4). The lizards diversified spectacularly, and this diversification was accompanied by a remarkable differentiation of hearing organ structure, evolving a family-, subfamily-, genus-, and in some cases even species-specific anatomy. This diversity is greater than that seen in any other group of amniotes. The snakes arose from a common ancestor with monitor lizards and are structurally much more uniform than are the lizards.

In extant lizards, the basilar papilla always has two clearly recognizable types of hair cell areas (Manley 2000a). Since no other lepidosaurs show this feature, it can be assumed to be apomorphic to the lizards and thus probably arose in the Jurassic period (Fig. 1.4). Electrophysiological studies have shown that one of these areas contains hair cells that respond to low frequencies; it is most sensitive to frequencies below 1 kHz (see below, section 5.1.2). This hair cell area probably derives from and corresponds to the entire papilla of stem reptiles and, incidentally, of the snakes (Manley, Chapter 7). In the other type of hair cell area, which is a synapomorphy of lizards, the hair cells respond best to frequencies above 1 kHz, generally up to 5 or 6 kHz. The upper limit is species-specific. Thus lizards, like mammals and archosaurs, also show two groups of hair cells. The hair cell populations of lizards, however, are not separated across the papilla and within the same frequency ranges, but rather in groups separated along the papilla and thus differentiated with respect to their frequency responses (Fig. 1.5). It is clear that they arose independently of the hair cell specializations of mammals and archosaurs.

Miller (1980, 1992) suggested that within the lizards, the most primitive structural pattern is represented by those groups (some iguanid, agamid, and anguid, lizards) in which the basilar papilla is in fact made up of three hair cell areas. The central area responds to low frequencies and corresponds to the plesiomorphic state. Here, all the hair cell stereovillar bundles have the same orientation (abneural, so-called unidirectional orientation). The other two hair cell areas flank the central area and are mirror symmetrical, at least primitively. The apomorphic, high-frequency area may thus have arisen at both ends of the stem-reptile papilla. These two areas contain roughly equally large groups of hair cells whose stereovillar bundles face each other (at least in less derived lizard papillae), a condition known as bidirectional orientation. These hair cell areas respond to higher frequencies.

From this basic morphological pattern, a remarkable array of evolutionary trends over time is discernible in the different lizard families (Köppl and Manley, 1992; Manley, Chapter 7). For example, in some lizard families, one of the bidirectional hair cell areas was lost in some or in all species. In some of these cases (e.g., skinks, cordylids), the apical bidirectional area disappears, and thus the unidirectional area lies at the apical end of the papilla. The frequency organization of such papillae, in which the lowest frequencies lie at the apical end of the papilla, thus (convergently) resembles that of birds and mammals. In contrast, in geckos and pygopods the basal bidirectional, high-frequency area disappears, so that—uniquely among amniotes—the reverse tonotopic organization is found.

In many lizard families (as, quite independently, in archosaurs and mammals), there was a strong tendency to elongation of the papilla over time, an effect that is most pronounced in the high-frequency area(s). The elongation is, however, less than seen in archosaurs and mammals, the maximal papillar length reached being only about 2 mm, in contrast to 11 mm in birds and 104 mm in mammals (Manley 1990) (Fig. 1.5). Papillar elongation is accompanied by an increase in the number of hair cells up to a maximum in lizards of about 2000 (Miller and Beck 1988; Miller 1992). The possible physiological meaning of such variations in structural patterns is discussed below and in detail in the following chapters.

4.4 The Synapsid Descendants, the Mammals

The earliest group to diverge from the stem reptiles was that of the mammal-like synapsids, of which only the mammals survived to the present (Fig. 1.4). Over a long period of time, these animals sequentially evolved the many features (more than 60; Hennig 1983) that we associate with the mammals. The transition to true mammals took place at some time during the Triassic period. Four features are typical for most, but not all, mammalian ears: the three-ossicle middle ear, the usually highly elongated sensory epithelium, the coiled cochlea, and the specialization of rows of hair cells into two clearly distinct populations across the papilla. It is likely that the specialization of hair cells into two distinct populations, the inner and outer hair cells, and arranged across the organ of Corti took place very early after the origin of true mammals, since all mammals show this feature. In Figure 1.4, the time of this development has been placed in the Triassic–Jurassic periods.

In contrast to the placental and marsupial mammals, however, the hearing organ of the monotreme mammals (duck-billed platypus and spiny anteater) is not coiled and the numbers of hair cell rows in a cross section of the hearing organ is much greater (Pickles 1992; Vater et al., Chapter 9). The time of the evolution of the coiled cochlea can thus be traced to the middle Cretaceous period, after these two main mammalian lineages diverged (Fig. 1.4). It is likely that at that time (middle to late Cretaceous), there was also a general reduction in the number of hair cell rows in the placental-marsupial line to the 1 + 3 hair cell row pattern (one row of inner, three rows of outer hair cells; Fig. 1.5). The plesiomorphic condition is still seen in the monotremes, the Echidna having three to five rows of inner and four to six rows of outer hair cells (Pickles 1992).

5. General Evolutionary Trends in the Physiology of Amniote Hearing Organs

Despite the obvious differences between the synapomorphic patterns of the mammalian organ of Corti and the basilar papillae of archosaurs and lepidosaurs, there are common trends observable in the evolution of their hearing organs, trends that developed parallel to each other over time.

5.1 Three Functional Principles Underlying Hearing in All Amniotes

The basic sensory unit of the hearing organ, the hair cell, represents an extremely old development. It can be traced back to an origin in the lateral line of the first vertebrates (Coffin et al., Chapter 3). The evolution of the hair cell to modern times, where it functions as the basic receptor cell type in lateral line, vestibular, and auditory systems, is a remarkable story of adaptability. Much of this adaptability can be traced back to three basic aspects of the function of these sensory cells. During the course of evolution, these three properties of hair cells and their surrounding structures were sculpted in response to selective pressures for both greater sensitivity and a greater frequency response range.

First, the passive and active electrical properties of the hair cell membrane can strongly influence the cell's ability to respond to stimuli, especially in the frequency domain. Cell membrane-potential oscillations resulting from the properties of different types of membrane channels are a good example of these influences. These led to the existence of the most primitive type of frequency selectivity, the so-called electrical tuning found in some types of hair cells (see below, section 5.1.1).

Second, as mechanoreceptors, hair cell sensory responses can be strongly influenced by the accessory structures around them. Thus the response pattern of a given cell will at least partly be the result of where exactly that cell is found in the epithelium. It will also be the result of changes in the characteristics of the stimulus incurred during its transmission through the accessory structures (e.g., middle ear, basilar membrane, and tectorial membrane) to the hair cell. Hair cells in auditory organs are generally connected to an accessory gelatinous membrane, the tectorial membrane, that together with the micromechanical properties of the cell's own stereovillar bundle, provides the basis for a micromechanical resonance system that underlies the second fundamental type of frequency selectivity (see below, section 5.1.2).

Third, hair cells are not just passive transducers of mechanical stimuli; they have an active process (known as the "cochlear amplifier") that increases the sensitivity of the hearing organ. There is evidence for the existence of two types of mechanism underlying the active process, an older one that was probably maintained in all lineages and a newer one that developed in the mammalian lineage (see below, section 5.1.3).

5.1.1 Electrical Tuning of the Hair Cell Membrane

As outlined above, the putative hearing epithelium of the stem reptiles contained several hundred hair cells that were covered by a thick tectorial membrane. It is likely that the membrane electrical properties of these hair cells, as in modern turtles, varied systematically along the epithelium. In turtles, the hair cell membrane contains at least two types of active channel, and the relative numbers and kinetics of these channels determine the best response frequency of each hair cell (see Coffin et al., Chapter 3; Manley, Chapter 7). If there are relatively few channels and their kinetics are slow, the cell membrane potential oscillates slowly between depolarization and hyperpolarization. If there are many channels and these have faster kinetics, these voltage swings take less time. Thus as long as the transduction channels are active, the cell membrane's potential oscillates at a rate depending on the number and the kinetics of the channels. As the number of channels of each type and the channel kinetics vary systematically in hair cells along the papilla, this membrane-channel mechanism forms the basis both for an electrical tuning of hair cells and for their tonotopic organization (Wu et al. 1995). In the case of modern turtles, the lowest frequency responses are found apically, the highest (about 1 kHz) basally. For each cell, there is a specific frequency at which the membrane electrical resonance is optimal. Above and below this frequency, the mechanism becomes increasingly inefficient, leading to a steady fall in the cell's response sensitivity, the further a frequency departs from the center, or characteristic frequency. Thus graphical representations of a cell's sensitivity across frequency yield V-shaped, so-called tuning curves. Thus even the most primitive basilar papillae show a clear frequency selectivity in each hair cell and a systematic distribution of frequencies along the hearing organ based mainly on the systematic trend in the membrane properties of hair cells. These two basic principles of hearing, a high-frequency selectivity and a tonotopic organization, can thus be regarded as plesiomorphic, deriving from the earliest hearing organs of stem reptiles (Manley and Köppl 1998).

One of the problems associated with electrical tuning is that it also has an inherent frequency limitation (Wu et al. 1995). Due for example to the fact that the membrane only has space for a finite number of channels of the different types, electrical tuning has an inherent upper frequency limit. Where exactly this limit lies depends on the temperature. Even at higher temperatures, however, it appears that the upper limit of operation of this membrane channel-based mechanism lies well below 10 kHz. This can be assumed to be the reason why all amniotes have developed other mechanisms for frequency selectivity, sometimes in addition to electrical tuning and sometimes apparently replacing it completely. So far, it is only clear that mammals have completely replaced it. The other groups appear to use it to a greater or lesser extent in the higher frequency ranges.

5.1.2 The Evolution of Micromechanical Tuning and High-Frequency Hearing

Compared to electrical tuning, a strong reliance on micromechanical tuning is a relatively new evolutionary development (see Manley, Chapter 7). We do not know, however, whether it was truly independently developed in the different groups. All hair cells have some degree of micromechanical tuning, if only because they have a stereovillar bundle that has a specific stiffness, etc. The micromechanical tuning results from the specific combination of stiffnesses (e.g., of the hair cell bundle or bundles in a group of hair cells) and masses (e.g., of the tectorial membrane) in a given resonant unit. A hair cell (or group of hair cells) with a shorter (i.e., stiffer) bundle and a small tectorial mass over it has a higher mechanical resonance frequency than a hair cell with a tall bundle and a larger mass. Depending on its size and properties, the tectorial membrane couples smaller or larger groups of hair cells.

Since even in primitive papillae such as that of the turtle, weak anatomical gradients in hair cell bundle structure are seen (Sneary 1988; Manley 1990), the basis for a micromechanical contribution to tuning can be assumed to have been present very early in hair cell evolution. Micromechanical tuning came to play a more important role during specializations for responses to increasingly high frequencies (lizards up to about 8 kHz, birds up to 14 kHz, and mammals up to more than 100 kHz). In birds, there is good evidence for electrical hair cell tuning, but since no clear anatomical transitions are seen, it has to be assumed that the weighting of electrical and micromechanical tuning gradually changes along the papilla from low to high response frequencies. Only in lizards do we find two groups of hair cells that are separated functionally into a low-frequency and a high-frequency group, perhaps based on these fundamentally different but interacting mechanisms of frequency selectivity.

5.1.3 The Principle of Active Amplification

Although our knowledge is not so far advanced in this respect, it is also likely that the principle of active amplification was also realized in the first amniote hearing organs. Considering the great resistance to movement incurred by hair cells in the inner-ear fluids, it is not surprising that an active amplification of weak stimuli evolved in vestibular hair cells (Hudspeth 1997).

One mechanism underlying active amplification in hair cells relies on the properties of the adaptation system of the hair cell stereovillar bundle (Hudspeth and Gillespie 1994). To maintain its exquisite sensitivity to stimuli (responding to movements of the bundle's tip of a few nanometers, despite much greater net displacements of the bundle during body motion, etc.), each tip link between hair cell stereovilli is equipped with an adaptation system. The tip link is apparently connected to a tiny motor system, in which myosins can actively climb up (or passively slide down) the actin matrix of the stereovilli. These movements are quite rapid, and, following a steady displacement, lead quickly to the restoration of the normal tension of the tip link that is necessary to maintain the hair cell's sensitivity. Active force generation in this adaptational system has been observed in various hair cell systems (including vestibular hair cells), and in some cases rapid twitching of the bundle occurs (Benser et al. 1996). It is likely that the speed of this system is such that at frequencies up to at least a few kilohertz, the adaptational motors could feed energy into an oscillating stereovillar bundle and thus increase the effective amplitude of the stimulus (Manley and Gallo 1997). As yet, however, there is no evidence linking such myosin-based systems to fast bundle motions. Instead, the newest studies point to a direct involvement of the transduction channel such that binding of calcium to open transduction channels induces their premature closure (Choe et al. 1998). This exerts a force on the bundle that, if the relative phase is appropriate, adds mechanical energy to the bundle's movement (see Hudspeth 1997 and Martin and Hudspeth 1999 for discussions of this issue).

In mammals, new data support the idea that outer hair cell bundles are also active like those of nonmammals (Kennedy et al. 2003). There is also strong evidence that at least at high frequencies, a new amplification system evolved.

This second mechanism in mammals is based on extremely rapid conformational changes of densely packed, integral membrane proteins in response to voltage changes at the cell's membrane (Dallos and Evans 1995). Taken together, new data indicate that all amniotes show evidence of one kind or another (e.g., spontaneous otoacoustic emissions) for the existence of this third basic principle underlying the operation of amniote hearing organs, that of active amplification (Manley 2001).

5.2 Elongation of the Basilar Papilla

During the evolution of most groups of amniotes, the basilar papilla elongated to a greater or lesser extent, and this occurred independently in all groups (Fig. 1.4). The elongation was most likely a response to the increasing importance of micromechanical tuning, since the fundamental mechanisms involved in this kind of tuning are more dependent on the amount of space available than is electrical tuning. Thus we observe that in elongated lizard papillae, it is always the high-frequency, micromechanically tuned hair cell area that became longer during evolution. The process of elongation was carried further in archosaurs, where papillar lengths of >10 mm are observed (in some owls), and even further in some mammals, where the organ of Corti is between 7 mm (in the mouse) and more than 100 mm long (e.g., 104 mm in the blue whale). The coiling of the cochlea of eutherian and metatherian mammals was an elegant solution for the problem of accommodating such highly elongated sensory epithelia. This coiling evidently occurred after the monotreme branch split off from the ancestral mammalian line, during the middle to late Cretaceous period.

This elongation of the auditory epithelium enabled an increase in the upper limit of hearing by providing space to accommodate higher octaves. To a first approximation, the upper frequency limit of hearing in amniotes is a function of the length of the papilla. It is, however, subject to other, partly group-specific effects, such as the animal's absolute size (Manley 1973). In addition to an increase in the upper frequency limit of hearing, however, papillar elongation was generally accompanied by an increase in the amount of space devoted to each octave. Thus the space per octave available in the smaller papillae of lizards is generally less than that in birds, which itself is generally less than that in mammals (Fig. 1.6) (Manley et al. 1988; Manley 1990). However, it should be emphasized that this is only a general rule, that there is overlap between the groups and that there are many exceptions. Even in mammals, there are great differences in emphasis for low- and high-frequency hearing (Heffner et al. 2001; see Vater et al., Chapter 9).

An increase in the space available for the analysis of sound can have several different consequences, since there is a trade-off between the frequency range that can be analyzed and the detail with which it can be analyzed. The space constant (expressed in millimeters devoted to each octave along the hearing epithelium) influences the degree to which an enhanced innervation of a narrow frequency region can provide increased input for the parallel processing of in-

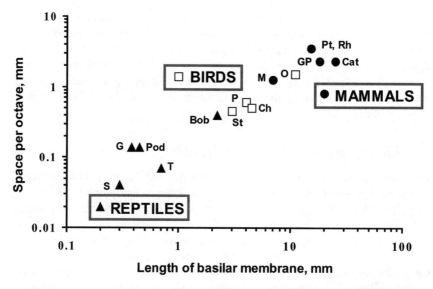

FIGURE 1.6. The space constant with which different frequencies are distributed along the hearing organ as a function of the length of the basilar membrane in a number of species of lizards, birds, and mammals. Regions of auditory foveae in bats and the barn owl have not been included. The species are as follows: REPTILES: S, *Sceloporus*, the granite spiny lizard; G, *Gerrhonotus*, an alligator lizard; Pod, *Podarcis*, the European wall lizard; T, turtle; Bob, *Tiliqua*, the bobtail skink. BIRDS: St, starling; Ch, chicken; P, pigeon; O, barn owl (average). MAMMALS: M, mouse; GP, guinea pig; Cat, cat; Pt, Pteronotus, a bat; Rh, Rhinolophus, another bat. (For references, see Manley 1990; Müller 1991 for the rat.)

formation (Fig. 1.6). Thus some species sacrifice a potentially greater breadth of frequency response for a greater degree of detail concerning specific frequency ranges. Therefore, although the distribution of frequencies along the auditory papilla is generally roughly exponential with frequency (that is, each octave—a doubling in frequency—occupies approximately the same absolute length of the papilla), there are a number of very important and interesting exceptions in specialist "hearers" such as some bats and owls (Köppl et al. 1998; see Gleich et al., Chapter 8; Vater et al., Chapter 9). Some of these species have strongly expanded the frequency representation of specific, behaviorally important frequency ranges.

5.3 Efficiencies Achieved Through Specialization in Hair Cell Function

The presence of two populations of hair cells in lizards, archosaurs, and mammals has already been noted. Whereas the two hair cell types in lizards are responsible for different frequency ranges, and are thus placed at different positions along the papilla, the situation in archosaurs and mammals is quite different. As we know too little about the *Crocodilia*, the archosaurs will be represented in this discussion by the birds. The hair cell specializations in birds and mammals that is manifest in the morphological differences between different populations is an interesting case of convergent evolution (Manley et al. 1989). In these two groups, the hair cell populations have specialized their function to a greater or lesser degree. Whereas in the auditory organ of stem reptiles (and many modern groups) every micromechanically tuned hair cell is both a receptor for mechanical stimuli and (through its active process) a microamplifier, both mammals and birds appear to have evolved hair cell populations specialized for one function or the other.

Since the evolution of specialized hair cell populations occurred independently in birds and mammals (both presumably during the Triassic–Jurassic periods, Fig. 1.4), these populations of hair cells are analogous and not homologous, and they do not represent a plesiomorphic feature. It is thus not unexpected that their detailed configuration shows very clear differences that are presumably related to the exact mechanisms involved in carrying out the specialized functions of each population. These differences are not, however, so great as to mask a number of obvious common features (Manley et al. 1989). Thus, the innervation patterns of the respective neurally and abneurally placed hair cell groups are similar in both birds and mammals. The neural group of hair cells receives a prominent afferent and a weak (birds) or no (mammals) efferent innervation, whereas the abneural hair cell group receives a prominent efferent innervation and a weak (mammals) or no (birds) afferent innervation (Fig. 1.5). Sensory cells in birds without an afferent innervation would not make any sense without the concept of active amplification in the hearing organ. Although the presence of a so-called cochlear amplifier is well established only for the mammalian cochlea, the remarkable morphological similarities to the bird cochlea and the absence of afferent innervation to avian short hair cells suggests that at least at higher frequencies, this principle is also realized in birds. Spontaneous otoacoustic emissions are generally accepted as unequivocal evidence for an active process, and they have been demonstrated in all land vertebrates (Köppl 1995; Manley 2001). Thus both the avian and mammalian cochleae contain normal receptor cells lying neurally and receptor cells modified as "effector" cells lying abneurally. It is presumed that this division of labor led to an increased efficiency of the hearing apparatus.

5.4 Physiological Gradients Across the Hearing Epithelium

Since the hair cell structure and their innervation patterns change across the width of the hearing epithelium in birds and mammals, it might be expected that there would be physiological differences in the responses of hair cells and their innervating afferent fibers across the width of the organ. It is not easy to study such phenomena, since it involves staining individual, physiologically characterized afferent nerve fibers and tracing these fibers back to their peripheral origin. For technical reasons, this can be done only for one or a few fibers per animal. In spite of this, information relevant to this important question is available for both birds and mammals.

In mammals, the afferent innervation is strongly concentrated on the inner hair cells (10 to 15 afferent fibers terminate on each inner hair cell) and as yet it has not proved possible to record responses to sound from outer hair cell afferent fibers. The afferent fibers on inner hair cells, however, are not all the same. They differ in their anatomy, in the position they occupy on the membrane of the inner hair cell, and in their physiological response properties. One of the most interesting differences between fibers is in their sensitivity to sound; the afferent fibers innervating the lateral surface of the hair cell are more sensitive to sound than those innervating the neural surface (Liberman and Oliver 1984) (Fig. 1.7). Other physiological differences (e.g., in the spontaneous activity and frequency selectivity) also indicate that, in some way, fibers that em-

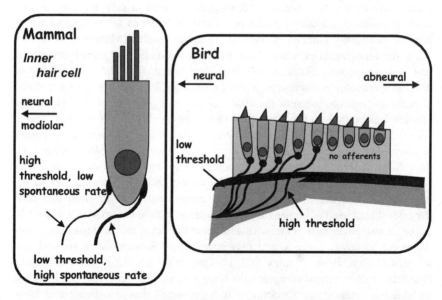

FIGURE 1.7. Some physiological gradients across the hearing epithelium in mammals and birds. In general, the input to the brain in mammals is derived from inner hair cells. On the left of this figure an inner hair cell is illustrated that is innervated by two afferent fibers. The fiber on the left, or neural side, is thinner; it has a poorer threshold and lower spontaneous activity (Liberman and Oliver 1984). In birds, such patterns across single hair cells are not seen. Instead, the more sensitive afferent fibers tend to innervate (tall) hair cells near the neural edge. Less sensitive afferents get their input from hair cells near the middle of the epithelium. More abneural (short) hair cells often receive no afferent terminals. (After Manley 2000c, © 2000 Wiley-VCH Verlag GmbH, with permission.)

anate from individual hair cells nonetheless differ systematically in their responses. The variety of response patterns increases the ability of the hearing organ to code different aspects of the sound stimuli in their input to the brain.

In birds, the innervation density on the innermost hair cells (tall hair cells) is generally not as high as in mammals (with the exception of some cells in the barn owl; Fischer 1994). Normally, avian tall hair cells contact less than three afferent nerve fibers. As in mammals, the sensitivity of afferents can differ, and the spread of thresholds to sound is in fact greater in birds. The evidence to date suggests that in birds the most sensitive fibers contact hair cells near the neural or inner edge of the auditory papilla, and the least sensitive afferents contact hair cells near the middle of the epithelium (Fig. 1.7) (Gleich and Manley 2000; see Gleich et al., Chapter 8, for a review). At least in higher frequency regions, there are in most cases no afferents to hair cells lying more abneurally than the middle of the papilla; thus no thresholds can be measured for these regions (Fig. 1.7). The large threshold spread in birds is correlated with the spontaneous activity, the frequency selectivity, and the position of the hair cell across the width of the papilla. These correlations are weaker or stronger depending on the frequency range and the species. Comparing mammals to birds, it appears that mammals achieve in single hair cells at least partly what birds achieve through cells distributed across the hearing epithelium. This suggests that the mechanisms underlying the interactions between hair cell populations, which also involve the active process, operate a little differently in mammals and birds (Yates et al. 2000).

It thus appears that during evolution, the selection pressures working on similar, large hair cell epithelia in the various groups of amniotes have led to some similar and to some different solutions to the problems of developing sensitive, high-frequency hearing. Many of the solutions found utilized preexisting abilities of hair cells that, through specialization, could be made more efficient. The end result of these trends was a remarkable variety of hearing organs whose sensitivities and frequency-response ranges differ in characteristic ways between the various modern groups of amniotes. The various chapters in this book describe these issues in detail for the different vertebrates groups.

6. Summary

In the context of the known paleontological history of vertebrates, it is clear that over time the various groups of vertebrates have utilized preexisting sensory epithelia for hearing, but have also developed new epithelia. These epithelia show a highly group-specific structure that results in a range of inner ears whose morphology and, to some extent, physiology show characteristic features. An understanding of these features is only possible in the context of evolutionary history.

We have provided a framework in which to integrate the contents of the chapters of this book. Studies of the evolution of vertebrate hearing demonstrate on the one hand the conservatism typical of the evolutionary process, in which for example in all vertebrates the basic unit of the hair cell has been retained as the sensory element. Nonetheless, this element has also been modified for its function within the various groups, with emphasis on electrical or micromechanical tuning, and on hair-bundle– or hair-cell-membrane–based active processes. Whereas as groups the fishes and the amphibians use more than one organ for hearing, the basic form of the basilar papilla of the amniotes was established right from their beginnings if not earlier. This basic form experienced formative evolution in many aspects of its structure and hair cell specializations, and these happened independently in the different amniote lineages. The most interesting research question for the future concerns the extent to which true apomorphies have occurred in the amniote lineages with respect to important mechanisms underlying the hearing process.

References

Benser ME, Marquis RE, Hudspeth AJ (1996) Rapid, active hair-bundle movements in hair cells from the bullfrog's sacculus. J Neurosci 16:5629–5643.

Carroll RL (1987) Vertebrate Paleontology and Evolution. New York: W. H. Freeman.

Choe Y, Magnasco MO, Hudspeth AJ (1998) A model for amplification of hair-bundle motion by cyclical binding of Ca2+ to mechanoelectrical-transduction channels. Proc Natl Acad Sci USA 95:15321–15326.

Dallos P, Evans B (1995) High-frequency motility of outer hair cells and the cochlear amplifier. Science 267:2006–2009.

Fischer FP (1994) General pattern and morphological specializations of the avian cochlea. Scanning Microscopy 8:351–364.

Fritzsch B (1987) Inner ear of the coelacanth fish *Latimeria* has tetrapod affinities. Nature 327:153–154.

Gleich O, Manley GA (2000) The hearing organ of birds and *Crocodilia*. In: Dooling R, Popper AN, Fay RR (eds) Comparative Hearing: Birds and Reptiles. New York: Springer-Verlag, pp. 70–138.

Hedges SB, Poling L (1999) A molecular phylogeny of reptiles. Science 283:898–901.

Heffner RS, Koay G, Heffner HE (2001) Audiograms of five species of rodents: implications for the evolution of hearing and the perception of pitch. Hear Res 157: 138–152.

Hennig W (1983) Stammesgeschichte der Chordaten. Hamburg: Verlag Paul Parey.

Hudspeth AJ (1997) Mechanical amplification of stimuli by hair cells. Curr Opin Neurobiol 7:480–486.

Hudspeth AJ, Gillespie PG (1994) Pulling springs to tune transduction: adaptation by hair cells. Neuron 12:1–9.

Janvier P (1996) Early Vertebrates. Oxford: Clarendon Press.

Kennedy HJ, Evans MG, Crawford AC, Fettiplace R (2003) Fast adaptation of mechanoelectrical transducer channels in mammalian cochlear hair cells. Nat Neurosci 6: 832–836.

Kitching IJ, Forey PL, Humphries CJ, Williams DM (1998) Cladistics, the Theory and

Practice of Parsimony Analysis, 2nd ed. Systematics Association publication 11. Oxford: Oxford University Press.

Köppl C (1995) Otoacoustic emissions as indicators of active cochlear mechanics: a primitive property of vertebrate auditory organs. In: Manley GA, Klump GM, Köppl C, Fastl H, Oeckinghaus H (eds) Advances in Hearing Research. New Jersey, London, Hong Kong: World Scientific, pp. 200–209.

Köppl C, Manley GA (1992) Functional consequences of morphological trends in the evolution of lizard hearing organs. In: Webster DB, Fay RR, Popper AN (eds) The Evolutionary Biology of Hearing. New York: Springer-Verlag, pp. 489–509.

Köppl C, Klump GM, Taschenberger G, Dyson M, Manley GA (1998) The auditory fovea of the barn owl—no correlation with enhanced frequency resolution. In: Palmer A, Rees A, Summerfield AQ, Meddis R (eds) Psychological and Physiological Advances in Hearing. London: Whurr, pp. 153–159.

Liberman MC, Oliver ME (1984) Morphometry of intracellularly labeled neurons of the auditory nerve: correlations with functional properties. J Comp Neurol 223:163–176.

Manley GA (1973) A review of some current concepts of the functional evolution of the ear in terrestrial vertebrates. Evolution 26:608–621.

Manley GA (1990) Peripheral Hearing Mechanisms in Reptiles and Birds. Heidelberg: Springer-Verlag.

Manley GA (2000a) The hearing organs of lizards. In: Dooling R, Popper AN, Fay RR (eds) Comparative Hearing: Birds and Reptiles. New York: Springer-Verlag, pp. 139–196.

Manley GA (2000b) Cochlear mechanisms from a phylogenetic viewpoint. Proc Natl Acad Sci USA 97:11736–11743.

Manley GA (2000c) Design plasticity in the evolution of the amniote hearing organ. In: Manley GA, Fastl H, Kössl M, Oeckinghaus H, Klump GM (eds) Auditory Worlds: Sensory Analysis and Perception in Animals and Man. Weinheim: Wiley-VCH, pp. 7–17.

Manley GA (2001) Evidence for an active process and a cochlear amplifier in non-mammals. J Neurophysiol 86:541–549.

Manley GA, Gallo L (1997) Otoacoustic emissions, hair cells and myosin motors. J Acoust Soc Am 102:1049–1055.

Manley GA, Köppl C (1998) Phylogenetic development of the cochlea and its innervation. Curr Opin Neurobiol 8:468–474.

Manley GA, Brix J, Gleich O, Kaiser A, Köppl C, Yates G (1988) New aspects of comparative peripheral auditory physiology. In: Syka J, Masterton RB (eds) Auditory Pathway—Structure and Function. London: Plenum Press, pp. 3–12.

Manley GA, Gleich O, Kaiser A, Brix J (1989) Functional differentiation of sensory cells in the avian auditory periphery. J Comp Physiol [A] 164:289–296.

Martin P, Hudspeth AJ (1999) Active bundle movements can amplify a hair cell's response to oscillatory mechanical stimuli. Proc Natl Acad Sci USA 96:14306–14311.

Miller MR (1980) The reptilian cochlear duct. In: Popper AN, Fay RR (eds) Comparative Studies of Hearing in Vertebrates. New York: Springer-Verlag, pp. 169–204.

Miller MR (1992) The evolutionary implications of the structural variations in the auditory papilla of lizards. In: Webster DB, Fay RR, Popper AN (eds) The Evolutionary Biology of Hearing. New York: Springer-Verlag, pp. 463–487.

Miller MR, Beck J (1988) Auditory hair cell innervational patterns in lizards. J Comp Neurol 271:604–628.

Pickles JO (1992) Scanning electron microscopy of the Echidna: morphology of a primitive mammalian cochlea. In: Cazals Y, Demany L, Horner K (eds) Auditory Physiology and Perception. Oxford: Pergamon Press, pp. 101–107.

Pough FW, Janis CM, Heiser JB (2002) Vertebrate Life, 6th ed. Upper Saddle River, NJ: Prentice Hall.

Sneary, MG (1988) Auditory receptor of the red-eared turtle: I. General ultrastructure. J Comp Neurol 276:573–587.

Wever EG (1978) The Reptile Ear. Princeton, NJ: Princeton University Press.

Wu Y-C, Art JJ, Goodman MB, Fettiplace R (1995) A kinetic description of the calcium-activated potassium channel and its application to electrical tuning of hair cells. Prog Biophys Mol Biol 63:131–158.

Yates GK, Manley GA, Köppl C (2000) Rate-intensity functions in the emu auditory nerve. J Acoust Soc Am 107:2143–2154.

2
Environmental Variables and the Fundamental Nature of Hearing

EDWIN R. LEWIS AND RICHARD R. FAY

1. Introduction

Comparative morphologists, physiologists, and neuroethologists (including the authors of this chapter) make observations at the level of phenotypes. It is natural for them to assume that observed phenotypic traits have been sculpted by evolution and therefore, somehow, have increased the fitnesses of the organisms in which they occur. Inferences about the pathways of evolution typically are drawn from formal or informal cladograms (Manley and Clack, Chapter 1, section 1, and Fig. 1.1; Ladich and Popper, Chapter 4, Fig. 4.1; Clack and Allin, Chapter 5, Fig. 5.1). In the past, such cladograms were based on morphological relationships. This is still the case for species long extinct, but for living species they now are based more and more on molecular relationships (e.g., see Gleich et al., Chapter 8, section 2). The inferences represented by cladograms surely will be strengthened greatly when investigators finally are able to relate them directly to underlying changes in genotype—to the genetic networks that through development give rise to the observed morphological, physiological, and behavioral traits (e.g., see Coffin et al., Chapter 3, section 5; Grothe et al., Chapter 10, section 2). No matter how strong one believes the inferences about evolutionary paths have become, however, one is left with uncertainty regarding the selective pressures and physical constraints that resulted in the taking of those paths, and the ways in which the phenotypic features associated with those pathways resolved those pressures and accommodated those constraints.

Many of the selective pressures would have been ecological. The analysis of ecological pressures in living systems is sufficiently difficult to make one believe that such analysis for ancient, long-extinct systems will always be speculative at best. What, then, can one do if one wishes to draw strong inferences about selective pressures? One approach is to consider ecological factors that must have been omnipresent and those that must have been invariant, and to construct inferences about how observed morphologies, physiologies, and behaviors of living organisms are especially adapted to cope with those factors or to take advantage of them. Among the omnipresent factors, throughout the evolution of the vertebrates, surely must have been predation, competition for resources, and competition for mates. Among the invariant factors surely must have been the physics of matter, including the omnipresence of random thermal motion (noise).

The evolutionary neurobiologist interested in hearing, for example, must be safe in the assumptions that throughout the course of evolution, matter was capable of conducting acoustic waves over long distances from their sources and that the physical natures of those waves bore implications regarding the identities and states of the sources as well as implications about the physical world irradiated by those sources. Thus, animals always had the potential to use acoustic waves to construct inferences about distant sources and about the world surrounding those sources and themselves. Realization of that potential would require the ability to sense the waves and then to compute the inferences from them. As the axiomatic basis of the evolutionary neurobiology of hearing, the authors assume that the selective advantage of the auditory nervous system is its ability to carry out those two functions—sensing and inferential computation. As clues to the sorts of things that might be inferred from acoustic waves, one can consider their physical nature and the changes imposed upon them as they travel about the physical world. As clues to the sorts of inferences that actually are computed by the auditory nervous systems of living animals, one can rely on neuroethology, including that informal branch of human neuroethology, introspection.

One strong conclusion that arises from introspection, as well as from experiment (e.g., Bregman 1990) is that a quintessential part of the human ability to compute inferences about the source of a sound is the ability to separate that sound from all the others simultaneously impinging on the ears, and to focus attention on that sound. For all vertebrate classes except agnathans, it has been shown that at least some extant species can hear (Fay 1988; see also section 3, below). Hearing probably arose among the early fishes. Whether it arose several times, independently, as different inner-ear organs moved to hearing (Ladich and Popper, Chapter 4), or just once, one wonders what it was like in its earliest stages. How did it differ from the sense of hearing of extant vertebrates? It is tempting to suggest that it was a simplified sense compared with the hearing of extant birds and mammals, primarily computing inferences, from sound, about predators or other dangers and prey or other resources. The neural encoding and processing necessary for constructing such inferences are often implicitly assumed to be far simpler than those required for sound communication. Examination of this assumption, however, leads one to conclude that constructing inferences from the sounds of predators and prey or, for that matter, sounds from any other sources (including those that would be innocuous or neutral), must be a highly complex function of *all* vertebrate auditory systems.

In an impoverished acoustic environment, the early hearing vertebrate might have confronted the sound from one source at a time. In that case, the presence of a sound would be a novelty and the inferential computations might have been limited to estimating the direction of the source in order to direct the gaze appropriately, allowing the visual system to take over. They also might have included identification of the source (potential danger, potential resource, or neutral) and decisions about whether or not to move into a higher state of alert, or to initiate a defensive (e.g., escape) response, or an aggressive response, or an exploratory response. In rich acoustic environments, which one must assume were faced by most early hearers, the presence of sound would not be a novelty, and sounds from several sources often would converge on the hearer at once (see also Ladich and Popper, Chapter 4, section 3). In that case, to be effective at all, inferential computations must have included separation of those sounds (Yost 1991; Lewis 1992; de Cheveigne 2000). For any one of the several sources, this process (called sound source segregation) must comprise at least two elements: (1) the construction of inferences regarding which parts of the total sound impinging on the hearer's ears should be attributed to that source and which should not; and (2) the focusing of attention on the former, effectively ignoring the latter. One concludes as well that source segregation is fundamentally required for effective intraspecific sound communication. For example, a communication sound in a normal, noisy environment must not be integrated with other, neutral sounds that happen to accompany it, but must be segregated to properly represent the biologically significant (communicating) source. We consider source segregation to be the quintessential element of auditory (and other sensory) systems in general, defining the biological significance of hearing in a most fundamental sense. We believe further that the requirement of source-segregation capability has produced the most important pressures shaping the evolution of the sense of hearing throughout vertebrate history.

In human listening, sound-source segregation is represented in the well-known cocktail-party effect, the ability of a listener to focus on the voice of a single speaker in a room full of people all speaking at once. It has been attributed largely to the ability of the nervous system to use binaural cues to infer which sound components belong to the speaker being attended and which do not. Human psychophysical studies have shown the neural computations for source segregation involve much more than binaural cues and spatial localization. Monaural cues also play important roles (Hartmann 1988). One can easily make a compelling case that a physicist considering only the properties of sound itself would have predicted precisely the monaural and binaural cues that psychophysical studies found to be important. Surely this must be true of cues for other computations carried out by the auditory nervous system. Therefore, an evolutionary neurobiologist interested in hearing could well begin by considering the invariant properties of physical sound and sound sources, and how those properties might be exploited by a nervous system to compute inferences about sources important to survival. In this context, it seems clear that the neural machinery necessary for source segregation must not be compromised or preempted by other functions or other inferential computations, including those associated with acoustic communication.

Included in the explosively evolving human technology over the past century or so have been astounding advances in the ability to design and construct acoustic sensors and systems that process the signals derived from those sensors. These advances have required deep understanding of the potentialities and limitations inherent in the physical elements from which acoustic sensors might be constructed, as well as of the fundamental potentialities and limitations of signal-processing schemes. In both areas, acoustic-sensor design and signal processing, technologists have developed repertoires of basic building blocks and repertoires of mathematical tools and concepts to aid in the design of systems they want to build with those blocks, taking advantage of the potentialities and accommodating the limitations. Animal auditory systems must have faced many of the same potentialities and limitations. Therefore, biologists considering the evolution of hearing should and do look to the technologists' repertoires of building blocks, tools, and concepts for inspiration. Similarly, technologists might look to hearing research for inspiration for some of their outstanding problems. To date, for example, they have not solved the general problem of source segregation.

2. The Circuit-Theory Metamodel

Among the technologists' concepts that already are invoked widely in the literature on the evolution of hearing, are active and passive, impedance matching, transformers, bidirectional transduction, and resonance (Manley and Clack, Chapter 1, section 5; Clack and Allin, Chapter 5, section 2.2; Smotherman and Narins, Chapter 6, section 4). All of these concepts arise in the mathematical tool that is most widely used for acoustical design. That tool is the circuit-theory metamodel. In its most basic and useful form, the circuit-theory metamodel uses the structures and concepts of nonequilibrium thermodynamics; and with those structures and concepts it leads one to models capable of incorporating the full range of classical physical processes as they are currently understood. The metamodel and conclusions relevant to the evolution of hearing that follow from simple applications of it are cited frequently in this chapter (see sections 4, 5, 6, and 8). Formal exposition of the metamodel, however, is placed in the Appendix, where the reader can refer to it at will, without being forced to interrupt the flow of thought from the rest of the text.

As its name suggests, the circuit-theory metamodel provides recipes for constructing models. Applied to a simple physical process, such as sound conduction in a uniform medium, the appropriate recipe always begins with selection of a single entity that is taken to be conserved and to move from place to place during the process. Common choices include charge, mass, and momentum. Having chosen the conserved entity, one next defines a model space (a model of a region of physical space or state space) in which the process is assumed to occur, and then divides that model space into places at which the conserved entity is assumed to accumulate. Once these steps have been taken, development of a model is exceedingly straightforward. Occasionally, a modeler wishes to consider more than one conserved entity. Each entity defines a physical realm (e.g., an electric realm in which charge is the conserved entity, a mechanical realm in which momentum is the conserved entity). In that case, through devices known as transducers, a model can represent the transfer of free energy back and forth between different realms. Transducers, of course, are key elements in models of peripheral hearing.

3. What Is Sound and What Is Hearing?

If one subscribes to conventional nomenclature, then one would label acoustical waves as *physical sound* if they were of sufficient amplitude and appropriate spectral composition to be capable of producing perceived sound in a human subject. Because human hearing is limited, approximately to the frequency range of 20 Hz to 20 kHz, acoustical waves at frequencies below 20 Hz commonly are defined to be *infrasound* (below sound), and those at frequencies above 20 kHz commonly are defined to be *ultrasound* (beyond sound). Although human hearing normally occurs in air, human subjects also can experience the sensation of hearing (*perceived sound*) in response to acoustic waves in that same frequency range in water (e.g., a human diver can hear the song of the humpback whale or the snap of a pistol shrimp) or in solids (e.g., a human hunter with his ear to the ground can hear the thunder of the hooves of a distant herd). Extending the definitions of *sound* and *hearing* to nonhuman animals occasionally has been difficult and controversial (Webster 1992). There is a common tendency, for example, to lump the seismic senses (sensitivity to substrate-borne vibration), whether mediated by specialized inner-ear organs, such as the frog saccule (Lewis et al. 1982), or by somatosensory organs, and to separate them from hearing. On the other hand, when mediated by the cochlea, or its homologs the reptilian and avian basilar papillae, sensitivities to waterborne acoustic waves and to airborne acoustic waves are both commonly included in hearing. In amphibians, by analogy, sensitivity to airborne acoustic waves by the two auditory papillae (whose homologies are uncertain) also is included in hearing. Also by analogy, inner-ear mediated sensitivity to waterborne acoustic waves in fishes and aquatic amphibians usually is included in hearing. In many instances, this inner-ear sensitivity to waterborne acoustic stimuli appears to be mediated in essentially the same manner as inner-ear seismic sensitivity (i.e., by the motion of an inertial element in the inner ear when the entire skull is subjected to vibration). Thus the distinctions between seismic sensitivity and hearing are blurred. Nonetheless, there is general agreement that physical sound includes both airborne and waterborne acoustic waves of the appropriate frequency range. There is also general agreement that physiological or behavioral response to physical sound of low to moderate intensity [e.g., less than 100 dB sound pressure level (SPL)] implies hearing. By that definition, hearing is present in all vertebrate classes (with the possible exception of agnathans) and in many invertebrate taxa (Fay 1988).

4. Sources of Physical Sound

By that same definition of sound, any physical phenomenon that causes a time-varying pressure gradient (and accompanying particle motion) with sufficient amplitude and appropriate spectral content to be heard is a source of physical sound. For air- and waterborne sound, such phenomena include the following: (1) turbulence in the medium, (2) explosions and implosions, (3) excitation of transient motion in solids or liquids in contact with air and water, and (4) excitation of resonances in bounded volumes of solids, liquids, or gases in contact with air and water. In considering the physical sounds produced by such phenomena, it may be instructive to look at some familiar examples related to humans listening in air and to consider the attributes known, through introspection, to be computed for the corresponding perceived sounds by the human auditory nervous system.

Turbulence occurs when the relative velocity between the medium and an object in the medium is sufficiently large. For example, air passing through an orifice or around an obstacle may become turbulent, and objects moving through otherwise still air may create turbulence. In some instances, turbulence produces physical sound with broad spectral content, as in a whispered voice. In that case, the human auditory system may compute only a vague or indefinite pitch from the perceptual version of the sound. The auditory system can be quite sensitive to the spectral shape, however, as in the differences among whispered vowels. In other instances, the turbulence may take the form of a periodic string of detached eddies. These can yield physical sound with narrow spectral distribution. In that case, the perceived version may be assigned a definite pitch by the auditory system. Such sounds have been labeled *aeolian tones*.

Explosions and implosions are sudden rushes of the medium that are transformed into propagating waves, usually with broad spectral content. An explosion is produced, for example, by the near-instantaneous conversion of a liquid or a solid into a gas, or by the near-instantaneous intense heating of air by an electrical spark (as in lightning). Less violent explosions, often called plosions, occur when pressure is allowed to build behind a closure in the vocal system, then suddenly released, as in the phonemes k, t, and p. Implosions can be created in the vocal tract by developing a negative pressure behind a closure, and then suddenly releasing the outside air to flow inward, as in a click. Phonemes actually labeled "implosives" by phoneticians, on the other hand, evidently do not involve much implosion, if any (Ladefoged 1982).

Probably the most common immediate causes of sound-producing transient motion of liquids are turbulent flow, implosion of cavitation bubbles, and impact (liquid colliding with liquid or liquid colliding with solid). In solids, the most common immediate causes probably are friction, fracture, percussion, and rapid deformation. Examples of sounds generated by these phenomena include the following: sound of a mountain stream (turbulent flow, possibly some cavitation); sound of rain (impact); sound of a jumping fish (impact); sound of a falling tree (friction, fracture, and percussion); sound of a twig snapping (fracture); or a dry leaf being crushed (fracture); sound of a footfall (percussion); a slap (impact or percussion), or a bell (percussion); sound of paper being crumpled (rapid deformation) or a sheet of metal being shaken (rapid deformation). Percussion is a common source of animal acoustic communications, including foot drumming by various mammals, and, among arthropods, drumming on the substrate with various body parts and striking body parts together (Muller-Preuss and Ploog 1983; Ewing 1989). Depending on the nature of the colliding objects, the physical sound produced may have narrow spectral distribution (e.g., around a fundamental and each of its overtones), in which case the human auditory system may compute a definite pitch from the perceptual version of the sound; or it may have a smoother spectral distribution, leading to the perception of indefinite pitch. Among percussion instruments used in Western music, xylophones, glockenspiels, chimes, bells, and kettledrums produce sounds leading to perception of definite pitch; triangles, base drums, snare drums, tambourines, cymbals, castanets, and gongs produce sounds leading to perception of indefinite pitch (Olson 1967). For sound produced by percussion in arthropods, the perception of pitch in the human listener also can arise from the frequency of collision, which can be well into the frequency range for physical sound.

Excitation of resonances also is a common source of animal acoustic communication signals, including music produced by humans (Appendix, see discussion around Eqs. 82 to 86). Usually it involves energy from some nonfluctuating source (e.g., a more-or-less constant pressure difference, more-or-less constant motion, a more-or-less constant temperature difference) being converted into rapid fluctuations in a resonant system. In many instances the nonfluctuating source appears to act directly on a relaxation process, leading to abrupt changes of airflow or mechanical motion (Olson 1967). Each relaxation (each abrupt change) excites the resonance. A key feature is feedback from the resonance to the relaxation process, influencing the timing of relaxations in such a way that the excitation that each of them puts into the resonance is nearly in phase with that already present. Thus, the energy contributed by each relaxation is largely added to that accumulated from previous relaxations. As the excitation of the resonance grows, its influence on the timing of relaxations becomes increasingly precise, and the addition of energy becomes increasingly efficient. In terms of the circuit-theory metamodel, it is the very low driving-point impedance (Appendix, Eqs. 82 and 83) at the resonant frequency of a circuit that allows the nonlinear system to settle easily into this mode of operation. For a resonance formed by a standing wave structure (Appendix, section 9.2), imagine the following sequence: (1) The first relaxation cycle initiates a wave at end number 1 of the structure. (2) The wave travels along the structure until it reaches end number 2, where it is reflected back toward end number 1. (3) The next relaxation cycle is triggered when the reflected wave reaches end number 1, leading to reinforcement. Repeated relaxation cycles will yield a growing standing wave, with a fundamental period equal to the round-trip travel time for the structure.

The action of the bow on a violin string is another example (Schellung 1978). With a more-or-less constantly moving bow, the string alternates between sticking to the resin-coated fibers of the bow and being released to slide along the fibers. While it is sticking, the string is being drawn away from its resting position and tension increases. Eventually, the tension breaks the grip of the fibers, suddenly releasing the string to slip back along the bow fibers toward its resting position; tension decreases. Subsequently, the grip of the fibers is restored and the string sticks again. If the string were not a resonant structure, the alternation between sticking and slipping would be somewhat random, only crudely periodic. The string *is* a resonant structure, however, and its resonance converts each stick-slip cycle of the string into a slightly damped oscillation of tension and motion in the string. This oscillation interacts with the force and motion produced by the static and dynamic friction between the steadily moving bow fibers and the string. Consider a single pass of a bow over a string with fixed fingering. After the first stick-slip cycle, the oscillatory wave produced in the string affects the timing of the second stick-slip cycle in such a way that the oscillation produced by it is nearly in phase with the oscillation produced by the first cycle. Thus, the oscillation amplitude is reinforced (i.e., the energy from the bow is converted into oscillatory energy in the string, and that oscillatory energy accumulates from one stick-slip cycle to the next). As the amplitude of oscillation builds, the timings of the sticks and slips become more and more precisely controlled by the energy of oscillating string. In steady state, the energy added by each stick-slip cycle of the bow is equal to the energy lost from the string by damping. Some of the damping is a consequence of conversion of mechanical energy in the string into heat. Some of it is a consequence of energy transfer to other resonances (e.g., Helmholtz resonances, see Appendix, Eq. 85) in the body of the violin, however, and subsequent conversion of that mechanical energy to airborne sound. Thus, a fraction of the energy from the steady pull of the bow is converted into physical sound, with narrowly distributed spectral components (fundamental plus harmonics) shaped by the violin string and body (Hutchins 1978). The bowing of a violin is a form of stridulation (Ewing 1989), one of the most common mechanisms of sound production in arthropods. Arthropod stridulation involves friction between a pair of body parts, with one member of the pair typically having a spatially periodic structure, as in the teeth of a comb (Elsner 1983; Ewing 1989). The spatial periodicity and the relative velocity of the two body parts determine the temporal periodicity of the slip-stick relaxation process. In some species, evidently including many aquatic species, the body parts involved in stridulation are in direct contact with a resonant system (Ewing 1989). For sound production in those species, the operation of the violin very likely provides an excellent analogy.

In brass instruments, the relaxation process is the opening and closing (buzzing) of lips held tightly together as air is expelled through them. In woodwinds with reeds, the relaxation process is the opening and closing of the valve formed by the one or two reeds. In air-jet instruments, such as the flute, the relaxation process is the switching on and off of airflow from a transverse stream into the opening of a tube. In each case, the resonance is created by the bounded volume of air in the instrument (Benade 1978a,b). In the trained human singer, the relaxation process is opening and closing the vocal folds of the glottis; the resonance is created by the volume of air bounded by the vocal tract (Sundberg 1978). One finds analogous sound production structures and operations in many vertebrate species, including mammals (Muller-Preuss and Ploog 1983) and birds (Seller 1983).

In the cases of the violin, the brass instruments, the woodwinds, and the trained human singer, the resonances are formed by standing-wave structures. Therefore, resonances occur for whole sets of harmonics. Because the waveforms generated by the relaxation processes are rich in harmonic content, the physical sounds generated by these various instruments also will be, and the human auditory system computes distinct timbres for the perceptual sounds from each of them.

5. Environmental Effects

Phenomena affecting sound propagation in air include refraction, absorption in air, scattering, reflection, boundary absorption, interference, and diffraction. Refraction (within the air itself) is a direct consequence of gradients of the speed of sound. Absorption in air is a consequence of the random motions of air particles. Scattering, reflection, and boundary absorption occur at boundaries between air and solids or liquids. Interference occurs when sound takes more than one path from source to listener. Diffraction is the spreading or bending that occurs in a wave front when it is partially blocked by an acoustic barrier. Although the following descriptions are based on sound transmission in air, the same basic principles apply to sound transmission in water.

5.1 Refraction

In a still, fluid medium, the speed (ϕ) of audible compressional waves is equal to the square root of the adiabatic bulk modulus divided by the square root of the density (Appendix, Eq. 44). At atmospheric pressures, the adiabatic bulk modulus of air is directly proportional to the pressure, and, according to the gas law, the density is directly proportional to the pressure divided by RT (the gas constant times the absolute temperature). Therefore, the speed of sound in still air is equal to $\phi_0(1 + \Delta T/273)^{1/2}$ where ΔT is the temperature in degrees centigrade and ϕ_0 is the speed of sound at 0°C. Thus, temperature gradients, which can be strong near surfaces illuminated by the sun or near cold surfaces, are sources of conspicuous gradients of ϕ. The gradient of ϕ is a vector. When it has a component perpendicular to the direction of sound propagation, then the direction of sound propagation tends to rotate away from the direction of the vector. To understand this process, consider a plane sound wave with the gradient of the speed of sound aligned parallel to the wave front (perpendicular to the direction of propagation). The wave front will move faster where the speed of sound is greater. Thus, as it moves, the wave front will bend away from the high-speed region, and the direction of sound propagation (which is perpendicular to the wave front) will bend more and more toward the region of low sound speed. Unless the speed gradient has a constant magnitude, the wave front will not remain planar. The steeper the gradient, the more rapidly the wave front will be bent, and the more rapidly the direction of propagation will change.

Owing to drag forces between moving air and the ground, wind typically has a velocity gradient perpendicular to the ground and directed away from it (i.e., upward movement takes one from a region of lower wind velocity to a region of higher wind velocity). Because the velocity of airflow (wind) adds (in a vectorial sense) to the velocity of sound propagating in the air, velocity gradients in a steady wind create direction-dependent gradients of the speed of sound and thus lead to refraction. For sound propagating in the direction of a steady wind, therefore, the gradient of ϕ is upward and the sound is bent downward, into the ground as it travels. Sound propagating in the opposite direction (into the wind) is bent upward. In gusty wind, gradients of the speed of sound are strong, irregular, and have horizontal as well as upward components. Sound passing through such wind tends to be scattered by refraction. The energy in the coherent wave front reaching the listener thus will be attenuated.

In warm, still air over cold bodies, such as lakes or snowfields, temperature gradients tend to be upward. Because the gradient of ϕ is aligned with the temperature gradient, sound propagating through those gradients is bent downward. In air over ground illuminated by the sun, temperature gradients may be downward; sound propagating through those gradients is bent upward.

5.2 Absorption in Air

Absorption of sound as it travels through air usually is attributed to three phenomena: viscosity, heat conduction, and molecular absorption. Viscosity in air is a consequence of the motion of the air molecules (Appendix, see discussion around Eq. 77). The particle velocity in the compressional wave model (Appendix, Eqs. 21 and 45) is superimposed on the random thermal motion of the particles. Components of random motion that are perpendicular to the particle velocity vector tend to transfer the momentum of particle velocity from layer to layer in the air, creating the equivalent of viscous drag between any two layers with different particle velocities. Thus, if there are particle velocity gradients normal to the particle velocity vector, as there are when sound passes through air close to a solid or liquid surface, there will be attenuation owing to viscosity.

Heat conduction in air is a consequence of the fact that hotter particles move faster than colder ones. Therefore, when a region occupied by hot particles is next to a region occupied by cold particles, the hot particles tend to move into the cold region at a greater rate than that at which the cold particles move into the hot region. In the range of audible frequencies, compression and rarefaction in the compressional wave model are adiabatic (Appendix, discussion around

TABLE 2.1. Absorption and speed of sound in still air ($T = 20°C$).

Frequency (Hz)	Absorption (dB/km) (humidity)			Speed (m/s)* (humidity)		
	(0%)	(50%)	(100%)	(0%)	(50%)	(100%)
20	0.51	0.02	0.01	343.48	344.06	344.68
40	1.07	0.07	0.04	343.51	344.06	344.68
50	1.26	0.11	0.06	343.53	344.06	344.69
63	1.43	0.17	0.09	343.54	344.06	344.69
100	1.67	0.38	0.22	343.55	344.06	344.69
200	1.84	1.03	0.77	343.56	344.07	344.69
400	1.96	1.85	2.02	343.56	344.07	344.69
630	2.11	2.41	3.05	343.56	344.08	344.70
800	2.27	2.79	3.57	343.56	344.08	344.70
1250	2.82	4.04	4.59	343.56	344.08	344.70
2000	4.14	7.14	6.29	343.56	344.08	344.70
4000	8.84	22.2	13.6	343.56	344.08	344.71
6300	14.9	51.3	27.7	343.57	344.08	344.71
10,000	26.3	123	63.5	343.57	344.08	344.71
12,500	35.8	187	96.6	343.57	344.09	344.71
16,000	52.2	294	155	343.57	344.09	344.71
20,000	75.4	455	238	343.57	344.10	344.71
40,000	267	1255	884	343.57	344.13	344.72
63,000	645	2147	1974	343.57	344.15	344.73
80,000	1032	2768	2913	343.57	344.16	344.74

*Computed for air as an ideal gas at 20°C, assuming adiabatic compression, the speed of sound is 343.37 m/s.
From Weast, RC (ed) 1986 Handbook of Chemistry and Physics. Boca Raton, FL: CRC Press.

Eq. 2). Therefore, regions of compression are warmer than the mean ambient air temperature and regions of rarefaction are cooler. Along the direction of travel, the sound wave comprises alternating regions of compression and rarefaction, thus alternating regions of warm and cool. Thermal conduction produces a flow of particles from the warm regions (compression) to the cool regions (rarefaction), tending to diminish the pressure difference and attenuate the wave amplitude. Because the pressure gradient (and thus the temperature gradient) increases with increasing frequency, so does the rate of attenuation by this mechanism.

Molecular absorption is attributed largely to polyatomic molecules (e.g., H_2O and CO_2), which have vibrational modes of energy storage not available in diatomic molecules (e.g., O_2 and N_2). As one would expect, it is strongly dependent on both the humidity and the frequency (Beranek 1954). Some of the effects of absorption by water molecules can be inferred from Table 2.1.

5.3 Reflection, Boundary Absorption, and Scattering

When airborne sound impinges on the interface between the air and a solid or liquid object, such as the ground, the surface of a lake, or the surface of a powdery snowfield, it undergoes varying degrees of reflection, absorption, and scattering. Scattering occurs when irregularities in the surface are comparable in size to a wavelength of the sound in air. Scattering tends to destroy the coherence of the wave front and thus to drastically attenuate the sound. Specular reflection occurs when the surface is smooth (no irregularities comparable in size to a wavelength) and not porous. Boundary absorption occurs when the surface is porous. Thus, rough ground tends to scatter sound, a smooth lake surface tends to reflect it, and a field of powdery snow tends to absorb it. Imagine the effect of a gentle wind or a thermal gradient over each of these surfaces. A surface that appears smooth relative to the long wavelengths of low-frequency sound may be extremely irregular relative to the short wavelengths of high-frequency sound. Therefore, the reflection and scattering from a surface are frequency dependent and thus will shape the spectral distribution of propagating sound.

5.4 Interference

According to the compressional wave model (see Appendix, especially discussion surrounding Eq. 45), the pressures of waves arriving at a point from different directions sum at that point, and the particle velocities of those waves add vectorially at that point:

$$P_{total}(x,y,z,t) = \sum_{i=1}^{N} P_i(x,y,z,t)$$

$$\vartheta_{total}(x,y,z,t) = \sum_{i=1}^{N} \vartheta_i(x,y,z,t)$$

$$\vartheta_{xtotal}(x,y,z,t) = \sum_{i=1}^{N} \vartheta_{ix}(x,y,z,t) \quad (2.1)$$

where $P_i(t)$ is the pressure of the ith wave at the point of convergence and ϑ_{ix} is the particle velocity component in the x-direction of the ith wave at the point of convergence. The particle velocity vector (ϑ_i) of a compressional wave always is parallel to the direction of propagation of that wave, being in the forward direction when the corresponding pressure, $P_i(t)$ is positive, in the reverse direction when $P_i(t)$ is negative. Because $P_i(t)$ and $\vartheta_{ix}(t)$ may be either positive or negative, the contributions to P_{total} and ϑ_{xtotal} from pairs of converging waves may be of the same sign and thus reinforce one another (*constructive interference*), or they may be of opposite sign and thus tend to cancel one another (*destructive interference*). For a pair of converging waves, particle velocities may undergo constructive interference in one direction and, at the same time, destructive interference in another.

5.5 Diffraction

Diffraction is a spreading of the direction of propagation that occurs whenever sound propagates around or between acoustic barriers. The usual model for diffraction of a wave is based on Huygen's principle, which states (1) that all points in the surface tangent to the wave front can be considered as point sources, each emitting spherical secondary wavelets; and (2) that after a period of time, the new position of the wave front is the surface forming an envelope tangent to all the secondary wavelets. Add to this Kirchhoff's constraint (that the wavelets have zero intensity in the reverse direction) and, with some simple geometric construction, it is relatively easy to show that according to this model, a complete, advancing spherical wave front in a uniform medium will remain spherical about the same center, and an advancing planar wave front will remain planar. On the other hand, if a barrier with a small aperture were placed in the path of the advancing wave, so that only the sound impinging on the aperture were allowed to pass, then the pattern of the original wave front would be broken. A model of the wave front emerging from the aperture can be constructed as the envelope of the Huygen's wavelets. If the aperture is sufficiently small, the emerging wave front will be approximately spherical, with the aperture being at the center of the sphere.

If the diffracted sound wave comprises sinusoidal components, and if the width of the aperture is not small compared to the wavelength of such a component, then an interference pattern for that component will be superimposed on the wave front. Thus, at some places along the advancing wave front interference will be predominantly constructive, and at other places it will be predominantly destructive. One can construct a model of the pattern by taking Huygen's model more literally and considering the time of travel (at the speed of sound) from the source (center) of each spherical wavelet to each point along the wave front. For sinusoidal waves, time of travel can be converted to phase, which in turn translates into instantaneous sign and amplitude for the wavelet from each source. Constructions of this sort usually are carried out at distances from the aperture that are large in comparison to the width of the aperture. This guarantees that, at a given point along the wave front, the particle-velocity vectors of all converging wavelets will be nearly parallel.

5.6 Combining the Environmental Effects

The seven effects described in this section (i.e., refraction, absorption in air, scattering, reflection, boundary absorption, interference, and diffraction) work together to reshape not only the amplitude and spectral content of physical sound as it propagates through the environment, but also its temporal structure. In the absence of any of these effects in a spherical wave front, which spreads radially outward from its source, the amplitude of the pressure declines as the reciprocal of the distance from the source (a *spreading attenuation* of 20 dB for every 10-fold increase in distance). In the absence of the other six effects, absorption in still air leads to additional attenuation. Figure 2.1, constructed from the data in Table 2.1, shows propagation distances in still air required for 20-dB attenuation (99% reduction in power level) due to absorption. For frequencies below 1 kHz, the required distances range upward from approximately 10 km. For frequencies greater than 10 kHz, they range downward from approximately 100 m. For

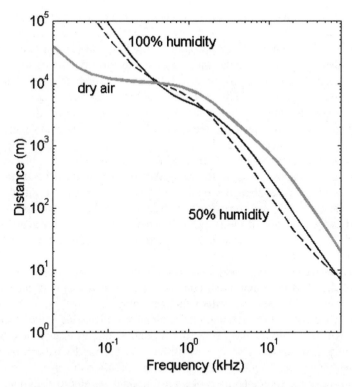

FIGURE 2.1. Distance required for airborne sound to lose 99% of its energy (20-dB loss) to absorption, plotted as a function of frequency for dry air (gray line), 50% humidity (dashed line), and 100% humidity (solid line) at 20°C. (Data from Table 2.1.)

humid air and frequencies greater than approximately 60 kHz, the required distance is less than 10 m (i.e., for every 10 m of propagation distance, the power would be reduced 99% by absorption alone). It is clear that frequencies in this range would be unsuitable for long-distance communication. Very low frequencies, on the other hand, would be quite suitable, being not only less susceptible to absorption in air, but also less susceptible to scattering and reflection by small to moderate-sized objects in the sound path (Poole et al. 1988). The high frequencies (approximately 40 kHz and above) would be suitable for very local, private communication—not apt to be overheard by distant predators. Signals at those frequencies also would not be subjected to interference by similar signals from distant sources. This could be an advantage for animals with high population densities. It is interesting to consider its importance for echolocation (Griffin 1971).

Small obstacles, such as leaves, in the path of the sound introduce an additional component of frequency-dependent attenuation (Aylor 1971; Marten and Marler 1977; Marten et al. 1977). This is largely a consequence of scattering, reflection, boundary absorption, and additional absorption in air owing to increased effects of viscosity (more surfaces creating particle-velocity gradients). These factors make the optimum choices of communication frequencies strongly dependent on the habitat (of course they continue to depend on other factors, such as the desirability of privacy). Upward-directed gradients of ϕ (the speed of sound) over scattering or absorbing surfaces tend to create acoustic shadows beyond a certain distance. Over powdery snow, for example, the refracted sound wave is absorbed. One can imagine the impact of this on the defensive strategies of a varying hare in lynx country in the winter. Over reflecting surfaces, upward-directed gradients of ϕ tend to create acoustic ducts, through which sound does not spread spherically and therefore suffers spreading attenuation conspicuously less than 20 dB for every 10-fold increase of distance from the source. An upward gradient of the speed of sound over a smooth lake surface, for example, would bend the wave into the lake, only to have it reflected upward again, then refracted downward, then reflected upward, and so on. The sound energy would be trapped in a layer over the surface of the lake and thus could carry over a long distance. The edge of a still lake would be an excellent station for a listener (e.g., a hunter) or a communicator. An analogous effect occurs for low-frequency sounds in the sea, producing a channel for very-long-distance communication (Mercado and Frazer 1999; Stafford et al. 1998). Marine vertebrates also have the option of local, private communication (Fine and Lenhardt 1983; see Ladich and Popper, Chapter 4, section 5).

Diffraction around large objects and reflection from large objects alter the direction of sound propagation and can create multiple paths, with different travel times from source to listener. This obscures the direction of the source and makes the temporal structure of the sound reaching the listener's ear different from that at the source. In enclosed volumes with reflecting walls, complicated reverberations occur, with decay times determined by the combination of surface absorption and absorption in air. These features place additional burdens on the computational machinery of the auditory central nervous system (CNS).

6. Temporal and Spectral Properties of Periodic Acoustic Waves

Although the sine and cosine functions are human mathematical inventions, those functions describe very well the dynamical behavior of simple resonances (Appendix, Eqs. 82 to 86) when they contain residual energy and are allowed to operate freely, with no further input of energy. The second law of thermodynamics tells us that the sinusoidal oscillations of freestanding macroscopic resonances will be damped and eventually will disappear.

Fourier's theory tells us that any periodic wave can be reconstructed as a sum of harmonically related sine waves, each of appropriate amplitude and phase. Thus, if one could adjust the amplitudes, frequencies, and relative phases of the outputs of many oscillators, one could combine them to synthesize the output of a single resonance-based musical instrument or a single resonance-based bioacoustic source. The reverse of this process (decomposing a periodic waveform into sine waves of various frequencies, amplitudes, and relative phases) is called *Fourier analysis*.

When confronted with a periodic waveform, modern researchers typically approximate Fourier analysis by capturing and sampling a segment of that waveform, shaping the sampled waveform with a window function, and computing the discrete Fourier transform of the result. The outcome is a Fourier analysis of a new periodic wave, comprising the sampled, truncated, windowed waveform repeated over and over again. The fundamental period of the resulting set of harmonically related sine waves is the duration of the truncated waveform. The same approach is applied as well to segments of aperiodic waveforms and to transient waveforms. Strictly speaking, transient waveforms and segments of aperiodic waveforms cannot be constructed as sums of sine waves. What one can construct, however, is a periodic waveform for which the shape of the wave during each period is a windowed approximation to the segment or the transient waveform.

Prior to the widespread availability of digital computers, researchers often used a spectrographic approach to estimate the harmonic content of a waveform. Typically, this involved one of two procedures. One could pass the waveform repeatedly through a single linear filter whose tuning was varied systematically. Alternatively, one could pass it once through a parallel array of linear filters, each tuned to give maximum sinusoidal steady-state response at a slightly higher frequency than that of its neighbor on one side and a slightly lower frequency than that of its neighbor on the other side. This alternative approach was especially convenient. The output of the array, called a *spectrogram* (or *sonogram*), is a three-dimensional function: amplitude versus time for each linear filter, versus frequency when taken over the entire tonotopic array of filters. Although the spectrum of a transient waveform is not well defined, the spectrogram gives an intuitively satisfying picture of frequency content versus time for such a wave.

6.1 Doppler Shifts

If a sound source is moving relative to the fluid (e.g., air or water) in which it is immersed, the spatial length of the emitted acoustic signal depends on the direction of propagation from the source. For example, if the duration of a very brief signal at a nonmoving source in a uniform fluid is T_s, the spatial length of the signal would be the same in every direction, namely ϕT_s, where ϕ is the speed of sound in the fluid. If source were moving with constant speed, s_s, then directly ahead of the object the spatial length of the signal would be shortened to $(\phi - s_s)T_s$, and directly behind the object the spatial length of the signal would be lengthened to $(\phi + s_s)T_s$. In other directions, the spatial length, L, of the signal would be intermediate, given by the more general expression

$$L = (\phi - s_s\cos(\theta))T_s \tag{2.2}$$

where θ is the angle of the direction of propagation relative to the direction of source motion. As the signal passes a motionless listener or target, its duration, T_L, would be given by its spatial length divided by the speed of sound in the fluid,

$$T_L = L/\phi \tag{2.3}$$

For a listener or target moving with speed s_L directly away from the source, the duration of the signal would be increased to $L/(\phi - s_L)$. For a listener or target moving in some other direction, the duration of the signal would be given by the more general expression

$$T_L = L/(\phi - s_L\cos(\chi)) \tag{2.4}$$

where χ is the angle of the listener's or target's motion relative to the direction of propagation (the direction of a line from the source to the listener). For the very brief signal being considered here, θ and χ would be expected to be constant. It is interesting to contemplate the effects of variations in θ and χ over prolonged signals. For the very brief signal, if the target or the listener also were a reflector, it could be treated as a source emitting a new signal of duration T_L, and the calculations could be continued from there rather easily.

If the signal contained a sinusoidal component at frequency f_s, the number of cycles contained in the signal would be f_sT_s. The listener would hear the same number of cycles spread over T_L. The frequency would be shifted to f_sT_s/T_L. The difference between f_s and f_sT_s/T_L is the *Doppler shift*.

6.2 Properties of Sounds from Relaxation Oscillators

As suggested above in section 4, periodic waveforms in nature, such as the vocalizations produced by many vertebrates, typically are produced by a relaxation process coupled to a resonance. Spectral analysis of such a periodic waveform would yield a very large number of harmonically related components, i.e., the frequencies of the components would be integral multiples of the same fundamental frequency. All of these components would begin together when the periodic waveform begins, and they would maintain constant phase relationships to one another. If the relative velocities of the source and the listener were changing, then all of the spectral components would be subject to the same, time-varying, Doppler-shift factor (i.e., to the same temporal patterns of frequency modulation). All of the spectral components also would face the same time-varying attenuating factors (e.g., changes in the orientation of the source, objects moving in and out of the sound path, etc.). The degree of amplitude modulation by such factors would depend on the frequency of the component, but the basic temporal pattern of modulation would be the same for all components. In a simple environment, all spectral components would arrive at the listener's position from the same direction, and thus exhibit the same interaural time difference.

6.3 Sorting Sounds from Different Sources

The acoustic environment of the listener will not always be a quiet one. To draw inferences about the source of a sound (is it a potential mate, an offspring in distress, a potential predator, a potential prey, a potentially dangerous nonbiological phenomenon?), the listener's brain must separate that source's sounds from the other sounds impinging on the listener's ears (see section 9, below). The higher level neural computations involved in this separation task have been studied only psychophysically and primarily in human listeners. Nonetheless, these psychophysical studies have been extraordinarily revealing. For periodic sounds, the physical features that the auditory brain uses to sort sounds are precisely those listed in the previous paragraph—harmonicity, common onset time, common amplitude-modulation pattern, common frequency-modulation pattern, and common interaural time difference (Hartmann 1988; Yost 1991).

To utilize those features, the auditory nervous system must decompose monaural auditory input (from many sources) into spectral components. Then it must identify a subset of those components that exhibited common onset time, common amplitude modulation, and common frequency modulation. At the same time, it must search for a similar set of components from the other ear with common interaural time differences. Finally, it must integrate all of the components meeting those criteria into an estimate of the acoustic source inferred to have given rise to the waveform. The first step in the neural computation of that inference would be spectral decomposition. Biological structures cannot perform true Fourier analysis, but they can perform spectrographic decomposition. All that would be required would be an appropriate tonotopic array of filters. To allow for the detection of harmonicity, such an array would have to possess spectral resolution that was a small fraction of an octave. For example, detection of the sixth harmonic would require spectral resolution much better than 0.26 octave, the difference between the fifth and sixth harmonic. To allow for the detection of common onset times, on the other hand, the array would have to respond very quickly in time. Therefore, one expects the individual filters in the array performing spectrographic analysis to provide both good frequency discrimination and rapid temporal response to incoming signals (Lewis 1987, 1990, 1992; Cortopassi and Lewis 1998).

6.4 Filter Requirements

There are two ways that signal-processing engineers can achieve good frequency discrimination in linear filters: (1) make the bandwidth of the filter extremely narrow, so that the filter responds preferentially to ongoing sine waves whose frequencies fall within a very narrow range; (2) make at least one band edge of the filter extremely steep, so that the responses of the filter to ongoing sine waves within its pass band are much greater than those to ongoing sine waves at frequencies beyond the steep band edge. The first method can be achieved with structures of low dynamic order, i.e., structures possessing a low number of interacting, independent energy storage elements (e.g., a low number of interacting masses and springs). Because low dynamic order implies a small number of structural elements, parsimony suggests that such a scheme would be favored in the evolution of inner-ear tuning structures. Unfortunately, in accomplishing tuning by reduction of bandwidth, one must sacrifice time resolution. In the context of the previous paragraph, this leads to two serious problems: (1) following the onset of a sine wave, the response of a narrow-band filter requires considerable time to develop, and that time increases as the bandwidth of the filter decreases, and (2) residual excitation in a narrow-band filter, such as that elicited by prior background noise or prior signals, requires considerable time to decay, and that time also increases as the bandwidth of the filter decreases. In other words, as one improves the frequency discriminability of a filter by narrowing its bandwidth, that filter becomes more sluggish in responding to new signals, and those signals must compete with increased interference by signals from the immediate past. These problems are not present in filters in which frequency discrimination is achieved by steep band edges rather than narrow bandwidth. In filters of this type, which require high dynamic order (large numbers of interacting, independent energy storage elements), there need be no trade-off between spectral acuity and time response. Therefore, for purposes of using both spectral and temporal cues to sort acoustic signals from one another (e.g., for tracking a given signal in an acoustic environment filled with noise and competing signals), an array of such filters would be the appropriate structure for performing spectrographic analysis (Lewis 1987, 1990, 1992; Cortopassi and Lewis 1998).

When the environment is quiet, the sounds of individual sources should stand out clearly in the vicinity of the ears. In that case, the segregation and integration of sounds from external sources is a less important aspect of the receiver system. The principal impediment to sensing and identifying the sound under those circumstances is the internal noise in the ear itself. This arises from thermal (Brownian) motion of the inner-ear fluids, random opening and closing of ion channels, noise (random voltage fluctuations) generated by electric currents passing through open ion channels, and the like (de Felice 1981; Bialek 1983; Bialek and Schweitzer 1985). Under those circumstances, spectrographic analysis with tonotopically arrayed, narrow-band filters would be thoroughly appropriate (Lewis 1987). Thus, considering only the physical properties of periodic sounds, the power of spectrographic analysis, and the nature of filters, one concludes that a good design for the auditory periphery would be narrow-band filtering at very low ambient sound levels, giving way to broader-band filtering (with steep band edges) at higher ambient sound levels. This indeed seems to be the design path followed by evolution in the terrestrial vertebrates (Evans 1977; Moller 1977, 1978, 1986; Carney and Yin 1988; Yu 1991; Lewis and Henry 1994).

6.5 Electrical Resonances

Over the past 20 years, electrical resonances have been found to be ubiquitous among acoustic hair cells of nonmammalian vertebrates (Manley and Clack, Chapter 1, section 5.1.1; Coffin et al., Chapter 3). These resonances exhibit dynamic order two, implying a pair of complementary energy storage elements (Lewis 1987; see also Appendix, section 9.1). With appropriate quality factor (Q), such resonances could serve as narrow-band filters for operation at very low ambient sound levels; but, for the reasons given above, they would be utterly maladaptive at higher levels. Electrical resonances have been observed largely in hair cells from which the tectorial structures have been removed. Their hair bundles thus have been separated from their normal mechanical milieus. A question raised by the maladaptive nature of low-order filters is the following: What would be required to make the electric elements of such a resonance (voltage-dependent calcium channels and calcium-dependent potassium channels) part of a tuning structure of high dynamic order? The most reasonable answer to this question is bidirectional transduction in the hair cell (Lewis 1987, 1988, 1990, 1992; Lewis and Lombard 1988). Under normal conditions (tectorium in place) this would allow the electrical elements of the hair cell to interact with the mechanical elements associated with the tectorium. It also would allow the electrical elements of neighboring hair cells to interact with each other, through the tectorium. The importance of bidirectional coupling and the presence of the tectorium can be argued convincingly from the observations made on turtle hair cells, the first in which electrical resonances were found (Crawford and Fettiplace 1980). When the tectorium is disconnected from the hair bundle, the quality factor of the resonance is approximately 10-fold greater than it is when the tectorium is connected (Art and Fettiplace 1987). This implies that when the tectorium is connected, during each cycle of resonance, the electrical elements transfer a considerable fraction of their stored energy to the tectorium (Sneary and Lewis 1989). This, in turn, would be transferred, through forward transduction, to the electrical elements of neighboring hair cells (Lewis and Lombard 1988).

7. How the Nervous System May Have Taken Advantage of Noise

So far, the presence of noise (along with competing signals) has been portrayed as an obstacle that had to be overcome by a sensory system. One might make a strong case, on the other hand, for the proposition that noise has made the peripheral nervous system, as it currently exists, possible. Noise arises from the seemingly random, ongoing process of energy redistribution among the members of populations of particles. In neurons, for example, this leads to (among other things) seemingly random motions of ions and seemingly random opening and closing of ion channels (de Felice 1981). Among neurobiologists it is widely believed that the peripheral nervous system in macroscopic organisms owes it existence to the ability of the neuron to generate and propagate the all-or-none spike (or action potential). In his classic theoretical studies of action potentials, FitzHugh (1955, 1961) argued that the all-or-none (threshold) property of neurons must arise from the presence of noise. Without noise, he argued, the neuron would lack its apparent threshold and would produce completely graded responses. Cole et al. (1970) showed that the all-or-none (threshold) properties of the squid giant (stellate) axon disappeared if the axon were warmed about 20° (C) above its normal operating temperature. Clay (1976) subsequently showed that this was strong confirmation of FitzHugh's theory—that the increase in the speed of channel kinetics brought about by the temperature increase would be expected to overcome the effects of noise. For a more detailed overview of these arguments, see MacGregor and Lewis (1977) and Lewis et al. (2000). In summary, the presently available theoretical and experimental evidence compellingly favors the notion that the nervous system uses noise to create the all-or-none spike and its apparent threshold.

At nearly the same time that FitzHugh was drawing his conclusions, Lowenstein (1956) was attempting to explain how peripheral sensory systems could overcome the all-or-none property of the neuron to encode very weak sensory signals—signals that, by themselves, would be far below the spike threshold. His solution was a suprathreshold DC biasing current applied to the spike trigger region of the sensory neuron. This would produce a background spike rate that could be modulated by sensory signals that would have been below threshold for the quiescent neuron. Applied to an idealized spike-trigger model, without noise, a constant suprathreshold current yields perfectly periodic spikes (Agin 1964; Stein 1967). Superimposed on such a spike train, even the tiniest perturbation of a spike period would, in principle, be detectable. Owing to the noise that must be present in any neuron and any sensor operating far from $0°K$, however, some degree of randomness is expected in spike periods. This is reflected in nonzero coefficient of variation (CV) of the spike period. If one takes the mean spontaneous spike period (the mean spike period in the absence of input stimulus) to be the parameter of the spike train that represents Lowenstein's idealized background level, then the random deviations from the mean spike interval would be considered background noise. The CV of the spike period in the absence of sensory input then becomes the ratio of the background noise amplitude to the idealized background level. Afferent axons of the vertebrate vestibular system exhibit CVs ranging from the neighborhood of 0.1 (relatively low background noise) to the neighborhood of 1.0 (relatively high background noise) (Goldberg and Fernandez 1971). Working largely on that system, Lowenstein must have encountered many low-CV axons. In such axons the implication of a DC background was clear.

The spontaneous spike activities in vertebrate auditory neurons, on the other hand, exhibit CVs in the neighborhood of 1.0. French and Stein (1970; see also Stein 1970; Stein et al. 1972) proposed that such background noisiness would allow a sensory neuron to encode the continuously varying amplitude of a sensory input in a continuous manner. In the scheme proposed by French and Stein, Lowenstein's DC biasing current becomes a random AC, which occasionally causes the neuroelectric potential at the spike trigger to exceed threshold, leading to random spike production. The resulting mean spike rate then is modulated by any additional current (response current) produced by the sensory input. When the response current is negative, the probability of exceeding threshold decreases in proportion to the strength of the current. The spike rate decreases accordingly. When the response current is positive, the effect is reversed. This is analogous to the process of dithering, which is used by auditory engineers in digital recording (Vanderkoy and Lipshitz 1984). In that case, low-amplitude, high-frequency noise is added to the analog sound waveform to remove the distortion imposed by the threshold effects during analog-to-digital conversion. Currently available evidence makes a compelling case for the presence of the French-Stein mode of operation in a large proportion of vertebrate primary acoustic afferent axons (Yu and Lewis 1989; Lewis et al. 2000). Thus, it apparently is noise that gives an axon its all-or-none threshold properties, and, in many instances, it apparently is noise that, through a dithering process, overcomes the strong nonlinearity imposed by the axonal threshold and allows nearly linear representations of low-level sensory signals.

8. Impedance Matching

Sensors operate by capturing some of the energy of the signal to be detected. In the case of the ear, that energy ultimately is used to gate or modulate neuroelectric signals, which subsequently are processed by the nervous system. When the site of the translation from acoustic signals to neuroelectric signals is internal, as it is in the case of the vertebrate inner ear, then one should expect to see pathways for acoustic signal energy from the outside of the body to that site of translation. Furthermore, especially for low signal intensities, one should expect to see structures in those pathways that enhance the transfer of energy from the outside to the inside. Specifically, for low signal intensities one should expect to find structures that reduce *reflection* of sound energy back to the outside world. This means that one should expect to see, in the pathway, devices (transformers) that provide impedance matching (see Appendix, section 5). This, in turn, means that one should look for chains of passive mechanical transducers, such as pistons (or diaphragms) and semilevers, as well as acoustic transformers, such as tubes that flare (analogous to a megaphone). Of course, one finds them (see Ladich and Popper, Chapter 4, section 5; Clack and Allin, Chapter 5; Smotherman and Narins, Chapter 6, section 3). The columella of modern frogs and the ossicular chain of mammals, as examples, both appear to be concatenations of semilevers, with piston-like devices on each end for example (the inverse of the hydraulic press, combined with one or more pairs of semilevers). This follows precisely the general recipe for a transformer (Appendix, see discussion around Eqs. 59 and 60). For terrestrial vertebrates, it has been said that these transformers serve to match the characteristic (acoustic) impedance of the air to that of the aqueous medium in the ear (i.e., to the characteristic impedance of water). The discussion around Equations 64 to 69 and 75 and 76 (Appendix) should raise serious doubt about that conclusion. The characteristic acoustic impedance of water is directly related to the adiabatic bulk modulus of water (Appendix, see discussion around Eq. 44). It seems highly unlikely that the bulk modulus of water has anything to do with any of the mechanical impedances faced by the acoustic signals traveling through the middle ears of terrestrial vertebrates.

9. Sound Source Segregation and Integration

Clearly, the sensing of a sound is computationally much simpler and more basic than discrimination, where time-varying differences in temporal and spectral patterns must be processed. It seems to be widely believed that one of the most primitive and fundamentally important original functions of hearing is the detection of predators (and other dangers) and prey (and other resources). But a little reflection reveals that what one could mean by predator and prey detection may be considerably more complex than simple signal detection and successive discrimination, as usually studied in the laboratory. First, predators and prey are objects that have locations, distances, sizes, masses, movements, etc. Second, these objects of potential interest are usually part of the collection of objects and events (sound sources and scatterers) making up the usual environments, and useful decisions about them would require that information be acquired and processed independently of the rest of the auditory ambience, or scene (Bregman 1990). Thus, the sounds from these objects not only need to be sensed and be discriminably different, but also must be segregated or sorted from one another (see section 6.2, above) when present simultaneously in order for appropriate behaviors to occur with respect to them (Yost 1991; Fay 1992, 1998, 2000; Lewis 1992; Cortopassi and Lewis 1998; de Cheveigne 2000).

A simple case of segregation would have to occur in the event of detecting the sounds from a single predator or prey source against a background of nonbiological noise (say, the environmental sounds caused by wind and rain). It would be tempting to suggest that such noise could be "filtered out" by components of the auditory system that are somehow "tuned" or especially selective to sounds of obvious biological significance, utterly rejecting or ignoring all other sounds. While one could argue that nonbiological environmental noise may not be insignificant to the organism's fitness in many situations (e.g., in the organism's determination of the general structure of the local environment), this does not alter the basic signal-processing requirements of the auditory system. Whether or not the nonbiological noise is relevant to the animal's survival, the problem for the animal is the same, namely the segregation of "noise" and "signal" as independent entities. This is not simply a matter of signal detection or signal-to-noise ratio. A serious mistake would be to confound the acoustic components of these two types of sources in a way that predicted some single, virtual source that had no single existence in reality. This is an example of sensory incompetence, and is probably very rare. Unless the "noise" and "signals" encountered by an organism are very well predictable (this does not seem likely), the most useful solution to source determination here is probably not modeled by a predator filter or a prey filter, but rather is likely to be a "separation machine" (de Cheveigne 2000) that parses mixtures of sounds from several sources into independent perceptual entities, each representing a single source. In this context, it is important to recognize that all audible sounds are of "biological significance" in the sense that all must be subject to this active, segregation operation. To the extent that certain sound sources (and scatterers) are of greater immediate importance than others with respect to survival, appropriate behavior can arise from decisions made *after* the segregation operation has taken place. The fundamental argument here is that source segregation with some degree of validity must take place before these hierarchical decisions can be made. This scenario is not necessarily simpler than that required for receiving communication sounds and behaving appropriately with respect to them. It seems likely that the structures and neural processing mechanisms used in basic sound communication are the same ones that are required for sound source determination and segregation.

Bregman (1990) and many others have demonstrated that sound source segregation is a hallmark of human hearing, and it is easy to recognize its operation by introspection. This aspect of human perception might appear to be in the realm of consciousness. Thus, it could be argued that its operation seems recently derived in animals like humans with comparatively large brains. But we believe that this fundamental segregation operation would be essential to all hearing organisms, and for all time. Sound source segregation has been demonstrated experimentally in European starlings using natural bird song signals (Hulse et al. 1997; Wisniewski and Hulse 1997) and more abstract pure tone sequences (MacDougall-Shackelton et al. 1998). In addition, behavior consistent with source segregation has been demonstrated in goldfish (Fay 1992, 1998, 2000). Thus sound source segregation and the neural encoding processes and representations at its foundation seem likely to be primitive or shared characters among vertebrates that originated quite early in vertebrate history, if not before the first vertebrates appeared. Among other things, this scenario suggests that these fundamental encoding and representation processes in the sense of hearing should be found widely among vertebrate animals, and there is no obvious reason to suggest that they suddenly arose in higher vertebrates. From the discussion in section 6.2, for example, it is clear that the prerequisite to the ability of the auditory nervous system to carry out the segregation operation is a peripheral sensor/processor that provides both good frequency discrimination and rapid temporal response to incoming signals. This ability indeed is found in the ears of all vertebrates investigated, including mammals (Patuzzi 1996), birds (Gleich and Manley 2000), reptiles (Manley 2000), amphibians (Lewis 1992), and fishes (Fay 1997). Its necessity may have been the principal selective pressure leading to bidirectional transduction at the auditory periphery (see section 6.4) and massive parallelism in the auditory CNS (Lewis 1987).

10. Summary

This chapter identified and discussed some of the important ecological and physical factors that would have been omnipresent throughout the history of vertebrate evolution and must have had an impact on the evolution of vertebrate hearing. Subsequent chapters in this text show how observed morphologies, physiologies, and behaviors of extant auditory systems may have been adapted to effectively cope with those factors. There are two factors considered sufficiently important to warrant elaboration in this chapter, however. Those are the omnipresence of random thermal motion (noise) in systems operating far from 0°K, and the occurrence in some environments of numerous, simultaneous acoustic signals from many sources. For a signal from an individual source among the latter, the other signals compose a formidable blanket of interference. We believe the following: In addition to acoustic transducers, the sine qua non for any auditory system to function in such a world is the ability to isolate the signals from an individual source. This quintessential task must be achieved largely by neural computation, and its requirement establishes essential constraints on the acoustic transducers and the signal-processing structures (such as peripheral filters) associated with them and on the brain itself.

Clearly, the random thermal motion (noise) inevitably associated with acoustic transducers and signal-processing structures must add to the blanket of interference that tends to mask the signal from a single source. In this chapter, however, the discussion did not focus on that consequence of noise. Instead, it was argued that the evolving nervous system actually derived advantage from noise, using it to create a threshold effect, making possible the familiar all-or-none spike, and using it as well to overcome the limitation imposed by a threshold on the coding of weak sensory signals.

This chapter discussed the nature of sound sources, the impact of the physical world on sound being transmitted from place to place, and the elements of human technology that seem to provide insight into evolutionary design, such as signal-processing theory, including the notion of spectrographic representation, and the circuit-theory metamodel, including the notions of impedance and resonance.

References

Agin D (1964) Hodgkin-Huxley equations: logarithmic relation between membrane current and frequency of repetitive activity. Nature 201:625–626.
Art JJ, Fettiplace R (1987) Variation of membrane properties in hair cells isolated from the turtle cochlea. J Physiol (Lond) 385:207–242.
Aylor D (1971) Noise reduction by vegetation and ground. J Acoust Soc Am 51:197–205.

Benade AH (1978a) The physics of woodwinds. In: Hutchins CM (ed) The Physics of Music. San Francisco: Freeman, pp. 34–43.
Benade AH (1978b) The physics of brasses. In: Hutchins CM (ed) The Physics of Music. San Francisco: Freeman, pp. 44–55.
Beranek, LL (1954) Acoustics. New York: McGraw-Hill.
Bialek W (1983) Thermal and quantum noise in the ear. In: De Boer E, Viergever MA (eds) Mechanics of Hearing. Delft: Delft University Press, pp. 185–192.
Bialek W, Schweitzer A (1985) Quantum noise and the threshold of hearing. Phys Rev Lett 54:725–728.
Bregman AS (1990) Auditory Scene Analysis: The Perceptual Organization of Sound. Cambridge, MA: MIT Press.
Carney LH, Yin TCT (1988) Temporal coding of resonances by low-frequency auditory nerve fibers: single-fiber responses and a population model. J Neurophysiol 60:653–677.
Clay JR (1976) A stochastic analysis of the graded excitatory responses of nerve membrane. J Theor Biol 59:141–158.
Cole KS, Guttman R, Bezanilla F (1970) Nerve membrane excitation without threshold. Proc Natl Acad Sci USA 65:884–891.
Cortopassi KA, Lewis ER (1998) A comparison of the linear tuning properties of two classes of axons in the bullfrog lagena. Brain Behav Evol 51:331–348.
Crawford AC, Fettiplace R (1980) The frequency selectivity of auditory nerve fibres and hair cells in the cochlea of the turtle. J Physiol (Lond) 306:79–125.
De Cheveigne A (2000) The auditory system as a "separation machine." In: Breebart D, Houtsma A, Kohlrausch A, Prijs V, Schoonhoven R (eds) Physiological and Psychophysical Bases of Auditory Function. Maastricht: Shaker, 2001.
De Felice LJ (1981) Introduction to Membrane Noise. New York: Plenum.
Elsner N (1983) Insect stridulation and its neurophysiological basis. In: Lewis B (ed) Bioacoustics. New York: Academic Press, pp. 69–92.
Evans EF (1977). Frequency selectivity at high signal levels of single units in cochlear nerve and nucleus. In: Evans EF, Wilson JP (eds) Psychophysics and Psychology of Hearing. London: Academic Press, pp. 185–196.
Ewing AW (1989) Arthropod Bioacoustics. Ithaca, NY: Cornell University Press.
Fay RR (1988) Hearing in Vertebrates: A Psychophysics Databook. Winnetka, IL: Hill-Fay Associates.
Fay RR (1992) Structure and function in sound discrimination among vertebrates. In: Webster D, Fay RR, Popper AN (eds) The Evolutionary Biology of Hearing. New York: Springer-Verlag, pp. 229–263.
Fay RR (1997) Frequency selectivity of saccular afferents of the goldfish revealed by revcor analysis. In: Lewis ER, Long GR, Lyon RF, Narins PM, Steele CR, Hecht-Poinar E (eds) Diversity in Auditory Mechanics. Singapore: World Scientific Publishers, pp. 69–75.
Fay RR (1998) Auditory stream segregation in goldfish (*Carassius auratus*). Hear Res 120:69–76.
Fay RR (2000) Frequency contrasts underlying auditory stream segregation in goldfish. J Assoc Res Otolaryngol 1:120–128.
Fine ML, Lenhardt ML (1983) Shallow-water propagation of the toadfish mating call. Comp Biochem Physiol [A] 76:225–231.
FitzHugh R (1955) Mathematical models of threshold phenomena in the nerve membrane. Bull Math Biophys 17:257–278.

FitzHugh R (1961) Impulses and physiological states in theoretical models of nerve membrane. Biophys J 1:445–466.
French AS, Stein RB (1970) A flexible neural analog using integrated circuits. IEEE Trans Biomed Eng 17:248–253.
Gleich O, Manley GA (2000) The hearing organ of birds and *Crocodilia*. In: Dooling RJ, Fay RR, Popper AN (eds) Comparative Hearing: Birds and Reptiles. New York: Springer-Verlag, pp. 70–138.
Goldberg JM, Fernandez C (1971) Physiology of peripheral neurons innervating semicircular canals of the squirrel monkey. I. Resting discharge and response to constant angular acceleration. J Neurophysiol 34:635–660.
Griffin DR (1971) The importance of atmospheric attenuation for the echolocation of bats (*chiroptera*). Anim Behav 19:55–61.
Hartmann WM (1988) Pitch perception and the segregation and integration of auditory entities. In: Edelman GM, Gall WE, Cowan WM (eds) Auditory Function: Neurological Bases of Hearing. New York: Wiley, pp. 623–645.
Hulse SH, MacDougall-Shackelton SA, Wisniewski B (1997) Auditory scene analysis by songbirds: stream segregation of birdsong by European starlings (*Sturnus vulgaris*). J Comp Psychol 111:3–13.
Hutchins CM (1978) The physics of violins. In: Hutchins CM (ed) The Physics of Music. San Francisco: Freeman, pp. 56–68.
Ladefoged P (1982) A Course on Phonetics. New York: Harcourt Brace Janovich.
Lewis ER (1987) Speculations about noise and the evolution of vertebrate hearing. Hear Res 25:83–90.
Lewis ER (1988) Tuning in the bullfrog ear. Biophys J 53:441–447.
Lewis ER (1990) Electrical tuning in the ear. Comm Theor Biol 1:253–273.
Lewis ER (1992) Convergence of design in vertebrate acoustic sensors. In Webster D, Fay RR, Popper AN (eds) The Evolutionary Biology of Hearing. New York: Springer-Verlag, pp. 163–184.
Lewis ER, Henry KR (1994) Dynamic changes in tuning in the gerbil cochlea. Hear Res 79:183–189.
Lewis ER, Lombard RE (1988) The amphibian inner ear. In: Fritzsch B, Ryan MJ, Wilczynski W, Hetherington TE, Walkowiak W (eds) The Evolution of the Amphibian Auditory System. New York: Wiley, pp. 93–123.
Lewis ER, Baird RA, Leverenz EL, Koyama H (1982) Inner ear. Dye injection reveals peripheral origins of specific sensitivities. Science 215:1641–1643.
Lewis ER, Henry KR, Yamada WM (2000) Essential roles of noise in neural coding and in studies of neural coding. Biosystems 58:109–115.
Lowenstein O (1956) Peripheral mechanisms of equilibrium. Br Med Bull 12:114–118.
MacDougall-Shackelton SA, Hulse SH, Gentner TQ, White W (1998) Auditory scene analysis by European starlings (*Sturnus vulgaris*): perceptual segregation of tone sequences. J Acoust Soc Am 103:3581–3587.
MacGregor RJ, Lewis ER (1977) Neural Modeling. New York: Plenum.
Manley G (2000) The hearing organs of lizards. In: Dooling RJ, Fay RR, Popper AN (eds) Comparative Hearing: Birds and Reptiles. New York: Springer-Verlag, pp. 139–196.
Marten K, Marler P (1977) Sound transmission and its significance for animal vocalization. I. Temperate habitats. Behav Ecol Sociobiol 2:271–290.
Marten K, Quine D, Marler P (1977) Sound transmission and its significance for animal vocalization. II. Tropical forest habitats. Behav Ecol Sociobiol 2:291–302.

Mercado E 3rd, Frazer LN (1999) Environmental constraints on sound transmission by humpback whales. J Acoust Soc Am 106:3004–3016.

Moller AR (1977) Frequency selectivity of single auditory-nerve fibers in response to broadband noise stimuli. J Acoust Soc Am 62:136–142.

Moller AR (1978) Frequency selectivity of the peripheral analyzer studied using broad band noise. Acta Physiol Scand 104:24–32.

Moller AR (1986) Systems identification using pseudorandom noise applied to a sensorineural system. Comp Math Appl 12A:803–814.

Muller-Preuss P, Ploog D (1983) Central control of sound production in mammals. In: Lewis B (ed) Bioacoustics. New York: Academic Press, pp. 125–146.

Olson HF (1967) Music, Physics and Engineering. New York: Dover.

Patuzzi R (1996) Cochlear micromechanics and macromechanics. In: Dallos P, Popper AN, Fay RR (eds) The Cochlea. New York: Springer-Verlag, pp. 186–257.

Poole J, Payne K, Langbauer WR Jr, Moss C (1988) The social content of some very low frequency calls of African elephants. Behav Ecol Sociobiol 22:385–392.

Schellung JC (1978) The physics of the bowed string. In: Hutchins CM (ed) The Physics of Music. San Francisco: Freeman, pp. 69–75.

Seller TJ (1983) Control of sound production in birds. In: Lewis B (ed) Bioacoustics. New York: Academic Press, pp. 93–124.

Sneary M, Lewis ER (1989) Response properties of turtle afferent fibers: evidence for a high-order tuning mechanism. In: Wilson JP, Kemp DT (eds) Cochlear Mechanisms. New York: Plenum, pp. 235–240.

Stafford KM, Fox CG, Clark DS (1998) Long-range acoustic detection and localization of blue whale calls in the northeast Pacific Ocean. J Acoust Soc Am 104:3616–3625.

Stein RB (1967) The frequency of nerve action potentials generated by applied currents. Proc R Soc Ser B 167:64–86.

Stein RB (1970) The role of spike trains in transmitting and distorting sensory signals. In: Schmidt FO (ed) The Neurosciences. New York: Rockefeller University Press, pp. 597–604.

Stein RB, French AS, Holden AV (1972) The frequency response, coherence, and information capacity of two neuronal models. Biophys J 12:295–322.

Sundberg J (1978) The acoustics of the singing voice. In: Hutchins CM (ed) The Physics of Music. San Francisco: Freeman, pp. 18–23.

Vanderkoy J, Lipshitz P (1984). Resolution below the least significant bit in digital systems with dither. J Audiol Eng Soc 32:106–113.

Webster DB (1992) Epilogue on the conference on the evolutionary biology of hearing. In: Webster D, Fay RR, Popper AN (eds) The Evolutionary Biology of Hearing. New York: Springer-Verlag, pp. 787–794.

Wisniewski AB, Hulse SH (1997) Auditory scene analysis in European starlings (*Sturnus vulgaris*): discrimination of starling song segments, their segregation from conspecific songs, and evidence for conspecific song categorization. J Comp Psychol 111:337–350.

Yost WA (1991) Auditory image perception and analysis: the basis for hearing. Hear Res 56:8–18.

Yu XL (1991) Signal Processing Mechanics in Bullfrog Ear Inferred from Neural Spike Trains. Ph.D. dissertation, University of California, Berkeley.

Yu XL, Lewis ER (1989) Studies with spike initiators: linearization by noise allows continuous signal modulation in neural networks. IEEE Trans Biomed Eng 36:36–43.

3
Evolution of Sensory Hair Cells

ALLISON COFFIN, MATTHEW KELLEY, GEOFFREY A. MANLEY, AND
ARTHUR N. POPPER

1. Introduction

The ears of all vertebrate species use sensory hair cells (Fig. 3.1) to convert mechanical energy to electrical signals compatible with the nervous system. However, although the basic structure of hair cells is ubiquitous among the vertebrates and hair cells are also found in the lateral line of fishes and aquatic amphibians, a growing body of literature has demonstrated considerable heterogeneity in morphology and physiology in different taxa and even within different end organs of the same species. Although far less is known about the functional diversity that accompanies the differences in structure and physiology, it is increasingly likely that these differences reflect the ability to respond to different types of signals and/or to process signals in different ways before a neurotransmitter is released and a signal is sent to the brain.

What is also clear is that the vertebrate sensory hair cell is likely to be very ancient, at least in terms of the origins of the vertebrates (see Manley and Clack, Chapter 1, for a discussion of vertebrate phylogeny). Although soft tissues such as sensory hair cells do not survive in the fossil record, several lines of evidence support an origin of hair cells at least as far back as the very earliest vertebrates and perhaps even in vertebrate (chordate) ancestors (Burighel et al. 2003). Such evidence includes the presence of hair cells in both modern lamprey (jawless fishes that represent a sister group to jawed vertebrates) and modern gnathostomes (jawed vertebrates), suggesting an origin in a common ancestor at least 430 million years ago (Fig. 3.2 for a phylogeny highlighting the taxonomic groups discussed here). The existence of an inner ear with hair cells in *Myxine* (hagfishes), craniates that are considered the sister taxa to vertebrates, is further evidence for the early evolution of inner-ear structures (Löwenstein and Thornhill 1970; Forey and Janvier 1993), as is the finding of very similar cells in tunicates (Burighel et al. 2003). Ears similar to those of the hagfishes, with only a single semicircular canal, were probably present at the dawn of vertebrate evolution (Jarvik 1980). Although there is no direct evidence of hair cells, one would imagine that the presence of a distinct ear means the presence of mechanosensory cells within that ear.

At the same time, ciliated mechanosensory cells are not unique to vertebrates,

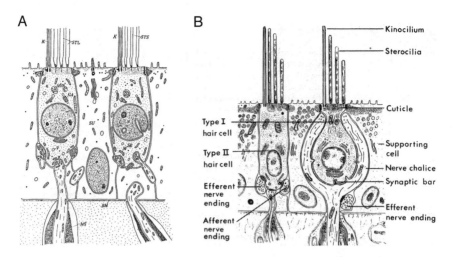

FIGURE 3.1. **A**: Schematic drawing of hair cells from the ear of the ray, *Raja clavata*. BM, basement membrane; CR, ciliary rod; CU, cuticular plate; GA, Golgi apparatus; K, kinocilium; M, mitochondria; MS, myelin sheath, MR, membrana reticularis; MV, microvillus; MVB, multivesicular body; N, nucleus; NE, nerve ending; Se, sensory cell; STL, "large" diameter stereocilia; STS, "small" diameter stereocilia; SU, supporting cell; V, vesicle. (From Löwenstein et al. 1964, with permission.) **B**: Drawing of type I and type II vestibular hair cells from the mammalian ear. (From Wersäll and Bagger-Sjöbäck 1974, with permission.)

and derivatives of vertebrate sensory hair cells are used in other detection paradigms such as electroreception (Jørgensen 1989). Mechanosensory "hair cells" in invertebrates have many features in common with those of vertebrates (Jørgensen 1989), although these similarities do not necessarily imply homology. Secondary sensory cells occur in tunicates (Bone and Ryan 1978; Burighel et al. 2003) and in a fossil organism that may be a vertebrate ancestor (Conway Morris 2000). Consequently, there has been discussion as to whether the vertebrate sensory hair cell arose de novo in the early evolution of the vertebrates (or chordates) or whether vertebrate hair cells and one or more of the even more ancient invertebrate mechanoreceptor cells share a common ancestry. This discussion has been amplified recently with the introduction of molecular techniques to study gene expression in specific cells in different taxa. As a result,

FIGURE 3.2. Phylogeny of species discussed in this chapter. **A**: The vertebrates. **B**: The phylogeny of the anamniote chordates discussed herein. The phylogeny shown here is closely allied with that for the vertebrates found in Manley and Clack (Chapter 1). **C**: Details of the ancestral invertebrates discussed in this chapter.

3. Evolution of Sensory Hair Cells   57

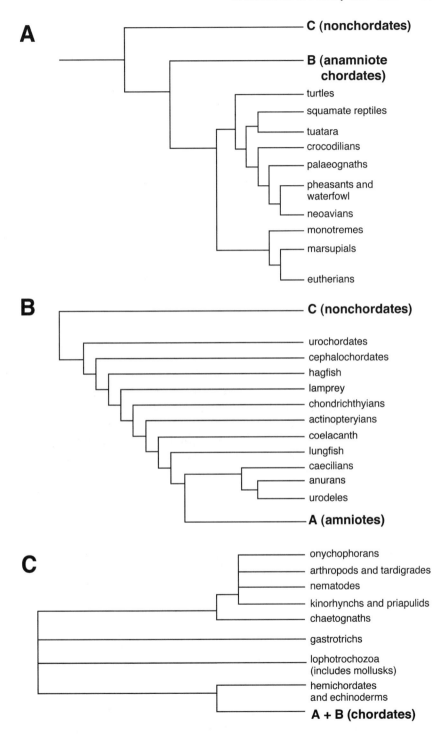

molecular markers provide additional data for the debate about hair cell relationships between vertebrates and invertebrates.

This chapter examines several issues. First, we examine the structure and, to a lesser degree, the function of sensory hair cells in different vertebrate taxa. We discuss the similarities and differences in hair cells in different taxa, between end organs within an organism, and within single end organs of particular species. This first section demonstrates the similarities of hair cells among vertebrates but also emphasizes that a good deal of heterogeneity exists that probably was built upon the basic vertebrate hair cell bauplan.

Second, we briefly examine hair cell types in different invertebrate taxa as a prelude to considering homology of cell types between invertebrates and vertebrates and asking whether the vertebrate hair cell may have a common ancestry with cells in an invertebrate. Third, we examine the arguments based on morphology and molecular biology that have led various investigators to argue that the vertebrate hair cell is homologous to some invertebrate sensory cell systems, including *Drosophila* mechanoreceptors. We conclude with considerations as to the current state of evidence regarding the origins of vertebrate sensory hair cells and consider the evolution of hair cells within the vertebrates.

2. The Comparative Biology of Sensory Hair Cells

The general structure of sensory hair cells does not vary much across the vertebrates, and the cells in the ear and the lateral line are strikingly similar. At the same time and within the basic architecture of the sensory hair cell, there are substantial differences in the details of structure and physiology that suggest that hair cells in different taxa, and even within different end organs of the ear of the same species, vary in the stimuli to which they respond and probably in the nature of this response.

2.1 General Morphology of Vertebrate Sensory Hair Cells

Hair cells are specialized epithelial cells. Apically, each cell has a sensory "hair" bundle that serves as the actual organelle of mechanotransduction. The bundle has a single, eccentrically placed true cilium, called a kinocilium, and many shorter actin-filled structures called stereocilia (reviewed in Hudspeth 1985). These stereocilia are rigid rods that are firmly embedded in the cuticular plate, a cytoskeletal meshwork found just under the apical plasma membrane. The kinocilium, with its $9 + 2$ arrangement of microtubules, extends deep into the cell, terminating in a basal body. In some mammalian and bird cochleae, hair cells lose the kinocilium as development proceeds, but the basal body remains as a marker of the original position of the kinocilium. Hair bundles are polarized, with the longest stereocilia positioned closest to the kinocilium and progressively shorter stereocilia oriented in a steplike fashion. Each stereocilium is connected to its nearest neighbors by thin, filamentous-tip links that play a crucial role in transduction and by ankle links that help maintain bundle cohesion (Pickles et al. 1989; García-Añoveros and Corey 1997). Generally speaking, the hair cell soma is cylindrical, with a centrally located nucleus and the usual cellular complement of organelles including mitochondria, endoplasmic reticulum, and Golgi bodies. Both afferent and efferent synapses are found along the basolateral surface of almost all hair cells. Interestingly, a set of bird cochlear hair cells lack afferent connections (Fischer 1992), and in mammalian inner hair cells, efferents connect almost exclusively to the afferent fibers and not directly to the hair cell itself. The primary afferent neurotransmitter is probably glutamate, whereas acetylcholine and γ-aminobutyric acid (GABA) serve as the efferent neurotransmitters (Klinke 1981; Cochran and Correia 1995; Schrott-Fischer et al. 2002). Mechanotransduction is initiated when a shearing force acts on the bundle along the axis of polarization, causing the stereocilia to pivot about their base (reviewed in Hudspeth et al. 2000). Force in the direction of the kinocilium (and longest stereocilia) leads to depolarization of the cell body, and force in the reverse direction causes hyperpolarization. Although the molecular structure of the transduction channel has not been determined, many of its properties may be inferred from other data (see Strassmaier and Gillespie 2002). The channels are located at the ends of the stereocilia in close proximity to the tip links. They are nonselective cation channels but may be modulated by calcium (reviewed in Strassmaier and Gillespie 2002).

Although this generalized description of hair cell structure and function is a good starting point for any hair cell discussion, it is also a great oversimplification. Both morphological and physiological studies reveal that hair cells exhibit astonishing diversity. The following sections outline some of this heterogeneity.

2.2 Amniote Vestibular Epithelia

The vestibular system of amniotes is a well-studied example of morphological hetereogeneity in hair cells. Amniote vestibular epithelia (canal cristae, saccule, utricle, and lagena in all but therian mammals) possess two distinct hair cell types, called type I and type II (e.g., Wersäll 1956; Wersäll and Bagger-Sjöbäck 1974) (Fig. 3.1B). Type II hair cells, which Wersäll and Bagger-Sjöbäck considered to be phylogenetically older than type I cells, have cylindrical cell bodies and bouton afferent terminals. Efferent synapses directly contact the cell body. Type I cells have flask-shaped somata and a distinct neck region and are enclosed by an afferent nerve calyx. Efferents contact the calyx rather than synapse directly on the hair cell. Bundle types also differ, with type I cells having more stereocilia per bundle (Wersäll and Bagger-Sjöbäck 1974; Peterson et al. 1996). Ultrastructurally, type I hair cells have a subnuclear stack of membranous structures that is possibly endoplasmic reticulum (Wersäll 1956).

Type I and type II hair cells differ considerably in their physiological response properties. Type I cells have a very negative resting potential and a considerable amount of current is activated at rest (Correia and Lang 1990). The dominant type I current is a large, noninactivating K$^+$ delayed rectifier that is activated at low voltages (Correia and Lang 1990; Rennie and Correia 1994; Rüsch and Eatock 1996). The delayed rectifier in type II cells has a smaller activation range (Rüsch and Eatock 1996). The very negative membrane potential of type I cells coupled to the low-voltage activated K$^+$ current could serve to give these cells a faster frequency response by keeping a significant fraction of current on at rest (Rennie and Correia 1994). However, the more depolarized membrane potential and smaller resting current of type II cells may make these cells more sensitive (Correia et al. 1996).

Much speculation surrounds the type I afferent calyx. Calyx fibers have a higher maximum response rate than bouton fibers, suggesting an expanded response range and increased gain to low-intensity stimuli (Brichta and Goldberg 1996; Correia et al. 1996). Transmission electron microscopic (TEM) studies of the calyx have identified synaptic vesicles within the nerve terminal, implying that the afferent terminal may release neurotransmitter on to the presynaptic hair cell membrane (Sans and Scarfone 1996). Because the calyx envelopes the entire cell body, neurotransmitter release from the calyx could promote feedback at the apical end of the hair cell (Sans and Scarfone 1996). Non–*N*-methyl-D-aspartate (NMDA) receptors have been identified in type I hair cells, providing further evidence for neurotransmitter feedback on these cells (Devau et al. 1993). Calcium imaging in type I cells of guinea pig cristae demonstrated that glycine affects intracellular calcium concentration through its action on both NMDA and glycine receptors (Devau 2000). Because glycine and glutamate are coagonists of NMDA receptors, this finding suggests an autofeedback mechanism by which glutamate is released by hair cells and accumulates in the synaptic cleft where it stimulates receptors on both the pre- and postsynaptic membranes. This presynaptic stimulation coupled to glycine coactivation of NMDA receptors causes an increase in intracellular calcium that then triggers further glutamate release (Devau 2000). Direct activation of glycine receptors may modulate this autofeedback process.

Several lines of evidence suggest that type II hair cells represent an earlier ontological form of vestibular hair cell. Type II hair cells are the first recognizable type during embryonic development, whereas type I cells in the mouse utricle are physiologically mature by postnatal day 8 (Rüsch et al. 1998). Hair cell regeneration studies in chickens find that, following ototoxic insult, type II cells are the first to reappear (Weisleder et al. 1995). Tritiated thymidine labeling of proliferating cells shows that some of these new type II cells undergo further differentiation and assume type I–like morphology and innervation (Weisleder et al. 1995). Patch-clamp recordings from regenerating canal cristae in birds show that type II current responses are present soon after ototoxic damage, whereas type I responses appear considerably later (Masetto and Correia 1997; Correia et al. 2001). Type II–like currents were recorded from cells that occupy the normal type I hair cell area and most likely become type I cells, thereby providing further evidence for the differentiation of some type II hair cells into type I (Masetto and Correia 1997).

2.3 Fish Inner Ear and Lateral Line

It was argued for many years that fish have only a single hair cell type in the inner ear and that this is similar to the type II vestibular cell of amniotes (Wersäll 1960; Flock 1964). Diversity of fish hair cells was first recognized as differences in bundle morphology in several species of bony fish. In 1977, Popper identified three major types of hair bundles (F1, F2, and F3) that differed in relative stereocilia and kinocilium lengths, although no specific functional differences were recognized. Despite these bundle differences, the cell bodies all appeared similar, lacking the characteristic shape and nerve calyx that distinguishes type I amniote hair cells (see Popper 1981).

Several studies in the 1990s provided evidence that classifying all fish hair cells as type II is an oversimplification (Fig. 3.3B). Work with an antibody to the calcium-binding protein S-100 identified two distinct populations of hair cells in the oscar, *Astronotus ocellatus* (Saidel et al. 1990). Hair cells in the striolar region of the utricle and most of the saccule were immunoreactive for

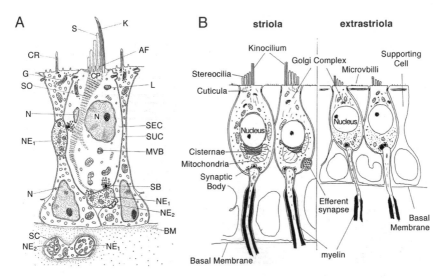

FIGURE 3.3. **A**: Drawing of a single sensory cell from a crista of the labyrinth of an ammocoete larva (lamprey). This cell has a large striated organelle that starts at the cuticular plate and runs to the base of the cell. AF, axial fiber; BM, basement membrane; CR, ciliary rod; CP, cuticular plate; G, Golgi apparatus; K, kinocilium; L, lysosome; MVB, multivesicular body; N, nucleus; NE_1, large nerve ending; NE_2, small nerve ending; S, stereocilia; SB, synaptic bar; SC, Schwann cell; SEC, sensory cell; SO, striated organelle; SUC, supporting cell. (From Lowenstein and Osborne 1964, with permission. © Nature Publishing Group.) **B**: Striolar (left) and extrastriolar (right) sensory hair cells from the utricle of a teleost fish. The cells on the left are called "type I–like," whereas those on the right are type II cells. (From Chang et al. 1992. © John Wiley & Sons. Reprinted by permission of Wiley-Liss, Inc. a subsidiary of John Wiley & Sons.)

S-100, whereas other cells were not. The presence of S-100 in only a subset of hair cells suggests a specific need for calcium regulation, implying functional differences between S-100–positive and S-100–negative hair cells. Enhanced cytochrome oxidase activity in subcuticular mitochondria of hair cells in the striolar region of the utricle in a number of fish species suggests that striolar hair cells have greater metabolic requirements (Saidel and Crowder 1997).

A study using the ototoxic drug gentamicin sulfate found that S-100–positive cells in the utricle of *A. ocellatus* and similar cells in the lagena were selectively sensitive to gentamicin (Yan et al. 1991). This was the first direct evidence for similarity of some fish hair cells and type I amniote hair cells because type I cells show greater gentamicin sensitivity than type II cells (Lindeman 1969).

Somewhat later, ultrastructural investigation of the utricle in *A. ocellatus* (Chang et al. 1992) showed that striolar hair cells have perinuclear cisternae similar to the subnuclear membranous structures of type I cells (Engström and Wersäll 1958) (Fig. 3.1B). Extrastriolar hair cells in the *A. ocellatus* utricle do not have these structures nor do type II cells in amniotes (see Table 2 of Chang et al. 1992 for a complete ultrastructural comparison). These findings prompted Chang et al. (1992) to suggest the names "type I–like" and "type II" for the classification of fish hair cells. The discovery of hair cells with enlarged, calyx-like afferent terminals in the goldfish, *Carassius auratus*, cristae provides additional support for this classification (Lanford and Popper 1996).

Fish hair cells also differ in their innervation patterns. Rostral hair cells in the saccule of *C. auratus* are innervated by thick (S1) afferent fibers, whereas caudal hair cells receive afferent innervation by thin diameter (S2) fibers (Furukawa and Ishii 1967). Hair cells in both regions receive efferent innervation as indicated by the position of acetylcholine esterase–positive fibers (Sugihara 2001). In *A. ocellatus*, however, only a subset of saccular hair cells receive efferent innervation, and efferents in the utricle are primarily restricted to striolar hair cells (Popper and Saidel 1990; Chang et al. 1992).

Physiological evidence further supports the notion of multiple hair cell types in fish sensory epithelia. Current-clamp recordings in the saccule of *C. auratus* show that ovoid-shaped hair cells found on the rostral end of the epithelium have a Ca^{2+}-activated outwardly rectifying K^+ current and exhibit damped oscillations between 40 and 220 Hz (Sugihara and Furukawa 1989). More caudally located hair cells generally have longer cylindrical cell bodies and longer ciliary bundles and do not show resonant behavior (Platt and Popper 1984; Sugihara and Furukawa 1989; Lanford et al. 2000). Instead, these cells spike in response to changes in membrane potential and are dominated by Ca^{2+} and Na^+ currents (Sugihara and Furukawa 1989). Rostrally located hair cells in this preparation responded to higher frequencies than caudally located cells, suggesting that there may be a crude tonotopic map in the goldfish auditory periphery and thus relevant functional distinctions between these cell types. Physiological differences in current type are also found in hair cells of the toadfish (*Opsanus tau*) horizontal canal crista (Steinacker et al. 1997), suggesting that physiological hair cell heterogeneity is a widespread feature of the fish inner ear, with functional significance for peripheral sensory processing. Significantly, there are also morphological differences between rostral and caudal hair cells in the *C. auratus* saccule that suggest that the caudal cells are type II, whereas those lying more rostrally are type I–like (Lanford et al. 2000).

The ear of elasmobranchs (sharks and rays—the cartilaginous fishes) are quite similar to those of bony fishes, suggesting that this basic type of inner ear was established before the divergence of these fish groups. Although there have been no recent studies on the hair cells of any elasmobranch, Chang et al. (1992) pointed out that in illustrations from earlier studies, both Löwenstein (1971) and Corwin (1977) had shown some hair cells that lack a subnuclear structure, whereas other hair cells have a distinct structure that resembles the subnuclear structure found in bony fishes (Fig. 3.1A). Although it is clear that considerable work remains to be done on hair cells in elasmobranchs, this observation coupled with the differentiation of cell types encountered in bony fishes leads to the suggestion that hair cell heterogeneity is likely to be present in cartilaginous as well as bony fishes and thus potentially in their ancestors.

The lateral lines of fishes also contain sensory hair cells that have the same basic structure as hair cells of the ear. Although there have been few studies on the physiology and ultrastructure of lateral line hair cells, Song et al. (1995) demonstrated that hair cells in the lateral line canals (canal neuromasts) are sensitive to gentamicin, whereas hair cells on the body outside the canals (free neuromasts) are not. This finding suggests that at least some of the differentiation in hair cell types encountered in the ear of fishes is also found in the lateral line. This idea is further supported by Ekström von Lubitz's (1981) discovery of both columnar and flask-shaped hair cells in the lateral line of a ratfish, *Chimaera monstrosa*, a cartilaginous fish. Although further work is needed, the finding of differentiation in hair cell physiology in the lateral line that is parallel to that found in the ear suggests that the origin of hair cell heterogeneity may have been exceedingly early in the origin of vertebrates, before the divergence of the cells that would give rise to the hair cells of the ear and lateral line (Song et al. 1995).

2.4 Jawless Fishes

Hair cells of lamprey sensory epithelia (a single macula communis and two canal cristae) bear a striking resemblance to those of gnathostomes (jawed vertebrates) (Löwenstein et al. 1968; Hoshino 1975; Popper and Hoxter 1987; Ladich and Popper, Chapter 4). Because lamprey are considered the sister group to the gnathostomes (Forey and Janvier 1993), this similarity suggests that ancestral vertebrates possessed an inner ear with sensory hair cells. Studies of several lamprey species identify two distinct types of hair cells in the lamprey inner ear (*Lampetra fluviatilis*, Löwenstein et al. 1968; *Entosphenus japonicus*, Hoshino 1975; *Petromyzon marinus*, Popper and Hoxter 1987). One type, called type A by Hoshino (1975), has striated organelles and a hair bundle similar to the F3 type classified by Popper (1977; Popper and Hoxter 1987) (Fig. 3.3A).

The second, referred to as type B (Hoshino 1975), lacks striated organelles and has an F2-type bundle (Popper 1977; Popper and Hoxter 1987). Despite these differences in organelle complement and bundle type, all lamprey hair cells studied by Hoshino (1975) had cylindrical cell bodies and were classified as type II based on the earlier observations of Wersäll (1960). Few studies (e.g., Löwenstein 1970) have examined the physiological characteristics of lamprey hair cells, and no data are available on the physiological heterogeneity of these cells.

Because lamprey and the rest of the vertebrates (Forey and Janvier 1993) have evolved separately for over 430 million years, it is not known whether these striated organelles are a plesiomorphic hair cell character or whether they are uniquely derived in lamprey. Because these organelles may be composed of thin endoplasmic reticulum (Hoshino 1975), Chang et al. (1992) suggested analogy (and possible homology) between the striated organelle of lamprey hair cells and the perinuclear cisternae of striolar hair cells in the oscar utricle and other fishes. This implies a homology between type A hair cells of lamprey and type I–like hair cells of fishes. However, the lack of fossilized soft tissue makes this hypothesis difficult to substantiate.

2.5 Amphibians

The amphibian inner ear provides a wealth of opportunities to study hair cell diversity (see Smotherman and Narins, Chapter 6). In anurans (frogs), the most widely studied amphibian group with respect to the auditory periphery, the ear has three canal cristae, three otolith organs, and two auditory papillae, each with its complement of hair cells (Fig. 3.4). Urodeles (newts and salamanders) and caecilians (legless amphibians) have a macula neglecta as well, and some orders of urodeles and all caecilians lack a basilar papilla (Wever 1975).

Wersäll and Bagger-Sjöbäck (1974) classified all frog hair cells as type II based on cell shape and afferent innervation. However, Lewis and Li (1975), using ciliary bundle morphology, identified at least six distinct cell types from the otolithic and papillar epithelia of *Rana catesbeiana* (bullfrog), although this study did not consider the same cell body characteristics used to differentiate type I and II hair cells in other vertebrates. The ciliary bundle types (A–F) found in frogs differ from one another in overall bundle length, relative lengths of the kinocilium to the longest stereocilia, and the presence or absence of a kinocilial bulb (Lewis and Li 1975) (Fig. 3.4). Two types (A and D) were only found in the saccule, amphibian papilla, and basilar papilla (all considered to be auditory end organs), whereas types B and C were restricted to the lagena and utricle, generally regarded as vestibular organs (but see Cortopassi and Lewis 1998). Based on the peripheral position of type A cells within the auditory organs, it was proposed that this type may represent an immature form (see also Lewis and Li 1973). Studies on *Ascaphus truei* (tailed frog), considered the most primitive extant anuran, showed that all hair cells of this species

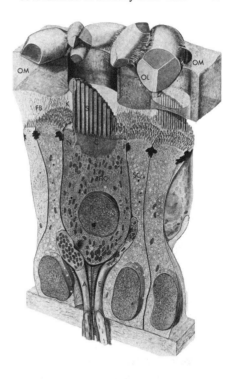

FIGURE 3.4. Amphibian hair cell shown surrounded by supporting cells. C, cuticular plate; FB, fibers connecting the otolithic membrane to the microvilli on supporting cells; K, kinocilium (note bulb); OL, otoconia; OM, otolithic membrane; S, stereocilia. (From Hillman 1976, with permission.)

lack kinociliary bulbs, as do hair cells of urodeles (Lewis 1981). These findings suggest that kinociliary bulbs are a derived feature of anuran hair cells that may serve to more tightly couple hair bundles to the overlying tectorial membrane (Lewis 1981). [Such bulbs commonly occur in lizard hair cells (Miller 1992), where they are presumably independently derived.] Additional studies using TEM to explore ultrastructural features of frog hair cells will be useful in the further classification of morphological cell types.

Extensive exploration of single-cell conductances in frog hair cells has revealed physiological diversity suggestive of functional differences (e.g., Holt and Eatock 1995; Smotherman and Narins 1999a). All frog hair cells studied to date possess at least one calcium conductance (I_{Ca}) and one Ca^{2+}-activated K^+ conductance ($I_{K(Ca)}$) along with additional potassium currents (Masetto et al. 1994; Holt and Eatock 1995; Smotherman and Narins 1999a,b). The calcium conductance stimulates synaptic vesicle fusion at the presynaptic membrane, whereas $I_{K(Ca)}$ repolarizes the cell membrane (Smotherman and Narins 1999b, 2000).

At least three distinct calcium currents, one L-type and two R-type, were identified in semicircular canal hair cells of *Rana esculenta* (grass frog) (Martini et al. 2000; Perin et al. 2000). Patch-clamp recordings in thin slices from cristae of this species also reveal multiple potassium conductances with distinct distributions. Centrally located hair cells possess an inwardly rectifying K^+ current and have a more negative resting potential than peripheral cells, which have a transient A-type K^+ conductance (I_A) (Masetto et al. 1994).

Saccular hair cells of *Rana pipiens pipiens* (leopard frog) also possess multiple, differentially distributed K^+ conductances that parallel differences in cell shape (Holt and Eatock 1995). Spherical hair cells have a slowly activating, slowly inactivating mixed K^+/Na^+ current (I_h). Cylindrical hair cells also have a faster, potassium-selective inward rectifier (I_{K1}) that may contribute to the more negative resting potential of these cells (Holt and Eatock 1995).

The amphibian papilla is a novel structure in amphibians that responds to low- and midfrequency airborne sounds (reviewed in Smotherman and Narins 2000; see also Smotherman and Narins, Chapter 6). Like all inner-ear epithelia discussed so far, hair cells of the amphibian papilla may be subdivided into distinct types. Cells in the low-frequency (rostral) end of the *R. pipiens* amphibian papilla show electrical resonance, whereas caudal cells appear to be only mechanically tuned (Smotherman and Narins 1999a). Electrically resonant cells also possess more Ca^{2+} channels and Ca^{2+}-dependent K^+ channels, with tonotopic variation in Ca^{2+}-dependent K^+ channel kinetics that may additionally tune the resonant frequency of individual hair cells (Fettiplace 1987; Smotherman and Narins 2000).

The frog basilar papilla, a higher frequency auditory organ, is also unique to amphibians (Wever 1985). This papilla is also distinct in that the hair cells appear to comprise a mostly homogeneous population whose electrical properties are dominated by two potassium conductances with differing activation kinetics (Smotherman and Narins 1999b). All hair cells of the frog basilar papilla are tuned to approximately the same frequency in a species- and size-specific manner (see Smotherman and Narins 1999b). Hair cells of the frog basilar papilla also lack efferent innervation, although efferents are present in the urodele basilar papilla (Fritzsch and Wahnschaffe 1987). This papillar homogeneity has prompted the suggestion that these hair cells represent the ancestral condition for the tetrapod auditory system (Smotherman and Narins 1999b). Because hair cell diversity is so widespread among anamniote ears, including other amphibian end organs, it seems more likely that the frog basilar papilla is a specialized receptor. In addition, fish auditory epithelia show clear tuning to a range of frequencies, suggesting that the amphibian basilar papilla does not represent a primitive condition.

2.6 Turtles and Squamate Reptiles

The reptilian basilar papilla is the earliest auditory end organ considered homologous to the mammalian cochlea (Wever 1978; but see Fritzsch 1987 for a different opinion). Manley (2000a), following Miller (1980), suggested that the turtle basilar papilla may be the closest extant representative of the stem reptile condition. The similarity of the basilar papilla of the rhynchocephalian *Sphenodon*, the most primitive extant squamate, to that of the turtles supports this notion. Therefore, hair cells on the turtle basilar papilla may be closest to the ancestral condition for amniote auditory receptors (but see Sneary 1988a for an opposing view) (Fig. 3.5).

Morphologically, turtle auditory hair cells represent a fairly uniform population of abneurally oriented, ultrastructurally similar cells with both afferent and efferent innervation (Sneary 1988a,b). Sneary (1988a) did, however, identify two different bundle types in the basilar papilla of the red-eared slider, *Pseudemys scripta elegans* (see also Hackney et al. 1993). The major type, seen in hair cells positioned over the basilar membrane, has a kinocilium no taller than the tallest stereocilia (similar to Popper's F1 bundle in fish; 1977) and a kinocilial bulb. The second type, seen at the ends of the papilla, has a long kinocilium lacking a bulb and many short stereocilia (Sneary 1988a; Hackney et al. 1993). This second type is similar to the F2 bundle in fishes (Popper 1977) and the type A bundle in frogs (Lewis and Li 1975) and therefore may represent an immature form of ciliary bundle (Hackney et al. 1993).

Physiologically, the basilar papilla of *P. scripta* is a well-studied system for understanding ionic currents and frequency tuning in auditory hair cells (e.g., Goodman and Art 1996; Jones et al. 1999; Ricci et al. 2000). Frequency tuning in turtle hair cells results mainly from electrical resonance (reviewed in Fettiplace 1987). Resonance in low-frequency hair cells results from the interplay between currents carried by voltage-dependent K^+ channels and voltage-dependent Ca^{2+} channels along with an inwardly rectifying K^+ current and the membrane capacitance of the cell (Art and Goodman 1996; Goodman and Art 1996). High-frequency resonance is driven by voltage-dependent Ca^{2+} and Ca^{2+}-dependent K^+ channels (BK channels), with the inward rectifier playing a minor role (Goodman and Art 1996). Electrical resonance within the complete frequency-response range is possible by alternative splicing and differential expression of BK channel subunits (Jones et al. 1999), resulting in channels with differing kinetics. BK channel differences in calcium affinity and differences in channel number per hair cell also aid in frequency tuning (Jones et al. 1999; Ricci et al. 2000).

Turtles represent a small and very unusual fraction of reptiles (see also Manley and Clack, Chapter 1), and consideration of auditory hair cells in squamate reptiles (lizards and snakes) provides additional examples of hair cell diversity and insight into hair cell evolution. Snake hair cells resemble those of turtles in terms of bundle type, polarization, and innervation (Miller 1978; Miller and Beck 1990). Although single-cell responses are not available for snakes, gross auditory responses suggest a similar frequency range as found in turtles and, therefore, electrical resonance for frequency tuning (Hartline and Campbell 1969; Hartline 1971a,b).

The lizard basilar papilla shows remarkable structural diversity (e.g., Miller 1985, 1992; Miller and Beck 1988). Although much of this diversity is represented by the overall papillar organization, two different hair cell types, termed unidirectional (UHC) and bidirectional (BHC) hair cells, are also present in the papillae of 10 lizard families (Miller 1985) (Fig. 3.5). Unidirectional hair cells

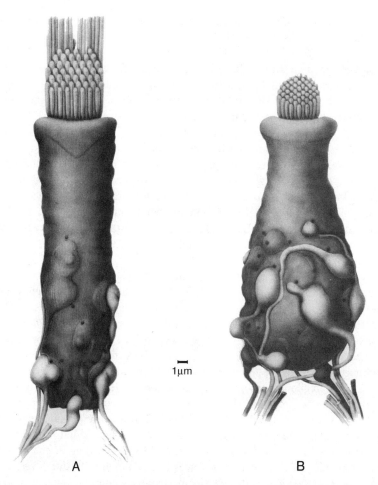

FIGURE 3.5. Hair cell from the auditory papilla of the western blue-bellied fence lizard *Sceloporus occidentalis* reconstructed from serial electron micrographs. The hair cell on the left (**A**) is from a unidirectional hair cell segment of the basilar papilla that is covered by tectorial membrane. The cell on the right (**B**) is from a bidirectional hair cell segment that has free standing stereocilia. The afferent nerve fiber endings are depicted with the synaptic body inside the cell. The unidirectional cell also has efferent endings. (Drawn by Dr. Catherine Jackson Miller for the unpublished doctoral dissertation of Dr. Paul Teresi and used with the kind permission of Dr. Teresi.)

have short hair bundles, large diameter cell bodies, and abneurally oriented hair bundles. These cells received both afferent and efferent innervation, although UHCs in agamid lizards may lack efferents (Miller and Beck 1988, 1990). Bidirectional hair cells have small cell bodies, fewer afferent synapses, and no efferent innervation and are in groups of oppositely polarized or oriented cells (Miller 1985; Miller and Beck 1988, 1990). To avoid confusion, because a single hair cell cannot be bidirectional, Manley (1990) suggested referring to these as unidirectional-type (UDT) or bidirectional-type (BDT) hair cells. Thus the terms can be used consistently in cases where, for example, basilar papillae contain oppositely oriented UDT-type hair cells, such as in skinks. The essential features of these hair cell types do not depend on their orientation in the papilla. Functional differences between UDT and BDT areas are described by Manley (Chapter 7).

Unidirectional-type hair cells in lizards resemble hair cells in turtles and snakes (Miller 1978, 1985) and respond to low-frequency sound. This response may involve electrical resonance (see Manley 2000b), suggesting that this is the ancestral hair cell type in reptiles and that extinct reptiles were probably low-frequency hearers (Clack and Allin, Chapter 5). Bidirectional-type hair cells respond to higher frequencies, above about 1 kHz (Manley 1981, 1990), and may have evolved as a response to selective pressures for an extended hearing range following the development of a tympanic middle ear (see also Clack and Allin, Chapter 5). The lack of efferent innervation to BDT hair cells is interesting. There is a loss of phase locking in BDT cells compared to UDT cells (Manley et al. 1990), but this is probably unrelated to the lack of efferent innervation because phase locking operates too quickly to require efferent feedback. Recent evidence indicates that BDT cells, which are always found in higher frequency areas, have bundles showing an active motility (Manley et al. 2001) similar to that documented for hair cells of the frog saccule and the turtle papilla (reviewed in Manley 2001). This commonality of mechanisms with vestibular hair cells suggests that these reptiles have retained a plesiomorphic active process that contributes to the high sensitivity of their hair cell systems.

2.7 Archosaurs (Birds and Crocodilians)

Crocodilians and birds make up a monophyletic group (Carroll 1988; Manley and Clack, Chapter 1) that shares many features including inner-ear structure. Hair cells in basilar papillae of this group may be divided into two morphologically distinct types termed "tall" (THCs) and "short" (SHCs) hair cells (Takasaka and Smith 1971; Leake 1977; Tanaka and Smith 1978; Gleich and Manley 2000) (Fig. 3.6). Tall hair cells have small efferent nerve terminals and sit over the neural limbus, whereas SHCs have larger efferent endings and rest over the free basilar membrane (Takasaka and Smith 1971). More recent study of basilar papillae from many species demonstrates that most papillae, in fact, show a continuous morphological gradient from THCs to SHCs, suggesting that this simple division is an insufficient description of papillar hair cells (e.g., Manley

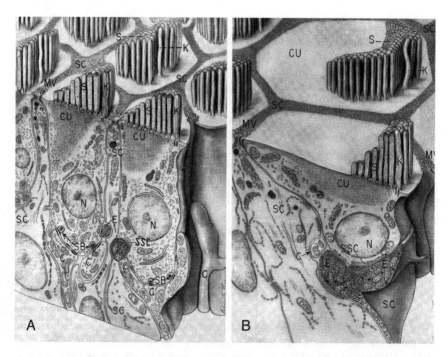

FIGURE 3.6. Tall (**A**) and short (**B**) sensory hair cells from the basilar papilla of a bird. C, cochlear afferent nerve; CU, cuticular plate; E, efferent ending; K, kinocilia; MV, microvilli on supporting cell; N, nucleus; S, stereocilia; SB, synaptic bodies; SSC, subsynaptic cisternae. (From Takasaka and Smith 1971. Reprinted from *Journal of Ultrastructure Research*. © 1971, with permission from Elsevier.)

et al. 1989; Fischer 1992, 1994b; Gleich et al., Chapter 8). Remarkably, a number of SHCs totally lack afferent innervation, leading Fischer (1994a,b) to suggest a differentiation of hair cell types based on innervation pattern rather than graded morphology. This is further supported by studies on *Dromaius novaehollandiae* (emu), a member of the Ratitae, the most primitive group of extant birds (Padian and Chiappe 1998). *D. novaehollandiae* also has a subset of abneural hair cells that lack afferent innervation, but most cell bodies are "tall" according to the usual morphological criteria (Fischer 1998; Köppl et al. 1998). Köppl et al. (1998) proposed that the evolution of functionally distinct SHCs in bird papillae occurred before their morphological differentiation. Loss of afferent innervation to hair cells in the abneural area of the avian basilar papilla appears to be a derived characteristic, with greater afferent loss in more advanced land birds such as the hearing specialist *Tyto alba* (barn owl) compared to the more primitive *D. novaehollandiae* (Gleich and Manley 2000; Köppl et al. 2000).

A few morphological characteristics of bird hair cells defy parsimonious evolutionary explanations. One interesting finding in *Gallus gallus domesticus* (domestic chicken) hair cells was the lack of a kinocilium on many SHCs and some THCs (Tanaka and Smith 1978). Because all *Columba livia* (pigeon) hair cells have a kinocilium (Takasaka and Smith 1971), it was suggested that kinocilium loss resulted from inbreeding during domestication (Tanaka and Smith 1978). Hair cells in *Anas platyrhynchos* (mallard duck), a member of the primitive water bird assemblage, all have kinocilia (Chandler 1984), but this structure is missing in some *D. novaehollandiae* hair cells (Fischer 1998). Kinocilia, therefore, have been lost multiple times during avian hair cell evolution.

Cuticular plate reduction follows a similar pattern. THCs have a centrally located hair bundle and a cuticular plate that covers most of the apical surface, whereas bundles of SHCs are small and eccentrically located, with a reduced cuticular plate in *C. livia* and *G. gallus* (Takasaka and Smith 1971; Tanaka and Smith 1978). The cuticular plate covers the entire apical surface in all hair cells in *A. platyrhynchos* (Chandler 1984) but does not in all *D. novaehollandiae* hair cells (Fischer 1998), suggesting that cuticular plate reduction may also have occurred multiple times.

Single-cell recordings in *G. gallus* hair cells show that at least the extreme forms of THCs and SHCs are physiologically distinct (Martinez-Dunst et al. 1997). THCs have voltage-dependent K^+ channels (K_v) and Ca^{2+}-dependent K^+ channels (BK) similar to those seen in turtles (Goodman and Art 1996; Martinez-Dunst et al. 1997). SHCs have BK channels and an A-type, rapidly inactivating K^+ current similar to I_A in frogs (Masetto et al. 1994; Martinez-Dunst et al. 1997).

Tuning to lower frequencies in *G. gallus* hair cells follows the same mechanism as in turtles, with alternative splicing and differential expression of BK-channel subunits presumably producing hair cells with different resonant frequencies (Ramanathan et al. 1999, 2000). High-frequency tuning may result from unidentified splice variants and/or by nonelectrically resonant mechanisms such as micromechanical tuning (Manley 1995; Ramanathan et al. 2000).

One great mystery of bird and crocodilian hair cells is SHC function (Gleich and Manley 2000). Because true SHCs lack afferent innervation (Fischer 1994b), their action must be restricted to the basilar papilla. Efferent terminals release acetylcholine onto the SHC basolateral membrane, activating a novel acetylcholine receptor with mixed nicotinic and muscarinic properties (Firbas and Müller 1983; Fuchs and Murrow 1992a,b). Efferent activation triggers a small inward Ca^{2+} current, then a large outward K^+ current that hyperpolarizes the cell (Fuchs and Murrow 1992a,b). Some evidence suggests that SHCs may serve a similar function to the outer hair cells (OHCs) of mammals, adding mechanical energy to the motion of the papilla (see Manley 1995). Not only are avian hearing systems highly sensitive and sharply frequency tuned, but at least some bird ears emit spontaneous otoacoustic emissions, thus manifesting properties typical of ears that use active processes (Manley 2001). This would explain the retention of SHCs in the papilla despite the lack of afferent innervation to these hair cells.

2.8 Mammals

The mammalian cochlea demonstrates considerable hair cell diversity. All mammalian cochleae studied to date contain two distinct hair cell types (e.g., Wever et al. 1971; Aitkin 1995; Ladhams and Pickles 1996; Vater et al., Chapter 9). Inner hair cells (IHCs) are considered to be the primary auditory receptors, whereas OHCs are thought to amplify low-intensity sounds (Fig. 3.7). OHCs also apparently play a role in the sharpening of frequency tuning curves (Dallos 1996). Morphologically, IHCs have flask-shaped cell bodies and central nuclei, whereas OHCs are cylindrical with basally lying nuclei (Forge et al. 1991; Slepecky 1996). Both cell types contain hair bundles with a few rows of stereocilia arranged in stereotyped patterns. A kinocilium is lacking in mature hair cells of eutherians (placental mammals) but present in some OHCs in *Tachyglossus aculeatus* (short-beaked echidna), a monotreme (Engström and Wersäll 1958; Ladhams and Pickles 1996). Ultrastructurally, IHCs resemble other vertebrate hair cells, whereas OHCs have abundant lateral cisternae and membrane-associated proteins indicative of their specialized function (Saito 1983; Forge 1991).

Inner hair cells receive extensive afferent innervation from myelinated radial fibers (Liberman et al. 1990). Unmyelinated efferent fibers from the lateral

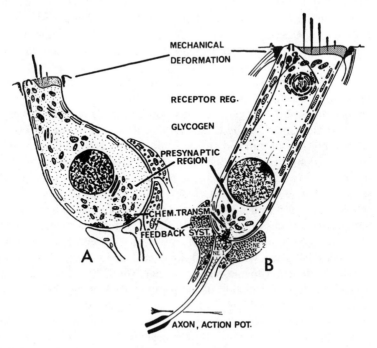

FIGURE 3.7. Inner (**A**) and outer (**B**) hair cells from the mammalian organ of Corti. (From Ades and Engström 1974, with permission.)

superior olive synapse on these afferents rather than directly on IHCs (Liberman et al. 1990). OHCs receive sparse afferent innervation from thin, unmyelinated fibers but ample efferent contact from the medial superior olive via thick, myelinated axons (Liberman et al. 1990). These differences in innervation reinforce the idea that IHCs transduce stimuli, whereas OHCs are involved in active feedback (Dallos 1996). However, this clear separation breaks down somewhat in the echolocating *Rhinolophus rouxi* (horseshoe bat), which lacks a medial olivary efferent system and therefore efferent control of OHCs (Bruns and Schmieszek 1980).

Outer hair cells are found only in mammals and have the unique property of electromotility. Brownell et al. (1985) noted that isolated guinea pig OHCs shorten when depolarized and elongate when hyperpolarized. Ashmore (1987) demonstrated that OHC motility was graded with membrane potential and independent of adenosine triphosphate (ATP) and calcium, suggesting that OHC motility differs from conventional motor-based cellular movement. Fast electromotility is driven by a series of densely packed prestin molecules in the basolateral OHC membrane (Dallos et al. 1991; Zheng et al. 2000). Intracellular anions serve as a voltage sensor, binding prestin and probably causing a conformational change in the protein (Oliver et al. 2001). This conformational change is thought to result in changes in cell surface area and, therefore, in cell length, and serves as the manifestation of electromotility (Zheng et al. 2000).

In addition to prestin-driven electromotility, OHC responses are modulated by efferent stimulation. Efferent activation releases acetylcholine onto OHCs and activates a novel, mostly nicotinic receptor similar to that seen in avian SHCs (Fuchs and Murrow 1992a,b; Erostegui et al. 1994). Receptor activation causes an initial, brief depolarization due to calcium influx, then a larger hyperpolarization caused by activation of Ca^{2+}-dependent K^+ channels (Erostegui et al. 1994). This hyperpolarization is thought to mediate basilar membrane stiffness. OHCs also exhibit a slower motile response following efferent activation that may protect the cochlea against acoustic trauma (Reiter and Liberman 1995; Sridhar et al. 1995). Slow motility is calcium dependent and results from phosphorylation of cytoskeletal proteins by myosin light chain kinase and calcium/calmodulin kinase II (Puschner and Schacht 1997; Szönyi et al. 2001). This phosphorylation results in changes in axial stiffness that reduces OHC force and therefore cochlear gain (Dallos et al. 1997).

2.9 Summary of Vertebrate Hair Cell Types

The lack of soft tissue such as hair cells in the fossil record makes it impossible to trace them in the evolution of vertebrate inner ears. However, comparative studies between hair cells of extant vertebrates shed considerable light on the evolutionary picture. The existence of hair cells in lamprey and hagfish provides strong evidence for an inner ear with hair cells in the earliest vertebrate ear over 470 million years ago (Forey and Janvier 1993).

Hair cell heterogeneity in otolithic epithelia and canal cristae occurs in bony fishes, as discussed in section 2.3. Although some of these hair cell features may be unique to fishes, ototoxic sensitivities and the presence of subnuclear cisternae in a subset of fish hair cells suggests that the ancestor to amniote type I hair cells arose first in primitive fishes. The finding by Lanford and Popper (1996) of enlarged afferent terminals similar to calyces in the *C. auratus* cristae further supports this hypothesis. Enlarged afferent terminals have also been seen in frog sensory epithelia (Wersäll and Bagger-Sjöbäck 1974), suggesting that anamniotes possess a rudimentary calyx that further developed during amniote hair cell evolution. Selective pressure for an expanded synaptic cleft, possibly to allow for afferent feedback (reciprocal synapses), could have driven the evolution of the type I nerve calyx (Sans and Scarfone 1996). Further functional studies of the calyx synapse should help uncover the selective pressures driving type I hair cell evolution.

The basilar and amphibian papillae of amphibians may be novel structures that evolved only in the lissamphibian lineage (Wever 1985), although there has been the suggestion that the basilar papilla evolved in tetrapod ancestors (Fritzsch 1987). Although the hair cells on these epithelia are surely homologous to other vertebrate hair cells and share most features (e.g., sensory hair bundle, mechanotransduction), unique hair cell specializations may have evolved during papillar evolution in this taxon (Smotherman and Narins 1999a,b).

The amniote auditory organ has undergone much divergence and specialization. The ancestral form may be best represented in turtles by a short, ovoid papilla containing a single hair cell type (Sneary 1988a). From this plesiomorphic condition, however, evolved several specialized papillae in parallel, all with morphologically and physiologically diverse hair cells (e.g., Tanaka and Smith 1978; Ladhams and Pickles 1996; Manley and Köppl 1998; Gleich and Manley 2000; Manley 2000a).

Perhaps the most striking case of parallel hair cell evolution is seen in the auditory organs of birds and mammals. Tall hair cells of the avian papilla, with both afferent and efferent innervation, may represent the ancestral form and are analogous to IHCs of the mammalian cochlea. Avian SHCs, devoid of afferent innervation, are more specialized and appear analogous to OHCs in mammals. Outer hair cell evolution was probably driven by selective pressure for amplitude gain at low stimulus levels and perhaps otoprotection at high intensities, and SHC evolution may have been driven by similar selective pressures. Comparative physiological studies between these analogous cell types should allow for a better understanding of this parallel relationship.

In summary, what probably started as a single hair cell type in the vertebrate ancestor has given rise to an astonishing array of hair cell types. Furthermore, the current classification system of anamniote hair cells insufficiently describes the diversity present in sensory epithelia of these taxa (e.g., Lewis and Li 1975; Chang et al. 1992; Masetto et al. 1994). Selective pressures to maximize stimulus detection and for the processing of stimuli that differ widely in such temporal characteristics as frequency and duration together with the selective distribution of cells with specialized active mechanisms to increase sensitivity have probably driven the evolution of hair cell diversity in the vertebrate inner ear.

3. Sensory Hair Cells in Nonvertebrate Chordates

Vertebrates are placed within the phylum Chordata (see Fig. 3.2). Within this phylum are found the craniates, the group that includes vertebrates and their sister group the hagfishes, and two invertebrate groups, the Urochordates (tunicates and relatives) and Cephalochordates (lancelets and relatives). The chordates all share at some time in their life cycle the presence of a notochord, a dorsal hollow nerve cord, and pharyngeal slits. The closest phyla to the Chordata are the Echinodermata, which includes the starfishes, and the lesser known Hemichordata. These phyla are grouped together because they share many embryonic features with one another.

3.1 Primary and Secondary Sensory Cells

The sensory hair cells in all vertebrates are considered to be secondary sensory cells in that they synapse with afferent (and often efferent) neurons. In contrast, many invertebrates such as *Drosophila* have sensory endings that resemble vertebrate sensory cells in many ways, but they are often primary sensory cells whereby the sensory ending is a modification of the terminal end of a neurite. This neurite then extends to other parts of the nervous system, carrying information generated in its mechanosensitive ending.

3.2 Hagfish

Hagfish (order Myxiniformes) had long been considered to be in the same vertebrate class as the lamprey (Agnatha). They are currently classified as being the only nonvertebrate craniate relatives of the vertebrates (e.g., Forey and Janvier 1993; Bardack 1998) but stay within the Craniata because they share a number of basic features with the vertebrates, including the presence of a cranial region, an enlarged brain, paired sensory organs on the head, and neural crest cells. They are not considered vertebrates because they lack other vertebrate characteristics such as the vertebral column.

The hagfish inner ear is basically a simple circular ring. Löwenstein and Thornhill (1970) described the ear as being toroidal in shape, with cristae anteriorly and posteriorly and with a single ventral macula communis (see also Jørgensen et al. 1998). Using both scanning electron microscopy (SEM) and TEM, Jørgensen et al. (1998) examined the inner ear of *Myxine glutinosa* and found sensory hair cells both in the cristae and on the macula communis. Those on the cristae do not appear to be covered by a cupula as in vertebrates, whereas there appear to be otoconial crystals associated with the epithelium (Jørgensen et al. 1998). Each hair cell has a single kinocilium that lacks the two central microtubules and terminates in a basal body but no basal foot. There are also a series of graded stereocilia to one side of the kinocilium, although there are some cells where the kinocilium is in the center of the bundle of stereocilia (Jørgensen et al. 1998). The latter type may represent an early developmental stage rather than a second type of mature cell.

The lateral line of hagfishes is simpler than that in lamprey and gnathostomes (Braun 1996; Braun and Northcutt 1997). It has two or three series of shallow epidermal grooves, each containing what appears to be a single type of sensory hair cell (Braun and Northcutt 1997), although there does not appear to have been a detailed analysis of the ultrastructure of these cells to determine whether there is the possibility of multiple cell types. Braun and Northcutt (1997), using SEM, identified sensory cells that have a single kinocilium surrounded by rings of microcilia without the length gradations found in vertebrates and without a clear orientation.

The hagfish lateral line appears, based on its development and innervation, to be homologous to the vertebrate system, leading Braun and Northcutt (1997) to conclude that the lateral line arose within or before the very earliest craniates and thus among the vertebrate ancestors. Still, they concluded that the current state of the lateral line in hagfish may be highly derived (see also Braun 1996). It is clear, however, that the presence of vertebrate sensory hair cells in the hagfish lateral line and ear suggests a very ancient origin to these cells.

3.3 Cephalochordates

The closest relatives of the craniates (including the vertebrates) are the cephalochordates and the urochordates. Neither of these groups contains evidence of having derivatives of dorsolateral placodes, embryonic structures that give rise to the craniate ear and lateral line systems (Braun and Northcutt 1997). Indeed, Northcutt and Gans (1983) have argued that neurogenic placodes and secondary sensory cells, the basis for the craniate ear and lateral line, were not present before the origin of the craniates.

Earlier investigations suggested that cephalochordates have two types of sensory cells called type I and type II (Fig. 3.8) (Schulte and Riehl 1977; Stokes and Holland 1995; Lacalli and Hou 1999). (It should be pointed out that the names of these cells in no way suggest a relationship with the similarly named hair cells in vertebrates.) As described in *Branchiostoma floridae* (lancelet) by Lacalli and Hou (1999), the type I cell has a single cilium with a 9 + 2 microtubule pattern that is encircled basally by a ring of shorter microcilia. Although Lacalli and Hou (1999) described two variants on the type I cell, the differences are small and are not considered here (see also Holland and Yu 2002). Based on serial TEM studies, Baatrup (1981) clearly demonstrated that the type I cell is a primary neuron and that it projects to the central nervous system with an extended axon.

The type II cell has a modified cilium that is packed with 20 to 30 microtubules rather than having a 9 + 2 or 9 + 0 microtubule pattern typically found

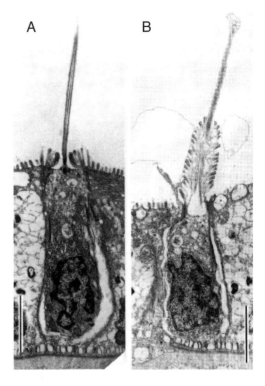

FIGURE 3.8. Transmission electron micrographs of sensory cells from *Amphioxus*. **A**: Cell referred to as type I. **B**, A type II cell. The function of these cells is not clear, but each has an elongate kinocilium-like structure. (From Lacalli and Hou 1999, with permission.)

in the type I cell in this species and in craniate hair cells (Lacalli and Hou 1999). The cilium is surrounded by what has been referred to as an extended collar of branched microvilli forming a brushlike structure around the lower region of the cilium. Lacalli and Hou (1999) suggested that the highly branched microvilli around the central cilium in type II cells is what one might expect from chemoreceptors as opposed to a more mechanoreceptive structure associated with type I cells. Baatrup (1981) found evidence that the type II cell is secondary and that it synapses with what appear to be Retzius biopolar cells (Holland and Yu 2002).

Although there are indeed some similarities between these cells and craniate sensory hair cells, it has been demonstrated by TEM that at least type I cells have basal axonic processes indicative of primary sensory cells (Baatrup 1981). As a consequence, it is not reasonable to assume that either cell type has given rise to the craniate sensory hair cell (Braun and Northcutt 1997).

3.4 Urochordates (Tunicates)

Tunicate adults have an atrial (exhalant) siphon extending out of the body, and in the region of the junction between the siphon and body there lies a set of cupular organs (Bone and Ryan 1978). Bone and Ryan reported the presence of ciliated cells in this region in *Ciona*. The mature cell has a central cilium that arises from the apical regions of the cells, and this cilium is surrounded by a corona of microvilli (Fig. 3.9A). Two other cell types exist in the same organs, but Bone and Ryan suggested these are developmental stages of the ciliated cell. Transmission electron microscopy studies demonstrated that this is a primary sensory cell.

The central cilium has a 9 + 2 tubular array toward its base, but at higher levels of the cilium, this is not apparent. Bone and Ryan (1978) argued that this change is not a fixation artifact but a true differentiation in the cilium. The cell body has an apical granular or fibrous zone that has some resemblance to the cuticular plate of vertebrate sensory hair cells. The sensory cells are surrounded by supporting cells that bear some resemblance to vertebrate supporting cells (Bone and Ryan 1978). Although there have been no direct physiological studies of the function of the tunicate sensory cells, Bone and Ryan did determine that the cells are sensitive to vibration, with best sensitivity from 25 to 400 Hz.

Recently, a new mechanoreceptive organ, the coronal organ, containing secondary sensory cells was discovered in two asidacian tunicates (Fig. 3.9B) *Botryllus schlosseri* and *Botrylloides violaceus* (Burighel et al. 2003). These cells have a single true cilium and multiple smaller structures that appear similar to the stereocilia found in vertebrate hair cells. Furthermore, these cells synapse

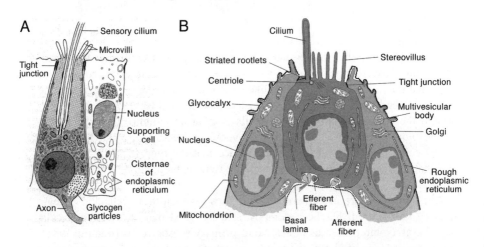

FIGURE 3.9. **A**: Primary sensory hair cell from the ascidian *Ciona* cupular organ. Note that the sensory area is an expansion of the afferent nerve fiber. (From Bone and Ryan 1978, with permission of Cambridge University Press.) **B**: Schematic drawing of a secondary sensory hair cell from the ascidian *Botryllus schlosseri* coronal organ. The organ is located in the siphon and appears to function for the detection of water motion, much as the vertebrate lateral line. Note the presence of a single cilium. No cuticular plate is present, however. (From Burighel et al. 2003. © John Wiley & Sons. Reprinted by permission of Wiley-Liss, Inc., a subsidiary of John Wiley & Sons.)

with both afferent and efferent fibers and are surrounded by nonsensory cells with characteristics of the supporting cells found in vertebrate hair cell epithelia (Burighel et al. 2003). Based on this finding, it is possible that vertebrate hair cells may have arisen in early chordates and may have predated the origin of vertebrates. However, further study is needed to test this suggestion.

4. Sensory Hair Cells in Other Invertebrate Taxa

A wide range of ciliated cells has been described in invertebrates (reviewed in Jørgensen 1989; Budelmann 1992a,b). In the vast majority of cases, the ciliated endings are part of primary sensory cells. There are exceptions in the cephalopod mollusks, where a single statocyst may contain both primary and secondary sensory cells (e.g., Budelmann 1988, 1992a,b; Neumeister and Budelmann 1997). Although there have been few physiological studies on the function of these mechanosensory cells themselves, Jørgensen (1989) made the interesting distinction between vertebrate sensory hair cells, where the major transduction process is associated with the stereocilia, and invertebrate systems, where the limited evidence suggests that the transduction process is a function of the single central (or displaced) cilium. Jørgensen goes on to conclude that on the basis of the transduction process itself, the vertebrate sensory hair cell is unique among animals.

4.1 Cephalopod Sensory Hair Cells

Of all the invertebrate mechanoreceptive systems, some well-studied examples are from various species of the derived mollusks known as cephalopods (e.g., octopods, decapods, chambered *Nautilus*) where there is a well-developed statocyst serving as an organ for the detection of angular and linear acceleration. Studies of the secondary sensory cells in the statocysts of *Nautilus* revealed the presence of two types of sensory cells, called type A and type B (Neumeister and Budelmann 1997). In both cases, the apical end of the cell has a large number of protruding kinocilia with the typical 9 + 2 microtubule pattern but no stereocilia. Both cell types are morphologically polarized and, presumably, physiologically polarized as well (Budelmann and Williamson 1994; Neumeister and Budelmann 1997). Type A cells in *Nautilus* have 10 to 15 kinocilia in a single row, whereas type B cells have 8 to 10 kinocilia. In contrast, the statocysts of faster moving cephalopods such as *Octopus vulgaris* include both primary and secondary sensory cells (e.g., Budelmann and Thies 1977; Budelmann et al. 1987).

4.2 Drosophila *Mechanoreceptors*

Although *Drosophila* only have primary mechanoreceptive cells (i.e., neurons), they are so well studied that we include them here. Known as type I mechanoreceptors (not related to vertebrate type I hair cells), these cells have a single true cilium (9 + 0) at the end of a single sensory dendrite (Fig. 3.10) (Eberl 1999; Jarman 2002). Bristle cells are type I cells that cover the fly and serve multiple functions such as olfaction, proprioception, and wind detection. Chordotonal organs, which include internal proprioceptors and the specialized auditory organ known as Johnston's organ, also contain monodendritic type I neurons (see Eberl 1999). In all cases, these are true neurons where a single cilium contains the transduction apparatus.

Despite the morphological differences, there are many physiological similarities between *Drosophila* mechanoreceptors and vertebrate hair cells. Mechanotransduction in bristle cells occurs with submillisecond latencies characteristic of a mechanically gated channel (Walker et al. 2000). The transduction current is highly polarized and adapts to prolonged stimulation. Recent work demonstrates that sensory neurons in Johnston's organ in *Drosophila* show active motility, another vertebrate hair cell character (Göpfert and Robert 2003). These similarities have led some investigators to suggest that vertebrate hair cells and *Drosophila* bristle cells are functionally analogous (Walker et al. 2000; Göpfert and Robert 2003).

Importantly, a putative mechanosensory channel, *nompC*, has been cloned in *Drosophila* (Walker et al. 2000). *nompC* is a member of the transient receptor potential (TRP) superfamily that also contains mechanoreceptive channels in the nematode *Caenorhabditis elegans* (Duggan et al. 2000; Walker et al. 2000). Because mechanosensation shares some similarities across animal phyla, these

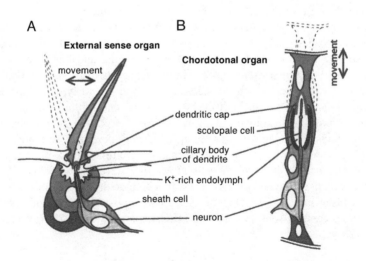

FIGURE 3.10. Type I mechanosensory neurons in *Drosophila*. **A**: Bristle organ. **B**: Chordotonal organ. Note that in both cases the transduction apparatus is at the tip of a neurite, whereas the entire structure is multicellular. (From Jarman 2002, with permission of Oxford University Press.)

findings in invertebrates may aid in the search for the vertebrate hair cell transduction channel.

4.3 Major Differences Between Invertebrate and Vertebrate Sensory Cells

Given the nature of the stimuli to be detected, it is not surprising that mechanosensory cells of different animal groups bear very similar apical segments. Only in the urochordates, however, do the secondary mechanosensory cells of invertebrate animals bear a complex of surface structures with the same organization as those of vertebrate hair cells (Burighel et al. 2003). The organization of these cells in nonchordate groups thus points more to convergent evolution rather than a common origin.

5. The Origin of Vertebrate Hair Cells

From what cell type did vertebrate hair cells evolve? As noted above, it has recently been suggested by some investigators that vertebrate hair cells are homologous with invertebrate mechanosensory neurons (e.g., Eddison et al. 2000; Fritzsch and Beisel 2001; Gillespie and Walker 2001). According to this view, ciliated mechanosensory cells have evolved once during animal evolution, with the specialized hair cells seen in the vertebrate ear representing modified versions of a primordial mechanoreceptor. However, at this point, the data to support this hypothesis are still fairly limited. In the following discussion, we argue that the evidence in support of a single ancestral mechanosensory cell is still lacking and that a multiphyletic origin for mechanosensory cells is still the most reasonable hypothesis.

The argument for mechanosensory homology is based on superficial morphology, physiological similarity, and molecular genetics. Superficially, the dendritic ciliary process of *Drosophila* mechanosensory neurons and the stereociliary bundles of vertebrate hair cells appear similar. Both structures represent a vibrationally sensitive projection arising from the apical portion of the cell. However, it is important to consider the structural nature of these projections. In vertebrates, the apical projection is composed of a group of actin-based stereocilia and a single tubulin-based true cilium, with the transductional components of the bundle restricted to the stereocilia. In *Drosophila*, the apical projection is a single true cilium (reviewed in Jarman 2002). In other invertebrates, mechanoreceptor structure varies greatly, with some containing many true cilia, some only a single cilium, and some an array of short microcilia (Jørgensen 1989). In cnidarians (sea anemones, jellyfish, and relatives), hair "cells" can be multicellular structures, with two to four neighboring cells each providing a portion of an apical bundle constituent (Mire and Watson 1997). The existence of superficially similar yet structurally distinct bundle types could imply convergence on a common design, suggesting that hairlike apical protrusions are probably the best way of creating a vibrationally sensitive superficial mechanoreceptive organelle. In fact, a recent symposium at the Society for Neuroscience found many examples of convergence in the central nervous system (CNS) and suggested that in the CNS, "some solutions seem easier to evolve" (Phelps 2002, p. 159). This argument is likely to apply to sensory systems.

Receptor cell type also differs between vertebrates and most invertebrates. In vertebrates, asidacians, and, to some extent, cephalopods, mechanosensory cells are secondary receptors that form synapses with afferent neurons (Jørgensen 1989; Burighel et al. 2003). In other invertebrates, including *Drosophila*, mechanoreceptors are primary sensory neurons that extend axons to central processing structures (Keil 1997). Fritzsch and Beisel (2001) suggested that the separation of mechanosensory neuron into two cell types (secondary receptor cell and primary afferent) has occurred twice during evolution but that this is a case of parallel evolution from the ancestral condition seen in some invertebrates, including *Drosophila*. We suggest that the existence of secondary mechanoreceptors in chordates and cephalopods could just as easily be a case of convergent evolution from independent ancestral cell types.

A second aspect of both vertebrate and invertebrate mechanosensation that has been examined in terms of homology is the physiological function of each cell type (Eddison et al. 2000; Gillespie and Walker 2001). *Drosophila* bristle cells and vertebrate hair cells have similar operating ranges. Vertebrate hair bundles are stimulated by a deflection of less than 1 nm (Hudspeth 1989) and bristle cells have similar sensitivity (Walker et al. 2000). However, maximal sensitivity is a common feature of sensory systems. Maximal sensitivity simply optimizes a sensory system to provide an organism with the most possible information about its environment, because greater sensitivity often translates into greater chances for survival. Viewed in this context, the similar operating range seen in hair cells and bristle cells is not unexpected because directional selection would act to converge on maximal sensitivity.

Adaptation during prolonged stimulation is also a common feature of sensory receptors, exhibited by *Drosophila* bristle cells, vertebrate hair cells, photoreceptors, and olfactory receptor neurons (Gillespie and Corey 1997; Walker et al. 2000; Zufall and Leinders-Zufall 2000; Fain et al. 2001). Just as a sensory system is optimized for maximal sensitivity, so is that system optimized for physiological adaptation, making it more sensitive to changes in stimulation. It is often these changes that are of biological relevance to the organism such as a sudden decrease in light intensity caused by the shadow of a passing predator. An important argument for homology between bristle cells and hair cells based on physiological adaptation could be built if both cell types shared a common mechanism for adaptation. In most vertebrate hair cells, the motor myosin Ic is hypothesized to drive adaptation by using structural aspects of the actin-based stereocilia to physically position the transduction channel along the stereociliary shaft (Gillespie and Corey 1997). However, this mechanism has not been confirmed in all vertebrate classes, and the role of myosin Ic in cochlear hair cells is still being debated (Kros et al. 2002). Moreover, mutations in *Drosophila* myosins have not, to date, been demonstrated to lead to defects in sensory neuron function (Caldwell and Eberl 2002). It should be noted that an unpublished reference has indicated that crinkle (*Drosophila* myosin VIIa mutant) results in defects in both chordotonal and sensory bristle function (cited in Caldwell and Eberl 2002), but the nature of these defects, in particular as it relates to adaptation, has not been determined. Therefore, at this point, although the phenomenon of adaptation clearly occurs in mechanosensory structures from multiple species, it is not clear that the mechanisms that mediate this adaptation are similarly conserved.

The final basis for the argument of homology between invertebrate sensory neurons and vertebrate hair cells has been the existence of conservation at the molecular genetic level, largely based on the observation that the basic helix-loop-helix (bHLH) transcription factor Atonal is required for both the formation of chordotonal organs in *Drosophila* and for the formation of hair cells in vertebrates (there called Math1) and that the Notch signaling pathway plays a role in the subsequent development of both structures. The advent of molecular biological techniques has significantly changed the methods that can be applied to the study of evolutionary biology. In particular, the discovery that specific genes have been conserved at both the genetic and functional level between most eukaryotes has provided a powerful tool for understanding cellular and genetic functions. However, it is important to consider that the expression of a stereotypic genetic signaling pathway in cells with analogous function is not necessarily an indication that two cell types are homologous. Despite the seemingly disparate morphology of diverse phyla, a high degree of commonality exists in the developmental events that must occur to generate an organism. For example, regulation of proliferation or sorting of multiple progenitor cells into different cell fates represents fundamental developmental events that must occur not only in all organisms but also repeatedly in the same organism. Considering the relatively limited size of most genomes in comparison with the number of developmental events that must occur, it is not surprising that the same genetic signaling pathway will be used to mediate the same type of developmental event. For instance, the progression of mitotically active cells through the cell cycle is universally controlled through cyclins, cyclin-dependent kinase, and cyclin-dependent kinase inhibitors (reviewed in Miller and Cross 2001).

Perhaps the most compelling example of the power of a molecular genetic argument in the face of a seemingly impossible spectrum of morphological diversity is the ongoing contention that *pax6* may act as a "master control gene" for the eye phenotype (Gehring and Ikeo 1999). The broad spectrum of eye morphologies has led to the conclusion that the eye is an example of convergent evolution in multiple phyla of animals. However, the demonstration that *pax6* homologs are present in multiple phyla as well and that invariably the primary role of this gene is to induce the formation of eyes has resulted in the formulation of the master control gene hypothesis (Gehring and Ikeo 1999). It is important to consider that this argument has been significantly strengthened by the demonstration that a *pax6* homolog is expressed in the eye of virtually every animal yet tested and by the demonstration that *pax6* activates a cascade of downstream genes that are similarly conserved (Fernald 2000).

Therefore, the question becomes: Do the same criteria apply to Atonal and the specification of mechanoreceptors? At this point, the answer is no, for several reasons. First, an *atonal* homolog has been identified in *C. elegans*, *lin-32*, but this gene is apparently not required for the development of any mechanosensory structures in this organism and instead plays a role in the specification of a different type of sensory neuron (Portman and Emmons 2000). Even within *Drosophila*, *atonal* is not required for the specification of all mechanosensory neurons (Jarman et al. 1993). In terms of downstream targets of Atonal, very little is known; however, *senseless*, a target of Atonal in *Drosophila* (Nolo et al. 2000), has recently been shown to not be regulated by Math1 in vertebrate hair cells (Wallis et al. 2003). Moreover, Atonal does not uniquely specify mechanoreceptors. In *Drosophila*, Atonal is also is required for the formation of the R8 photoreceptor (Jarman et al. 1994), whereas in vertebrates Math1 is also required for the specification of cerebellar granule cells and neurons within the proprioceptive pathway (Ben-Arie et al. 1997; Bermingham et al. 2001). Finally, a recently discovered *Drosophila* bHLH gene, *amos*, which is required for the formation of odorant receptors, is as closely related to *math1* as is *atonal* (Goulding et al. 2000). This brings into question the true identity of the *math1* ortholog in *Drosophila* (Jarman 2002).

Finally, it remains to be seen whether the ultimate target of mechanosensory development, the formation of a mechanosensing channel, is conserved. NompC, a member of the TRP family of ion channels, has been shown to be required for mechanotransduction in *Drosophila* (Walker et al. 2000). A homolog, Ce-NompC, is expressed in ciliated mechanoreceptors in *C. elegans* but has not been shown to be required for mechanoreceptor transduction (Walker et al. 2000). In vertebrates, the identity of the transduction channel has not been determined. In mammals, there are at least 13 genes that code for TRP or TRP-related channels (Minke and Cook 2002), and one of these may turn out to be the transduction channel, but at present the channel remains unknown.

Similar arguments can be made for the Notch pathway, only on a much larger scale. The Notch pathway appears to be an ancestral pathway that has evolved, for the most part, to nonspecifically regulate the commitment of a subset of cells within an equivalence group (reviewed in Baron et al. 2002). In both *Drosophila* and mice, genomic defects in the Notch pathway typically lead to defects so profound that the embryos are aborted at very early stages (Atravanis-Tsakonas et al. 1983; Swiatek et al. 1994). These include significant defects in events as profound as the size of the primary neural plate. Based on its broad pattern of expression, it is difficult to draw any insights on cellular homology based on the expression of the Notch pathway.

6. Summary

We put forth the argument that vertebrate sensory hair cells and invertebrate mechanoreceptive neurons have not yet been shown to be homologous as mechanoreceptors. Under this view, similarities such as bundle morphology, physiological operating range, and bundle adaptation result from convergent evolution due to similar requirements. Molecular homologies suggest co-opting of genes and the use of existing genetic pathways during evolution (e.g., Notch) where precise developmental patterning is necessary. This view does not preclude the possibility that all mechanosensors arose from a common primordial cell type, but at this point the data to support that possibility are insufficient, and it seems as likely that similarities between vertebrate hair cells and invertebrate mechanosensory neurons arose by convergent evolution due to selective pressure to optimize sensory systems.

At the same time, we recognize that we are still at an early stage in our understanding of the molecular biology of sensory systems, and even our knowledge of the molecular biology of vertebrate hair cells is still limited. We would not be surprised if the views expressed here evolve over the next several years, but at the same time, we will not predict the direction in which this evolution will take place.

References

Ades HW, Engström H (1974) Anatomy of the inner ear. In: Keidel WD, Neff WD (eds) Handbook of Sensory Physiology: Auditory System: Anatomy Physiology (Ear). Berlin: Springer-Verlag, pp. 125–158.
Aitkin LM (1995) The auditory neurobiology of marsupials: a review. Hear Res 82:257–266.
Art JJ, Goodman MB (1996) Ionic conductances and hair cell tuning in the turtle cochlea. Ann NY Acad Sci 781:103–122.
Artavanis-Tsakonas S, Muskavitch MA, Yedvobnick B (1983) Molecular cloning of Notch, a locus affecting neurogenesis in *Drosophila melanogaster*. Proc Natl Acad Sci USA 80:1977–1981.
Ashmore JF (1987) A fast motile response in guinea-pig outer hair cells: the cellular basis of the cochlear amplifier. J Physiol 388:323–347.
Baatrup E (1981) Primary sensory cells in the skin of amphioxus (*Branchiostoma lanceolatum* (P)). Acta Zool 62:147–157.
Bardack D (1998) Relationships of living and fossil hagfishes. In: Jørgensen JM, Lomholt JP, Weber RE, Malte H (eds) The Biology of Hagfishes. London: Chapman and Hall, pp. 3–14.
Baron M, Aslam H, Flasza M, Fostier M, Higgs JE, Mazaleyrat SL, Wilkin MB (2002) Multiple levels of Notch signal regulation (review). Mol Membr Biol 19:27–38.
Ben-Arie N, Bellen HJ, Armstrong DL, McCall AE, Gordadze PR, Guo Q, Matzuk MM, Zoghbi HY (1997) *Math1* is essential for genesis of cerebellar granule neurons. Nature 390:169–172.
Bermingham NA, Hassan BA, Wang VY, Fernandez M, Banfi S, Bellen HJ, Fritzsch B,

Zoghbi HY (2001) Proprioceptor pathway development is dependent on Math1. Neuron 30:411–422.

Bone Q, Ryan KP (1978) Cupular sense organs in *Ciona* (Tunicata: Ascidiacea). J Zool Lond 186:417–429.

Braun CB (1996) The sensory biology of the living jawless fishes: a phylogenetic assessment. Brain Behav Evol 48:262–276.

Braun CB, Northcutt RG (1997) The lateral line system of hagfishes (Craniata: Myxinoida). Acta Zool 78:247–268.

Brichta AM, Goldberg JM (1996) Afferent and efferent responses from morphological fiber classes in the turtle posterior crista. Ann NY Acad Sci 781:183–195.

Brownell WE, Bader CR, Bertrand D, de Ribaupierre Y (1985) Evoked mechanical responses of isolated cochlear outer hair cells. Science 227:194–196.

Bruns V, Schmieszek E (1980) Cochlear innervation in the greater horseshoe bat: demonstration of an acoustic fovea. Hear Res 3:27–43.

Budelmann BU (1988) Morphological diversity of equilibrium receptor systems in aquatic vertebrates. In: Atema J, Fay RR, Popper AN, Tavolga WN (eds) Sensory Biology of Aquatic Animals. New York: Springer-Verlag, pp. 757–782.

Budelmann BU (1992a) Hearing in crustaceans. In: Webster, DB, Fay RR, Popper AN (eds) The Evolutionary Biology of Hearing. New York: Springer-Verlag, pp. 131–139.

Budelmann BU (1992b) Hearing in nonarthropod invertebrates. In: Webster DB, Fay RR, Popper AN (eds) The Evolutionary Biology of Hearing. New York: Springer-Verlag, pp. 141–155.

Budelmann BU, Thies G (1977) Secondary sensory cells in the gravity receptor system of the statocysts of *Octopus vulgaris*. Cell Tissue Res 182:93–98.

Budelmann BU, Williamson R (1994) Directional sensitivity of hair cell afferents in the *Octopus* statocysts. J Exp Biol 187:245–259.

Budelmann BU, Sachse M, Staudigl M (1987) The angular acceleration receptor system of *Octopus vulgaris*, *Sepia officinalis,* and *Loligo vulgaris*. Brain Res 56:25–41.

Burighel P, Lane NJ, Fabio G, Stefano T, Zaniolo G, Carnevali MDC, Manni L (2003) Novel, secondary sensory cell organ in ascidians: in search of the ancestor of the vertebrate lateral line. J Comp Neurol 461:236–249.

Caldwell JC, Eberl DF (2002) Towards a molecular understanding of *Drosophila* hearing. J Neurobiol 53:172–189.

Carroll RL (1988) Vertebrate Paleontology and Evolution. New York: Freeman.

Chandler JP (1984) Light and electron microscopic studies of the basilar papilla in the duck, *Anas platyrhynchos*. I. The hatchling. J Comp Neurol 222:506–522.

Chang JSY, Popper AN, Saidel WM (1992) Heterogeneity of sensory hair cells in a fish ear. J Comp Neurol 324:621–640.

Cochran SL, Correia MJ (1995) Functional support of glutamate as a vestibular hair cell transmitter in an amniote. Brain Res 670:321–325.

Conway Morris S (2000) The Cambrian "explosion": slow-fuse or megatonnage? Proc Natl Acad Sci USA 97:4426–4429.

Correia MJ, Lang DG (1990) An electrophysiological comparison of solitary type I and type II vestibular hair cells. Neurosci Lett 116:106–111.

Correia MJ, Ricci AJ, Rennie KJ (1996) Filtering properties of vestibular hair cells: an update. Ann NY Acad Sci 781:138–149.

Correia MJ, Rennie KJ, Koo P (2001) Return of potassium ion channels in regenerated hair cells: possible pathways and the role of intracellular calcium signaling. Ann NY Acad Sci 942:228–240.

Cortopassi KA, Lewis ER (1998) A comparison of the linear tuning properties of two classes of axons in the bullfrog lagena. Brain Behav Evol 51:331–348.

Corwin JT (1977) Morphology of the macular neglecta in sharks of the genus *Carcharhinus*. J Morphol 152:341–362.

Dallos P (1996) Overview: cochlear neurobiology. In: Dallos P, Popper AN, Fay RR (eds) The Cochlea. New York: Springer-Verlag, pp. 1–43.

Dallos P, Evans BN, Hallworth R (1991) Nature of the motor element in electrokinetic shape changes of cochlear outer hair cells. Nature 350:155–157.

Dallos P, He DZZ, Lin X, Sziklai I, Mehta S, Evans BN (1997) Acetylcholine, outer hair cell electromotility, and the cochlear amplifier. J Neurosci 17:2212–2226.

Devau G (2000) Glycine induced calcium concentration changes in vestibular type I sensory cells. Hear Res 140:126–136.

Devau G, Lehouelleur J, Sans A (1993) Glutamate receptors on type I vestibular hair cells of guinea pig. Eur J Neurosci 5:1210–1217.

Duggan A, García-Añoveros J, Corey DP (2000) Insect mechanoreception: what a long, strange TRP it's been. Curr Biol 10:R384–R387.

Eberl DF (1999) Feeling the vibes: chordotonal mechanisms in insect hearing. Curr Opin Neurobiol 9:389–393.

Eddison M, Le Roux I, Lewis J (2000) Notch signaling in the development of the inner ear: lessons from *Drosophila*. Proc Natl Acad Sci USA 97:11692–11699.

Ekström von Lubitz DKJ (1981) Ultrastructure of the lateral-line sense organs of the ratfish, *Chimaera monstrosa*. Cell Tissue Res 215:651–665.

Engström H, Wersäll J (1958) The ultrastructural organization of the organ of Corti and of the vestibular sensory epithelia. Exp Cell Res Suppl 5:460–492.

Erostegui C, Norris CH, Bobbin RP (1994) In vitro pharmacologic characterization of a cholinergic receptor on outer hair cells. Hear Res 74:135–147.

Fain GL, Matthews HR, Cornwall MC, Koutalos Y (2001) Adaptation in vertebrate photoreceptors. Physiol Rev 81:117–151.

Fernald RD (2000) Evolution of eyes. Curr Opin Neurobiol 10:444–450.

Fettiplace R (1987) Electrical tuning of hair cells in the inner ear. Trends Neurosci 10:421–425.

Firbas W, Müller G (1983) The efferent innervation of the avian cochlea. Hear Res 10:109–116.

Fischer FP (1992) Quantitative analysis of the innervation of the chicken basilar papilla. Hear Res 61:167–178.

Fischer FP (1994a) General pattern and morphological specializations of the avian cochlea. Scan Microsc 8:351–364.

Fischer FP (1994b) Quantitative TEM analysis of the barn owl basilar papilla. Hear Res 73:1–15.

Fischer FP (1998) Hair cell morphology and innervation in the basilar papilla of the emu (*Dromaius novaehollandiae*). Hear Res 121:112–124.

Flock Å (1964) Structure of the macula utriculi with special reference to directional interplay of sensory responses as revealed by morphological polarization. J Cell Biol 22:413–431.

Forey P, Janvier P (1993) Agnathans and the origin of jawed vertebrates. Nature 361:129–134.

Forge A (1991) Structural features of the lateral walls in mammalian cochlear outer hair cells. Cell Tissue Res 265:473–483.

Forge A, Davies S, Zajic G (1991) Assessment of ultrastructure in isolated cochlear hair cells using a procedure for rapid freezing before freeze-fracture and deep-etching. J Neurocytol 20:471–484.

Fritzsch B (1987) Inner ear of the coelacanth fish *Latimeria* has tetrapod affinities. Nature 327:153–154.

Fritzsch B, Beisel KW (2001) Evolution and development of the vertebrate ear. Brain Res Bull 55:711–721.

Fritzsch B, Wahnschaffe U (1987) Electron microscopical evidence for common inner ear and lateral line efferents in urodeles. Neurosci Lett 81:48–52.

Fuchs PA, Murrow BW (1992a) Cholinergic inhibition of short (outer) hair cells of the chick's cochlea. J Neurosci 12:800–809.

Fuchs PA, Murrow BW (1992b) A novel cholinergic receptor mediates inhibition of chick cochlear hair cells. Proc R Soc Lond B 248:35–40.

Furukawa T, Ishii Y (1967) Neurophysiological studies on hearing in goldfish. J Neurophysiol 30:1377–1403.

García-Añoveros J, Corey DP (1997) The molecules of mechanosensation. Annu Rev Neurosci 20:567–594.

Gehring WJ, Ikeo K (1999) Pax 6: mastering eye morphogenesis and eye evolution. Trends Genet 15:371–377.

Gillespie PG, Corey DP (1997) Myosin and adaptation by hair cells. Neuron 19:955–958.

Gillespie PG, Walker RG (2001) Molecular basis of mechanosensory transduction. Nature 413:194–202.

Gleich O, Manley GA (2000) The hearing organ of birds and Crocodilia. In: Dooling RE, Fay RR, Popper AN (eds) Comparative Hearing: Birds and Reptiles. New York: Springer-Verlag, pp. 70–138.

Goodman MB, Art JJ (1996) Variations in the ensemble of potassium currents underlying resonance in turtle hair cells. J Physiol 497:395–412.

Göpfert MC, Robert D (2003) Motion generation by *Drosophila* mechanosensory neurons. Proc Natl Acad Sci USA 100:5514–5519.

Goulding SE, zur Lage P, Jarman AP (2000) *amos*, a proneural gene for *Drosophila* olfactory sense organs that is regulated by lozenge. Neuron 25:69–78.

Hackney CM, Fettiplace R, Furness DN (1993) The functional morphology of stereociliary bundles on turtle cochlear hair cells. Hear Res 69:163–175.

Hartline PH (1971a) Physiological basis for detection of sound and vibration in snakes. J Exp Biol 54:349–371.

Hartline PH (1971b) Mid-brain responses of the auditory and somatic vibration systems in snakes. J Exp Biol 54:373–390.

Hartline PH, Campbell HW (1969) Auditory and vibratory responses in the midbrains of snakes. Science 163:1221–1223.

Hillman DE (1976) Morphology of the peripheral and central vestibular systems. In: Llinás R, Precht W (eds) Frog Neurobiology. Berlin: Springer-Verlag, pp. 452–480.

Holland ND, Yu J-K (2002) Epidermal receptor development and sensory pathways in vitally stained amphioxus (*Branchiostoma floridae*). Acta Zool 83:309–319.

Holt JR, Eatock RA (1995) Inwardly rectifying currents of saccular hair cells from the leopard frog. J Neurophysiol 73:1484–1502.

Hoshino T (1975) An electron microscopic study of the otolithic maculae of the lamprey (*Entosphenus japonicus*). Acta Otolaryngol 80:43–53.
Hudspeth AJ (1985) The cellular basis of hearing: the biophysics of hair cells. Science 230:745–752.
Hudspeth AJ (1989) How the ear's works work. Nature 341:397–404.
Hudspeth AJ, Choe Y, Mehta AD, Martin P (2000) Putting ion channels to work: mechanoelectrical transduction, adaptation, and amplification by hair cells. Proc Natl Acad Sci USA 97:11765–11772.
Jarman AP (2002) Studies of mechanosensation using the fly. Hum Mol Genet 11:1215–1218.
Jarman AP, Grau Y, Jan LY, Jan YN (1993) *atonal* is a proneural gene that directs chordotonal organ formation in the *Drosophila* peripheral nervous system. Cell 73:1307–1321.
Jarman AP, Grell EH, Ackerman L, Jan LY, Jan YN (1994) *atonal* is the proneural gene for *Drosophila* photoreceptors. Nature 369:398–400.
Jarvik E (1980) Basic Structure and Evolution of Vertebrates, vol 1. London: Academic Press.
Jones EMC, Gray-Keller M, Fettiplace R (1999) The role of Ca^{2+}-activated K^+ channel spliced variants in the tonotopic organization of the turtle cochlea. J Physiol 518:653–665.
Jørgensen JM (1989) Evolution of octavolateralis sensory cells. In: Coombs S, Görner P, Münz H (eds) The Mechanosensory Lateral Line. New York: Springer-Verlag, pp. 115–146.
Jørgensen JM, Shichiri M, Geneser FA (1998) Morphology of the hagfish inner ear. Acta Zool 79:251–256.
Keil TA (1997) Functional morphology of insect mechanoreceptors. Microsc Res Tech 39:506–531.
Klinke R (1981) Neurotransmitters in the cochlea and the cochlear nucleus. Acta Otolaryngol 91:541–554.
Köppl C, Gleich O, Schwabedissen G, Siegl E, Manley GA (1998) Fine structure of the basilar papilla of the emu: implications for the evolution of hair-cell types. Hear Res 126:99–112.
Köppl C, Wegscheider A, Gleich O, Manley GA (2000) A quantitative study of cochlear afferent axons in birds. Hear Res 139:123–143.
Kros CJ, Marcotti W, van Netten SM, Self TJ, Libby RT, Brown SD, Richardson GP, Steel KP (2002) Reduced climbing and increased slipping adaptation in cochlear hair cells of mice with Myo7a mutations. Nat Neurosci 5:41–47.
Lacalli TC, Hou S (1999) A reexamination of the epithelial sensory cells of amphioxus (*Branchiostoma*). Acta Zool 80:125–134.
Ladhams A, Pickles JO (1996) Morphology of the monotreme organ of Corti and macula lagena. J Comp Neurol 366:335–347.
Lanford PJ, Popper AN (1996) Novel afferent terminal structure in the crista ampullaris of the goldfish, *Carassius auratus*. J Comp Neurol 366:572–579.
Lanford PJ, Platt C, Popper AN (2000) Structure and function in the saccule of the goldfish (*Carassius auratus*): a model of diversity in the non-amniote ear. Hear Res 143:1–13.
Leake PA (1977) SEM observations of the cochlear duct in *Caiman crocodilus*. Scan Electron Microsc II:437–444.

Lewis ER (1981) Evolution of inner-ear auditory apparatus in the frog. Brain Res 219: 149–155.

Lewis ER, Li CW (1973) Evidence concerning the morphogenesis of saccular receptors in the bullfrog (*Rana catesbeina*). J Morphol 139:351–362.

Lewis ER, Li CW (1975) Hair cell types and distributions in the otolithic and auditory organs of the bullfrog. Brain Res 83:35–50.

Liberman MC, Dodds LW, Pierce S (1990) Afferent and efferent innervation of the cat cochlea: quantitative analysis with light and electron microscopy. J Comp Neurol 301: 443–460.

Lindeman HH (1969) Regional differences in sensitivity of the vestibular sensory epithelia to ototoxic antibiotics. Acta Otolaryngol 67:117–189.

Löwenstein O (1970) The electrophysiological study of the responses of the isolated labyrinth of the lamprey (*Lampetra fluviatilis*) to angular acceleration, tilting, and mechanical vibration. Proc R Soc Lond B 174:419–434.

Löwenstein O (1971) Functional anatomy of the vertebrate gravity receptor system. In: Gordon SA, Cohen MJ (eds) Gravity and the Organism. Chicago: University of Chicago Press, pp. 253–261.

Löwenstein O, Osborne MP (1964) Ultrastructure of the sensory hair cells in the labyrinth of the ammocete larva of the lamprey, *Lampetra fluviatilis*. Nature 204:97.

Löwenstein O, Thornhill RA (1970) The labyrinth of *Myxine*, anatomy, ultrastructure and electrophysiology. Proc R Soc Lond B 176:21–42.

Löwenstein O, Osborne MP, Wersäll J (1964) Structure and innervation of the sensory epithelia of the labyrinth in the thornback ray (*Raja clavata*). Proc R Soc Lond B 160:1–12.

Löwenstein O, Osborne MP, Thornhill RA (1968) The anatomy and ultrastructure of the labyrinth of the lamprey (*Lampetra fluviatilis* L.). Proc R Soc Lond B 170:113–134.

Manley GA (1981) A review of the auditory physiology of the reptiles. Prog Sens Physiol 2:49–134.

Manley GA (1990) Peripheral Hearing Mechanisms in Reptiles and Birds. Heidelberg, Germany: Springer-Verlag.

Manley GA (1995) The avian hearing organ: a status report. In: Manley GA, Klump GM, Köppl C, Fastl H, Oeckinghaus H (eds) Advances in Hearing Research. Singapore: World Scientific, pp. 219–229.

Manley GA (2000a) The hearing organs of lizards. In: Dooling RE, Fay RR, Popper AN (eds) Comparative Hearing: Birds and Reptiles. New York: Springer-Verlag, pp. 139–196.

Manley GA (2000b) Cochlear mechanisms from a phylogenetic viewpoint. Proc Nat Acad Sci USA 97:11736–11743.

Manley GA (2001) Evidence for an active process and a cochlear amplifier in nonmammals. J Neurophysiol 86:541–549.

Manley GA, Köppl C (1998) Phylogenetic development of the cochlea and its innervation. Curr Opin Neurobiol 8:468–474.

Manley GA, Gleich O, Kaiser A, Brix J (1989) Functional differentiation of sensory cells in the avian auditory periphery. J Comp Physiol A 164:289–296.

Manley GA, Yates GK, Köppl C, Johnstone BM (1990) Peripheral auditory processing in the bobtail lizard *Tiliqua rugosa*: IV. Phase locking of auditory-nerve fibres. J Comp Physiol A 167:129–138.

Manley GA, Kirk D, Köppl C, Yates GK (2001) In vivo evidence for a cochlear amplifier in the hair-cell bundle of lizards. Proc Nat Acad Sci USA 98:2826–2831.

Martinez-Dunst C, Michaels RL, Fuchs PA (1997) Release sites and calcium channels in hair cells of the chick's cochlea. J Neurosci 17:9133–9144.

Martini M, Rossi ML, Rubbini G, Rispoli G (2000) Calcium currents in hair cells isolated from semicircular canals of the frog. Biophys J 78:1240–1254.

Masetto S, Correia MJ (1997) Ionic currents in regenerating avian vestibular hair cells. Int J Dev Neurosci 15:387–399.

Masetto S, Russo G, Prigioni I (1994) Differential expression of potassium currents by hair cells in thin slices of frog crista ampullaris. J Neurophysiol 72:443–455.

Miller ME, Cross FR (2001) Cyclin specificity: how many wheels do you need on a unicycle? J Cell Sci 114:1811–1820.

Miller MR (1978) Scanning electron microscope studies of the papilla basilaris of some turtles and snakes. Am J Anat 151:409–436.

Miller MR (1980) The reptilian cochlear duct. In: Popper AN, Fay RR (eds) Comparative Studies of Hearing in Vertebrates. New York: Springer-Verlag, pp. 169–204.

Miller MR (1985) Quantitative studies of auditory hair cells and nerves in lizards. J Comp Neurol 232:1–24.

Miller MR (1992) The evolutionary implications of the structural variations in the auditory papilla of lizards. In: Webster DB, Fay RR, Popper AN (eds) The Evolutionary Biology of Hearing. New York: Springer-Verlag, pp. 463–488.

Miller MR, Beck J (1988) Auditory hair cell innervational patterns in lizards. J Comp Neurol 271:604–628.

Miller MR, Beck J (1990) Further serial transmission electron microscopy studies of auditory hair cell innervation in lizards and a snake. Am J Anat 188:175–184.

Minke B, Cook B (2002) TRP channel proteins and signal transduction. Physiol Rev 82:429–472.

Mire P, Watson GM (1997) Mechanotransduction of hair bundles arising from multicellular complexes in anemones. Hear Res 113:224–234.

Neumeister H, Budelmann BU (1997) Structure and function of the *Nautilus* statocysts. Philos Trans R Soc Lond B 352:1565–1588.

Nolo R, Abbott LA, Bellen HJ (2000) Senseless, a Zn finger transcription factor, is necessary and sufficient for sensory organ development in *Drosophila*. Cell 102:349–362.

Northcutt RG, Gans C (1983) The genesis of neural crest and epidermal placodes: a reinterpretation of vertebrate origins. Q Rev Biol 58:1–58.

Oliver D, He DZZ, Klöcker N, Ludwig J, Schulte U, Waldegger S, Ruppersberg SP, Dallos P, Fakler B (2001) Intracellular anions as the voltage sensor of prestin, the outer hair cell motor protein. Science 292:2340–2343.

Padian K, Chiappe LM (1998) The origin and early evolution of birds. Biol Rev 73:1–42.

Perin P, Soto E, Vega R, Botta L, Masetto S, Zucca G, Valli P (2000) Calcium channels functional roles in the frog semicircular canal. Neuroreport 11:417–420.

Peterson EH, Cotton JR, Grant JW (1996) Structural variation in ciliary bundles of the posterior semicircular canal. Quantitative anatomy and computational analysis. Ann NY Acad Sci 781:85–102.

Phelps SM (2002) Like minds: evolutionary convergence in nervous systems. TREE 17:158–159.

Pickles JO, Brix J, Comis SD, Gleich O, Köppl C, Manley GA, Osborne MP (1989) The organization of tip links and stereocilia on hair cells of bird and lizards basilar papillae. Hear Res 41:31–41.

Platt C, Popper AN (1984) Variation in length of ciliary bundles on hair cells along the macula of the sacculus in two species of teleost fishes. Scan Electron Microsc 4:1915–1924.

Popper AN (1977) A scanning electron microscopic study of the sacculus and lagena in the ears of fifteen species of teleost fishes. J Morphol 153:397–417.

Popper AN (1981) Comparative scanning electron microscopic investigations of the sensory epithelia in the teleost sacculus and lagena. J Comp Neurol 200:357–374.

Popper AN, Hoxter B (1987) Sensory and nonsensory ciliated cells in the ear of the sea lamprey, *Petromyzon marinus*. Brain Behav Evol 30:43–61.

Popper AN, Saidel WM (1990) Variations in receptor cell innervation in the saccule of a teleost fish ear. Hear Res 46:211–228.

Portman DS, Emmons SW (2000) The basic helix-loop-helix transcription factors LIN-32 and HLH-2 function together in multiple steps of a *C. elegans* neuronal sublineage. Development 127:5415–5426.

Puschner B, Schacht J (1997) Calmodulin-dependent protein kinases mediate calcium-induced slow motility of mammalian outer hair cells. Hear Res 110:251–258.

Ramanathan K, Michael TH, Jiang G, Hiel H, Fuchs PA (1999) A molecular mechanism for electrical tuning of cochlear hair cells. Science 283:215–217.

Ramanathan K, Michael TH, Fuchs PA (2000) β subunits modulate alternatively spliced, large conductance, calcium-activated potassium channels of avian hair cells. J Neurosci 20:1675–1684.

Reiter ER, Liberman MC (1995) Efferent mediated protection from acoustic overexposure: relation to "slow" effects of olivocochlear stimulation. J Neurophysiol 73:506–514.

Rennie KJ, Correia MJ (1994) Potassium currents in mammalian and avian isolated type I semicircular canal hair cells. J Neurophysiol 71:317–329.

Ricci AJ, Gray-Keller M, Fettiplace R (2000) Tonotopic variations of calcium signalling in turtle auditory hair cells. J Physiol 524:423–436.

Rüsch A, Eatock RA (1996) A delayed rectifier conductance in type I hair cells of the mouse utricle. J Neurophysiol 76:995–1004.

Rüsch A, Lysakowski A, Eatock RA (1998) Postnatal development of type I and type II hair cells in the mouse utricle: acquisition of voltage-gated conductances and differentiated morphology. J Neurosci 18:7487–7501.

Saidel WM, Crowder JA (1997) Expression of cytochrome oxidase in hair cells of the teleost utricle. Hear Res 109:63–77.

Saidel WM, Presson JC, Chang JS (1990) S100 immunoreactivity identifies a subset of hair cells in the utricle and saccule of a fish. Hear Res 47:139–146.

Saito K (1983) Fine structure of the sensory epithelium of guinea pig organ of Corti: subsurface cisternae and lamellar bodies in the outer hair cells. Cell Tissue Res 229:457–481.

Sans A, Scarfone E (1996) Afferent calyces and type I hair cells during development. A new morphofunctional hypothesis. Ann NY Acad Sci 781:1–12.

Schrott-Fischer, A, Kammen-Jolly K, Scholtz AW, Gluckert R, Eybalin M (2002) Patterns of GABA-like immunoreactivity in efferent fibers of the human cochlea. Hear Res 174:75–85.

Schulte E, Riehl R (1977) Elektronmikroskopische Untersuchungen an der Oralcirren und der Haut von *Branchiostoma lanceolatum*. Helgoländer wissenschaftliche Meeresuntersuchungen 29:337–357.

Slepecky NB (1996) Structure of the mammalian cochlea. In: Dallos P, Popper AN, Fay RR (eds) The Cochlea. New York: Springer-Verlag, pp. 44–129.

Smotherman MS, Narins PM (1999a) The electrical properties of auditory hair cells in the frog amphibian papilla. J Neurosci 19:5275–5292.

Smotherman MS, Narins PM (1999b) Potassium currents in auditory hair cells of the frog basilar papilla. Hear Res 132:117–130.

Smotherman MS, Narins PM (2000) Hair cells, hearing, and hopping: a field guide to hair cell physiology in the frog. J Exp Biol 203:2237–2246.

Sneary MG (1988a) Auditory receptor of the red-eared turtle: I. Afferent and efferent synapses and innervation patterns. J Comp Neurol 276:588–606.

Sneary MG (1988b) Auditory receptor of the red-eared turtle: II. General ultrastructure. J Comp Neurol 276:573–587.

Song J, Yan HY, Popper AN (1995) Damage and recovery of hair cells in fish canal (but not superficial) neuromasts after gentamicin exposure. Hear Res 91:63–71.

Sridhar T, Liberman MC, Brown MC, Sewell WF (1995) A novel cholinergic "slow effect" of efferent stimulation on cochlear potentials in the guinea pig. J Neurosci 15: 3667–3678.

Steinacker A, Monterrubio J, Perez R, Mensinger AF, Marin A (1997) Electrophysiology and pharmacology of outward potassium currents in semicircular canal hair cells of toadfish, *Opsanus tau*. Hear Res 109:11–20.

Stokes MD, Holland ND (1995) Embryos and larvae of a lancelet, *Branchiostoma floridae*, from hatching through metamorphosis: growth in the laboratory and external morphology. Acta Zool 76:89–176.

Strassmaier M, Gillespie PG (2002) The hair cell's transduction channel. Curr Opin Neurobiol 12:380–386.

Sugihara I (2001) Efferent innervation in the goldfish saccule examined by acetylcholinesterase histochemistry. Hear Res 153:91–99.

Sugihara I, Furukawa T (1989) Morphological and functional aspects of two different types of hair cells in the goldfish sacculus. J Neurophysiol 62:1330–1343.

Swiatek PJ, Lindsell CE, del Amo FF, Weinmaster G, Gridley T (1994) Notch1 is essential for postimplantation development in mice. Genes Dev 8:707–719.

Szönyi M, He DZZ, Ribári O, Sziklai I, Dallos P (2001) Intracellular calcium and outer hair cell electromotility. Brain Res 922:65–70.

Takasaka T, Smith CA (1971) The structure and innervation of the pigeon's basilar papilla. J Ultrastruct Res 34:20–65.

Tanaka K, Smith CA (1978) Structure of the chicken's inner ear: SEM and TEM study. Am J Anat 153:251–272.

Walker RG, Willingham AT, Zuker CS (2000) A *Drosophila* mechanosensory transduction channel. Science 287:2229–2234.

Wallis D, Hamblen M, Zhou Y, Venken KJ, Schumacher A, Grimes HL, Zoghbi HY, Orkin SH, Bellen HJ (2003) The zinc finger transcription factor Gfi1, implicated in lymphomagenesis, is required for inner ear hair cell differentiation and survival. Development 130:221–232.

Weisleder P, Tsue TT, Rubel EW (1995) Hair cell replacement in avian vestibular epithelium: supporting cell to type I hair cell. Hear Res 82:125–133.

Wersäll J (1956) Studies on the structures and innervation of the sensory epithelium of the cristae ampullares in the guinea pig. Acta Otolaryngol Stockh Suppl 126: 1–85.

Wersäll J (1960) Vestibular receptor cells in fish and mammals. Acta Otolaryngol Stockh Suppl 163:25–29.
Wersäll J, Bagger-Sjöbäck D (1974) Morphology of the vestibular sense organs. In: Kornhuber HH (ed) Handbook of Sensory Physiology: Vestibular System, Part 1. Berlin: Springler-Verlag, pp. 123–170.
Wever EG (1975) The caecilian ear. J Exp Zool 191:63–72.
Wever EG (1978) The Reptile Ear. Princeton, NJ: Princeton University Press.
Wever EG (1985) The Amphibian Ear. Princeton, NJ: Princeton University Press.
Wever EG, McCormick JG, Palin J, Ridgway SH (1971) The cochlea of the dolphin, *Tursiops truncatus*: hair cells and ganglion cells. Proc Natl Acad Sci USA 68:2908–2912.
Yan HY, Saidel WM, Chang JS, Presson JC, Popper AN (1991) Sensory hair cells of a fish ear: evidence of multiple cell types based on ototoxic sensitivity. Proc R Soc Lond B 245:133–138.
Zheng J, Shen W, He DZZ, Long KB, Madison LD, Dallos P (2000) Prestin is the motor protein of cochlear outer hair cells. Nature 405:149–155.
Zufall F, Leinders-Zufall T (2000) The cellular and molecular basis of odor adaptation. Chem Senses 25:473–481.

4
Parallel Evolution in Fish Hearing Organs

FRIEDRICH LADICH AND ARTHUR N. POPPER

1. Introduction

Fishes, as broadly defined to include agnathans (jawless fishes), cartilaginous fishes, and bony fishes, are the earliest vertebrates (Fig. 4.1). Because an inner ear is found in the fossil record of the most primitive jawless vertebrates (Forey and Janvier 1994), it is reasonable to assume that the ear, and possibly hearing, arose quite early in this group or was present in their ancestral chordates. Although there has been some suggestion that vertebrate inner-ear sensory hair cells may be derived from a statocyst-like system invertebrate mechanoreceptive cell, this is very much open to question (reviewed in Coffin et al., Chapter 3). More importantly for this chapter, it is highly likely that the vertebrate ear arose de novo in this group or perhaps in craniate ancestors (see van Bergeijk 1967 and Wever 1974 for a discussion of the origin of the vertebrate ear and Lewis and Fay, Chapter 2, for a discussion of the origin of hearing).

In this chapter we draw some conclusions about the evolution of fish ears from what we know about the perception and generation of sound in extant groups of fishes. We sort out which evolutionary constraints may have resulted in the development of the auditory system and to what extent this is linked to acoustic communication. From an anthropocentric point of view, hearing seems to be primarily bound to the transfer of acoustic information within a species, although there are several indications that this was not the major selective pressure for the origin of hearing in the earliest vertebrates (Fay and Popper 2000; see also Lewis and Fay, Chapter 2). We elaborate on the auditory tasks that should be fulfilled by any acoustic sensory system and describe the evolution of the inner-ear structure represented in extant fish taxa. Although hearing is based on the inner ear (or labyrinth), the specific structures involved in sound detection vary in different anamniote groups. Moreover, unlike the amniotes (birds, reptiles, mammals; see Smotherman and Narins, Chapter 6; Manley, Chapter 7; Gleich et al., Chapter 8) where there is a single clearly defined auditory end organ, the individual end organs in fishes, like those in some amphibians (Lewis and Narins 1999; Smotherman and Narins, Chapter 6), may exhibit both auditory and vestibular functions. Moreover, the relative contributions to hearing of different end organs may differ between different fish species.

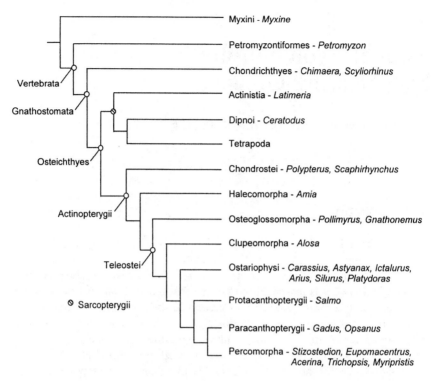

FIGURE 4.1. Cladogram of taxa and the species mentioned throughout the chapter. This cladogram follows that of Pennisi (2003) and Nelson (1994).

It is likely that the most primitive vertebrate ear was a motion detector that sensed changes in head position relative to gravity (van Bergeijk 1967). Van Bergeijk (1967) proposed that over time, this structure evolved into an accelerometer-type of system that responded to low-frequency particle motion. This would have provided fish with the detection of relatively low-frequency sounds, with the specific response characteristics of the system probably dependent on the size of the mass that was associated with the system. The limitations of this system were that it could only be low frequency (e.g., perhaps up to 500 Hz). Although we have no living (or even fossil) examples of such early ears, it is likely that selective pressures independently resulted in the evolution of increased hearing bandwidth in different fish groups. This most often resulted from the development of an acoustic coupling between a higher frequency pressure impedance-matching system, generally the swim bladder, and the inner ear. Such coupling not only would have broadened hearing bandwidth but also would have increased sensitivity.

2. Evolutionary Perspective

Among the chordates, the detection of acoustic vibrations is only known among the vertebrates. There is no indication of an ear in the early chordates (Wever 1974), but sensory cells with similarities to vertebrate hair cells are known from other chordates (Burighel et al. 2003; see also Coffin et al., Chapter 3). This may partly be explained by the life history of these organisms and their lack of a complex nervous system associated with living as either sessile or planktonic organisms or buried in the substrate (e.g., *Amphioxus*). All these groups are filter feeders and thus are not required to actively search for food. Therefore, they possess only a limited ability to orient and move in their environment, and this would limit their need to respond to positional or acoustical stimuli.

Fossil agnathans (jawless vertebrates), including extinct "ostracoderms" that represent the earliest vertebrates, had a vestibular labyrinth. The majority of the known extinct species appear to have had one or two semicircular canals (Forey and Janvier 1994). There is no evidence to suggest that any of these species detected sound, and it is not clear whether extant agnathans detect sound. However, it should be noted that the lack of knowledge as to whether jawless vertebrates detect sound may be more a function of lack of experimental study rather than detection actually not occurring in any species.

When did hearing first evolve in vertebrates? Considering that a labyrinth is present in all known extinct and extant vertebrates and that there exists no particular structure or end organ exclusively devoted to sound detection in fishes, the vestibular labyrinth of early agnathans could potentially have served for detection of acoustic vibrations. Given the structure of vestibular epithelia, with dense, calcareous otoliths overlying the hair cells, motion of such structures in the presence of low-frequency sound would have been unavoidable and thus available for detection by the sensory epithelium (de Vries 1950). Thus it is possible that sound detection might have evolved in jawless fishes in the Cambrian period more than 400 million years ago. Alternatively, if hearing is linked to a higher diversification of the labyrinth and sensory epithelium that took place in early gnathostomes (jawed vertebrates), then hearing may have evolved in the Ordovician period when gnathostomes first appeared (Nelson 1994).

Why did hearing first evolve? Hearing in vertebrates is always linked to a vestibular labyrinth that is also required for active movement and orientation in a three-dimensional space. In addition to various equilibrium and gravitational tasks, sensory epithelia might initially have responded to large oscillations of the water such as those caused by wind, waves, and currents and perhaps low-frequency acoustic vibrations backscattered from objects in their environment. A sensory system helping in the determination of the magnitude of such vibratory cues, their patterns of change, and their direction is potentially very useful, especially in an environment with limited light and where the spread of olfactory cues over any significant distance is subject to the direction and uniformity of water currents. Additional selective pressures toward the evolution of hearing may have involved eavesdropping on sound unintentionally emitted by predators, prey, and, perhaps, conspecifics (e.g., "interception" as defined by Myrberg 1981). There is no evidence that agnathans and sharks produce communication sounds, and thus it appears that the detection of sounds of nonbiological origin or unintentionally produced sounds is likely to have been the major reason for the evolution and further refinement of hearing in bony fishes and more "advanced" vertebrates.

Acoustic communication, which is the detection of sounds produced by conspecifics for information transfer, most likely evolved much later in evolution. The "oldest" group in which sound production has been described are representatives of the genus *Polypterus* or bichirs (Ladich and Tadler 1988), a group that is thought to be an early line of actinopterygians (ray-finned fishes), the group that comprises the majority of modern fishes (Nelson 1994). *Polypterus* and its relatives (the Polypterids) resemble Devonian fishes, and thus one might speculate that acoustic communication in vertebrates arose some 300 million years ago.

3. Auditory Tasks

Any auditory system must solve complex problems in the interpretation of incoming sounds, including analyzing the frequency composition and intensity as well as temporal patterns of sounds and segregating mixtures of sounds (see Lewis and Fay, Chapter 2). All of these tasks have to be performed in the presence of other sounds.

It is reasonable to suggest that the earliest vertebrate sound-detecting ear would have responded to low-frequency, large-magnitude signals. This means that hearing was presumably restricted to short distances and thus to the kinetic component of sound (the acoustic "near field" and hydrodynamic fields) (Lewis and Fay, Chapter 2). The acoustic near field is usually defined as the region around a source where a hydrodynamic flow and particle motion generated by the vibrating object exceeds that accompanying the sound pressure of a progressive sound wave. This field declines rapidly with distance by $1/r^2$ or $1/r^3$ (where r is the distance from the source depending on the specific characteristics of the source) (van Bergeijk 1967; Kalmijn 1988; Rogers and Cox 1988; Hawkins 1993). The region beyond the near field is called the acoustic "far field." However, because the boundary between far field and near field depends on the frequency and the pattern of motion of objects (which are usually more complex than simple physical models such as monopoles or dipoles), the transition zones can only be roughly estimated as being from one sixth up to one half of a wavelength from the source. This means that for a 50-Hz pure tone, the near field extends up to 15 m and for a 1-kHz tone up to 0.75 m and that it varies for broadband signals such as feeding or splashing noises or certain communication sounds.

Sensitivity and frequency bandwidth of the auditory system most likely increased during phylogenetic development, although the extent of this increase is not clear from extant species because the majority of species without hearing specializations (section 5) all have the same general frequency range and sensitivity (Fay 1988; Popper and Fay 1999; Popper et al. 2003). The limit on bandwidth and sensitivity is thus a function of the structure of the otolithic end organs of the ear and any accessory structures, but we still have only the most limited understanding of the features of the ear that specifically limit bandwidth (Lychakov and Rebane 2002).

A widening of the frequency bandwidth and sensitivity no doubt occurred in teleosts, and particularly in hearing specialists, with the evolution of structural connections between the inner ear and gas-filled cavities. A pulsating gas bubble produces a considerable near-field displacement in an alternating acoustic pressure field and enables animals to detect sound pressure in the acoustic near and far fields. The further evolution of hearing organs in vertebrates, and especially in mammals, consisted of a 10- or even 50-fold increase in the frequency bandwidth detectable, whereas sensitivities in hearing specialists among fishes approached those for some birds and mammals (Poggendorf 1952). Interestingly, a small number of fish species (some herringlike fishes, the clupeids) can detect ultrasound using hearing (Mann et al. 1998, 2001; Plachta and Popper 2002).

Another important aspect of the identification and separation of sound sources is the ability to discriminate between different temporal patterns of sounds. This ability seems to be of major importance in fishes, a conclusion drawn from sound propagation tests in shallow-water environments and from the acoustic structure of communication sounds (the majority of which are built up from a series of pulses) (Myrberg et al. 1978b). Mann and Lobel (1997) demonstrated that except for the pulse period, the variance in sound characteristics increased with increasing distance from the sound source. Temporal parameters such as sound duration, pulse length, pulse periods, or duty cycles were frequently found to differ between closely related species and individuals and thus seem to be important in species recognition, mate choice, and assessment of opponents (sunfishes: Gerald 1971; damselfishes: Spanier 1979; labyrinth fishes: Ladich et al. 1992; mormyrids: Crawford et al. 1997, Marvit and Crawford 2000; toadfishes: McKibben and Bass 1998). However, from an evolutionary point of view, it is unclear whether the temporal resolution ability down to or even below 1 millisecond (ms) evolved in ancestral fishes or if it improved during the evolution of teleosts. Despite the fact that this ability was primarily investigated in teleosts that communicate acoustically, studies on the nonvocal goldfish (*Carassius auratus*) (Fay 1982, 1985; Fay et al. 1983) and comparative studies including vocal and nonvocal cyprinids and gouramis (labyrinth fishes) showed clear similarities in temporal resolution in sound-producing and non–sound-producing species (Wysocki and Ladich 2002). This suggests that the ability to do refined temporal resolution evolved before the origin of sound communication. Although the selective pressures for such a resolution capability are not clear, they are likely to have been in response to environmental constraints and/or they reflect the necessity within the auditory pathways of the central nervous system to resolve patterns as part of an analysis of the "acoustic scene."

Sensory components that arise from distinct environmental events have to be segregated into separate perceptual representations, which are referred to as auditory streams (Bregman 1990). Auditory stream segregation is based on the grouping of acoustical events by temporal patterns, frequency composition (e.g., harmonicity), loudness, and the location from which sounds are coming. This grouping results in the segregation and possibly the localization and identification of each sound generator. Analyzing the auditory scene by separating out important sound producers such as predators, prey, competitors, or mates from unimportant events such as background noise caused by currents or other animals such as insects or crustaceans is a major task of the auditory system and a prerequisite for appropriate behavioral responses. Auditory stream segregation has been demonstrated behaviorally in goldfish (Fay 1998, 2000). However, analyzing the auditory scene could be impaired because several acoustic components are less audible in the presence of others in the same or adjacent frequency spectrum, an effect called masking. Masking is likely to occur in all but the most quiet environments (Tavolga 1967; Fay 1974; Fay et al. 1978; Popper and Clarke 1979). Intense broadband noise or pure tones can result in temporary hearing loss over the entire frequency range or at particular frequencies in species possessing well-developed hearing abilities such as cyprinids and catfishes (Popper and Clarke 1976; Scholik and Yan 2000; Amoser and Ladich 2003). If and to what degree ambient sounds in the environment influenced the evolution of sound detection or source segregation are unknown. But there are some indications that such a connection exists (Schellart and Popper 1992).

4. Evolution of Inner-Ear Structure

The acousticolateralis hypothesis suggested that the ear arose as an invagination of the lateral line (Ayers 1892; van Bergeijk 1967). This suggestion was primarily based on presumed structural, developmental, and functional similarities between the ear and lateral line such as the utilization of hair cells for perception of mechanical stimuli and presumed similarities of neural innervation. However, no trace of an intermediate stage between the lateral line and the vestibular labyrinth can be found in any extant or extinct fish species (reviewed in Popper et al. 1992). Indeed, it is now known that there are significant differences between the ear and lateral line both developmentally and in their innervation, and this has led to the octavolateralis hypothesis that suggests that the two systems arose independently.

4.1 Agnathans

The two groups of jawless fishes possess somewhat different types of labyrinths. The labyrinth of hagfishes (genus *Myxine*), a group that is now considered to be a sister group of the vertebrates and not a true vertebrate (Bardack 1998), has one semicircular canal and two cristae ampulares (Löwenstein and Thornhill 1970) (Fig. 4.2). *Myxine* also has one sensory epithelium (macula communis) with no obvious differentiation (de Burlet 1934) (Fig. 4.2A).

In contrast, lampreys (genera *Petromyzon* and *Lampetra*), true vertebrates and a group that is more closely related in the cladistic sense to gnathostomes than to hagfishes (Nelson 1994), differ from hagfishes in various ways (Fig. 4.2B).

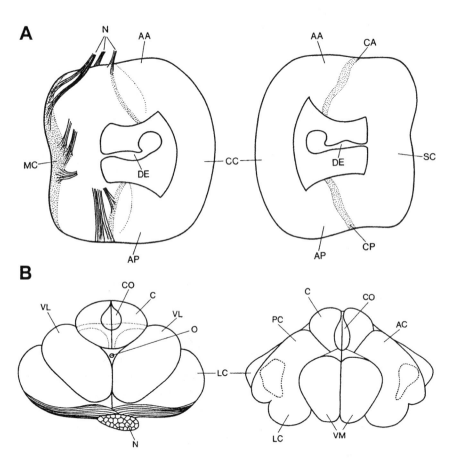

FIGURE 4.2. Medial (left) and lateral (right) views of the ears of (**A**) a craniate from a group that is sister group to the vertebrates (*Myxine glutinosa*) and (**B**) a primitive vertebrate, the lamprey (*Petromyzon fluviatilis*). AA, ampulla anterior; AC, anterior semicircular canal; AP, ampulla posterior; C, commissure; CA, crista anterior; CC, canalis communis; CO, commissural cone; CP, crista posterior; DE, ductus endolymphaticus; LC, lateral compartment of ampulla trifida; MC, macula communis; N, vestibular and auditory branches of the eighth nerve; O, opening of the ampulla trifida to the vestibulum; PC, posterior semicircular canal; SC, saccus communis; VL, vestibulum lateral part; VM, vestibulum medial part. (Redrawn from Retzius 1881.)

Their labyrinth consists of two semicircular canals, whereas the sensory epithelium, the macula communis, is partially differentiated into several regions that are connected by epithelial bridges (Löwenstein et al. 1968; Popper and Hoxter 1987). Although these regions were named the maculae neglecta, lagenae, sacculi, and utriculi (de Burlet 1934), it has been questioned whether they are homologous to structures with similar names in gnathostomes (Popper and Hoxter 1987). In addition, the lamprey labyrinth possesses ciliary chambers that appear to cause fluid flow, a feature that is unknown in other vertebrates (Popper and Hoxter 1987). However, it is not known whether this fluid flow affects the motion of the endolymph within the semicircular canals or at the sensory epithelia.

The function of the lamprey sensory epithelium and its various regions is not known. So far, it has to be assumed that the epithelium serves in the detection of linear acceleration, including gravity (Löwenstein and Osborne 1964; Löwenstein 1970). Whether the epithelium is also used for the detection of acoustic vibrations cannot be deduced from morphological findings. Sensory epithelia of lampreys have been observed to respond to sudden loud sounds produced in the water nearby (Löwenstein 1970), but it remains to be clarified whether this reaction is mediated by the labyrinth (or perhaps by the lateral line) and, if so, by which region and thus if it functions as a hearing organ.

Although the agnathan labyrinths may be highly derived, unlike the earliest vertebrate sensory epithelia, it is also possible that they represent the phylogenetic succession from a very basic labyrinth possessing one tilted, vertical semicircular canal and one sensory epithelium through a more developed labyrinth with two vertical canals and a slightly differentiated sensory epithelium and shallow recesses and finally to the stage currently found in all jawed vertebrates that possess three canals and several separated sensory epithelia. This latter notion is supported by the observation that two semicircular canals are also found in early "ostracoderm" and therefore do not seem to be the result of reduction tendencies (Romer and Parsons 1983; Forey and Janvier 1994). Furthermore, the embryonic development of vestibular labyrinths and inner ears always starts with a single sensory epithelium before the latter begins to separate into an anterior and posterior field of nerve endings that later differentiate into various cristae and epithelia (or papillae) (de Burlet 1934). At the same time, a recent neuroanatomical study indicated that the horizontal canal sensory epithelium in lampreys segregates from an already existing sensory epithelium (Fritzsch et al. 2001).

4.2 Chondrichthyes

Chondrichthyans, like all jawed vertebrates, possess three semicircular canals with ampullary enlargements, two of them oriented vertically and one horizontally (Fig. 4.3). The addition of a third horizontal semicircular canal in jawed vertebrates most likely improves motion detection in the horizontal plane. This would allow faster and more accurate orientation in a three-dimensional space

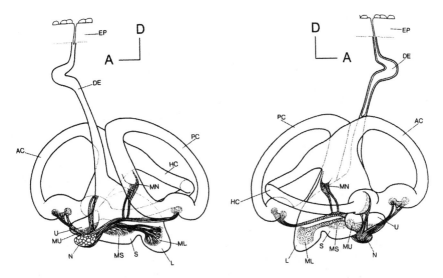

FIGURE 4.3. Medial (left) and lateral (right) views of the right ear of an elasmobranch (*Scyliorhinus canicula*). A, anterior; AC, anterior semicircular canal; DE, ductus endolymphaticus; D dorsal; EP, skin; HC, horizontal semicircular canal; L, lagena; ML, macula lagenae; MN, macula neglecta; MS, macula sacculi; MU, macula utriculi; N, vestibular and auditory branches of the eighth nerve; PC, posterior semicircular canal; PL, papilla lagenae; S, saccule; U, utricle. (Redrawn from Retzius 1881.)

than in lampreys, an important prerequisite for hunting moving prey or for escaping from predators.

The ventral part of the labyrinth consists of an otic sac containing the sensory epithelia of the saccule, lagena, and utricle as well as the macula neglecta (Retzius 1881; Tester et al. 1972; Corwin 1981) (Fig. 4.3). In holocephalans (genus *Chimaera*), the most ancient chondrichthyan group, the regions of the sensory epithelia of the ear are not fully separated from one another (de Burlet 1934). The lagenar and saccular epithelia form one sensory field in Holocephali, whereas a full separation seems to be the case in elasmobranchs, its sister group.

There is ample evidence that sharks and rays are sensitive to sound and that this sensitivity is mediated via the labyrinth (Löwenstein and Roberts 1950, 1951; Fay et al. 1974; Corwin 1981). Behavioral observations suggest that sharks are attracted to sounds that resemble those of struggling prey and that they have the best sensitivity between 40 and 300 Hz (Nelson and Johnson 1976; Fay 1988). Sharks can also localize source direction, and oriented responses to intense sound were observed when sharks were 250 m from a sound source (Myrberg et al. 1972, 2001). At the same time, it must be recognized that there is only the most limited behavioral data on shark hearing (Fay 1988; Myrberg 2001) and that considerably more data are needed before broad generalizations about hearing by this group can be made.

This sensitivity to sounds may be based on two different sound transmission and detection channels within the head and inner ears of sharks that could potentially provide at least some groups of sharks (charcharinids) with mechanisms for near-field and far-field hearing. It has been suggested that there is a nonotolithic sound transmission path in which acoustic vibrations are transmitted through the fenestra ovalis, a membranous region that covers an endolymph-filled region of the dorsal part of the skull and that connects via a tube to the region of the macula neglecta (Tester et al. 1972; Fay et al. 1974; Corwin 1981). The macula neglecta consists of one or two sheets of sensory hair cells that are often quite large in many sharks and in which the gelatinous cupula lacks calcareous inclusions. In contrast, the apical gelatinous membranes of the saccule, utricle, and lagena in elasmobranchs (and other vertebrates) contain fusiform crystals (otoconia), and, therefore, this pathway is called otolithic. Otolithic sound transmission is mainly based on sound detection by the saccular epithelia, although there are some indications that the utricle and lagena play a role in hearing in some species of elasmobranchs (Corwin 1989).

Corwin has suggested that chondrichthyans hear using parallel pathways that include the otolithic and nonotolithic sound channels. This indicates that hearing may have arisen 400 million years ago because the oldest remains of chondrichthyans are from the Silurian era (Nelson 1994). Based on the similarity of chondrichthyans to placoderms, it can be assumed that hearing may have evolved somewhat earlier (Nelson 1994). Possibly different ways of sound transmission within the head suggests the likelihood that cartilaginous fishes (and perhaps bony fishes as well) found several ways to improve hearing sensitivity via the presence of a pressure-detecting discontinuity (e.g., a gas bubble in the head or trunk) that allows sound pressure detection and far-field hearing in teleosts.

4.3 Osteichthyes

The ears of bony fishes (sarcopterygian and actinopterygian) differ from those in cartilaginous fishes in several ways (Figs. 4.4 and 4.5). First, the macula neglecta is diminutive in bony fishes and is not present in many species (Retzius 1881). Second, the utricle is not in a separate recess (except in lung fishes) but is in the common crus of the horizontal and anterior vertical semicircular canals (Fritzsch 1999). Third, bony fishes most likely do not have a nonotolithic sound detection mechanism similar to that described in elasmobranchs by Corwin (1981). Direct openings of the labyrinth to the outer medium and sensory epithelia without calcareous inclusions (other than the macula neglecta) have not been found in bony fishes. In all groups, otoconia or otolith-bearing sensory end organs use the inertia of the denser mineral deposits to create a shearing force with respect to the lagging otoconia otolith (de Vries 1950; reviewed in Popper and Fay 1999).

In ray-finned bony fishes (Actinopterygii), there is a single, often large, otolith in each otic chamber in contrast to the mass of otoconia in other fishes and in

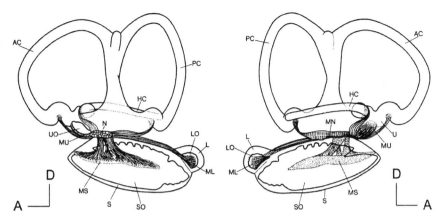

FIGURE 4.4. Medial (left) and lateral (right) views of the ears of a bony fish, the pike-perch (*Stizostedion lucioperca*). This species is a typical hearing generalist. A, anterior; AC, anterior semicircular canal; D, dorsal; HC, horizontal semicircular canal; L, lagena; LO, lagena otolith; ML, macula lagenae; MN, macula neglecta; MS, macula sacculi; MU, macula utriculi; N, vestibular and auditory branches of the eighth nerve; PC, posterior semicircular canal; S, saccule; SO, saccular otolith; U, utricle; UO, utricular otolith. (Redrawn from Retzius 1881.)

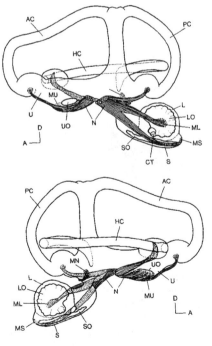

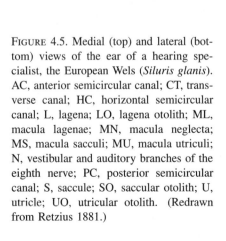

FIGURE 4.5. Medial (top) and lateral (bottom) views of the ear of a hearing specialist, the European Wels (*Siluris glanis*). AC, anterior semicircular canal; CT, transverse canal; HC, horizontal semicircular canal; L, lagena; LO, lagena otolith; ML, macula lagenae; MN, macula neglecta; MS, macula sacculi; MU, macula utriculi; N, vestibular and auditory branches of the eighth nerve; PC, posterior semicircular canal; S, saccule; SO, saccular otolith; U, utricle; UO, utricular otolith. (Redrawn from Retzius 1881.)

tetrapods. The otolith is separated from the ciliary bundles of hair cells by a thin otolithic membrane. The functional significance of a single otolith compared with otoconia remains unclear because there is no known auditory or vestibular advantage in bony fishes in having a single structure versus the otoconial "paste" found in chondrichthyans (Tester et al. 1972) and other nonteleost vertebrates (Carlström 1963).

One major feature of the osteichthyan labyrinth is the enormous structural diversity of the saccule and, in a few species, of the lagena. The utricle and semicircular canals are generally more conservative and vary to a smaller degree, although, as discussed in section 4.2, there are a few species in which the utricle is highly derived and in which it clearly plays a major role in audition. The functional consequences of the diversity in end-organ size and shape is not known. Nonteleost groups of bony fishes such as chondrosteans, halecomorphs, and brachiopterygii (Polypteriformes) have the same general inner-ear gross morphology as teleosts. However, the shovel-nose sturgeon *Scaphirhynchus*, the bowfin *Amia calva*, and the bichir *Polypterus* all have both saccular and lagenar epithelia in a single chamber (Popper 1978; Popper and Northcutt 1983). The same has been demonstrated for lungfish (Retzius 1881; Platt, Jørgensen, and Popper 2004) (Fig. 4.6) and the coelacanth *Latimeria chalumnae* (Millot and Anthony 1965; Platt personal communication) (Fig. 4.7).

In *Polypterus* and *Amia,* the sensory region of the lagena is larger than that of the saccular region, whereas in many teleosts, the lagena is smaller than the

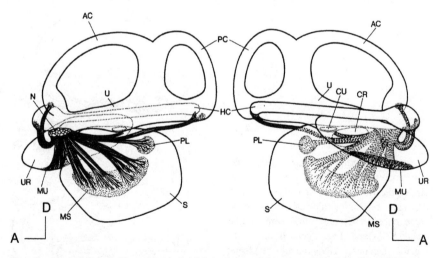

FIGURE 4.6. Medial (left) and lateral (right) views of the ears of the lungfish (*Ceratodus forsteri*). AC, anterior semicircular canal; CU, canalis utriculo-saccularis; CR, canalis recessu-saccularis; HC, horizontal semicircular canal; MN, macula neglecta; MS, macula sacculi; MU, macula utriculi; N, vestibular and auditory branches of the eighth nerve; PC, posterior semicircular canal; PL, papilla lagena; S, saccule; U, utricle; UO, utricular otolith; UR, utricular recess. (Redrawn from Retzius 1881.)

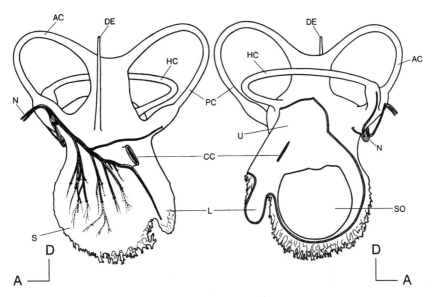

FIGURE 4.7. Medial (left) and lateral (right) views of the ears of the coelacanth (*Latimeria chalumnae*). AC, anterior semicircular canal; CC, canalis communis; DE, ductus endolymphaticus; HC, horizontal semicircular canal; L, lagena; N, vestibular and auditory branches of the eighth nerve; PC, posterior semicircular canal; PL, papilla lagena; S, saccule; SO, saccular otolith; U, utricle. (Redrawn from Millot and Anthony 1965.)

saccule and forms a separate recess that is often only connected to the saccule by a small opening) (Ladich and Popper 2001) (Fig. 4.4). The size relationship between the saccule, lagena, and semicircular canals varies widely among teleosts. Although in gobies the saccular chamber is huge and reaches almost the upper edge of the vertical canals, with the lagena being a small caudal appendix, the situation is quite different in the lumpsucker *Cyclopterus lumpus* (Retzius 1881) where both the saccule and lagena are tiny compared with the semicircular canals.

A series of modifications of this scheme are found in otophysan fishes, which, together with anotophysi, comprise the taxon Ostariophysi. These modifications are most likely linked to the development of the Weberian apparatus and the connection it provides between the inner ear and the swim bladder (Weber 1820). The otophysan saccule is an elongated compartment, whereas the lagena is larger, round in shape, and located dorsal to the saccule (Retzius 1881) (Fig. 4.5). The saccular otolith in otophysans is sticklike and has a fluted anterior and a rodlike posterior part, whereas the lagenar otolith is quite enlarged and round to ovoid in shape (Retzius 1881; von Frisch 1936) (Fig. 4.5). The saccules of the two ears are connected by a transverse canal (canalis communicans transversus; Chranilov 1927) that has a caudally situated sac (sinus endolymphaticus Weberianus). This sinus is attached either directly to the Weberian ossicles (catfish family Loricariidae) or to the unpaired perilymphatic space that itself is connected to the most anterior of the Weberian ossicles via two caudal extensions (Weber 1820; Chranilov 1929; von Frisch 1936). A ventral, thinned portion of the saccular wall, called the release membrane by von Frisch (1936), separates the saccular lumen from the periotic channel that terminates in contact with the cerebrospinal fluid. This release membrane was suggested by von Frisch (1936) to facilitate movements of the endolymphatic fluid in the saccule and lagena in a manner analogous to the round window in mammals (Jenkins 1977). There are, however, no experimental data to support this suggestion. This condition is found in certain siluriform families such as Siluridae, Ictaluridae, Doradidae, and Aspredinidae (Jenkins 1977).

A major deviation from the typical otophysan pattern of inner-ear morphology is found in the marine catfish (family Ariidae). In *Arius felis*, the utricle and its otolith are very large compared with the other otolithic end organs in this species (Popper and Tavolga 1981). In addition, the utricular sensory epithelium does not cover the whole lower hemisphere of the chamber as it does in most other vertebrates but instead consists of a band around the equatorial region. *Arius* has better low-frequency hearing than any other fish studied to date (Popper and Tavolga 1981; Fay 1988), although unlike other otophysans that generally hear to above 3,000 Hz, *Arius* only detects sounds to about 1,500 Hz. Popper and Tavolga (1981) suggested that the enlarged utricle, and particularly the very large otolith, serves as a low-frequency detector useful for detection of self-generated sounds that *Arius* uses to determine the presence of objects in its environment (Tavolga 1976).

The unusual form of the saccular otolith of otophysans (sticklike with flutes) suggests differences in the function of the inner ear compared with other taxa. It is assumed that the flutes, which are located near the opening of the canalis transversus, serve in registering movements of the endolymph within this channel caused by the oscillations of the Weberian ossicles–swim bladder complex (von Frisch 1936). These oscillations of the sagitta cause bending of the ciliary bundles. This seems to be in contrast to the basic hearing process in fishes, especially in those species possessing no hearing specializations. Basically, the fish ear is a particle motion detector that responds to oscillations of the whole fish within a sound field in relation to the denser otolith that lags behind (Hawkins and MacLennan 1976). Furthermore, it is unclear which functional role the large lagena has in otophysans. Some experimental evidence indicates that the lagena may play a role in directional hearing comparable to that generally accepted for the saccule of the hearing nonspecialists (Popper and Fay 1999; Popper et al. 2003).

5. Hearing Specializations

A number of fish species have evolved ways to enhance hearing sensitivity up to several kilohertz (kHz). This enhancement occurs by combining the accelerometer system with the input to the otolith end organs from gas-filled spaces that are likely to undergo volume changes during the passage of pressure waves. This results in transmission of movements of the wall of these spaces to the inner ear. Thus fishes were able to perceive acoustic sound pressure and detect sound sources in the far field (Rogers and Cox 1988; Ladich and Bass 2003a).

Gas-filled spaces can contact the inner ear either directly or indirectly via a chain of bony ossicles and perilymphatic spaces (otophysans), or the contact is mediated by the tissue between the swim bladder and the inner ear. The latter species such as the damselfish *Eupomacentrus partitus* and the cod *Gadus morhua* have a less extended frequency range, which might be due to the lack of a particularly efficient connection between the gas-filled cavity and the ear (Chapman and Hawkins 1973; Myrberg and Spires 1980). Sometimes pressure-sensitive fishes are called *hearing specialists*, whereas at other times this term is limited to those groups possessing (known) morphological adaptations for hearing enhancement (Hawkins and Myrberg 1983) (Fig. 4.8).

The best-known specializations for hearing are the Weberian ossicles, the functional significance of which remained unknown for many years (for a review, see Ladich and Bass 2003b). Weber (1820), who first described these ossicles and after whom they were named, originally stated that the "ossicles are conductors of sound (from the air-bladder) to the ear." Several investigators including Sagemehl (1885) and Bridge and Haddon (1889, 1892) questioned this function due to the supposed inelasticity of the swim bladder wall and the assumed compressibility of ligaments between ossicles. They assumed that the Weberian ossicles, together with the swim bladder, primarily functioned in the detection of hydrostatic pressure. Von Frisch (1936; von Frisch and Stetter 1932) finally proved by ablation experiments the importance of the swim bladder and its connection to the inner ear for increasing the acuity of hearing.

The Weberian apparatus typically consists of four ossicles: the claustrum, scaphium, intercalarium, and tripus (Fig. 4.9). These are derived from the anterior vertebrae and are connected to one another by ligaments. Oscillations of the dorsoanterior wall of the swim bladder are transmitted via these bones to the perilymphatic space of the ear. Functionally, the anterior swim bladder wall can be compared with the tympanum of the tetrapod ear and the Weberian ossicles to the middle ear bones (Chranilov 1927; Alexander 1964). Interestingly, anatomical studies revealed a rich diversity in the structure of the siluriform swim bladder and Weberian ossicles (Bridge and Haddon 1889, 1892; Sörensen 1895; Chranilov 1927, 1929; Chardon 1968). This diversity mainly originates from the reduction and bony encapsulation of the swim bladder as well as a reduction in ossicles among many species, especially siluriforms. In loricariid catfish, the swim bladder is medially separated in two lateral parts, each of which is surrounded by bone except for the areas below the skin. The left and right swim bladders are connected to the sinus impar via the Weberian ossicles (Chranilov 1929; Bleckmann et al. 1991). Furthermore, the caudal (tripodes) and cranial ossicles (scaphia) have lost their processes in some species, whereas the intermediate ossicles (intercalaria) either form nodules or are absent. In callichthyids, the tripodes and scaphia fuse to form a single Weberian ossicle (Alexander 1964; Coburn and Grubach 1998). The reduction in the swim blad-

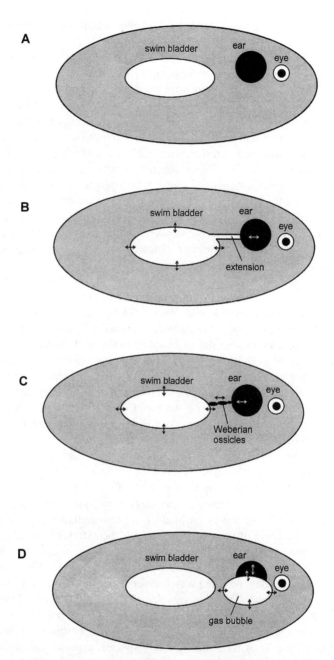

FIGURE 4.8. Schematic relationships between ears and air-filled cavities in hearing generalists such as cichlids (no specialization) (**A**), a fish with direct connection between the swim bladder and the ear via a swim bladder extension such as the more specialized holocentrid *Myripristis kuntee* (**B**), otophysans possessing a Weberian apparatus such as cyprinids (**C**), and anabantoids (**D**) having an air-breathing cavity (suprabranchial chamber) laterally of the ears (gas bubble). Arrows indicate volume changes of the air-filled cavities in a sound pressure field and their transmission to the inner ear endolymphs.

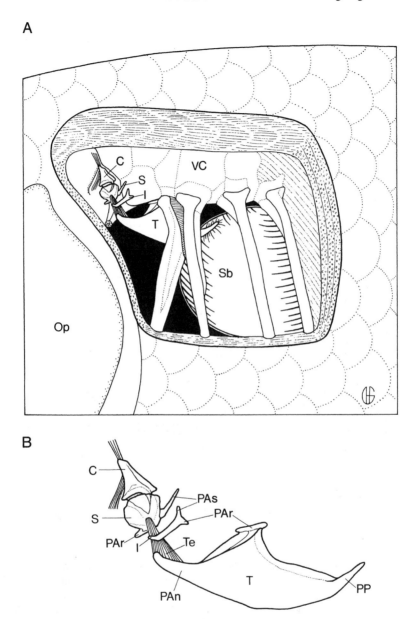

FIGURE 4.9. **A**: Lateral view of the goldfish showing the postcranial region illustrating the position of all four Weberian ossicles (C, claustrum; I, intercalarium; S, scaphium; T, tripus), the anterior ribs, and the swim bladder (Sb). **B**: Enlarged view of the four Weberian ossicles. Op, operculum; PAn, anterior process; PAr, articulation process; PAs, ascending process; PP, posterior process; PR, articulation process; Te, tendon; VC, vertebral column. (Graphic by H.C. Grillitsch, redrawn partly after Ladich and Wysocki 2003.)

der is regarded as an adaptation to a bottom-living habit, but contrary to other bottom-living taxa such as gobies and sculpins, the otophysans always retain some swim bladder, even if it is reduced.

The auditory sensitivity of otophysans extends up to several kilohertz (Fig. 4.10). Poggendorf (1952) reported an upper frequency hearing limit of 13 kHz in a small number of brown bullhead catfish *Ictalurus nebulosus*, and Popper (1970) observed upper limits in the characids genus *Astyanax* of 6.4 and 7.5 kHz. Upper hearing limits in otophysans of 3 to 5 kHz were measured in cypriniforms, siluriforms, characiforms, and gymnotiforms (Jacobs and Tavolga 1967; Weiss et al. 1969; Popper 1970, 1971; Fay 1988; Ladich 1999). An exceptionally low-frequency hearing range (to 1.5 kHz) was found in the marine ariid catfishes, which seem to be paralleled by differences in the inner ear anatomy (Popper and Tavolga 1981) (see section 4.3).

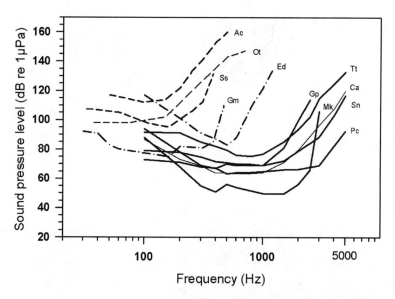

FIGURE 4.10. Auditory thresholds for hearing generalists [Ac, *Gymnocephalus cernuus* (*Acerina cernus*); Ot, *Opsanus tau*; Ss, *Salmo salar*] and for hearing specialists [Ca, *Carassius auratus*; Gp, *Gnathonemus petersii*; Mk, *Myripristis kuntee*; Pc, *Platydoras costatus*; Sn, *Pygocentrus* (*Serrasalmus*) *nattereri*; Tt, *Trichogaster trichopterus*]. The damselfish *Eupomacentrus dorsopunicans* (Ed) and the cod *Gadus morhua* (Gm) have a less extended frequency range and the swim bladder is not connected to the ear, though it is believed to play a role in hearing. There has been some discussion in the literature as to whether these fishes should be considered hearing "specialists" or "nonspecialists," but no definition is provided here. [Audiograms after Wolff 1968 (Ac), Chapman and Hawkins 1973 (Gm), Fish and Offutt 1972 (Ot), Hawkins and Johnstone 1978 (Ss), Coombs and Popper 1979 (Mk), Myrberg and Spires 1980 (Ed), McCormick and Popper 1984 (Gp), Ladich and Yan 1998 (Tt), and Ladich 1999 (P_c).]

In a series of extirpation experiments, several authors demonstrated the importance of the swim bladder for increasing auditory sensitivity in otophysans. Removal of the swim bladder in the minnow *Phoxinus phoxinus* weakened the hearing acuity, especially at higher frequencies (von Frisch and Stetter 1932). Further experiments in the channel catfish *Ictalurus punctatus,* the goldfish *Carassius auratus,* and the roach *Rutilus rutilus* confirmed that deflation of the swim bladder resulted in a considerable loss of sensitivity at frequencies above 100 Hz (Kleerekoper and Roggenkamp 1959; Fay and Popper 1974, 1975; Laming and Morrow 1981; Yan et al. 2000). The role of the Weberian ossicles was studied in experiments in which the swim bladder was left intact. Removal of the tripodes, the most caudal Weberian ossicles, resulted in a frequency-independent decrease in the auditory sensitivity of the brown bullhead catfish *I. nebulosus* (Poggendorf 1952) and, in contrast, a frequency-dependent hearing loss in the goldfish (Ladich and Wysocki 2003). Ladich and Wysocki (2003) observed that unilateral tripus extirpation did not affect hearing thresholds. The likely explanation for this is that the paired ossicles transmit swim bladder oscillations to the same unpaired perilymphatic sinus. Results of the tripodes removal experiments in the catfish and goldfish indicate that swim bladders play some role in hearing even after interruption of the Weberian ossicular chain.

The size of the swim bladder and the degree of encapsulation as well as the reduction in the Weberian ossicles, in particular in catfishes, seem to influence hearing sensitivity. There is evidence of differences in hearing capabilities among representatives of all four otophysan orders (Ladich 1999). For example, at 4 kHz, the auditory sensitivity of the pimelodid catfish *Pimelodus pictus* and the doradid *Platydoras costatus* was more than 50 dB lower than in the callichthyid *Corydoras paleatus*. The poor sensitivity in the latter seems to be due to their encapsulation of the paired lateral swim bladder sacs and a reduced ability of the wall to oscillate and perhaps the limited mobility of the single Weberian ossicle (Coburn and Grubach 1998; Ladich 1999). Thus, the evolution of the Weberian apparatus can be constrained by the tendency of bottom-dwelling catfishes to reduce their buoyancy and thus the swim bladder.

Accessory hearing structures are not limited to otophysans but also evolved in numerous nonrelated taxa. Two collaborators of von Frisch, Schneider (1941) and Stipetić (1939) showed that hearing acuity in fishes can be enhanced by air-filled cavities other than the swim bladder. Mormyrids, a family of weakly electric fishes from tropical Africa, possess small tympanic bladders, gas-filled cranial derivatives of the swim bladder that are directly attached to the saccule and improve hearing sensitivity up to 3.1 kHz (Stipetić 1939; McCormick and Popper 1984; Fletcher and Crawford 2001) (Fig. 4.8D). Anabantoids (labyrinth fishes or gouramis) possess a suprabranchial chamber (labyrinth) dorsally of the gills that is utilized for air breathing (Bader 1937). Because these labyrinths lie laterally of the inner ears and are only separated by a thin bony sheet or epithelium from the saccule, they enhance hearing sensitivity considerably (Fig. 4.8D). Labyrinth fishes such as the paradise fish *Macropodus opercularis* and the blue gourami *Trichogaster trichopterus* can detect sound up to 5 kHz

(Schneider 1941; Ladich and Yan 1998). Again, removal of the air from these bladders or chambers resulted in a drop in the upper hearing limit and in a decrease in the sensitivity by 20 to 30 dB (Schneider 1941; Yan et al 2000; Fletcher and Crawford 2001).

In some species of holocentrids (squirrel fish or soldier fish), the swim bladder is connected via an elongated anterior extension to the ear (Fig. 4.8B), and this enhances their hearing capacities (Tavolga and Wodinsky 1963; Coombs and Popper 1979). The audiograms within this family vary depending on the degree of the association between the swim bladder and the ear (Coombs and Popper 1979). The greatest sensitivity is found in the genus *Myripristis,* where an anterior lobe makes substantial contact with a thin membrane in the wall of the auditory bulla lateral to the saccule of the ear. Holocentrids and otophysines evolved very similar sensitivities based on different morphological structures (Hawkins and Myrberg 1983).

Among all fishes studied to date, perhaps the greatest variability is found within the family Sciaenidae, where there is extensive diversity in saccular structure, in the relationship between the swim bladder and the inner ear, in the size and shape of the saccular otolith, and in many other aspects of the structures generally associated with hearing (Ramcharitar et al. 2001; Ramcharitar and Popper unpublished data). These differences are associated with both sound production and differences in hearing capabilities and lead to the suggestion that this group may be excellent for future studies on the relationship between structure and function in auditory systems of fishes (Ramcharitar and Popper unpublished data).

What selective pressures led to the improvement of hearing in fishes? Interestingly, specializations are often found in taxa that initially arose in shallow freshwater habitats such as lakes, rivers, and flood plains, and they are seldom found in taxa from the upper regions of rivers or in fishes that evolved in marine environments (Fay and Popper 2000). The specializations in these species may be related to the limited range of travel of low-frequency sound in shallow water (Rogers and Cox 1988). As a result, to detect sound that comes from any distance, there would have been selective pressure for detection of the higher frequencies.

Morphological studies also lead to the suggestion that hearing specializations may have evolved in some deep-sea habitats (Popper 1980; Deng et al. 2002). Although there are limited data due to the difficulties in obtaining deep-sea fish and there are no behavioral or physiological data, a number of species of myctophids and gadids have specializations in the saccular sensory epithelium that are very similar to those found in hearing specialists but never in fishes that are not specialists (Popper and Coombs 1982). There is also evidence that some of these fishes have connections between the inner ear and swim bladder as well as sound-producing mechanisms (Marshall 1967).

The hearing specializations are seen throughout freshwater fish groups. This includes the otophysans, which include about 6,500 species and make up about

64% of all freshwater species (Nelson 1994). Other freshwater groups, including the mormyrids (elephantnose fishes) that live in tropical Africa and the Nile and the anabantoids that inhabit mostly still to slowly flowing waters with a high content of vegetation in eastern and southeast Asia as well as in southern parts of Africa (Richter 1988), also show specializations for hearing.

The physical preconditions such as no sound propagation below a certain frequency (cutoff frequency) in shallow waters most likely provided strong selective pressures in the ancestors of hearing specialists for the evolution of peripheral accessory structures for the exploitation of acoustical information of lower intensity and higher frequencies. Lowering the threshold was most likely linked to broadening of the hearing bandwidth. These improvements in hearing may have started out as a side effect of other functions such as air breathing in anabantoids or of the hydrostatic function of swim bladders. The selective pressures for the improvement of hearing were large enough to induce morphological changes such as anterior extensions of the swim bladder that touch the inner ear, as in holocentrids. The connections between the auditory and hydrostatic chambers of the swim bladder might have become very thin, as in clupeids, or obliterated totally, as found in the mormyrids that have an otic bladder directly attached to the saccule (Stipetić 1939; Blaxter et al. 1981) (Fig. 4.8D). Another way was exploited by otophysans that modified vertebral derivatives for sound conduction between the ear and bladder (Fig. 4.8C). The immediate advantage of this transformation of organs for better sound detection is less well understood and is discussed in section 6.

6. Functional Significance of Sound Detection

6.1 Acoustic Communication

Fishes are a vocal group of vertebrates, and it is intriguing to assume that many hearing refinements evolved in connection with adaptations for acoustic communication. However, the ability to produce sounds and to communicate acoustically is not limited to the hearing specialists or to freshwater fishes. Numerous hearing generalists such as bichirs (polypterids), toadfishes (batrachoidids), sunfishes (centrarchids), gobies (gobiids), cichlids, and drums (sciaenids) emit sounds in various behavioral situations (Tavolga 1958; Myrberg et al. 1965; Gerald 1971; Ladich and Tadler 1988; Connaughton and Taylor 1996; Lugli et al. 1997). These fish are obviously able to compensate for their relatively poor hearing thresholds by increasing the intensity of their sounds or by shortening the effective distance over which they communicate under given environmental noise levels. Acoustic communication often takes place over distances of 1 to 10 cm, where even faint sounds such as those emitted by tiny gobies are detectable by mates (Tavolga 1956; Ladich and Kratochvil 1989; Lugli and Fine 2003).

In addition, sound production is not a general feature of all hearing specialists, and it is not known whether the ancestors of those taxa were vocal. Among otophysines, cypriniforms are the most primitive group (Fink and Fink 1996). Interestingly, only a few representatives of the large order of cypriniforms are known to be vocal (Stout 1963; Ladich 1988, 1999; Johnston and Johnson 2000), and in no case has a specialized sonic organ been described. This indicates that sound-generating structures are less specialized within cypriniforms. Specialized sound-producing structures such as swim bladder drumming muscles only evolved in related groups. Many characids as well as numerous catfish families possess extrinsic swim bladder muscles (Schaller 1967; Markl 1971; Ladich and Bass 1998; Fine and Ladich 2003).

Similarly, among labyrinth fishes, only a few genera seem to be vocal, and representatives of only one genus evolved a highly specialized pectoral sound-producing mechanism (genus *Trichopsis*; Kratochvil 1985). Analysis of the fine structure of the saccular epithelia of several vocal and nonvocal species of anabantoids showed that the epithelia do not differ in shape, in hair cell density, in ciliary bundle type, or in orientation pattern (Ladich and Popper 2001). Data correlated with physiological results showed similar auditory sensitivity in anabantoids (Ladich and Yan 1998). Similar observations were made on mormyrids by Fletcher and Crawford (2001), who found a remarkable similarity in the best sensitivity in sound-producing weakly electric fish *Pollimyrus isidori* and the nonvocal *Brienomyrus niger*.

Although fossil evidence proving that sound communication was absent in ancestors of hearing specialists is lacking, the aforementioned data suggest that acoustic communication was not the main selective pressure for the evolution of hearing. The main energy of sounds is located within the optimal hearing range in pomacentrids of the genus *Eupomacentrus* (Myrberg and Spires 1980), the weakly electric fish *Pollimyrus isidori* (Crawford 1993), and the croaking gourami *Trichopsis vittata* (Ladich and Yan 1998). However, Ladich (1999) argued that the main sound energy is not always located within the optimal hearing range in particular species as comparative studies in otophysans and anabantoids revealed (Ladich 2000) (as Konishi showed for birds in 1970).

In summary, differences in auditory sensitivity between hearing generalists and specialists do not seem to be related to their ability to produce sound, and hearing sensitivity curves do not always match the frequency spectra of communication sounds.

6.2 Noncommunicative Tasks of the Auditory System

Which evolutionary constraints might have caused the ancestors of otophysans and other hearing specialists to enhance their hearing abilities? The ability to detect predators or prey over larger distances must have improved the chance of survival of many fish species. Clearly, our knowledge of the detection of conspecific sounds is considerable compared with what we know of the detection of other sound sources. Nevertheless, we should direct our attention in the future more to these evolutionary questions.

Many fish species respond to underwater sounds accompanying attacks by predators with a characteristic startle behavior so that the fish turns away from the direction of the attacks of predators (C-start) (Canfield and Eaton 1990; Canfield and Rose 1996). This escape response is mediated by the two bilateral Mauthner neurons in the spinal chord. Mauthner cells are present in basic fishes and predate the evolution of the swim bladder. Thus the successful expression of Mauthner-mediated escape could have influenced the evolution of a swim bladder for acoustic pressure transduction. Although the transition from buoyancy control to hearing is unknown, Canfield and Eaton (1990) suggested that sound pressure evoked Mauthner cell activation and predator evasion coevolved with the addition of hearing to swim bladder function. Thus fish with hearing specializations increased their effective acoustic range for escape initiation that might have played a crucial role during their evolution. Perhaps this explains the enormous success of otophysines, which constitute 64% of all freshwater species.

Predator avoidance through the development or improvement of specific hearing abilities has been the major selective pressure in the evolution of ultrasonic hearing in many flying insects such as moths and praying mantises (Hoy 1992). A similar response has been described in several species of Clupeiformes (American shad *Alosa sapidissima*, gulf menhaden *Brevoortia patronus*, and relatives) (Mann et al. 1997, 1998, 2001). Although all Clupeiformes appear to be able to detect sounds to over 3 kHz (Mann et al. 2001), at least the shads, herrings, and menhaden are able to detect ultrasound to well over 100 kHz (Ross et al. 1996; Mann et al. 1997, 1998, 2001; Wilson and Dill 2002; Plachta and Popper 2003). Moreover, both herring and American shad show escape responses to dolphinlike ultrasonic pulses (Wilson and Dill 2002; Plachta and Popper 2003), with the reactions increasing as the sound level of the pulses increases, indicating a predator that is closing in on the fish (Plachta and Popper 2003). Although a few non-Clupeiform species have been tested for ultrasound detection (Mann et al. 2001), the only other nonclupeid species shown to be able to detect ultrasound to date is the cod *Gadus morhua* (Astrup and Møhl 1993). However, it is possible that the cod was detecting the stimulus using touch receptors that were overdriven by very intense fish-finder sonar emissions (Astrup 1999).

Another important aspect improving survival is prey or food detection. Feeding or swimming noises of conspecifics as well as all kinds of noises generated by prey species might be very important for the localization of prey and food items. Markl (1972) observed that the piranha *Pygocendrus (Serrasalmus) nattereri*, representative of the otophysine order Characiformes, attacked prey items that make splashing noises significantly more often than silent prey items.

Although this behavior has seldom been investigated systematically, the attractive nature of certain underwater sounds for fishes has been documented for hundred of years. Acoustical fishery techniques based on the generation of underwater noise to attract fishes have been applied and exploited by indigenous people all over the world (for reviews, see Steinberg 1957; Wolff 1966). Hitting rods on the water surface attracts piranhas in South America as well as cichlids in Lake Tanganyika in Africa. Rattle fishery, known in the Baltic regions since the Medieval Ages, is based on rattling a plank underwater (Wolff 1966). It was reported that "large" numbers of perciforms were caught. The main energy of these sounds is up to 300 Hz, a frequency range that can be detected by almost all fish species (Wolff 1966). Rattling coconut shells underwater is very attractive for sharks ("shark rattle") and was used in the South Seas. Although in these and several other techniques, it is not fully understood whether catch success is due to attracting or startling fishes, the available data do point to the importance of hearing for orienting to or away from sounds other than those produced by conspecifics for acoustic communication.

It has also been shown that at least some fish species evolved the ability to detect infrasound to as low as 0.1 Hz. This has been described in representatives of four taxa including perch, plaice cod, and salmon (Sand and Karlsen 1986; Karlsen 1992a,b; Knudsen et al. 1992), although there is some doubt whether infrasound detection is significant beyond a few meters from the receiving animal (Sand and Karlsen 2000; Popper et al. 2003). This is unlikely because the level of ambient noise in the sea increases considerably toward lower frequencies (Urick 1983), and so infrasonic sounds would be well within the background noise except close to the source. Among the suggested sources of marine infrasound are turbulence due to ocean currents, seismic motions of the ocean floor, and moving schools of fishes. Ambient infrasound in the sea may be used as a cue for orientation during migration or even for prey detection (Sand and Karlsen 1986).

7. Comparison Between Fishes and Amniotes

Over the course of the evolution of the ear and hearing, fishes do not appear to have evolved an inner-ear structure entirely dedicated to acoustic function. This has been clearly demonstrated in amphibians (for reviews, see Lewis and Narins 1999; Smotherman and Narins, Chapter 6) and all other vertebrate classes. This difference, however, could be due to our gaps in knowledge on the particular function of end organs in fishes and less so to a physiological difference between vertebrate classes.

On the other hand, fishes independently evolved a number of characteristics that are found in mammals and other amniotes, although there is no reason to think that there is homology between structures in fishes and amniotes. A few fish groups evolved a "tympanum," a thin membrane in close contact with an air-filled space (most often the swim bladder), which oscillates in the sound field. The anterior wall of the swim bladder can be regarded as a functional tympanum. In this way, fishes are able to detect sound pressure as do amniotes. The oscillations of the "tympanum" in otophysines and amniotes are transmitted to the inner ear through a series of bones. Weber (1820; also see von Frisch 1936) used the term *middle ear ossicles*, although fishes do not possess middle ears and their ossicles have a different embryonic origin than middle ear bones in terrestrial vertebrates (see Clack and Allin, Chapter 5); they are thought to function analogously in that they probably transmit large oscillations of a membrane close to an air-filled space to oscillations of a fluid, the endolymph of the inner ear.

As mentioned in section 6.2, some fishes have evolved ultrasonic hearing to above 180 kHz and infrasonic hearing. Ultrasonic hearing is also a specialization in some mammals, but it is not known in amphibians, reptiles, or birds (Fay 1988).

Finally, some pimelodid catfishes seem to possess a mechanism to protect the inner ear from over stimulation. In mammals, the tensor tympani muscle contracts in response to intense sound that draws the malleus inward, increases tension on the tympanic membrane, and limits motion of the auditory ossicles (Yost 1994). In pimelodids (e.g., *Pimelodus blochi*), a small conical muscle (tensor tripodis) originates on the occipital region of the skull and inserts via a long tendon on the dorsal surface of the swim bladder near the tripus, the caudal Weberian ossicle (Bridge and Haddon 1889, 1892; Schachner 1977; Ladich and Fine 1994; Ladich 2001; Ladich and Bass 2003b). Although the action of this muscle has not been examined physiologically, Bridge and Haddon (1892) and Schachner (1977) suggested on anatomical grounds that it dampens oscillations of the tripus during swim bladder vibrations. This action would be particularly important during sound production when contractions of the massive drumming muscles cause large amplitude oscillations of the rostral swim bladder wall.

8. Summary

Ears constitute a basic element of the vertebrate Bauplan. There are several reasons to assume that ears in fishes evolved primarily as vestibular organs to facilitate orientation in a three-dimensional space. Auditory functions apparently appeared secondarily, utilizing the same morphological substrate as the vestibular system. However, we do not know when and in which taxa this happened first. From our current knowledge, it is reasonable to suggest that the majority of the extant cartilaginous and bony fishes are able to detect acoustic signals, but this ability remains questionable in agnathans and has not been tested in the majority of basic fish taxa. Early vertebrates, and fishes in particular, did not evolve a characteristic end organ for sound detection, and therefore we cannot deduce from anatomical findings whether and to what degree hearing is present. Hearing abilities improved in some teleost families (hearing specialists), so that the bandwidth of hearing approaches that of many amphibians and birds. The reasons for the evolution of hearing, and in particular hearing specializations, remain unclear and are open for speculation. It is unlikely that acoustic communication was a driving force in both evolutionary processes because basic fish groups such as cartilaginous fishes or chondrosteans are not known to be vocal. Hearing could have evolved in connection with the detection of inanimate sound sources advantageous for orientation such as wind, waves, tides, currents, or seismic events or for living sound sources such as predators, prey, or conspecifics. The selection of hearing specializations followed similar rules but obviously under lower background noise levels. However, in both cases, we lack sufficient experimental evidence, especially field data, establishing the importance of environmental noise of inanimate or animate origin in the life of fishes.

Acknowledgments. We thank Heidemarie Grillitsch for the drawings in Figures 4.2 through 4.7 and 4.9 and Helen A. Popper for editing the manuscript. Dr. Ladich's research was supported by the Austrian Science Fund (FWF No. 15873).

References

Alexander RMcN (1964) The structure of the Weberian apparatus in the Siluri. Proc Zool Soc Lond 142:419–440.
Amoser S, Ladich F (2003) Diversity in noise-induced temporary hearing loss in otophysine fishes. J Acoust Soc Am 113:2170–2179.
Astrup J (1999) Ultrasound detection in fish—a parallel to the sonar-mediated detection of bats by ultrasound-sensitive insects. Comp Biochem Physiol 124:19–27.
Astrup J, Møhl B (1993) Detection of intense ultrasound by the cod *Gadus morhua*. J Exp Biol 182:71–80.
Ayers H (1892). Vertebrate cephalogenesis. J Morphol 6:1–360.
Bader R (1937) Bau, Entwicklung und Funktion des akzessorischen Atmungsorgans der Labyrinthfische. Z wiss Zool Leipzig 149:323–401.
Bardack D (1998) Relationships of living and fossil hagfishes. In: Jørgensen JM, Lomholt JP, Weber RE, Malte H (eds) The Biology of Hagfishes. London: Chapman and Hall, pp. 3–14.
Blaxter JH, Denton EJ, Gray JAB (1981) Acousticolateralis system in clupeid fishes. In: Tavolga WN, Popper AN, Fay RR (eds) Hearing and Sound Communication in Fishes. New York: Springer-Verlag, pp. 39–56.
Bleckmann H, Niemann U, Fritzsch B (1991) Peripheral and central aspects of the acoustic and lateral line system of a bottom dwelling catfish, *Ancistrus* sp. J Comp Neurol 314:462–466.
Bregman AS (1990) Auditory Scene Analysis. The Perceptual Organisation of Sound. Cambridge: MIT Press.
Bridge TW, Haddon AC (1889) Contributions to the anatomy of fishes. I. The airbladder and Weberian ossicles in the Siluridae. Proc R Soc Lond 46:309–328.
Bridge TW, Haddon AC (1892) Contributions to the anatomy of fishes. II. The airbladder and Weberian ossicles in the Siluridae. Proc R Soc Lond 184:65–324.
Burighel P, Lane NJ, Fabio G, Stefano T, Zaniolo G, Carnevali MDC, Manni L (2003) Novel, secondary sensory cell organ in ascidians: in search of the ancestor of the vertebrate lateral line. J Comp Neurol 461:236–249.
Canfield JG, Eaton RC (1990) Swimbladder acoustic pressure transduction initiates Mauthner-mediated escape. Nature 347:760–762.
Canfield JG, Rose GJ (1996) Hierarchical sensory guidance of Mauthner-mediated escape

response in goldfish (*Carassius auratus*) and cichlids (*Haplochromis burtoni*). Brain Behav Evol 48:137–156.
Carlström D (1963) A crystallographic study of vertebrate otoliths. Biol Bull 125:441–463.
Chapman CJ, Hawkins AD (1973) A field study of hearing in the cod, *Gadus morhua* L. J Comp Physiol [A] 85:147–167.
Chardon M (1968) Anatomie comparee de l'appareil de Weber et des structures connexes chez les Siluriformes. Musee Royal de l'Afrique Centrale, Tervuren, Belgique. Annales, Serie in 8, Sciences Zoologiques 169:1–273.
Chranilov NS (1927) Beiträge zur Kenntnis des Weber'schen Apparates der Ostariophysi 1. Vergleichend-anatomische Übersicht der Knochenelemente des Weber'schen Apparates bei Cypriniformes. Zool Jb Anat 49:501–597.
Chranilov NS (1929) Beiträge zur Kenntnis des Weber'schen Apparates der Ostariophysi: 2. Der Weber'sche Apparat bei Siluroidea. Zool Jb Anat 51:323–462.
Coburn MM, Grubach PG (1998) Ontogeny of the Weberian apparatus in the armored catfish *Corydoras paleatus* (Siluriformes: Callichthyidae). Copeia 301–311.
Connaughton MA, Taylor MH (1996) Drumming, courtship, and spawning behavior in captive weakfish, *Cynoscion regalis*. Copeia 195–199.
Coombs S, Popper AN (1979) Hearing differences among Hawaiian squirrelfish (family Holocentridae) related to differences in the peripheral auditory system. J Comp Physiol 132:203–207.
Corwin JT (1981) Audition in elasmobranchs. In: Tavolga WN, Popper AN, Fay RR (eds) Hearing and Sound Communication in Fishes. New York: Springer-Verlag, pp. 81–102.
Corwin JT (1989) Functional anatomy of the auditory system in sharks and rays. J Exp Zool Suppl 2:62–74.
Crawford JD (1993) Central auditory neurophysiology of a sound-producing fish: the mesencephalon of *Pollimyrus isidori* (Mormyridae). J Comp Physiol [A] 172:139–152.
Crawford JD, Cook AP, Heberlein AS (1997) Bioacoustic behavior of African fishes (Mormyridae): potential cues for species and individual recognition in *Pollimyrus*. J Acoust Soc Am 102:1–13.
de Burlet HM (1934) Vergleichende Anatomie des stato-akustischen Organs. a) Die innere Ohrsphäre. In: Bolk L, Göppert E, Kallius E, Lubosch W (eds) Handbuch der vergleichenden Anatomie der Wirbeltiere. Berlin: Urban and Schwarzenberg, pp. 1293–1380.
Deng X, Wagner H-J, Popper AN (2002) Messages from the bottom of the Atlantic Ocean: comparative studies of anatomy and ultrastructure of the inner ears of several Gadiform deep-sea fishes. Abst Assoc Res Otolaryngol 25:101.
de Vries HL (1950) The mechanics of the labyrinth otoliths. Acta Oto-Laryngol 38: 262–273.
Fay RR (1974) Masking of tones by noise for the goldfish (*Carassius auratus*). J Comp Physiol Psychol 87:708–716.
Fay RR (1982) Neural mechanisms of an auditory temporal discrimination by the goldfish. J Comp Physiol [A] 147:201–216.
Fay RR (1985) Temporal processing by the auditory system of fishes. In: Michelsen A (ed) Time Resolution in Auditory Systems. New York: Springer-Verlag, pp. 28–57.
Fay RR (1988) Hearing in Vertebrates: A Psychophysics Databook. Winnetka, IL: Hill-Fay Associates.

Fay RR (1998) Auditory stream segregation in goldfish (*Carassius auratus*). Hear Res 120:69–76.
Fay RR (2000) Frequency contrasts underlying auditory stream segregation in goldfish. J Assoc Res Otolaryngol 1:120–128.
Fay RR, Popper AN (1974) Acoustic stimulation of the ear of the goldfish (*Carassius auratus*). J Exp Biol 61:243–260.
Fay RR, Popper AN (1975) Modes of stimulation of the teleost ear. J Exp Biol 62:379–387.
Fay RR, Popper AN (2000) Evolution of hearing in vertebrates: the inner ears and processing. Hear Res 149:1–10.
Fay RR, Kendall JI, Popper AN, Tester AL (1974) Vibration detection by the macula neglecta of sharks. Comp Biochem Physiol 47A:1235–1240.
Fay RR, Ahroon WA, Orawski AA (1978) Auditory masking patterns in the goldfish (*Carassius auratus*): psychophysical tuning curves. J Exp Biol 74:83–100.
Fay RR, Yost WA, Coombs SL (1983) Psychophysics and neurophysiology of repetition noise processing in a vertebrate auditory system. Hear Res 12:31–55.
Fine ML, Ladich F (2003). Sound production, spine locking, and related adaptations. In: Arratia G, Kapoor BG, Chardon M, Diogo M (eds) Catfishes, vol 1. Enfield, NH: Science Publisher, Inc., pp. 249–290.
Fink SV, Fink WL (1996) Interrelationships of ostariophysan fishes. In: Stiassny MLJ, Pasenti LR, Johnson GD (eds) Interrelationships of Fishes. San Diego, CA: Academic Press, pp. 209–249.
Fish MP, Offutt GC (1972) Hearing threshold from toadfish, *Opsanus tau*, measured in the laboratory and field. J Acoust Soc Am 51:1318–1321.
Fletcher LB, Crawford JD (2001) Acoustic detection by sound-producing fishes (Mormyridae): the role of gas-filled tympanic bladders. J Exp Biol 204:175–183.
Forey P, Janvier P (1994) Evolution of the early vertebrates. Am Sci 82:554–565.
Fritzsch B (1999) Hearing in two worlds: theoretical and actual adaptive changes of the aquatic and terrestrial ear for sound reception. In: Fay RR, Popper AN (eds) Comparative Hearing: Fish and Amphibians. New York: Springer-Verlag, pp. 15–42.
Fritzsch B, Signore M, Simeone A (2001) *Otx1* null mutant mice show partial segregation of sensory epithelia comparable to lamprey ears. Dev Genes Evol 211:388–396.
Gerald JW (1971) Sound production in six species of sunfish (Centrarchidae). Evolution 25:75–87.
Hawkins AD (1993) Underwater sound and fish behaviour. In: Pitcher TJ (ed) Behaviour of Teleost Fishes. London: Chapman and Hall, pp. 129–169.
Hawkins AD, Johnstone ADF (1978) The hearing of the Atlantic Salmon, *Salmon salar*. J Fish Biol 13:655–673.
Hawkins AD, MacLennan DN (1976) An acoustic tank for hearing studies on fish. In: Schuijf A (ed) Sound Reception in Fish. Amsterdam: Elsevier, pp. 149–169.
Hawkins AD, Myrberg AA (1983) Hearing and sound communication underwater. In: Lewis B (ed) Bioacoustics: A Comparative Approach. London: Academic Press, pp. 347–405.
Hoy RR (1992) The evolution of hearing in insects as an adaptation to predation from bats. In: Webster DB, Fay RR, Popper AN (eds) The Evolutionary Biology of Hearing. New York: Springer-Verlag, pp. 115–129.
Hoy RR (1998) Acute as a bug's ear: an informal discussion of hearing in insects. In: Hoy RR, Popper AN, Fay RR (eds) Comparative Hearing: Insects. New York: Springer-Verlag, pp. 1–17.

Jacobs DW, Tavolga WN (1967). Acoustic intensity limens in the goldfish. Anim Behav 15:324–335.
Jenkins DB (1977) A light microscopic study of the saccule and lagena in certain catfish. Am J Anat 150:605–630.
Johnston CE, Johnson DL (2000) Sound production in *Pimephales notatus* (Rafinesque) (Cyprinidae). Copeia 567–571.
Kalmijn AJ (1988) Hydrodynamic and acoustic field detection. In: Atema J, Fay RR, Popper AN, Tavolga WN (eds) Sensory Biology of Aquatic Animals. New York: Springer-Verlag, pp. 83–130.
Karlsen HE (1992a) Infrasound sensitivity in the plaice (*Pleuronectes platessa*). J Exp Biol 171:173–187.
Karlsen HE (1992b) The inner ear is responsible for detection of infrasound in the perch (*Perca fluviatilis*). J Exp Zool 171:163–172.
Kleerekoper H, Roggenkamp PA (1959) An experimental study on the effect of the swimbladder on hearing sensitivity in *Ameiurus nebulosus* (Lesueur). Can J Zool 37: 1–8.
Knudsen FR, Enger PS, Sand O (1992) Awareness reactions and avoidance responses to sound in juvenile Atlantic salmon, *Salmo salar* L. J Fish Biol 40:523–534.
Konishi M (1970) Comparative neurophysiological studies of hearing and vocalizations in songbirds. Z Vergl Physiol 66:257–272.
Kratochvil H (1985) Beiträge zur Lautbiologie der Anabantoidei—Bau, Funktion und Entwicklung von lauterzeugenden Systemen. Zool Jb Anat 89:203–255.
Ladich F (1988) Sound production by the gudgeon, *Gobio gobio* L., a common European freshwater fish (Cyprinidae, Teleostei). J Fish Biol 32:707–715.
Ladich F (1999) Did auditory sensitivity and vocalization evolve independently in otophysan fishes? Brain Behav Evol 53:288–304.
Ladich F (2000) Acoustic communication and the evolution of hearing in fishes. Philos Trans R Soc Lond 355:1285–1288.
Ladich F (2001) The sound-generating (and -detecting) motor system in catfishes: design of swimbladder muscles in doradids and pimelodids. Anat Rec 263:297–306.
Ladich F, Bass AH (1998) Sonic/vocal motor pathways in catfishes: comparison with other teleosts. Brain Behav Evol 51:315–330.
Ladich F, Bass AH (2003a) Underwater sound generation and acoustic reception in fishes with some notes on frogs. In: Collin SP, Marshall NJ (eds) Sensory Processing in Aquatic Environments. New York: Springer-Verlag, pp. 173–193.
Ladich F, Bass AH (2003b) Audition. In: Arratia G, Kapoor BG, Chardon M, Diogo R (eds) Catfishes, vol 2. Enfield, NH: Science Publishers, Inc., pp. 701–730.
Ladich F, Fine ML (1994) Localization of swimbladder and pectoral motoneurons involved in sound production in pimelodid catfish. Brain Behav Evol 44:86–100.
Ladich F, Kratochvil H (1989) Sound production by the marmoreal goby, *Protherorhinus marmoratus* (Pallas) (Gobiidae, Teleostei). Zool Jb Physiol 93:501–504.
Ladich F, Popper AN (2001) Comparison of the inner ear ultrastructure between teleost fishes using different channels for communication. Hear Res 154:62–72.
Ladich F, Tadler A (1988) Sound production in *Polypterus* (Osteichthyes: Polypteridae). Copeia 1076–1077.
Ladich F, Wysocki LE (2003) How does tripus extirpation affect auditory sensitivity in goldfish? Hear Res. 182:119–129.
Ladich F, Yan HY (1998) Correlation between auditory sensitivity and vocalization in anabantoid fishes. J Comp Physiol [A] 182:737–746.

Ladich F, Bischof C, Schleinzer G, Fuchs A (1992) Intra- and interspecific differences in agonistic vocalization in croaking gouramis (genus: *Trichopsis*, Anabantoidei, Teleostei). Bioacoustics 4:131–141.

Laming PR, Morrow G (1981) The contribution of the swimbladder to audition in the roach (*Rutilus rutilus*). Comp Biochem Physiol [A] 69:537–541.

Lewis ER, Narins P (1999) The acoustic periphery of amphibians: anatomy and physiology. In: Fay RR, Popper AN (eds) Comparative Hearing: Fish and Amphibians. New York: Springer-Verlag, pp. 101–154.

Löwenstein O (1970) The electrophysiological study of the responses of the isolate labyrinth of the lamprey (*Lampetra fluviatilis*) to angular acceleration, tilting and mechanical vibration. Proc R Soc Lond B 174:419–434.

Löwenstein O, Osborne MP (1964) Ultrastructure of the sensory hair cells in the labyrinth of the ammocete larva of the lamprey, *Lampetra fluviatilis*. Nature 204:97.

Löwenstein O, Roberts TDM (1950) The equilibrium function of the otolith organs of the thornback ray (*Raja clavatra*). J Physiol (Lond) 110:392–415.

Löwenstein O, Roberts TDM (1951) The localization and analysis of the responses to vibration from the isolated elasmobranch labyrinth. A contribution to the problem of the evolution of hearing in vertebrates. J Physiol (Lond) 114:471–489.

Löwenstein O, Thornhill RA (1970) The labyrinth of *Myxine*, anatomy, ultrastructure and electrophysiology. Proc R Soc Lond B 176:21–42.

Löwenstein O, Osborne MP, Thornhill RA (1968) The anatomy and ultrastructure of the labyrinth of the lamprey (*Lampetra fluviatilis* L.). Proc R Soc Lond B 170:113–134.

Lugli M, Fine ML (2003) Acoustic communication in two freshwater gobies: ambient noise and short-range propagation in shallow waters. J Acoust Soc Am 114:512–521.

Lugli M, Torricelli P, Pavan G, Mainardi D (1997) Sound production during courtship and spawning among freshwater gobiids (Pisces, Gobiidae). Mar Fresh Behav Physiol 29:109–126.

Lychakov DV, Rebane YT (2002) Otoliths and modelling ear function. Bioacoustics 12: 125–128.

Mann DA, Lobel PS (1997) Propagation of damselfish (Pomacentridae) courtship sounds. J Acoust Soc Am 101:3783–3791.

Mann DA, Lu Z, Popper AN (1997) Ultrasound detection by a teleost fish. Nature 389: 341.

Mann DA, Lu Z, Hastings MC, Popper AN (1998) Detection of ultrasonic tones and simulated dolphin echolocation clicks by a teleost fish, the American shad (*Alosa sapidissima*). J Acoust Soc Am 104:562–568.

Mann DA, Higgs DM, Tavolga WN, Souza MJ, Popper AN (2001) Ultrasound detection by clupeiform fishes. J Acoust Soc Am 109:3048–3054.

Markl H (1971) Schallerzeugung bei Piranhas (Serrasalminae, Characidae). Z Vergl Physiol 74:39–56.

Markl H (1972) Aggression und Beuteverhalten bei Piranhas (Serrasalminae, Characidae). Z Tierpsychol 30:190–216.

Marshall NB (1967) Sound-producing mechanisms and the biology of deep-sea fishes. In: Tavolga WN (ed) Marine Bio-Acoustics. Oxford, UK: Pergamon Press, pp. 123–133.

Marvit P, Crawford JD (2000) Auditory discrimination in a sound-producing electric fish (*Pollimyrus*): tone frequency and click rate difference detection. J Acoust Soc Am 108:1819–1825.

McCormick CA, Popper AN (1984) Auditory sensitivity and psychophysical tuning curves in the elephantnose fish, *Gnathonemus petersii.* J Comp Physiol [A] 155:753–761.

McKibben JR, Bass AH (1998) Behavioral assessment of acoustic parameters relevant to signal recognition and preference in a vocal fish. J Acoust Soc Am 104:3520–3533.

Millot J, Anthony J (1965) Anatomie de *Latimeria chalumnae.* II. Systeme Nerveux et Organes de Sense. Paris: Centre National de la Recherche Scientifique.

Myrberg AA (1981) Sound communication and interception in fishes. In: Tavolga WN, Popper AN, Fay RR (eds) Hearing and Sound Communication in Fishes. New York: Springer-Verlag, pp. 395–426.

Myrberg AA (2001) The acoustical biology of elasmobranchs. Environ Biol Fishes 60:31–45.

Myrberg AA, Spires JY (1980) Hearing in damselfishes: an analysis of signal detection among closely related species. J Comp Physiol 140:135–144.

Myrberg AA, Kramer E, Heinecke P (1965) Sound production by cichlid fishes. Science 149:555–558.

Myrberg AA, Ha SJ, Walewski S, Banbury JC (1972) Effectiveness of acoustic signals in attracting epipelagic sharks to an underwater sound source. Bull Mar Sci 22:926–949.

Myberg AA, Spanier E, Ha SJ (1978) Temporal patterning in acoustical communication, In: Reese ES, Lighter FJ (eds) Contrasts in Behavior. New York: Wiley and Sons, pp. 137–179.

Myrberg AA, Mohler M, Catala JD (1986) Sound production by males of a coral reef fish (*Pomacentrus partitus*): its significance to females. Anim Behav 34:913–923.

Nelson DR, Johnson RH (1976) Some recent observations on acoustic attraction of Pacific reef sharks. In: Schuijf A, Hawkins AD (eds) Sound Reception in Fishes. Amsterdam: Elsevier, pp. 229–239.

Nelson JS (1994) Fishes of the World, 3rd ed. New York: John Wiley and Sons.

Pennisi E (2003) Modernizing the tree of life. Science 300:1692–1697.

Plachta DTT, Popper AN (2003) Evasive responses of American shad (*Alosa sapidissima*) to ultrasonic stimuli. J Assoc Res Otolaryngol 4:25–30.

Platt C, Jørgensen JM, Popper AN (2004) The inner ear of the lungfish *Protopterus.* J Comp Neurol 471:277–278.

Poggendorf D (1952) Die absolute Hörschwelle des Zwergwelses (*Amiurus nebulosus*) und Beiträge zur Physik des Weberschen Apparates der Ostariophysen. Z Vergl Physiol 34:222–257.

Popper AN (1970) Auditory capacities of the Mexican blind cave fish (*Astyanax jordani*) and its eyed ancestor (*Astyanax mexicanus*). Anim Behav 18:552–562.

Popper AN (1971) The effects of size on auditory capacities of the goldfish. J Aud Res 11:239–247.

Popper AN (1978) Scanning electron microscopic study of the otolithic organs in the bichir (*Polypterus bichir*) and shovel-nose sturgeon (*Scaphirhynchus platorhynchus*). J Comp Neurol 181:117–128.

Popper AN (1980) Scanning electron microscopic studies of the sacculus and lagena in several deep-sea fishes. Am J Anat 157:115–136.

Popper AN, Clarke NL (1976) The auditory system of the goldfish (*Carassius auratus*): effects of intense acoustic stimulation. Comp Biochem Physiol 53A:11–18.

Popper AN, Clarke NL (1979) Non-simultaneous auditory masking in the goldfish, *Carassius auratus.* J Exp Biol 83:145–158.

Popper AN, Coombs S (1982) The morphology and evolution of the ear in Actinopterygian fishes. Am Zool 22:311–328.

Popper AN, Fay RR (1999) The auditory periphery in fishes. In: Fay RR, Popper AN (eds) Comparative Hearing: Fish and Amphibians. New York: Springer-Verlag, pp. 43–100.

Popper AN, Hoxter B (1987) Sensory and nonsensory ciliated cells in the ear of the sea lamprey, *Petromyzon marinus*. Brain Behav Evol 30:43–61.

Popper AN, Northcutt RG (1983) Structure and innervation of the inner ear of the bowfin, *Amia calva*. J Comp Neurol 213:279–286.

Popper AN, Tavolga WN (1981) Structure and function of the ear in the marine catfish, *Arius felis*. J Comp Physiol 144:27–34.

Popper AN, Platt C, Edds PL (1992) Evolution of the vertebrate inner ear: an overview of ideas. In: Webster DB, Fay RR, Popper AN (eds) The Evolutionary Biology of Hearing. New York: Springer-Verlag, pp. 49–57.

Popper AN, Fay RR, Platt C, Sand O (2003) Sound detection mechanisms and capabilities of teleost fishes In: Collin SP, Marshall NJ (eds) Sensory Processing of the Aquatic Environment. New York: Springer-Verlag, pp. 3–38.

Ramacharitar J, Higgs DM, Popper AN (2001) Sciaenid inner ears: a study in diversity. Brain Behav Evol 58:152–162.

Retzius G (1881) Das Gehörorgan der Wirbelthiere. I. Das Gehörorgan der Fische und Amphibien. Stockholm: Samson and Wallin.

Richter HJ (1988) Gouramis and Other Anabantoids. Neptune City, NJ: TFH Publications.

Rogers PH, Cox H (1988) Underwater sound as a biological stimulus. In: Atema J, Fay RR, Popper AN, Tavolga WN (eds) Sensory Biology of Aquatic Animals. New York: Springer-Verlag, pp. 131–149.

Romer AS, Parsons TS (1983) Vergleichende Anatomie der Wirbeltiere. Hamburg: Paul Parey Verlag.

Ross QE, Dunning DJ, Menezes JK Jr, Kenna MJ, Tiller GW (1996) Responses of alewives to high-frequency sound at a power plant intake in Ontario. North Am J Fish Manage 16:548–559.

Sagemehl M (1885) Beiträge zur vergleichenden Anatomie der Fische. III. Das Cranium der Characiniden nebst allgemeinen Bemerkungen über die mit einem Weber'schen Apparat versehenen Physostomen-Familien. Gegenbaur's Morphol Jahrb 10:1–119.

Sand O, Karlsen HE (1986) Detection of infrasound by the Atlantic cod. J Exp Biol 125:197–204.

Sand O, Karlsen HE (2000) Detection of infrasound and linear acceleration in fishes. Philos Trans R Soc Lond 355:1295–1298.

Schachner G (1977) Mechanismen und biologische Bedeutung der Schallerzeugung und-wahrnehmung beim südamerikanischen Antennenwels (*Pimelodus* sp. Pimelodidae). Doctoral thesis, University of Vienna.

Schaller F (1967) Die Lauterzeugung des Jaraqui, *Prochilodus insignis* Schomburgh 1841 (Pisces, Characoidei, Anastomidae). Verh Dt Zool Ges 1967:365–370.

Schellart NAM, Popper AN (1992) Functional aspects of the evolution of the auditory system of actinopterygian fish. In: Webster DE, Fay RR, Popper AN (eds) The Evolutionary Biology of Hearing. New York: Springer-Verlag, pp. 295–322.

Schneider H (1941) Die Bedeutung der Atemhöhle der Labyrinthfische für ihr Hörvermögen. Z Vergl Physiol 29:172–194.

Scholik AR, Yan HY (2000) Effects of underwater noise on auditory sensitivity of cyprinid fish. Hear Res 152:17–24.
Sörensen W (1895) Are the extrinsic muscles of the air-bladder in some Siluroidae and the "elastic spring" apparatus of others subordinate to the voluntary production of sounds? What is, according to our present knowledge, the function of the Weberian ossicles? J Anat Physiol 29:399–423, 205–229, 518–552.
Spanier E (1979) Aspects of species recognition by sound in four species of damselfish, genus *Eupomacentrus* (Pisces: Pomacentridae). Z Tierpsychol 51:301–316.
Steinberg R (1957) Unterwassergeräusche und Fischerei. Prot Fischereitech 4:216–249.
Stipetić E (1939) Über das Gehörorgan der Mormyriden. Z Vergl Physiol 26:740–752.
Stout JF (1963) The significance of sound production during the reproductive behaviour of *Notropis analostanus* (Family Cyprinidae). Anim Behav 11:83–92.
Tavolga WN (1956) Visual, chemical and sound stimuli as cues in the sex discriminatory behavior of the gobiid fish, *Bathygobius soporator*. Zoologica 41:49–64.
Tavolga WN (1958) Underwater sounds produced by males of the bleniid fish, *Chasmodes bosquianus*. Ecology 39:759–760.
Tavolga WN (1967) Masked auditory thresholds in teleost fishes. In: Tavolga WN (ed) Marine Bio-Acoustics II. Oxford, UK: Pergamon Press, pp. 233–245.
Tavolga WN (1976) Acoustic obstacle detection in the sea catfish (*Arius felis*). In: Schuijf A, Hawkins AD (eds) Sound Reception in Fish. Amsterdam: Elsevier, pp. 185–204.
Tavolga WN, Wodinsky J (1963) Auditory capacities in fishes. Pure tone thresholds in nine species of marine teleosts. Bull Am Mus Nat Hist 126:177–240.
Tester AL, Kendall JI, Milisen WB (1972) Morphology of the ear of the shark genus *Carcharhinus*, with particular reference to the macula neglecta. Pac Sci 26:264–274.
Urick RJ (1983) Principles of Underwater Sound, 3rd ed. Los Altos, CA: Peninsula.
van Bergeijk WA (1967) The evolution of vertebrate hearing. In: Neff WD (ed) Contributions to Sensory Physiology. New York: Academic Press, pp. 1–49.
von Frisch K (1936) Über den Gehörsinn der Fische. Biol Rev 11:210–246.
von Frisch K, Stetter H (1932) Untersuchungen über den Sitz des Gehörsinnes bei der Elritze. Z Vergl Physiol 17:687–801.
Weber EH (1820) De Aure et Auditu Hominis et Animalium. Pars I. De Aure Animalium Aquatilium. Leipzig: Gerhard Fleischer.
Weiss BA, Strother WF, Hartig GH (1969) Auditory sensitivity in the bullhead catfish (*Ictalurus nebulosus*). Proc Nat Acad Sci USA 64:552–556.
Wever EG (1974) The evolution of vertebrate hearing. In: Keidel WD, Neff WD (eds) Handbook of Sensory Physiology, vol V/1. Auditory System. Berlin: Springer-Verlag, pp. 423–454.
Wilson B, Dill LM (2002) Pacific herring respond to simulated odontocete echolocation sounds. Can J Fish Aquat Sci 59:542–553.
Wolff DL (1966) Akustische Untersuchungen zur Klapperfischerei und verwandter Methoden. Z Fisch Hilfswiss 14:277–315.
Wolff DL (1968) Das Hörvermögen des Kaulbarsches (*Acerina cernua L.*) und des Zanders (*Lucioperca sandra* Cuv. und Val.). Z Vergl Physiol 60:14–33.
Wysocki LE, Ladich F (2002) Can fishes resolve temporal characteristics of sounds? New insights using auditory brainstem response. Hear Res 169:36–46.
Yan HY, Fine ML, Horn HS, Colon WE (2000) Variability in the role of the gasbladder in fish audition. J Comp Physiol [A] 186:435–445.
Yost WA (1994) Fundamentals of Hearing, 3rd ed. San Diego: Academic Press.

5
The Evolution of Single- and Multiple-Ossicle Ears in Fishes and Tetrapods

JENNIFER A. CLACK AND EDGAR ALLIN

1. Fossil Record of the Ostariophysan Hearing System

The Ostariophysi are the hearing specialists of the ray-finned world. They are placed cladistically within the euteleosts and include the gonorynchiforms and a much larger group called the Otophysi (see Fig. 5.1 for a phylogeny of the taxa referred to in this chapter). The otophysans are distinguished by having some of the anterior neural arches and supraneurals modified into the "Weberian apparatus." This system of interarticulating elements is intimately connected to the swim bladder, and forms a unique sound transmission system. Although lacking the Weberian apparatus, other euteleosts nonetheless have hearing capabilities, if not always as acute as those of the otophysans. Weberian ossicles, however, represent a hearing apparatus whose evolution can be traced in the fossil record.

The gonorynchiforms or milkfish show an expanded third rib that contacts the swim bladder. This character is considered to be an example of a formative stage in the development of the Weberian apparatus. Gonorynchiformes are known from the Lower Cretaceous, and certainly they are the earliest teleosts to show indications of the system (Carroll 1987).

Otophysans, in which the Weberian system is fully developed, include many modern freshwater fishes such as carp, goldfish, characoids, cyprinoids, and catfish (Fig. 5.2). A few fossil genera that show some elements of the Weberian apparatus are also included within the group. These earliest otophysans were marine and date from the Late Cretaceous. They include some genera that have the supraneural of neural arch three enlarged and in contact with those anterior and posterior to it, but no other modifications. A form from the Eocene of Italy (*Chanoides*, Fig. 5.2A) shows a more fully developed Weberian apparatus that nevertheless has some unique features suggesting that it functioned in a slightly different manner from those of modern otophysans (Patterson 1984).

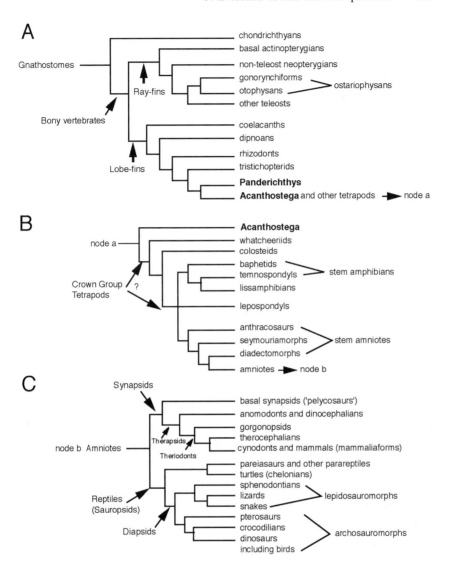

FIGURE 5.1. Consensus cladogram of the major groups referred to in this chapter (see text for sources).

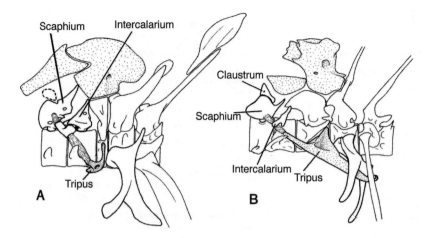

FIGURE 5.2. Weberian apparatus of the Eocene genus *Chanoides* from Monte Bolca, Italy (**A**) and the extant cypriniform *Opsariichthys* (**B**). Note that the narrow third centrum and minute tripus are unique to *Chanoides*, and suggest that the Weberian apparatus functioned in a different manner from that in modern forms. (From Patterson 1984.)

2. Tetrapods and Their Lobe-Finned Relatives

2.1 Lobe-Finned "Fishes"

The hearing capabilities of the tetrapods' closest living relatives are poorly known. It is not possible at this time to examine those of *Latimeria* experimentally, and little is known about those of lungfishes. Research on the soft tissue anatomy of lungfish and *Latimeria* ears shows a number of apparently plesiomorphic features in each (Platt 1994; Popper and Platt 1996). Lungfish and *Latimeria* are nonetheless very specialized, so that it is difficult to infer from them what conditions might have been like in the fossil osteolepiforms, which are actually the tetrapods' closest relatives.

There is some information about the fossil history of coelacanth and lungfish otic morphology. Some fossil lungfishes from the Middle Devonian (Miles 1977) have excellently preserved braincases from which some of the inner-ear gross morphology can be reconstructed, and it appears much like that of recent forms. There is a single sacculo-lagenar chamber and a large utricular chamber. Reconstructions of the ear in the primitive Late Devonian coelacanth *Diplocercides* (as *Nesides*, Jarvik 1980) show a separate lagenar pouch already present by this stage in coelacanth history. Otoliths are known in fossil and recent coelacanths (Clack 1996), while none has been described in any fossil lungfish. Recent lungfishes are described as having either solid (Shepherd 1914; Gauldie et al. 1986) or "pasty" statoliths (Retzius 1881; Carlström 1963) composed of many crystals held in a gelatinous matrix that disintegrates on death, as in most tetrapods.

The best known of the "osteolepiforms" (a paraphyletic group of the tetrapods' closest relatives) is *Eusthenopteron* (Jarvik 1980, Fig. 5.3A). In osteolepiforms, the braincase was divided by a hinge, with the otic region posterior to the hinge. The side wall carried paired facets for the hyomandibular attachment, and it is this region that underwent the most profound change during the "fish–tetrapod transition." Some record of the inner-ear structure in osteolepiforms can be seen in exceptionally well-preserved fossils. For example, *Eusthenopteron* had no separate lagenar pouch, and the later *Ectosteorhachis* (as *Megalichthys*, Romer 1937) had a large otolithic mass in an undivided sacculo-lagenar chamber.

2.2 Early Tetrapods

Fossil evidence of the ear region in tetrapods is abundant, though its interpretation is not straightforward and has undergone radical reassessment in recent years. Lack of clear evidence of the hearing abilities in osteolepiforms is only one of the problems. Another has been the changed appreciation of the interrelationships of early tetrapod groups; the advent of cladistics has changed perceptions about which characters were shared among all groups and which were derived separately in different lineages. This change has had an impact on the interpretation of the ear region in tetrapods as a whole.

Parts of the ear region that may be preserved include the otic capsules of the braincase, the fenestra vestibuli (FV) or oval window, the stapes, and bones surrounding the middle ear cavity. Features such as how much bone separates the otic capsules from the brain, the course of the perilymphatic system, the size of the lagena or other parts of the vestibular system, and the existence or development of the eustachian system in some animals may be accessible from fossil braincases. Bones of the skull roof and palate, such as the quadrate, squamosal and pterygoid, give clues to the size of middle ear cavity or how a tympanic membrane, if present, was attached. The components of a tympanic ear (i.e., a middle ear specialized for the reception of airborne sound, with a tympanum and air-filled cavity) that are readable from bony remains include a slender stapes passing across the middle ear cavity and free to vibrate within it, as well the size and shape of the middle ear cavity and tympanic membrane, inferred from the form of the quadrate. The assembly acts as an impedance matching device (see, for example, Smotherman and Narins, Chapter 6; Manley, Chapter 7; Gleich et al., Chapter 8). The size and orientation of the stapes may suggest whether it was likely to have been part of a tympanic ear, while the degree and style of attachment between the braincase, palate, and skull roof may also give clues about the role of the stapes in other functions not directly connected with hearing.

2.2.1 Palaeozoic Tetrapods

The earliest tetrapods come from the Devonian, the best known being *Acanthostega gunnari* from East Greenland. In *Acanthostega* good information on the ear region is preserved (Fig. 5.3B–D) (Clack 1998). In some features of the braincase it is very similar to that of *Eusthenopteron*. *Acanthostega*, however,

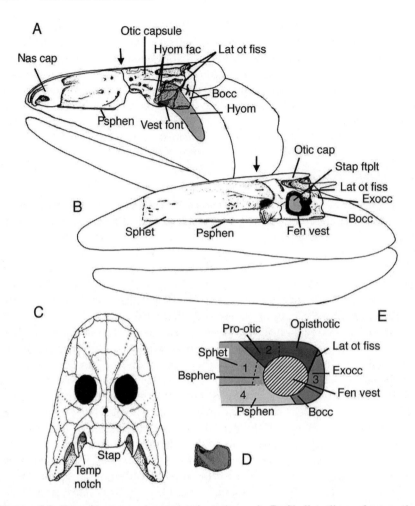

FIGURE 5.3. Devonian tetrapodomorph otic regions. **A, B**: Skull outlines of tetrapodomorphs with braincase shown in position, the otic regions indicated, and the position of the stapes/hyomandibula and vestibular fontanelle/fenestra vestibuli shown. A, *Eusthenopteron foordi*; B, *Acanthostega gunnari*. **C**: Skull reconstruction of *Acanthostega* (skull length about 150 mm). **D**: Stapes in dorsal view (not to scale). **E**: Diagram indicating the position of the fenestra vestibuli in early tetrapods with respect to the main embryogenic tissues. Area 1, sphenethmoid/basisphenoid (neural crest origin); area 2, pro-otic and opisthotic (otic capsule, cephalic mesoderm origin); area 3, basi- and exoccipital (occipital arch, somitic mesoderm origin); area 4, parasphenoid (dermal bone, neural crest origin). The metotic fissure (which remains open in stem tetrapods) is known as the lateral otic fissure in the adult animals shown in A and B. Bocc, basioccipital; Bsphen, basisphenoid; Exocc, exoccipital; fac, facet; Fen vest, fenestra vestibuli; Ftplt, footplate; Hyom, hyomandibula; cap, nasal capsule; Psphen, parasphenoid; Sphet, sphenethmoid; Stap, stapes/stapedial; Temp, temporal; Vest font, vestibular fontanelle.

shows several crucial tetrapod characters of the ear region, including a hole in the braincase wall, the FV with the footplate of the stapes inserted into it. One of the main functions of the fish hyomandibula is to move the bony opercular flap during breathing. The opercular series is lost in tetrapods, so the hyomandibula no longer has to operate in this way. At some point during the fish–tetrapod transition, the hyomandibula became redeployed as the tetrapod stapes (sometimes called the columella in nonmammalian tetrapods), though its initial function may have been as a structural brace rather than having a primary role in hearing. The stapedial shaft in *Acanthostega* is short and broad and is still in contact with the palatal bones. Notches on each side in the rear margin of the skull roof, once interpreted as supporting a tympanic membrane, are now envisaged as possibly having housed functional spiracles (Clack 1989). *Acanthostega* was essentially an aquatic animal, so its ear is unlikely to show adapations for terrestrial hearing. By contrast, it has recently been shown that another Devonian tetrapod, *Ichthyostega*, had an ear that was specialized for underwater audition (Clack et al. 2003).

Changes to the otic region during the fish–tetrapod transition appear to have been rapid, and one of their main outcomes was the incorporation of the head of the hyomandibula (stapes) into the braincase wall, as well as the elimination of the osteolepiform braincase hinge (Ahlberg et al. 1996). The FV into which the stapes fits has a margin positioned at the junction between four tissues of different embryogenic origins: the otic capsule dorsally is formed from cephalic mesoderm; the occipital arch posteriorly is formed from somitic mesoderm; the parasphenoid ventrally is a dermal bone formed from neural crest; and the sphenethmoid-basisphenoid anteriorly is also formed from neural crest tissue but is endoskeletal in origin. The junctions between these regions can be seen in the conspicuous embryonic fissures of the braincase such as the metotic and ventral cranial fissures (Fig. 5.3A, B, E). The hyomandibula or stapes is also a neural crest–derived element. The rapid braincase changes during the transition affected the neural crest derived tissues most profoundly, including the elimination of the ventral cranial fissure, and the incorporation of the stapes into the gap between the occipital arch, otic capsule, and para- and basisphenoid. Why these changes occurred is not clear, but the neural crest is a highly active tissue and is responsible for most of the major innovations found in vertebrates (Clack 2001).

The inner parts of the braincase of *Acanthostega* were poorly ossified, and the semicircular canals were not fully enclosed in bone. Because of this, there are many questions about early tetrapod ears that we cannot answer. Many of these are fundamental to the structure of the ear in tetrapods. Here are some examples. Was there a separate lagenar pouch, absent in both osteolepiforms and lungfishes? Was the perilymphatic system differentiated into a perilymphatic canal and cistern as in more derived tetrapods? If the perilymphatic canal were present, did it pass posterior to the sacculo-lagenar complex as modern amphibians or anterior to it as in amniotes? Was there a perilymphatic foramen (PLF) allowing the perilymphatic sac room to compensate for auditory vibrations? We cannot tell in early tetrapods whether the internal wall of the capsule was formed in cartilage, stiff membrane, or flexible membrane. This would no doubt make a difference to the functioning of any perilymphatic system and possible presence of a PLF.

During the Carboniferous, tetrapods radiated into many diverse groups, including the two major lineages, which eventually gave rise to modern amphibians and amniotes. All had essentially the conditions seen in *Acanthostega*. The FV was large and its margin involved the contribution of bones from the same four regions. The stapes was stout and broad; the otic capsules were poorly ossified internally; the PLF, recessus scalae tympani, semicircular canal tubes, and vestibular pouches are not clearly identifiable, and there was apparently no separate pressure-relief window (PRW) in the braincase wall, a feature that develops in later tetrapods to give the inner ear chamber more flexibility to accommodate the vibrations of the incompressible inner ear fluids (Figs. 5.3 and 5.4).

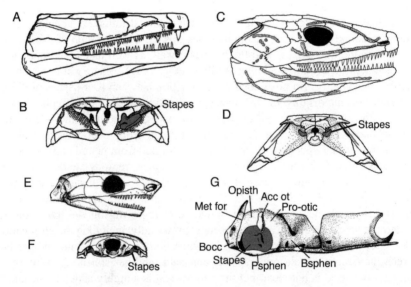

FIGURE 5.4. Carboniferous tetrapod otic regions. **A, B**: *Greererpeton burkemorani*, a colosteid, showing the lack of an otic notch and the presence of a large, flattened stapes. Skull length about 200 mm. A, right lateral view; B, posterior view. (From Smithson 1982.) **C, D**: *Palaeoherpeton decorum*, an embolomere, showing apparent otic notch, but with a stapes similar in shape to that of *Greererpeton*. Skull length about 200 mm. C, right lateral view; D: posterior view. **E–G**: *"Goniorhynchus"* (= *Rhynchonkos*) *stovalli*, a microsaur, showing lack of otic notch, and button-like stapes. Skull length about 17.5 mm. E, right lateral view; F, posterior view, G, enlarged view of right side of braincase. (From Carroll and Currie 1975.) The braincase shows the large stapedial footplate in contact with the four regions of the embryonic braincase. Note also the accessory otic element which has sometimes been interpeted as equivalent to the lissamphibian operculum. Acc ot, accessory otic element; Met for, metotic foramen; Opisth, opisthotic. (For other abbreviations, see Figure 5.3.)

Large forms such as whatcheeriids (Clack 2002a), baphetids (Beaumont and Smithson 1998; Clack 2003), and anthracosaurs had a deep notch at the rear of the skull, in the past interpreted as "otic" and bearing a tympanum. Before the stapes of these forms was found, this arrangement was supposed to be part of a so-called "labyrinthdont middle ear." However, it is now known that they all had stapes much like *Acanthostega*, albeit that the stapedial shaft and footplate were usually relatively smaller. In some forms such as colosteids, the stapes was proportionately larger than in *Acanthostega* and the notch was absent (Figs. 5.3 and 5.4), (Smithson 1982). Studies of the ear region in colosteids, anthracosaurs, and *Acanthostega* over about the past 20 years have helped to change the picture of early ear evolution. Small forms such as microsaurs and aïstopods, also quite common in the late Palaeozoic (Fig. 5.4E–G) (Carroll and Currie 1975), had no notch at the back of skull. The stapes (where known) had a very large footplate and a short blunt shaft.

This brief survey also shows that some diversity of otic morphology was present among early tetrapods. Nonetheless, considering the phylogeny of these animals and comparing them with later fossil forms as well as modern animals, it now seems improbable that they had evolved tympanic ears, and their hearing capabilities were possibly confined to low frequencies or use in water. The evolution of crucial features of the bony ear region can be traced in the fossil record of several tetrapod groups. The story seems largely one of independent development of most of those associated with terrestrial audition, mirroring very closely the pattern inferred from soft tissue anatomy.

3. The Amphibian Lineage: Temnospondyls and Lissamphibians

The first group interpreted as having a terrestrially adapted ear are the temnospondyls. It was the otic region in this group that originally suggested the concept of a labyrinthodont middle ear. However, this group is now considered to be part of a completely separate radiation of tetrapods from that which gave rise to amniotes. The earliest known temnospondyl, *Balanerpeton*, derives from the Early Carboniferous and has a large embayment at the back of the skull and a stapes with narrow shaft (Milner and Sequeira 1994); a slightly later (Late Carboniferous) but more primitive form, *Dendrerpeton*, has a similar embayment, though the stapes is relatively stouter (Godfrey et al. 1987) (Fig. 5.5). Braincase preservation in these forms is not good enough to give details of the otic capsules, but later and better preserved ones show that they had the primitive, complex margin to the FV and that the internal structure of the otic capsule was poorly differentiated. Most temnospondyls (Fig. 5.5) show a skull embayment presumed to be tympanic with a rod-like stapes directed to the tympanic notch, except where the notch has been lost in obviously secondarily aquatic forms. One family (the Dissorophidae) shows stapes with detailed similarities to those of frogs (Bolt and Lombard 1985). Temnospondyls evolved speciali-

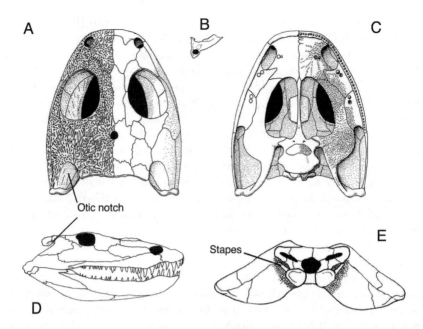

FIGURE 5.5. Temnospondyls. **A–C**: Skull and stapes of *Balanerpeton woodi*, the earliest known temnospondyl. A, skull in dorsal view, B, stapes in posterior view, C, skull in ventral view (skull length up to 48 mm). Note the large otic notch. The braincase of this species is very poorly known. (From Milner and Sequeira 1994. Reproduced by permission of the Royal Society of Edinburgh and A.R. Milner and S.E.K. Sequeira.) **D, E**: Skull reconstruction of *Eryops megacephalus*. D, skull in right lateral view (skull length about 400 mm), E, skull in posterior view. This large Permian temnospondyl was formerly taken as a model for a primitive labyrinthodont and its otic region established ideas about what the labyrinthodont ear was like. The stapes can be seen in the posterior view.

zations of the palate (large palatal vacuities) that may be associated with buccal breathing and that may have freed the stapes from its role as bracing member early in the group's evolution as a result (Clack 1992).

Many phylogenetic assessments place temnospondyls as the ancestral group giving rise to frogs or to all lissamphibians (Milner 1988; Trueb and Cloutier 1991). Amphibians use buccal pumping for ventilating the lungs, and frogs exploit this in a unique sound-production mechanism. If the frog–temnospondyl link is correct, this could imply an association between the development of the specialized palate in temnospondyls and the early development of hearing ability. Sound is used in communication by frogs, and it could be that temnospondyls were the first group of vertebrates to produce significant sounds. Under this hypothesis, urodeles and caecilians would show secondary loss of a tympanic ear. Supporting this hypothesis is the presence in some urodeles and caecilians of a receptor for relatively high-frequency hearing, a feature shared with frogs.

A competing phylogenetic hypothesis places lissamphibians (as well as amniotes) closer to the small tetrapods collectively known as "lepospondyls," including microsaurs (Fig. 5.4) and aïstopods (Laurin and Reisz 1997). These show no evidence of terrestrial hearing specializations, and are much like salamanders in the form of the stapes and their lack of a tympanic notch. If this hypothesis is correct, frogs represent a separate development of the tympanic ear from that presumed in temnospondyls, with urodeles and caecilians showing the primitive lissamphibian condition. The earliest frog- or salamander-like animal, *Triadobatrachus*, from the Late Triassic, would then be the first known lissamphibian with possible evidence of a tympanic ear (Rage and Rocek 1989). Each of these hypotheses is consistent with an independent development of a basilar papilla in amphibians and amniotes.

One consistent feature of the ear region of frogs and salamanders is a bony operculum inserted into the FV in addition to the stapedial footplate. It is associated with reception of ground-borne vibrations (Hetherington et al. 1986). However, no obvious equivalent exists in any nonlissamphibian fossil, with the possible exception of an "accessory otic element" in some microsaurs (Fig. 5.4).

4. The Amniote Lineage

4.1 Reptiliomorph Groups and Stem Amniotes

Late in the Carboniferous, the amniote lineage of tetrapods, also called the reptiliomorphs, diversified into numerous forms, of which two or three have living descendants. These are the synapsids from which mammals evolved, and the diapsids from which lepidosaurs (snakes and lizards) and archosaurs (crocodiles and birds) evolved. The third group is the parareptiles, the lineage from which turtles (chelonians) are widely believed to derive (Lee 1997a). However, some analyses suggest that turtles are modified diapsids (Rieppel and DeBraga 1996; DeBraga and Rieppel 1997; Hedges and Poling 1999).

The advent of full terrestriality among these forms brought in train changes to skull structure consequent upon the development of terrestrial feeding and breathing. These affected lower jaw construction, skull shape, and the formation of the occiput, and these changes are reflected in the changing role of the stapes. The development of aerial hearing in amniotes was one of the suite of features whose beginnings can be traced to adaptations for terrestrial feeding and loss of the muscular apparatus for operating the branchial system during buccal pumping (Clack 2002b,c).

Given this early separation between modern amniote groups, one might predict the evolution of differences in many ear characters. Although one might infer that birds and crocodiles would share many characters, lizards might be expected to show differences from crocodiles and birds, and turtles should either be different from any of the diapsid groups or share some basic features with them. The mammalian lineage, being one of the first amniote groups to become established, should show differences from all other amniote groups. These predictions are borne out by the fossil record, which gives reasonably good information on how and when such differences arose, and by the ear structure in modern representatives.

Early members of the reptiliomorphs include some terrestially adapted forms that have no living relatives, but apparently show specializations of the ear region. Seymouriamorphs, medium-sized forms from the Early Permian, show a unique form of otic capsule with the FV at the end of a long lateral extension of the capsule. A strongly embayed notch bounded by the squamosal and skull table combined with an apparently rod-like stapes suggests that a tympanic ear was present (Fig. 5.6) (White 1939; Laurin 1996). Once "lumped" with "labyrinthodonts," it is now obvious that their ear regions show differences from those of temnospondyls that were overlooked because of their misunderstood phylogenetic relationships.

Diadectomorphs were large herbivores also equipped with a unique otic region. The stapes was many-processed and attached to a thin sheet of bone usually interpeted as an ossified tympanic membrane fitting in a broadly embayed cheek margin (Olson 1966). The phylogenetic relationships of these groups are controversial, though both are considered primitive members of the amniote stem lineage. Again, we see diversity of otic morphology in primitive members of a lineage, but its significance is unclear.

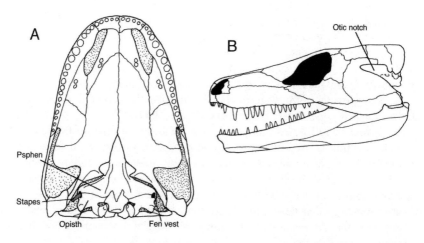

FIGURE 5.6. *Seymouria baylorensis*. **A**: Skull in ventral view. (skull length about 150 mm). **B**: Skull in left lateral view. (From White 1939.) The deeply embayed otic notch and the very small rod-like stapes were taken to be part of a labyrinthodont ear when this animal was considered to be a member of the labyrinthodonts. However, the form of the notch, and the otic region of the braincase are unique to seymouriamorphs. Note the extension of the parasphenoid and opisthotic forming a tube to carry the fenestra vestibuli laterally. The stapes (shaded) is tiny. (Abbreviations as in Figures 5.3 and 5.4.)

4.2 Early Amniotes

The earliest true (crown group) amniotes, the protorothyridids, date from the Late Carboniferous; the earliest synapsid (the lineage leading to mammals) and earliest diapsid come from the same period. All are more or less identical otically, and resemble Permian forms such as captorhinids, fortunately known from very good material (Fig. 5.7) (Heaton 1979).

There is no notch at the rear of the skull, the stapes is long and ventrally or horizontally aligned. It forms the major link between the back of the braincase and the skull roof via the quadrate. The paroccipital process, a strut of bone formed from bones of the otic capsule, does not reach the skull roof in these early forms. In later amniotes it secures the braincase to the skull roof, relieving the stapes from this duty. In these early amniotes, a very large stapedial footplate fitted into an FV like that of other early tetrapods, in which the occipital arch and ventral braincase elements were involved as well as the otic capsule. Usually there was a dorsal process of the stapes contacting the otic capsule. Permian diapsids such as *Araeoscelis* and synapsids like *Sphenacodon* and *Dimetrodon* have similar ear regions to *Captorhinus*. The conclusion must be that none of the animals representing the origins of the major amniote lineages shows any specializations for terrestrial hearing.

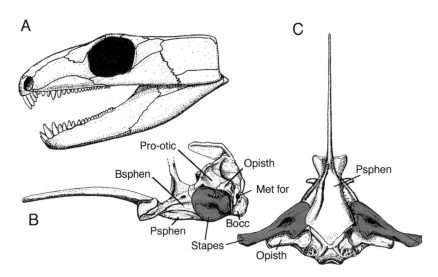

FIGURE 5.7. *Captorhinus aguti*, an early amniote. **A**: Skull in left lateral view (skull length up to 80 mm), showing lack of an otic notch common to all early amniotes. **B**: Braincase in left lateral view. **C**: Braincase in ventral view (length of braincase about 47 mm). Note the large stapedial footplate contacting the four embryonic braincase regions, as in stem tetrapods (stapes shaded). The metotic foramen is clearly defined on the metotic fissure, but is small and undivided. (From Heaton 1979.) (Abbreviations as in Figures 5.3 and 5.4.)

4.3 Neodiapsids

4.3.1 Neodiapsids I: *Youngina* and Basal Neodiapsids

The earliest diapsids resemble *Captorhinus* in the ear region, but within diapsids we see major changes in the two lineages, squamates and rhynchocephalians (lepidosauromorphs) forming one group, crocodiles, pterosaurs, dinosaurs, and birds (archosauromorphs) the other (Fig. 5.8). The base of this split lies among

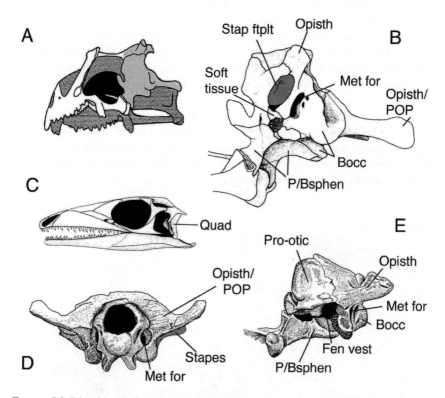

FIGURE 5.8. Diapsids. **A, B**: The basal lepidosauromorph *Sphenodon punctatus*. A: Skull outline in left posterolateral oblique view. Right parts of the skull are shown in dark shading; light shading shows part of the skull enlarged in B. B: Left posterolateral oblique view of braincase, stapes shaded. Note the area of soft tissue filling a gap in the ossification of the fenestra vestibuli ventral margin, so that the occipital arch and para-basisphenoid still approach the margin quite closely. The metotic foramen is large but undivided. (From Gower and Weber 1998.) **C–E**: The basal archosauromorph *Prolacerta broomi*. C: Skull in left lateral view showing embayed quadrate presumably for an early form of tympanic membrane (skull length about 50 mm). (From Gow 1975.) D, E: Braincase. D, left lateral view, E, posterior view. (Modified from Evans 1986.) There is a large undivided metotic foramen. The margins of the fenestra vestibuli are incompletely ossified and probably resembled that of *Sphenodon*. Note the general resemblance to the basal lepidosauromorph *Sphenodon*. POP, paroccipital process; Quad, quadrate. (Other abbreviations as in Figures 5.3 and 5.4.)

Late Permian or Early Triassic forms but the phylogeny of primitive diapsids is not well understood. Neodiapsids seem to form a clade based on the development of characters of the postcranium and quadrate region. There may be a link here with modifications to the basilar papilla.

At the base of the clade is *Youngina,* from the Late Permian of South Africa. This animal is critical to the story of ear evolution, because it is not clear whether it is an archosauromorph, lepidosauromorph, or falls below the split between them (Laurin and Reisz 1995). It has a relatively slender stapes with the footplate reduced in diameter compared with early nondiapsid amniotes, but still large compared with the diameter of the stapes (Gow 1975; Evans 1987). It apparently retains the primitive condition of the FV, although the walls of the otic capsule are still poorly ossified internally. It has a laterally projecting paroccipital process, but this contacts the supratemporal (a skull-table bone) rather than the squamosal (in the cheek) as in other neodiapsids. The quadrate shows no embayment for the possible location of a tympanic membrane or middle ear cavity. No clear evidence of hearing specializations are present except for the possible reduction in the stapedial shaft. Until the position of *Youngina* is resolved, and/or more basal neodiapsids are discovered, it will not be clear which of the otic characters we see in lepidosaurs and archosaurs were developed early in the clade before the split, and which arose independently.

4.3.2 Neodiapsids II: Lepidosauromorphs

It seems that even founder members of the lepidosauromorphs had some otic specializations suggesting hearing ability. A deeply embayed "quadrate conch," presumably for a tympanic membrane and middle ear cavity, appears among the most primitive lepidosauromorphs. However, it is not known whether any pressure relief mechanism existed for use in a tympanic ear. All lepidosauromorphs have a slender stapes and a paroccipital process tying the braincase to the skull roof via the squamosal at the posterodorsal corner, though in some this junction acted as part of the kinetic mechanism found in the skulls of these animals.

Rhynchocephalians (sphenodontids and related families) have an FV in which the bony margin is often incomplete ventrally; instead they probably had a membranous wall bounding the fenestra at that point. Rhynchocephalians cladistically below *Sphenodon* have a quadrate conch much like that of squamates, but they do not have a separate pressure relief window (see Clack 1997 and references therein) (Fig. 5.8A,B). It is not clear what the acuity of a system like this might have been, but it seems that lack of a tympanic ear and poor acuity in *Sphenodon* is to some extent a reversal from an earlier lepidosauromorph condition in which a tympanic membrane or its precursor may have been present.

Squamates show great variation in the condition of the inner-ear soft anatomy, and this is not easy to explain (Miller 1992; Manley and Gleich 1992; Manley 2002). The earliest squamates from the Middle and Late Jurassic may eventually help resolve the history of squamate otic characters. It may have been at this point that the specialization of the lepidosaur basilar papilla began (see Manley, Chapter 7). A tympanic ear is found in some Cretaceous squamates, and seems plesiomorphic at this level, with the FV small and bounded only by otic capsule bones. Most have a separate PRW, except where the tympanic ear was secondarily lost (e.g., chamaeleons). Manley (Chapter 7) has suggested possible evolution sequences for the basilar membrane and associated structures among lizard families.

In snakes, the stapes is long but delicate and contacts the quadrate, giving the arrangement a plesiomorphic appearance. However, some analyses place snakes as relatives of varanoid lizards, along with aquatic mosasaurs of the Cretaceous (Caldwell and Lee 1997; Lee 1997b; Caldwell 1999). Mosasaurs had a tympanic ear; there is evidence of a quadrate conch and also an ossified tympanic membrane preserved in some. Alternatively, this may be an ossified or calcified expanded extrastapes that met the conch, preventing water pressure from driving the stapes into the inner ear during dives. This proposal implies that if snakes evolved from swimming varanoids, they lost the tympanic ear during a long aquatic history.

4.3.3 Neodiapsids III: Archosauromorphs

Archosauromorphs are a very diverse group, having produced several lineages of conspicuous extinct forms, such as dinosaurs and pterosaurs, as well as two modern survivors, the crocodiles and birds. Archosauromorph phylogeny is the subject of much current controversy (Gower and Wilkinson 1996). Resolving these basal relationships is critical for working out the history of the ear region among the archosaur lineages.

Two genera are generally agreed to be among the most primitive known archosauromorphs, *Prolacerta* (Fig. 5.8C–E) and *Euparkeria*, both from the Early Triassic of South Africa (Evans 1986; Gower and Weber 1998). In both, the quadrate is shallowly embayed, and has usually been interpreted as serving for attachment of a tympanic membrane. In *Prolacerta* the quadrate has a small pocket on the mesial surface, possibly representing a middle ear cavity. The stapes of *Euparkeria* is unknown, but that of *Prolacerta* is slender and has lost its stapedial foramen, as in later archosaurs. The paroccipital process is horizontal and slender and contacts the skull roof at the squamosal, as seen in lepidosauromorphs. It seems that the stapedial footplate became reduced in diameter among archosaurs before the bony margins defining the FV came to match it. In both forms, the metotic foramen was large, again resembling rhynchocephalians, and no dedicated PRW was present. The latter, associated with tympanic hearing, appears only in later archosaurs. Significantly, though superficially like that in lepidosaurs, it is formed differently.

The otic region has been described in a number of dinosaurs, though not in the context of ear evolution. It is known that many dinosaurs had a separate PRW and a delicate stapes. However, the earliest dinosaurs had ear regions that were essentially like those of primitive archosauromorphs without a separate PRW (Sereno 1991). Evolution of ear specializations in dinosaurs would be an interesting study, to follow the acquisition of such features as the PRW. Among hadrosaurs, for example, peculiar cranial appendages are thought to have been employed in sound production (Weishampel 1981). The implication is that these later dinosaurs had sensitive hearing capable of discriminating intraspecific sounds, although it seems that early ones did not.

Archaeopteryx, considered to be the earliest known bird, has a relatively poorly known otic region, though it does have some bird-like features (Walker 1985). However, it also shows primitive features with no separate PRW. Other similarities between birds and crocodiles may be general archosaur synapomorphies giving a baseline for interpretation of early archosaurs and dinosaurs. The relationship of birds to other archosaurs has been disputed, though the majority view is that their closest relatives are small theropod dinosaurs. Palaeontological evidence strongly favors this hypothesis (Padian and Chiappe 1998), especially with new and spectacular feathered dinosaur material coming from China in recent years. The implication for ear evolution is that birds and *Archaeopteryx* show many reversals to more primitive otic conditions that may be paedomorphic in origin.

Although the flying reptiles, pterosaurs, are popularly recognized, almost nothing is known about their otic region, nor is it even clear where they fit among the archosaurs. There is no evidence yet for a specialized hearing apparatus, despite the appearance of an active and possibly warm-blooded lifestyle.

Modern crocodilians display many unique features of the otic region, including the configuration of the middle ear cavity and the anatomical relationships of the quadrate and quadratojugal to the squamosal. This arrangment encloses a narrow V-shaped slot that houses the tympanic membrane. The squamosal provides an overhanging ledge on which the upper "ear-lid" is supported in modern aquatic forms. The FV, PRW, and the ossifications of the inner ear chambers are also characteristic in structure. The cochlea in crocodiles has a "knee-bend" or geniculation reflected in the bony structure of the otic capsule. Crocodiles also have a complex suite of chambers and passages penetrating the braincase, associated with their unique eustachian system. Controversy over the phylogeny of the crocodile lineage means uncertainty over the sequence of acquisition of otic characters. Primitive and usually terrestrial Triassic crocodylomorphs show otic regions that resemble more generalized archosaurs rather than those of true crocodilians (Parrish 1993). The evolution of the modern crocodilian system can be traced through fossil material of these Triassic forms, some of which are represented by extremely well-preserved otic material (e.g., *Sphenosuchus*, Walker 1990). In successively earlier forms, however, it becomes increasingly difficult to distinguish the ear structure from that of other more generalized archosaurs (see also Clack 1997).

4.4 Turtles and Parareptiles

The origin of turtles (chelonians) has long been a contentious issue (Rieppel and DeBraga 1996; DeBraga and Rieppel 1997; Lee 1997a; Hedges and Poling 1999), but its solution is critical to understanding the origin of the ear in the group. Many modern phylogenetic analyses of turtles place them in a group called parareptiles, which are placed on the amniote tree above mammals but below diapsids, and are thus part of the sauropsid group (Fig. 5.1). Parareptiles are largely a Permian radiation of diverse forms, most of which show little evidence of otic specializations. Soft tissue anatomy of the ear strongly suggests a separate history for the chelonian ear from that of other amniotes (Wever 1978), and the bony anatomy shows many unique features. These include the middle ear cavity and its relationships to the quadrate bone, as well as the relationships of the quadrate to the braincase and otic capsule. Turtles have no PRW for the inner ear fluids, but have a "reentrant fluid system," surrounded by bony walls (Wever 1978). These features are amenable to preservation and can be seen in the long and well-documented fossil record of the group.

The only fossil turtle that does not show the characteristic ear structure is *Proganochelys*, the earliest, from the Late Triassic of Germany (Fig. 5.9) (Gaffney 1990). *Proganochelys* has a shallowly embayed quadrate, a relatively stout, imperforate stapes, with an FV that is larger than that of modern forms, but that is entirely confined within the otic capsule. It does not have the specialized reentrant fluid system, and the perilymphatic sac may have used the large metotic foramen as a PRW as in many early diapsids. In many respects, it bears comparison with generalized neodiapsids in ear structure. It is not clear what the basal conditions for the turtle radiation might have been, but if they arose from

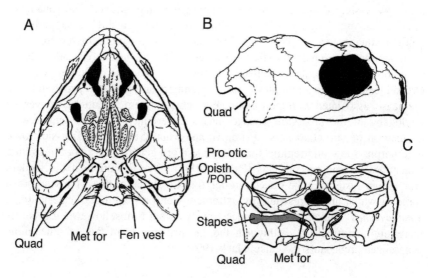

FIGURE 5.9. Skull of the Triassic turtle *Proganochelys quenstedti*. **A**: Ventral view showing the fenestra vestibuli, entirely within the otic capsule, and large undivided metotic foramen. **B**: Right lateral view showing embayed quadrate. **C**: Posterior view showing relatively slender stapes (shaded). (From Gaffney 1990.) (Abbreviations as in Figures 5.3, 5.4, and 5.8.)

among parareptiles, it seems likely that early members of the clade did not have a tympanic ear. If that is correct, the tympanic ear of turtles is, as soft anatomy suggests, a separate development. However, the analyses that have placed turtles among the diapsids are consistent with the otic structure of *Proganochelys*.

5. Synapsids and Hypotheses on Ear Evolution in This Clade

5.1 Synapsid Relationships and Structure

Synapsida comprises a paraphyletic cluster of very primitive forms, informally termed pelycosaurs, and a monophyletic group of more advanced forms termed therapsids. Therapsida consists of the theriodonts and other clades such as dinocephalians and anomodonts. Theriodontia includes Cynodontia, which includes Mammaliaformes, which in turn includes Mammalia (Rowe 1988) (Fig. 5.10). For a general survey of the nonmammalian synapsids see Kemp (1982),

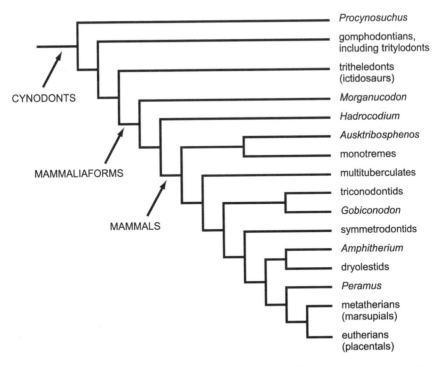

FIGURE 5.10. Cladogram of synapsid relationships (see text for sources). Cynodont phylogeny after Hopson and Kitching (2001) and Luo et al. (2001b). Some authors (e.g., Rowe 1988) consider tritylodonts to be closer to trithelodonts and mammaliaforms than to gomphodonts.

and for an extensive account of synapsid ear evolution see Allin and Hopson (1992). Novacek (1993) surveyed the development and evolution of the mammalian skull, including the auditory apparatus.

All nonmammalian tetrapods have a compound mandible consisting of several bones. One of these (the dentary bone) becomes the entire mandible in mammals, whereas others that project behind it (the four postdentary bones: angular, surangular, articular, and prearticular) become separated from the jaw as components of the mammalian middle ear, and other accessory jaw bones (the paradentary bones: coronoid and splenial) that initially assisted in attaching the postdentary bones to the dentary, among other functions, have been lost in extant mammals (Fig. 5.11). The angular bone becomes the tympanic annulus (tympanic or ectotympanic bone) of the mammalian embryo and fetus. This loosely suspended C-shaped element largely surrounds the eardrum and may remain

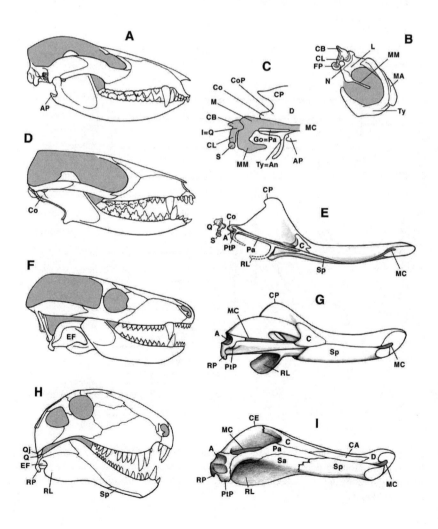

loosely suspended in the adult (as in opossums) or become synostosed to the squamosal and petrosal bones as part of a composite temporal bone. The surangular bone vanishes, except perhaps for a tiny vestige in a few mammals. The articular bone, formed by ossification of the posterior end of Meckel's cartilage, becomes the main part of the malleus, whereas the prearticular fuses to it as the anterior process. The anterior part of Meckel's cartilage degenerates in present-day mammalian fetuses or neonates, whereas the middle part becomes bonded to the skull and differentiates into the anterior ligament of the malleus and the sphenomandibular ligament. The original jaw joint, between the articular bone and the quadrate bone of the skull, persists as the malleo-incudal joint (the quadrate being renamed the incus), and a new jaw articulation forms between a process of the dentary and the squamosal bone. The primitive direct articulation between stapes and quadrate, lost in many other groups, is retained in all synapsids (Fig. 5.12), including mammals (as the incudostapedial joint). These remarkable homologies, initially recognized by Reichert on the basis of comparative embryology (Goodrich 1930; Sanchez-Villagra et al. 2002), have since been confirmed by detailed paleontological evidence showing progressively increasing similarity of form between the postdentary elements, quadrate, and stapes of premammalian therapsids (especially cynodonts) and their homologues in mammals, over the course of geologic time. Although the nature of this

FIGURE 5.11. Morphological series of synapsids in stratigraphic order to illustrate homologies and changes leading to the mammalian middle ear. **A**: Skull of the opossum *Didelphis* (about 120 mm long). **B**: Lateral view of tympanic annulus and middle ear ossicles of *Didelphis* (pars tensa of tympanum shaded). **C**: Medial view of corresponding structures or their primordia in a fetal marsupial, the bandicoot *Perameles* (cartilaginous structures shaded). **D, E**: Skull (about 40 mm long) and medial side of mandible of the Late Triassic mammaliaform (stem mammal) *Morganucodon*. **F, G**: Skull (about 120 mm long) and medial side of mandible of the Late Permian premammalian cynodont *Procynosuchus*. **H, I**: Skull (about 300 mm long) and medial side of mandible of the Early Permian sphenacodontid pelycosaur *Dimetrodon*. A, articular bone (homolog of main part of malleus); An, angular bone (homolog of tympanic bone); AP, angular process of dentary; C, coronoid bone; CB, crus breve of incus; CE, coronoid eminence; CL, crus longum of incus; Co, condyle of dentary; CP, coronoid process; D, dentary bone; E, external fossa; FP, footplate (base) of stapes; Go, goniale (homolog of prearticular); L, lamina of malleus; M, cartilaginous primordium of malleus; MA, anterior process of malleus, largely formed of goniale (prearticular); MC, Meckel's cartilage; MM, manubrium of malleus; N, neck of malleus; Pa, prearticular bone; PtP, process for insertion of pterygoideus muscle (homolog of tensor tympani); Q, quadrate bone (homolog of incus); QJ, quadratojugal bone; RP, retroarticular process of articular; RL, reflected lamina of angular bone; S, stapes primordium; Sa, surangular bone; Sp, splenial bone; Ty, tympanic or ectotympanic bone. (A, D, E, F, and G from Allin and Hopson 1992, with minor modifications; stapes and quadrate of E after Kermack et al. 1981; B after Allin 1986; C and I after Allin 1975, slightly modified.)

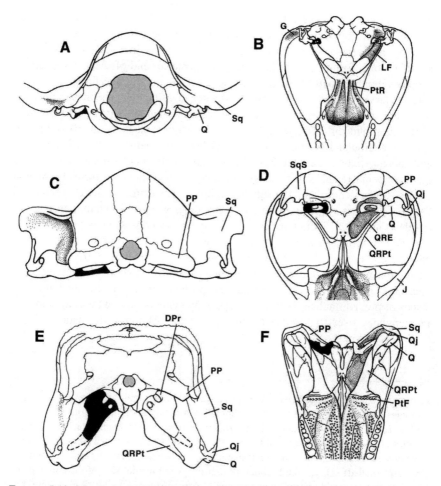

FIGURE 5.12. Morphological series of synapsid skulls in stratigraphic order, left stapes black. **A, B**: Posterior and ventral views of *Morganucodon*. **C, D**: Posterior and ventral views of *Procynosuchus*. **E, F**: Posterior and ventral views of *Dimetrodon*. DPr, dorsal process of stapes; LF, lateral flange of pro-otic; PP, paroccipital process of opisthotic; PtF, pterygoid flange; PtR, pterygopalatine ridge; QRE, quadrate ramus of epipterygoid (alisphenoid); QRPt, quadrate ramus of pterygoid; SqS, squamosal sulcus (probably for external auditory meatus). (Other abbreviations as in Figure 5.11.) (From Allin and Hopson 1992, slightly modified.)

radical structural transmogrification has been quite well established at the descriptive level, much remains to be understood as to the underlying functional reasons and developmental mechanisms (Mallo 2001).

Therapsids are thought to have arisen from sphenacodontian pelycosaurs (Figs 5.11H,I and 5.12E,F), the earliest animals to show a characteristic configuration of the postdentary part of the mandible that becomes more elaborate in therapsids and is recognizable in homologous ear stuctures of existing mammals. Part of this complex is a flange of the angular bone called the reflected lamina. Seen in no other vertebrates, this flange partly overlies a smooth-surfaced depression on the lateral side of the body of the angular bone called the external fossa. Posteroventrally, the fossa extends onto a ventrally projecting process of the articular bone termed the retroarticular process, although it may not be homologous with the posteriorly projecting process by that name in sauropsids that serves as a lever for a jaw-opening muscle, the depressor mandibulae (Presley 1989). This muscle is absent in mammals and is probably not homologous with the posterior belly of the digastric, as Parrington (1979) believed. There is a gap separating the posterodorsal margin of the reflected lamina from the margin of the external fossa, which has been termed the angular gap by Allin (1975). In cynodonts and early mammals it becomes clear that the reflected lamina is the homolog of the posteroventral limb of the mammalian tympanic annulus and the body of the angular bone corresponds to its anterodorsal limb, so the pars tensa of the mammalian tympanic membrane spans the homolog of the angular gap. It is often stated that the retroarticular process is homologous with the manubrium of the malleus, but it may be better considered to correspond to the neck of the malleus and manubrial base, the manubrium itself being a newer outgrowth.

Various opinions have been expressed as to what occupied the space delimited by the reflected lamina laterally and the external fossa medially, which Allin (1975) called the angular cleft. Romer and Price (1940) and Barghusen (1968) showed that no jaw muscle is likely to have entered the cleft from above, but that a good case can be made for an adductor mandibulae internus ("reptilian pterygoideus") muscle wrapping around into the fossa from below in sphenacodonts. However, the skeletal configuration in many therapsids does not favor such a muscle filling the cleft. Other authors have proposed that the cleft was occupied, in therapsids at least, by an air-filled diverticulum (recessus mandibularis) from a preexisting middle ear cavity or directly from the pharynx, and that overlying soft tissues became the mammalian eardrum, at least its pars tensa. Figure 5.13 illustrates this hypothesis (B) and a less likely alternative (A).

For many decades it was generally believed that pelycosaurs had a tympanic membrane homologous with that of labyrinthodonts, frogs, and sauropsids, located posterior to the quadrate. The arguments can be followed in the work of Westoll (1945), Gregory (1951), Watson (1953), Shute (1956), Hopson (1966), and Parrington (1979). However, with the idea that labyrinthodonts do not comprise a natural group, that the ear structure seen in temnospondyls is not primitive for tetrapods, and that primitive amniotes (including synapsids), like the earliest tetrapods, did not possess a tympanic membrane, many of these former scenario-based ideas have been superseded. However, the origin of the mammalian tympanum still remains to be explained.

Allin (1975) suggested that the thin tissues overlying the recessus mandibularis would have served for reception of low-frequency aerial sound early in synapsid history, incipiently in pelycosaurs but much more effectively in the-

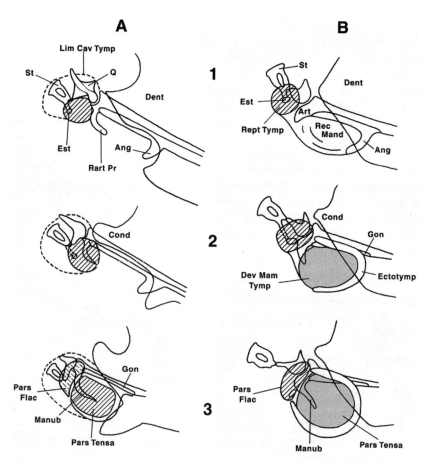

FIGURE 5.13. Contrasting theories of evolution of the mammalian middle ear from an advanced premammalian cynodont. **A**: Hopson's (1966) reconstruction for a tritylodont (stage 1), *Morganucodon* (2), and a Jurassic mammal (3). A postquadrate tympanum is visualized as the source of the pars tensa of the mammalian eardrum, becoming attached to the articular bone and setting it into vibration (as a malleus) following the evolution of a new (dentary-squamosal) jaw joint and later becoming supported by the reflected lamina of the angular bone when the latter shifted posteriorly; the pars flaccida is seen as a mammalian novelty. **B**: Westoll's (1945) theory represented by comparable stages. A postquadrate tympanum is visualized as the source of the pars laccida, whereas the pars tensa evolved by modification of the tissues external to an air space associated with the angular bone. Allin (1975 and later) agrees with Westoll as to the origin of the pars tensa, but favors a much earlier auditory role for the recessus mandibularis. Ang, angular; Art, articular; Cond, condyle of dentary; Dent, dentary; Dev Mam Tymp, developing mammalian (postdentary) tympanum; Est, extrastapes or tympanic process; Gon, goniale (prearticular); Lim Cav Tym, limits of cavum tympani; Rart Pr, retroarticular process; Rec Mand, recessus mandibularis; Rept Tymp, "reptilian" (postquadrate) tympanum; St, stapes. (From Hopson 1966.)

rapsids, and that selection for auditory improvement was a significant factor in the evolution of the mandible and jaw suspension, especially in cynodonts. He considered the large, thin reflected lamina of many therapsids to be part of a functional tympanum, its vibrations being transmitted to the rest of the postdentary unit, quadrate, and stapes. He also questioned the existence of a postquadrate eardrum in therapsids. The strongest evidence for such a tympanum had been the depression or groove on the posterolateral aspect of the therapsid squamosal (which he called the squamosal sulcus) that ended posterior to the quadrate, often at a concave ventral lip of the squamosal or paroccipital process. Allin made the tentative suggestion that the sulcus may have housed a depressor mandibulae muscle rather than an external auditory meatus, as a similar sulcus does in turtles. However, he subsequently accepted that the squamosal sulcus did, indeed, contain a meatus and that some sort of postquadrate tympanum may have been present, either separate from the postdentary tympanum (and receiving a different range of frequencies) or continuous with it as part of a large sound-receptive surface (Allin 1986, 1999; Allin and Hopson 1992). A tympanic process of the stapes may have projected across the postquadrate drum to its center, or an anterior hyoid cornu that retained its primitive continuity with the stapes may have served this role. The persisting dorsal process of the early cynodont stapes is consistent with a persisting hyostapedial connection. In mammals the dorsal process of the stapes and the hyoid process become separated from the stapes ontogenetically, bonding to the petrosal as the tympanohyal and stylohyal and probably assisting in support of the tympanic annulus during its detachment from the mandible (Goodrich 1930; Sanchez-Villagra et al. 2002).

5.1.1 Basal Synapsids (Pelycosaurs)

Pelycosaurs (Figs. 11H,I and 5.12E,F) have no evident skeletal attachment site for a tympanic membrane. Their stapes is massive, with a proximal dorsal process, and served a mechanical role as a supportive strut between the quadrate and braincase, in addition to whatever acoustic role it may have played. Furthermore, it continued to support the hyoid apparatus, as shown by a ventral rugosity for attachment of the anterior hyoid cornu (Romer and Price 1940). As is true of all nonmammalian synapsids, there is no skeletal auditory bulla so the ventral limit of the tympanic cavity, if any cavity was present, cannot be determined. There is no evidence of a stapedial process that may have projected to an eardrum, if such existed, and a mammal-like tympanic membrane confined to the small angular gap of sphenacodonts would be functionally implausible. Finally, there is no indication of cochlear enlargement suggestive of enhanced sensitivity, frequency range, and discrimination of sound frequencies. As in archaic sauropsids, it is unlikely that a true tympanic ear existed, although the ability to hear low-frequency substrate-borne sounds can safely be assumed and it is likely that these animals could detect loud low-frequency airborne sounds (probably well below 10,000 Hz, more likely nearer 1,000 Hz, and most sensitively in small species and juveniles) even in the absence of a refined tympanic membrane.

5.2 Therapsids

The most primitive therapsids, such as the biarmosuchids, are much like sphenacodont pelycosaurs, but the thin reflected lamina is larger and overlaps the external fossa to a greater extent. The external fossa and angular cleft are also enlarged. There is some simplification of the articulation of the postdentary bones with the rest of the mandible, and of the quadrate to the rest of the skull. Also, the quadrate is somewhat reduced in height and mass. Later therapsids carry these trends further.

The horizontally oriented stapes of nonmammalian therapsids retains a dorsal process, which has shifted distally (and has been mistaken for a tympanic process in some instances). In some genera there is evidence of a persistent connection to the anterior hyoid cornu. A foramen for the stapedial artery is retained and enlarged in many therapsids but lost in others. A squamosal sulcus is present, as mentioned, probably for an external auditory meatus associated with a postquadrate eardrum comparable to that of reptiles and birds but attached to the margin of the paroccipital process or the squamosal, rather than to the quadrate (which usually has no suitable concavity). A laterally open depression seems more probable than a tubular meatus, for primitive therapsids. There is no convincing evidence of a tympanic process of the stapes in any synapsid, but being cartilaginous such a process would not be likely to fossilize.

The dinocephalians were large herbivores and carnivores, whereas the anomodonts were herbivores of various sizes. The group includes the dicynodonts. In most of these animals the reflected lamina and external fossa are both enlarged, the rim of the lamina generally being separated from the margin of the fossa by an elongate, narrow angular gap, but in some dicynodonts the dorsal part of the lamina is reduced and the gap is wide. The angular cleft between lamina and fossa is usually narrow mediolaterally but in some dicynodonts it is deep. The short but stout stapes of dinocephalians and anomodonts generally lacks a foramen for the stapedial artery and meets the quadrate abruptly (without the cartilaginous distal part of the stapes that appears to have been present in pelycosaurs), and in at least one dicynodont an ossified hyoid cornu (ceratohyal) articulates with it.

5.3 Nonmammalian Theriodonts

The noncynodont theriodonts are the gorgonopsians and therocephalians. The former are rather large saber-toothed carnivores, as are primitive therocephalians. Advanced therocephalians are small to medium-sized mammals that were carnivores and herbivores (bauriamorphs). All have a mandible consisting of distinct dentary and postdentary moieties; a prominent coronoid process of the dentary bone; a broad, thin, ridged or corrugated reflected lamina; and a large external fossa. The angular gap is long and rather narrow in therocephalians, broader in gorgonopsians. The latter had a very mobile quadrate bone (Kemp 1982) and a long, forwardly curved retroarticular process.

Primitive cynodonts such as *Procynosuchus* (Figs. 5.11F,G and 5.12C,D) are similar in jaw structure to therocephalians, with a very large angular cleft, but three or more lineages of later premammalian cynodonts independently evolved a smaller postdentary region, reduced in height until it is rod-like in the later representatives, supported dorsally by an enlarged posterior process of the dentary. The reflected lamina is reduced in height and the angular gap greatly widened. The coronoid process is much higher, and a large, newly evolved masseter muscle is inserted over much of the posterior part of the dentary. Reorientation of muscle forces probably had the effect of reducing pressures borne by the jaw joint even as bite forces underwent enhancement by enlargement of muscles (Crompton 1963, 1994). Jaw muscles inserting on the postdentary moiety of the mandible became reduced in size (the adductor mandibulae internus, precursor of the tensor tympani and tensor veli palatini in mammals), or transferred to the dentary (the precursor of the medial pterygoid muscle, and the adductor mandibulae externus, precursor of the mammalian temporalis, masseter, and lateral pterygoid muscles) (Barghusen 1968, 1986). The postdentary unit is ensconced in a smooth-walled trough on the dentary, and the dentary-postdentary interface was clearly flexible in life. The quadrate also is diminished in size and, along with the small adjoined quadratojugal bone, flexibly articulated with notches in the squamosal. A new supplementary jaw articulation is present between a boss on the surangular and a facet on the squamosal (Crompton 1972), reducing forces borne by the quadrate. The stapes remains fairly long but is light, with a large foramen (much larger than necessary for the stapedial artery) and a dorsal process. In some genera, such as *Diademodon* and *Exaeretodon*, the squamosal sulcus is very large, and may have housed an external acoustic meatus that extended to a postdentary tympanum (unlike that of noncynodont therapsids). The parasphenoid ceases to participate in forming the fenestra vestibuli, but the round window remains confluent with the jugular foramen. An enlarged cochleolagenar recess (incipient cochlear canal) is present (Quiroga 1979; Allin 1986; Luo 2001), but no bulging cochlear promontory. The most derived nonmammalian cynodonts, tritylodonts and tritheledonts, are in most relevant regards much like stem mammals, tritheledonts even having a dentary-squamosal jaw articulation (of unique form) lateral to the primitive joint (Crompton 1963; Allin and Hopson 1992). Because the articular-quadrate joint of tritylodonts is oriented obliquely (Sues 1986), there must have been movement of the quadrate at its articulation with the paroccipital process, the postdentary unit at its interface with the dentary, or both, as the mandible was depressed and elevated. Tritylodonts and tritheledonts have narrow lateral grooves posterior to the choanae, bounded medially by prominent pterygopalatine ridges. These grooves may have represented the anterior parts of eustachian passages (Barghusen 1986; Allin and Hopson 1992). Similar grooves and ridges are present in early mammaliaforms but are absent in later mammals.

5.4 Basal Mammaliaforms (Stem Mammals)

Early mammaliaforms (Rowe 1988) such as *Morganucodon* (Figs. 5.10, 5.11D,E, and 5.12A,B), are tiny insectivorous creatures of the Late Triassic and

Early Jurassic. Most fossils of these and later Mesozoic mammals are isolated teeth or jaw fragments with the postdentary bones or ossicles missing. Fortunately, a few more complete mandibles and skulls have been discovered. A large dentary-squamosal joint, apparently displacing the pre-existing surangular-squamosal joint, is present lateral to the articular-quadrate joint. The diminutive postdentary elements and quadrate of *Morganucodon*, the best known genus (Kermack et al. 1981), show striking similarities to homologous entities in extant mammals (especially fetal stages). The postdentary bones almost certainly supported a tympanic membrane and participated in reception of aerial sound (Kermack 1983). The long, delicate reflected lamina is medial to a crescentic depression on the inner aspect of the dentary bone, perhaps marking the terminus of an external acoustic meatus. Probably the entire postdentary unit vibrated as a unit, the angular (tympanic) bone functioning as an ossicle rather than a stationary support for the eardrum, as appears also to be true of the tympanic bone in living monotremes (platypus and echidnas), which is synostosed to a sturdy anterior process of the malleus (Novacek 1993). It is not known whether the retroarticular process was continued forward by a slender manubrium. As in earlier synapsids, a groove for a persistent Meckel's cartilage runs forward between the ventral border of the surangular and the adjacent prearticular and angular, then continues along the medial side of the dentary to the mandibular symphysis. The quadrate has lost contact with the squamosal and is supported by the rounded posterolateral end of the paroccipital process. It articulates with the stapes by a new process homologous with the long crus of a modern incus (similar features are seen in tritylodonts) (Luo and Crompton 1994). The opisthotic and pro-otic bones have synostosed to form a unified petrosal or periotic bone as in later mammals, rigidifying the bony surroundings of the labyrinth, and the fenestra cochleae (round window) are fully separated from the jugular foramen. The bony cochlea has a finger-like shape, similar to that of a bird, and a conspicuous cochlear promontory bulges ventrolaterally (Luo et. al. 1995).

Most living mammals have a long external auditory meatus and a mobile pinna, unlike other vertebrates. The pinna is valuable for detecting the direction from which sounds emanate, by diffraction of sound waves. For an object to diffract sound of a given wavelength significantly, its width must be at least one-tenth of the wavelength. Airborne sound of 1,048 Hz (two octaves above middle C) cannot be significantly diffracted by a pinna less than 30 mm in diameter. At 10,000 Hz, roughly the upper limit audible to living reptiles, the minimum useful auricular width would be about 3 mm (about the size of the pinna of a housemouse). Thus, for mouse-sized animals, pinnae are most effective for sounds of higher frequencies than most nonmammals can readily hear, especially frequencies too high for most of them to detect at all. *Morganucodon* may well have been able to hear sounds of high enough frequencies that pinnae of reasonable size would have been useful (Rosowski 1992). No fossil pinnae of Mesozoic mammals or cynodonts are known, regrettably.

Hadrocodium, a tiny Early Jurassic mammaliaform with a surprisingly capacious braincase, lacks a postdentary trough on the mandible (Luo et al. 2001a).

The postdentary bones may have persisted as mandibular components with simplified attachments to the dentary, or they may have become fully detached from the jaw as purely auditory structures many millions of years earlier than in any other known mammal. Luo et al. (2001a) favor the latter interpretation and suggest that during development separation from the dentary may have been encouraged by traction exerted by the expanding brain, an epigenetic mechanism proposed for later mammals by Rowe (1996).

At the opposite extreme, the dentary of *Chronoperates*, described by Fox et al. (1992) as a nonmammaliaform cynodont that survived in the Paleocene, retains markings suggestive of persistent attachment of postdentary elements, many millions of years later than in any other known case.

5.5 Mammals

In addition to the extant monotremes and therian mammals (marsupials or metatherians and placentals or eutherians), there are several clades of largely or entirely Mesozoic mammaliaforms that are generally considered to be true mammals. These include the triconodontids, amphilestids, gobiconodontids, multituberculates, symmetrodonts, and dryolestoids. Some of these groups retained the Meckelian groove on the dentary but not the postdentary trough, indicating that the postdentary elements had either vanished or, more probably, become separated from the mandible as purely auditory structures. Retention as ossicles is strongly supported for triconodontids by cranial evidence, although the ossicles themselves have not been found. In the case of the multituberculates, actual ossicles quite similar to those of monotremes have now been described (Meng and Wyss 1995; Hurum et al. 1996; Rougier et al. 1996). Archaic therians such as *Amphitherium* (Allin and Hopson 1992) and *Tendagurutherium* (Heinrich 1998) have markings on the dentary that suggest incomplete detachment of postdentary bones. Wang et al. (2001) have made the unexpected discovery of entire skulls of two Early Cretaceous gobiconodontids (*Gobiconodon* and *Repenomamus*) with an ossified middle portion of Meckel's cartilage. This is a rather substantial rod that has a tapered anterior extremity occupying a matching groove on the dentary bone and a somewhat expanded posterior end that appears to have articulated with the ventral rim of the lateral flange of the petrosal. Whether this accessory jaw articulation was coaxial with the dentary-squamosal joint is unclear; if not, some flexibility of the articulation of the rod with the dentary would be necessary. Ossicles are not preserved but there is an apparent fossa incudis (crus breve attachment site). Presumably the anterior ligament of the malleus connected to the posterior end of the ossified cartilage. It is possible that a similar Meckelian bone existed during a transitional stage in the evolution of extant mammals, later being replaced by the sphenomandibular ligament.

By one line of reasoning, the trends observed in several cynodont lineages that culminated in the separation of the postdentary elements from the mandible were impelled by selective forces unrelated to hearing, such as increasing masticatory capabilities, and only when these elements had become essentially vestigial could they be actuated by airborne sound waves, assume an auditory role, and enter the middle ear (Parrington 1979). By another line of reasoning (Allin 1975), the postdentary elements served as middle ear components throughout cynodont history, and auditory adaptation was an important causal factor in the observed trends (reduction in mass of the postdentary bones and quadrate, loosening of their attachments to the dentary and cranium, enlargement and reconfiguration of the dentary, shifting of muscle insertions from the postdentary bones onto the dentary, and reorientation of muscle forces), and contact between the dentary and squamosal permitted ultimate full segregration of jaw functions from auditory functions. Separation of the postdentary bones from the mandible as components of the definitive mammalian middle ear may have happened independently in two or more mammalian lineages, including those leading to living therians and monotremes. The cladistic analysis of Luo et al. (2001b) implies this possibility, if detachment was incomplete in certain early relatives (*Henkelotherium* and *Peramus*), which are closer to the therians, and *Ausktribosphenos*, which is closer to the monotremes (Rich et al. 1997), as suggested by markings on the dentary. This conclusion is strongly supported by a newly discovered Early Cretaceous monotreme dentary with a well-defined trough and markings indicative of persisting attachment of postdentary bones (T. H. Rich et al. in preparation). However, Rougier et al. (1996) favor a single detachment event in the common ancestor of all mammals having the definitive mammalian middle ear.

The middle ear of Cretaceous marsupials and placentals resembled that of existing mammals with a loosely suspended ectotympanic, such as opossums and shrews. Several specimens have the tympanic annulus preserved but not the ossicles. However, a good malleus with a lamina, a long thin anterior process that had extensive contact with the tympanic, a forwardly directed manubrium, and a small head has been described for a Late Cretaceous eutherian (McKenna et al. 2000). In most later therians the tympanic bone fuses to the petrosal and squamosal and the anterior process of the malleus becomes much reduced. In some of these the head of the malleus becomes bulbous. Many therian mammals evolved an auditory bulla (bony or cartilaginous ventral wall of the middle ear cavity), using various different skeletal sources, presumably to reduce fluctuations in the volume and pressure of the middle ear cavity (Novacek 1993).

Extant therians differ from monotremes and archaic mammalian groups in having a much longer, fully coiled cochlea with primary and secondary bony spiral laminae to provide rigid support for the basilar membrane. Late Cretaceous marsupials and placentals already had these traits (Meng and Fox 1995a–c; Fox and Meng 1997), indicating enhanced auditory sensitivity (increased numbers of hair cells) as well as the ability to hear and discriminate an expanded range of frequencies, including those far above 10,000 Hz (in contrast to monotremes and all other tetrapods). Specialists in echolocation, bats and odontocete whales such as porpoises, have an exceedingly narrow proximal space between primary and secondary spiral laminae and this was already true in early fossil odontocetes (Geisler and Luo 1996). The petrosals and ectotympanics of modern whales are very large, dense, fused to one another, and isolated from the rest of the skull. Their ossicles are large and uniquely modified in shape. There is a good fossil record of these and associated structures (Oelschlager 1987; Luo and Marsh 1996; Luo and Gingerich 1999). Monotremes retain a lagena at the distal end of their cochlear duct but living therians do not. The slightly expanded end of the bony cochlear canal of the multituberculate *Nemegtbaatar* (Hurum 1998) hints of the retention of a lagena.

6. Summary

Among amniotes, a number of threads can be followed, showing changes to the inner- and middle-ear structures presumably associated with increased terrestrial hearing acuity:

- From its origin as a large hole in the braincase sidewall involving not only otic but occipital, sphenoidal, and dermal components, the FV reduces in size until it lies entirely surrounded by otic elements. This can be followed independently in synapsids, neodiapsids, and parareptiles. In neodiapsids, though the process had begun in the early forms, it does not appear to have been completed by the time lepidosaurs and archosaurs had split from each other. Each of these two groups seems to have completed it separately. By contrast, in amphibians the FV sometimes retains the parasphenoid in its margin, though the otic bones are often difficult to separate from the occipital ones.
- A pressure-relief window is critical to the efficient functioning of a tympanic ear. Among modern forms, a tympanic ear is almost always associated with a dedicated hole for the PRW. In early forms the perilymphatic sac appears to have shared a single foramen (known as the metotic foramen) with one or more cranial nerves and may have used this opening for pressure relief. The mammals, squamates, crocodiles, birds, and dinosaurs developed a dedicated PRW independently. The only other modern amniote group with a tympanic system, the turtles, use an alternative pressure-relief mechanism, the reentrant fluid system.
- Some other features of the inner ear are also amenable to study in fossils. The medial wall of the otic capsule becomes increasingly well ossified among mammals, squamates, and crocodiles. Ossification of the walls surrounding the vestibular system undergoes parallel increase starting with full enclosure of the semicircular canals, and in mammals, for example, it ends with the coils of the cochlea being echoed by their bony housing. It seems likely that this ossification increase may be associated with improved hearing acuity. The presence of bony spiral laminae and the narrowness of the space between primary and secondary laminae in the basal part of the cochlea provide evidence that Cretaceous marsupials and placentals already were capable of hearing higher sound frequencies than sauropsids, and that early odontocete whales

already could hear ultrasonic frequencies, indicating the use of echolocation. Some early archosaurs and lepidosaurs, despite lacking a separate PRW and having otic capsules that remained incompletely ossified, nevertheless appear to have had a middle ear cavity and tympanic membrane, judging from the quadrate bone. The acuity of such a system may not have been great, and *Sphenodon* may provide a good model not only for its capability but also its soft tissue anatomy.

- Reduction of the diameter of the stapedial shaft may correlate with a reduced role as a structural brace between quadrate and braincase. In diapsids, the paroccipital process develops between the otic capsule and the posterodorsal corner of the skull, and it may be significant that the stapes becomes a slender rod in concert with the outgrowth of the paroccipital process. In synapsids, the occipital plate becomes sutured to the skull roof, and later an additional connection develops between the palate and braincase, both factors possibly reducing reliance on the stapes as a structural element.
- Soft tissue structures rarely fossilize, but reasonable inferences can often be made about them on the basis of associated skeletal structures and development. A true tympanic membrane probably originated convergently within the sauropsids and synapsids, by modification of primitively unspecialized tissues behind and below the jaw that served as a crude tympanum, with vibrations reaching the stapes via the anterior hyoid cornu and also via the mandible and quadrate. For unknown reasons, different areas were emphasized for sound reception in sauropsids and synapsids.

Though it is true that the neurological, physiological, and behavioral aspects of hearing in tetrapods cannot be seen in fossil material, this chapter shows that the evolution of more structural aspects of the ear regions can be traced than has previously been realized. Combined with a well-supported phylogeny and neontological studies, the story of the evolution of tetrapod ears and hearing can be seen as a rich and complex one.

References

Ahlberg PE, Clack JA, Luksevics E (1996) Rapid braincase evolution between *Panderichthys* and the earliest tetrapods. Nature 381:61–64.

Allin EF (1975) Evolution of the mammalian middle ear. J Morph 147:403–437.

Allin EF (1986) The auditory apparatus of advanced mammal-like reptiles and early mammals. In: Hotton N III, MacLean PD, Roth JJ, Roth E (eds) The Ecology and Biology of Mammal-like Reptiles. Washington, DC: Smithsonian Institution Press, pp. 283–294.

Allin EF (1999) Hearing and positional sense. In: Singer R (ed) Encyclopedia of Paleontology, vol 1. Chicago: Fitzroy Dearborn, pp. 554–561.

Allin EF, Hopson JA (1992) Evolution of the auditory system in Synapsida ("mammal-like reptiles") as seen in the fossil record. In: Webster DB, Popper AN, Fay RR (eds) The Evolutionary Biology of Hearing. New York: Springer-Verlag, pp. 587–614.

Barghusen HR (1968) The lower jaw of cynodonts (Reptilia, Therapsida) and the evolutionary origin of mammal-like adductor jaw musculature. Postilla 116:1–49.

Barghusen HR (1986) On the evolutionary origin of the therian tensor veli palatini and tensor tympani muscles. In: Hotton N III, Mac Lean PD, Roth JJ, Roth EC (eds) The Ecology and Biology of Mammal-like Reptiles. Washington, DC: Smithsonian Institution Press, pp. 253–262.

Beaumont EI, Smithson TR (1998) The cranial morphology and relationships of the aberrant Carboniferous amphibian *Spathicephalus mirus* Watson. Zool J Linn Soc 122: 187–209.

Bolt JR, Lombard RE (1985) Evolution of the amphibian tympanic ear and the origin of frogs. Biol J Linn Soc 24:83–99.

Caldwell MW (1999) Squamate phylogeny and the relationships of snakes and mosasaurs. Zool J Linn Soc 125:115–147.

Caldwell MW, Lee MSY (1997) A snake with legs from the marine Cretaceous of the Middle East. Nature 386:705–709.

Carlström D (1963) A crystallographic study of vertebrate otoliths. Biol Bull 125:124–138.

Carroll RL (1987) Vertebrate Paleontology and Evolution. New York: W.H. Freeman and Co.

Carroll RL, Currie PJ (1975) Microsaurs as possible apodan ancestors. Zool J Linn Soc 57:229–247.

Clack JA (1989) Discovery of the earliest-known tetrapod stapes. Nature 342:425–427.

Clack JA (1992) The stapes of *Acanthostega gunnari* and the role of the stapes in early tetrapods. In: Webster D, Fay R, Popper AN (eds) Evolutionary Biology of Hearing. New York: Springer-Verlag, pp. 405–420.

Clack JA (1996) Otoliths in fossil coelacanths. J Vert Paleont 16:168–171.

Clack JA (1997) The evolution of tetrapod ears and the fossil record. Brain Behav Evol 50:198–212.

Clack JA (1998) The neurocranium of *Acanthostega gunnari* and the evolution of the otic region in tetrapods. Zool J Linn Soc 122:61–97.

Clack JA (2001) The otoccipital region—origin, ontogeny and the fish-tetrapod transition. In: Ahlberg PE (ed) Major Events in Early Vertebrate Evolution. London: Systematics Association Symposium Volume, pp. 392–505.

Clack JA (2002a) An early tetrapod from Romer's gap. Nature 418:72–76.

Clack JA (2002b) Gaining Ground: The Origin and Evolution of Tetrapods. Bloomington, IN: Indiana University Press.

Clack JA (2002c) Patterns and processes in the early evolution of the tetrapod ear. J Neurobiol 53:251–264.

Clack JA (2003) A new baphetid (stem tetrapod) from the Upper Carboniferous of Tyne and Wear, UK, and the evolution of the tetrapod occiput. Can J Earth Sci 40:483–498.

Clack JA, Ahlberg PE, Finney SM, Dominguez Alonso P, Robinson J, Ketcham RA (2003) A uniquely specialized ear in a very early tetrapod. Nature 425:65–69.

Crompton AW (1963) On the lower jaw of *Diarthrognathus* and the origin of the mammalian lower jaw. Proc Zool Soc Lond [B] 140:697–753.

Crompton AW (1972) The evolution of the jaw articulation of cynodonts. In: Joysey KA, Kemp TS (eds) Studies in Vertebrate Evolution. Edinburgh: Oliver and Boyd, pp. 231–251.

Crompton AW (1994) Masticatory function in non-mammalian cynodonts and early mammals. In: Thomason JJ (ed) Functional Morphology in Vertebrate Paleontology. Cambridge: Cambridge University Press, pp. 55–75.
DeBraga M, Rieppel O (1997) Reptile phylogeny and the interrelationships of turtles. Zool J Linn Soc 120:281–354.
Evans SE (1986) The braincase of *Prolacerta broomi* (Reptilia: Triassic). Neues Jahrb Geol Palaontol Abh 173:181–200.
Evans SE (1987) The braincase of *Youngina capensis* (Reptilia: Diapsida; Permian). Neues Jahrb Geol Palaontol Mh 1987:193–203.
Fox RC, Meng J (1997) An X-radiographic and SEM study of the osseous inner ear of multituberculates and monotremes (Mammalia): implications for mammalian phylogeny and the evolution of hearing. Zool J Linn Soc 121:249–291.
Fox RC, Youzwyshyn GP, Kraus DW (1992) A post-Jurassic mammal-like reptile from the Paleocene. Nature 358 (6383):233–235.
Gaffney ES (1990) The comparative osteology of the Triassic turtle *Proganochelys*. Bull Am Mus Nat Hist 194:1–263.
Gauldie RW, Dunlop D, Tse J (1986) The remarkable lungfish otolith. N Z J Mar Freshw Res 20:81–92.
Geisler JH, Luo Z (1996) The petrosal and inner ear of *Herpetocetus* (Mammalia, Cetacea) and their implication for the phylogeny and hearing of archaic mysticetes. J Paleont 70:1045–1066.
Godfrey SJ, Fiorillo AR, Carroll RL (1987) A newly discovered skull of the temnospondyl amphibian *Dendrerpeton acadianum* Owen. Can J Earth Sci 24:796–805.
Goodrich ES (1930) Studies on the Structure and Development of Vertebrates. London: Macmillan.
Gow CE (1975) The morphology and relationships of *Youngina capensis* Broom and *Prolacerta broomi* Parrington. Paleont Afr 18:89–131.
Gower DJ, Weber E (1998) The braincase of *Euparkeria*, and the evolutionary relationships of birds and crocodiles. Cambridge University Press Biol Rev 73:367–412.
Gower D, Wilkinson M (1996) Is there any consensus on basal archosaur phylogeny? Proc R Soc Lond [B] 263:1399–1406.
Gregory WK (1951) Evolution Emerging. New York: Macmillan.
Heaton MJ (1979) The cranial anatomy of primitive captorhinid reptiles from the Pennsylvanian and Permian of Oklahoma and Texas. Bull Okl Geol Surv 127:1–84.
Hedges SB Poling L (1999) A molecular phylogeny of reptiles. Science 283:898–901.
Heinrich W-D (1998) Late Jurassic mammals from Tendaguru, Tanzania, east Africa. J Mamm Evol 5(4):269–290.
Hetherington TE, Jaslow AP, Lombard RE (1986) Comparative morphology of the amphibian opercularis system: 1. General design features and functional interpretation. J Morph 190:43–61.
Hopson JA (1966) The origin of the mammalian middle ear. Am Zool 6:437–450.
Hopson JA, Kitching JW (2001) A probainognathian cynodont from South Africa and the phylogeny of non-mammalian cynodonts. Bull Mus Comp Zool Harv 156:5–35.
Hurum JH (1998) The inner ear of two Late Cretaceous multituberculate mammals, and its implications for multituberculate hearing. J Mamm Evol 5:65–93.
Hurum JH, Presley R, Kielan-Jaworowska Z (1996) The middle ear in multituberculate mammals. Acta Paleont Polon 41:253–275.
Jarvik E (1980) Basic Structure and Evolution of Vertebrates, vols 1 and 2. New York: Academic Press.

Kemp TS (1982) Mammal-Like Reptiles and the Origin of Mammals. London: Academic Press.

Kermack KA (1983) The ear in mammal-like reptiles and early mammals. Acta Paleont 28:147–158.

Kermack KA, Mussett F, Rigney HW (1981) The skull of *Morganucodon*. Zool J Linn Soc 71:1–158.

Laurin M (1996) A redescription of the cranial anatomy of *Seymouria baylorensis*, the best known seymouriamorph (Vertebrata: Seymouriamorpha). Paleobios 17:1–16.

Laurin M, Reisz RR (1995) A re-evaluation of early amniote phylogeny. Zool J Linn Soc 113:165–223.

Laurin M, Reisz RR (1997) A new perspective on tetrapod phylogeny. In: Sumida S, Martin KLM (eds) Amniote Origins—Completing the Transition to Land. London: Academic Press, pp. 9–59.

Lee MSY (1997a) Pareiasaur phylogeny and the origin of turtles. Zool J Linn Soc 120: 197–280.

Lee MSY (1997b) The phylogeny of varanoid lizards and the affinities of snakes. Philos Trans R Soc Lond [B] 352:53–91.

Luo Z-X (2001) Inner ear and its bony housing in tritylodonts and implications for evolution of the mammalian ear. Bull Mus Comp Zool Harv 156(1):81–97.

Luo Z-X, Crompton AW (1994) Transformation of the quadrate (incus) through the transition from nonmammalian cynodonts to mammals. J Vert Paleont 14(3): 341–374.

Luo Z-X, Gingerich PD (1999) Transition from terrestrial ungulates to aquatic whales: transformation of the braincase and evolution of hearing. Pap Paleont Mus Paleont Univ Mich 31:1–98.

Luo Z-X, Marsh K (1996) Petrosal (periotic) and inner ear of a Pliocene Kogine whale (Kogiinae, Odontoceti): implications on relationships and hearing evolution in toothed whales. J Vert Paleont 16(2):328–348.

Luo Z-X, Crompton AW, Lucas S (1995) Evolutionary origins of the mammalian promontorium and cochlea. J Vert Palaeont 15:113–121.

Luo Z-X, Crompton AW, Sun A-L (2001a) A new mammaliaform from the Early Jurassic and evolution of mammalian characteristics. Science 292:1535–1540.

Luo Z-X, Cifelli RL, Kielan-Jaworowska Z (2001b) Dual origin of tribosphenic mammals. Nature 409:53–57.

Mallo M (2001) Formation of the middle ear: recent progress on the developmental and molecular mechanisms. Dev Biol 231:410–419.

Manley GA (2002) Evolution of structure and function of the hearing organ of lizards. J Neurobiol 53:202–211.

Manley G, Gleich O (1992) Evolution and specialisation of function in the avian auditory periphery. In: Webster D, Fay R, Popper AN (eds) Evolutionary Biology of Hearing. New York: Springer-Verlag, pp. 561–580.

McKenna MC, Kielan-Jaworowska Z, Meng J (2000) Earliest eutherian mammal skull, from the Late Cretaceous (Coniacian) of Uzbekistan. Acta Paleont Polon 45(1):1–54.

Meng J, Fox RC (1995a) Therian petrosals from the Oldman and Milk River Formation (Late Cretaceous), Alberta, Canada. J Vert Paleont 15:122–130.

Meng J, Fox RC (1995b) Evolution of the inner ear from non-therians to therians during the Mesozoic: implications for mammalian phylogeny and hearing. In: Sun AL, Wang Y (eds) Sixth Symposium on Mesozoic Terrestrial Ecosystems and Biota, Short Papers. Beijing: China Ocean Press, pp. 235–242.

Meng J, Fox RC (1995c) Osseous inner ear structures and hearing in early marsupials and placentals. Zool J Linn Soc 115:47–71.

Meng J, Wyss AR (1995) Monotreme affinities and low frequency hearing suggested by the multituberculate ear. Nature 377:141–144.

Miles RS (1977) Dipnoan (lungfish) skulls and the relationships of the group: a study based on new species from the Devonian of Australia. Zool J Linn Soc 61:1–328.

Miller MR (1992) The evolutionary implications of the structural variations in the auditory papilla of lizards. In: Webster DB, Popper AN, Fay RR (eds) Evolutionary Biology of Hearing. New York: Springer-Verlag, pp. 463–488.

Milner AR (1988) The relationships and origin of the living amphibians. In: Benton MJ (ed) The Phylogeny and Classification of the Tetrapods, vol 1. Oxford: Clarendon Press, pp. 59–102.

Milner AR, Sequeira SEK (1994) The temnospondyl amphibians from the Viséan of East Kirkton, West Lothian, Scotland. Trans R Soc Edinb Earth Sci 84:331–362.

Novacek MJ (1993) Patterns of diversity in the mammalian skull. In: Hanken J, Hall BK (eds) The Skull, vol 2. Chicago: University of Chicago Press, pp. 438–546.

Oelschlager HA (1987) *Pakicetus inachus* and the origin of whales and dolphins (Mammalia, Cetacea). Gegenb Morphol Jahrb 133:673–685.

Olson EC (1966) The relationships of *Diadectes*. Fieldiana Geol 14:199–227.

Padian K, Chiappe LM (1998) The origin and early evolution of birds. Biol Rev 73:1–42.

Parrington FR (1979) The evolution of the mammalian middle and outer ears: a personal review. Biol Rev 54:369–387.

Parrish JM (1993) Phylogeny of the Crocodylotarsi, with reference to archosaurian and crurotarsan monophyly. J Vert Paleont 13(3):287–308.

Patterson C (1984) *Chanoides*, a marine Eocene otophysan fish (Teleostei: Ostariophysi). J Vert Paleont 4:430–456.

Platt CJ (1994) Hair cells in the lagenar otolith organ of the coelacanth are unlike those in amphibians. J Morph 220:381.

Popper AN, Platt C (1996) Sensory hair cell arrays in lungfish inner ears suggest retention of the primitive patterns for bony fishes. Soc Neurosci, vol 22, Washington, DC, abstracts, p. 1819.

Presley R (1989) Ontogeny and the evolution of the mammalian jaw complex. In: Wake DB, Roth G (eds) Complex Organismal Function: Integration and Evolution in Vertebrates. New York: Wiley, pp. 53–61.

Quiroga JC (1979) The inner ear of two cynodonts (Reptilia, Therapsida) and some comments on the evolution of the inner ear from pelycosaurs to mammals. Gegenb Morphol Jahrb 125:178–190.

Rage J-C, Rocek Z (1989) Redescription of *Triadobatrachus massinoti* (Piveteau 1936) an anuran amphibian from the Early Jurassic. Palaeontographica 206:1–16.

Retzius G (1881) Das Gehörorgan der Wirbelthiere. Morphologisch-histologische Studien. I. Das Gehörorgan der Fische und Amphibien. Stockholm, Sweden: Centraldruckerei.

Rich TH, Vickers-Rich P, Constantine A, Flannery TF, Kool L, Klaveren N (1997) A tribosphenic mammal from the Mesozoic of Australia. Science 278:1438–1442.

Rieppel O, DeBraga M (1996) Turtles as diapsid reptiles. Nature 384:453–455.

Romer AS (1937) The braincase of the Carboniferous crossopterygian *Megalichthys nitidus*. Bull Mus Comp Zool Harv 82:1–73.

Romer AS, Price LI (1940) Review of the Pelycosauria. Geol Soc Am Spec Pap 28: 1–538.

Rosowski JA (1992) Hearing in transitional mammals:predictions from the middle-ear anatomy and hearing capabilities of extant mammals. In: DB Webster, RR Fay, Popper AN (eds) The Evolutionary Biology of Hearing. New York: Springer-Verlag, pp. 615–631.

Rougier GW, Wible JR, Novacek MJ (1996) Middle-ear ossicles of the multituberculate *Kryptobaatar* from the Mongolian Late Cretaceous: implications for mammaliamorph relationships and the evolution of the auditory apparatus. Am Mus Nat Hist Novit 3187:1–42.

Rowe T (1988) Definition, diagnosis and origin of Mammalia. J Vert Paleont 8:241–264.

Rowe T (1996) Coevolution of the mammalian middle ear and neocortex. Science 273: 651–654.

Sanchez-Villagra MR, Gemballa S, Nummela S, Smith KK, Maier W (2002) Ontogenetic and phylogenetic transformations of the ear ossicles in marsupial mammals. J Morph 251(3):219–238.

Sereno P (1991) *Lesothosaurus*, "fabrosaurids" and the early evolution of Ornithisichia. J Vert Palaeont 11:168–197.

Shepherd CE (1914) On the location of the sacculus and its contained otoliths in fishes. Zoologist 4:103–109, 131–146.

Shute CD (1956) The evolution of the mammalian eardrum and tympanic cavity. J Anat 90:261–281.

Smithson TR (1982) The cranial morphology of *Greererpeton burkemorani* (Amphibia: Temnospondyli). Zool J Linn Soc 76:29–90.

Sues H-D (1986) The skull and dentition of two tritylodontid synapsids from the Lower Jurassic of western North America. Bull Mus Comp Zool Harv 151(4):217–268.

Trueb L, Cloutier R (1991) A phylogenetic investigation of the inter- and intrarelationships of the Lissamphibia (Amphibia: Temnospondyli). In: Schulte H-P, Trueb L (eds) Origins of the Higher Groups of Tetrapods. Ithaca: Cornell University Press, pp. 223–314.

Walker AD (1985) The braincase of *Archaeopteryx*. In: Hecht MK, Ostrom JH, Viohl G, Wellnhofer P (eds) The Beginnings of Birds. Eichstatt: Freunde des Jura-Museums, pp. 123–134.

Walker AD (1990) A revision of *Sphenosuchus acutus* Haughton, a crocodylomorph reptile from the Elliot Formation (Late Triassic or Early Jurassic) of South Africa. Philos Trans R Soc Lond [B] 330:1–120.

Wang Y, Hu Y, Meng J, Li C (2001) An ossified Meckel's cartilage in two Cretaceous mammals and origin of the mammalian middle ear. Science 294:357–361.

Watson DMS (1953) Evolution of the mammalian ear. Evolution 7:159–177.

Weishampel DB (1981) Acoustic analysis of potential vocalisation in lambeosaurine dinosaurs. Paleobiology 7:252–261.

Westoll TS (1945) The mammalian middle ear. Nature 155:114–115.

Wever EG (1978) The Reptile Ear. Princeton: Princeton University Press.

White TE (1939) Osteology of *Seymouria baylorensis* Broili. Bull Mus Comp Zool Harv 85:325–409.

6
Evolution of the Amphibian Ear

MICHAEL SMOTHERMAN AND PETER NARINS

1. Introduction

Most amphibians have within their ears the substrate to hear efficiently underwater, underground, and in air, a talent few if any other vertebrates can lay claim to. They have achieved this by being very conservative in the nature of novel additions and specialized adaptations to their ears. Indeed, regressive events appear to be just as common as progressive trends in the evolution of the amphibian ear. As a result, the amphibian ear reflects a diverse array of basically simple, presumably reliable mechanisms of auditory transduction; mechanisms that have in essence served as the foundations of the more sophisticated hearing apparati seen in other terrestrial vertebrates. Understanding of the mechanisms of hearing in amphibians can thus offer many insights into the physical and ecological forces that have fueled the evolution of hearing in terrestrial vertebrates.

The earliest vertebrates suspected of having the capacity to hear airborne sounds were still primarily aquatic animals (Coates and Clack 1991; Clack 1997, 2002). However, as the earliest terrestrial vertebrates came to spend increasing amounts of time out of water, more specialized adaptations for hearing on land were needed, though not necessarily at the expense of continuing to hear underwater. Most modern amphibians remain similarly attached to their need for hearing underwater. Amphibians today differ chiefly from their purely aquatic ancestors in that they possess inner-ear organs specifically receptive to sound pressure, and the anatomy of the middle and inner ear has been modified to provide more efficient mechanisms for directing sound pressure changes to these organs. Amphibians differ from reptiles, birds, and mammals in terms of the complexity of the acoustic processing performed both at the auditory periphery and throughout the entire auditory system.

Collectively, the ears of amphibians exhibit an array of unique adaptations that have contributed to the enduring success of this diverse vertebrate class. Differences in the physiology of hearing can be seen within every branch of the amphibian phylogenetic tree, and range from such gross morphological differences as the absence of a middle ear down to subtle differences in hair cell stereovillar architecture. Key differences include variations in the sound transfer pathway, and the presence, position, and architectures of the various receptor organs within the ear. In many instances we can assign specific ecological significance to auditory specializations, made possible largely by the fact that amphibians occupy a diverse array of habitats and because acoustic communication has played a key role in amphibian reproductive behavior and hence their evolution.

The physiology of the amphibian auditory system has been the focus of several reviews (Capranica 1976; Wilczynski and Capranica 1984; Wever 1985; Lewis and Narins 1998; Smotherman and Narins 2000), and throughout this chapter we refer the reader to many classical and some more recent descriptions of the physics and physiology of hearing in the amphibians and other vertebrates. This chapter highlights those aspects of hearing in amphibians that offer special insight into the evolution of hearing both within the amphibians and within the context of hearing among all vertebrates.

2. Origins and Evolution of the Amphibians

The living amphibians, or Lissamphibia, should be considered all highly derived from their earliest ancestors (Fig. 6.1). The fossil record for tetrapods begins with the aquatic Acanthostega (Coates and Clack 1991). The fossil remains of Acanthostega include a primitive stapes that, if indeed auditory in function, probably evolved in support of aquatic hearing (Coates and Clack 1991; Clack 2002; Clack et al. 2003). However, its presence suggests that an ear already capable of capturing and transmitting airborne sound could have already existed in related animals around the same time. Amphibians emerged as a separate lineage from the other tetrapods sometime during the Paleozoic era, roughly 300 million years ago (Duellman and Trueb 1994). It is now generally agreed that modern amphibians, the Caudata (salamanders and newts), the Gymnophiona (caecilians), and the Anura (frogs and toads) share a monophyletic origin (Duell-

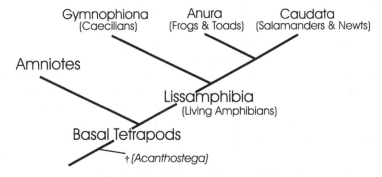

FIGURE 6.1. A phylogenetic tree of the amphibians († extinct).

man and Trueb 1994; Zardoya and Meyer 2001) with respect to the other tetrapods, and that the Caudata and Anura constitute sister groups within the amphibians (the so-called Batrachia hypothesis). Assuming a common origin, these groups apparently diverged from one another at least as early as the Jurassic period, some 200 million years ago, and the divergence of all three groups probably occurred within a very narrow time window (Zardoya and Meyer 2000). Of these, the salamanders and newts probably most closely resemble ancestral amphibian form (Duellman and Trueb 1994).

Among the Caudata may be found species that are, to varying degrees, aquatic, fossorial, or terrestrial. However, the anurans are the most abundant of the living amphibians, and have succeeded in invading almost every ecological niche, save the most inhospitable climates, through a broad array of behavioral and physiological adaptations. The tropical caecilians have adopted a long, legless trunk and seceded to either a burrowing or entirely aquatic lifestyle. With few exceptions the members of these three orders all share a common dependence on a moist or watery environment for reproduction. Some members live their entire lives underwater, some underground, some manage to survive in harshly arid environments, and some live at the tops of trees. The extent to which each species relies on an aquatic larval stage varies considerably: the well-known tadpole represents a significant stage in the life history of most frogs, whereas some terrestrial frogs and salamanders, the so-called direct-developers (Callery et al. 2001), have succeeded in nearly bypassing the free-swimming larval stage altogether by confining metamorphosis to within the egg. Some species require vocal communication to attract or compete for mates, others are entirely voiceless. All are sensitive to substrate-borne vibrations, whether in water or soil or on plants. The acoustic sensitivity of the amphibian ear is closely associated with each animal's dependence on acoustic communication, especially in the anurans. Indeed, that irrepressible *ribbit* sound, having proven itself equally attractive to physiologists as well as female frogs, has yielded much of our current insight into the interplay between the evolution of communication behavior and sensory physiology.

3. Basic Anatomy of the Amphibian Ear

The anatomy of the frog inner ear (Fig. 6.2) (Fritzsch 1998; Lewis and Narins 1998) reflects all the basic adaptations required by terrestrial vertebrates for hearing sounds traveling in air, although in many amphibians these have been modified or replaced by equally effective adaptations for hearing underwater or underground. For most species of anurans, sound enters the animal through tympanic membranes, or eardrums, located on both sides of the head. Movement of the tympanic membrane drives the middle ear apparatus, which in amphibians was long thought to consist of a single ossicle called the columella or stapes (we will use the term *stapes*), but has recently been shown also to include a small extrastapes located at the distal end of the stapes (Mason and Narins

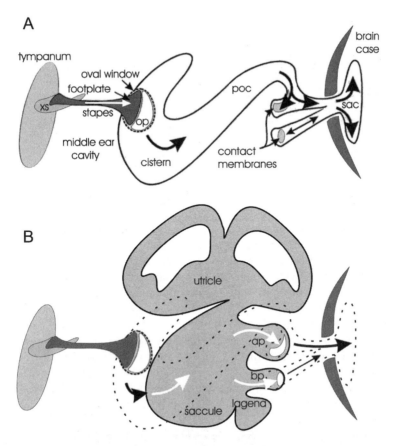

FIGURE 6.2. A schematic diagram of probable sound pathways through the ear of a frog, *Rana pipiens*. Amphibians do not have an external ear (or pinna) such as seen in mammals. Airborne sounds enter the ear via the middle-ear apparatus, which begins with the tympanum. The extrastapes (XS) is attached to the inner surface of the tympanum and is connected to the stapes at an articulated joint. A stapedial footplate and an operculum (op) impinge upon the periotic space through a lateral opening in the otic capsule, the oval window. From the lateral-most chamber of the periotic space, the cistern, sound travels through the inner ear by two separate pathways. **A**: Sound is passing along the periotic canal (POC) by the black arrows, bypassing the endolymphatic space (illustrated in B as gray area) and exiting from the otic capsule into the periotic sac within the brain case. From the sac, sound energy dissipates to the soft tissues and spinal column. In some amphibians, extensions of the sac may go so far as to provide a pathway to the contralateral ear. **B**: Sound passes into the endolymphatic space (white arrows) through the membranes of the saccular chamber, continues through to the auditory recesses, and reemerges into the periotic canal at specialized contact membranes [illustrated as white spaces within the amphibian papilla (AP) and basilar papilla (BP)] found at the distal ends of the recesses. Evidence suggests that the AP and BP receive their principal inputs via the endolymphatic pathway, whereas the POC serves as a shunt for high-intensity, low-frequency sounds that might otherwise overstimulate the auditory end organs (Purgue and Narins 2000b).

2002b). The stapes is attached medially to a membrane-covered opening on the lateral side of the otic capsule, called the oval window, through which sound first passes into the periotic space, and more specifically into a region known as the periotic cistern. For most amphibians both the stapes footplate and a second bony element called the operculum covers the oval window. The periotic cistern constitutes the beginning of two putative sound pathways through the inner ear (Wever 1985; Purgue and Narins 2000a). In the first pathway (Fig. 6.2A), sound waves propagate through the periotic space to the periotic canal, a tubular duct that passes medially around the saccule and communicates with the contact membrane of each of the auditory receptor organs just before passing out through the medial wall of the otic capsule. The other putative sound pathway (Fig. 6.2B) also begins in the periotic cistern, but here sound waves are presumed to pass directly into the endolymphatic space, specifically by entering the neighboring saccule through the thin flexible membranes separating the cistern from the saccule. From here the sound waves spread throughout the saccular chamber and enter the recesses of the auditory receptor organs, which in most amphibians are little more than evaginations of the medial wall of the saccule. By this pathway sound may directly stimulate the interior of the auditory receptor organs and, upon passing through them, reenter the periotic space through thin contact membranes. These two pathways should be considered interdependent rather than alternate possibilities, and the properties of each are suspected to dynamically influence energy flow within the other (Purgue and Narins 2000a,b). Collectively they may be considered either homoplastic or analogous to the pathways of sound transmission in other terrestrial vertebrates. However, as we will show, not a single component of these sound-transfer pathways has proven immune to modification during the evolution of amphibians. Everything from the tympanic membrane to the periotic sac exhibits at least one alternate form among the amphibians. Furthermore, alternate modes of inner-ear stimulation are not only possible but also quite common in the amphibians.

3.1 A Structural Basis for Diversity at Many Levels

Major variations in the organization of the amphibian inner ear can be categorized as concerning either how sounds enter the ear or which auditory sense organs are present to receive it. Since the very first amphibians or their immediate ancestors must be credited with the de novo creation of a middle ear and periotic canal, both essential for hearing on land, it would appear that for all living amphibians having a middle ear is the primitive condition. The stapes, or a remnant of the stapes, an oval window, and a periotic canal extending to at least one acoustic sense organ and ultimately the periotic sac have been found in all amphibians so far examined (Retzius 1881; Sarasin and Sarasin 1890; Burlett 1934; Villiers 1934; Lombard 1977; Jarvik 1980; Wever 1985; Jaslow et al. 1988; Bolt and Lombard 1992). However, the tympanum, much of the stapes, and the middle ear cavity are secondarily lost in all caecilians, many salamanders, and some frogs. Likewise, there is evidence that some of the putative acoustic sense organs may have been secondarily lost. Although all amphibians have at least one dedicated acoustic sense organ, two structures, the papilla neglecta and the basilar papilla (see below), appear to have been reduced or entirely lost in many of the Gymnophiona and Caudata (Wever 1975, 1985; Wever and Gans 1976; Lombard 1977; White and Baird 1982). Within each of the auditory sense organs the organization of the sensory epithelia, the physiology of the hair cells, and the mechanical means by which those hair cells come to be stimulated also show significant variability (Smotherman and Narins 2000).

3.2 Auditory Receptor Organs of the Inner Ear

Perhaps the most conserved auditory structure in the amphibian inner ear is the saccule, which is an essential part of the auditory system of both fish and amphibians, but on land also serves importantly as a highly sensitive seismic detector (Capranica and Moffat 1975; Narins and Lewis 1984; Christensen-Dalsgaard and Jørgensen 1988; Yu et al. 1991). The saccule is a fundamental component of the vertebrate sense of balance, but in amphibians as well as in teleost fish it is essential for the detection of approaching predators, which may be why this auditory organ exhibits the least variability of all the amphibian inner-ear sense organs.

The nomenclature of the more specialized amphibian auditory sense organs has a somewhat convoluted history. All amphibians so far studied apparently possess an auditory receptor unique to amphibians, the aptly named amphibian papilla (AP) (Retzius 1881; Sarasin and Sarasin 1890, 1892; Burlett 1934; Villiers 1934; Lombard 1977; Lewis 1978, 1984; White and Baird 1982; Wever 1985; Lewis et al. 1992; Lewis and Narins 1998). In the most primitive amphibians the AP is a small round outpocket near the utriculosaccular duct containing a patch of hair cells located on the chamber's roof, whereas in the more derived anurans the sensory epithelium is elongated rostrocaudally (Fig. 6.3). It has at times been confused with another structure, the papilla neglecta (Fig. 6.4), which among amphibians has only been found in the Gymnophiona (Sarasin and Sarasin 1890; White and Baird 1982). A structure resembling the papilla neglecta of amphibians has been described for some teleost fishes and elasmobranchs (Retzius 1881; Corwin 1977). The term *macula neglecta* may be more appropriate than *papilla neglecta*, since this sensory epithelium does not reside within a dedicated cavity or space in the same way as other auditory sense organs (White and Baird 1982).

The inner ear of all vertebrates can be divided into two halves; the highly conserved vestibular portion known as the pars superior, and the more variable auditory portion, the pars inferior. The *pars superior* normally includes the semicircular canals and the utricle, and when present the macula neglecta is

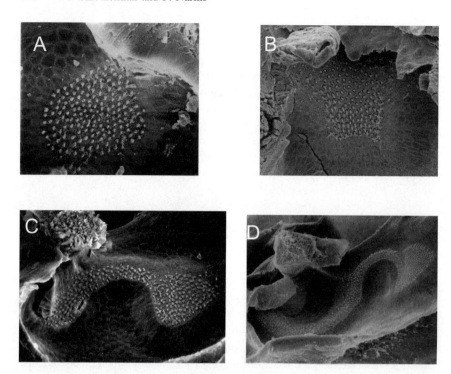

FIGURE 6.3. Scanning electron microscope (SEM) photos of the AP inner chambers from four different amphibians, including a salamander (**A**: *Ambystoma maculata*), and examples of a primitive frog (**B**: *Ascaphus truei*), an intermediate frog (**C**: *Atelopus varius*) and a frog for whom the AP sensory epithelium appears highly derived (**D**: *Kassina senegalensis*). Notice in particular the addition of the long thin extension that provides the substrate for higher frequency audition in the more derived animals. Tectorial membranes were removed to reveal the hair bundles. (Photos courtesy of E.R. Lewis.)

found within the utricular chamber. The standard definition of the AP is based on its position within the ear; as with all terrestrial vertebrate acoustic sense organs, the AP derives from an evagination from within that part of the ear known as the *pars inferior*. The AP is generally defined as a saccular out-pocket located very near the utriculosaccular duct (i.e., near the border of the pars superior and inferior), although in some salamanders the AP actually opens into the utriculosaccular duct rather than the medial wall of the saccule (Fig. 6.4) (Lombard 1977; Wever 1985). The macula neglecta of caecilians appears strikingly similar to the AP of caecilians and salamanders (Wever 1975), which has led some researchers to conclude that the AP may be derived directly from the more primitive macula neglecta (Wever 1975, 1985; Wever and Gans 1976; White and Baird 1982; Fritzsch and Wake 1988; Fritzsch 1992).

One indication that the AP has evolved as a sound pressure receiver is the presence of a thin contact membrane at the distal end of the papilla where it

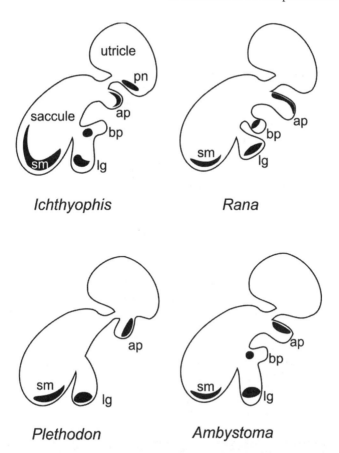

FIGURE 6.4. Major variations in the organization of the inner ear sensory structures among the amphibians. *Ichthyophis* (Gymnophiona, Ichthyophiidae) possesses a papilla neglecta (pn) and an AP (ap) that evaginate from the saccular space and a BP (bp) that evaginates from the lagenar (lg) recess. A typical advanced frog such as *Rana* (example Anura, Ranidae) possesses a notably elongated AP recess and sensory epithelium. Among the Caudata, *Plethodon* (Plethodontidae) is unusual in that its AP extends ventrally from the floor of the utricular chamber, and it appears to lack a BP. The inner ear of the salamander *Ambystoma* (Ambystomatidae) is similar to frogs except that its reduced BP extends from the lagenar recess. Saccular macula (sm). [Figures based on descriptions presented in Wever (1975, 1985), Wever and Gans (1976), Lombard (1977), White and Baird (1982), and White (1986b).]

meets the periotic canal. This contact membrane provides a low-resistance exit for the pressure waves, which is necessary for the efficient propagation of sound energy through the organ. Without such an exit, any mechanical displacements caused by sound pressure waves would depend on the compressibility of the fluid medium within the auditory organ, which is presumed to be minimal. No such connection between the macula neglecta and the periotic system has been observed.

A third auditory receptor organ known as the basilar papilla (BP, Fig. 6.5) can be found in all frogs, some salamanders, and in the basic but not the more derived caecilians (Sarasin and Sarasin 1890, 1892; Burlett 1934; Villiers 1934; van Bergeijk and Witschi 1957; Geisler et al. 1964; Wever 1975, 1985; Capranica 1976; Wever and Gans 1976; Lombard 1977; White and Baird 1982; Wilczynski and Capranica 1984; White 1986a,b; Fritzsch and Wake 1988; Fritzsch 1992). The architecture of the amphibian BP is surprisingly simple: the sensory epithelium typically consists of a patch of 50 or so small hair cells located beneath a thin tectorial membrane that is stretched like a sail across the tubular pathway of the BP recess (Fig. 6.5). Beyond the sensory epithelium lies a contact membrane through which sound may pass to or from the periotic space. The BP of caecilians and salamanders is present as a second sensory epithelium located within the lagenar chamber, or the epithelium may exist within a distinct evagination of the lagena (White and Baird 1982; White 1986a,b; Fritzsch and Wake 1988). The salamander and caecilian BP recesses extend from near where the lagena itself opens into the saccule, but nevertheless clearly open into the lagenar recess, and not the saccular chamber. In anurans a structure very similar in design to the salamander BP is found next to and tightly abutting the lagena, but in frogs the BP always appears as a separate evagination opening directly into the saccule (Lewis 1977a, 1978). Unlike the amphibian papilla, the anuran BP shows relatively little anatomical variation between different taxa. The apparent association of the amphibian BP with the lagena led Retzius (1881), de

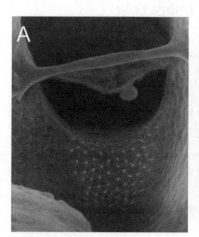 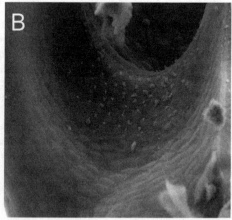

FIGURE 6.5. SEM photos of the BP sensory epithelium for two anurans, *Bufo punctatus* (**A**) and *Kassina senegalensis* (**B**). In both photos the tectorial membrane has been damaged and torn away from the sensory surface by the fixation process. (Photos courtesy of E.R. Lewis.)

Burlett (1934), and others to postulate a homology between it and the previously named BP of reptiles and birds. More recent developmental (see below) and molecular evidence tends to support this view as well (Fritzsch and Neary 1998; Fritzsch 2001; Fritzsch et al. 2002).

4. The Middle Ear: Terrestrial, Aquatic, and Fossorial Adaptations

For the first terrestrial vertebrates, hearing airborne sounds required the creation of specialized anatomical structures that could facilitate the transference of sound pressure waves from air into the animal (Lewis and Fay, Chapter 2). Normally, an enormous impedance mismatch between air and any solid body prohibits such a transfer, since the vast majority of airborne sounds are reflected by the animal's body surface. The evolution of a tympanic membrane and middle-ear ossicle directly addressed the need to provide some form of compensation for this impedance mismatch (Clack and Allin, Chapter 5). Many frogs have conspicuous, and in some cases exceptionally large, tympanic membranes. The anuran tympanic membrane exhibits considerable interspecific variations in size, and in some species, such as the bullfrog *Rana catesbeiana*, males and females may possess differently sized tympanic membranes (Noble 1931; Purgue 1997; Mason et al. 2003). The chief function of the tympanic membrane is to capture and transmit airborne sounds to the stapes; however, not all amphibians, and not even all frogs, spend enough time above ground to benefit from this. For these fossorial amphibians it appears to have been necessary to either remove the tympanic membrane completely or cover it under a dense, protective external epithelium. Apparently the presence of a tympanic membrane has been selected against in the subterranean and aquatic caecilians, salamanders, and some of the more primitive frogs, although the reason for this is unknown.

The tympanum and stapes shown in Figure 6.2 represent a typical middle ear for most anurans; however, several interesting variations have arisen among amphibians. Among the most primitive families of frogs, members of the Leiopelmatidae (for example, *Ascaphus truei*) lack all evidence of a tympanum or stapes (Stejneger 1899; Wever 1985). In *Ascaphus,* only an operculum rests against the oval window. However, in the equally primitive sister family Discoglossidae (for example *Bombina orientalis*), a tympanum, stapes, and air-filled middle ear cavity are present. Frogs in another primitive family, Pipidae, possess no external sign of an ear. Yet peeling away the skin at the expected tympanum position reveals an otherwise normal frog middle ear apparatus, including a tympanic "disk" (Wever 1985) attached to a stapes that passes through an air-filled middle ear cavity to the oval window. The Pipidae notably include the purely aquatic African clawed frog *Xenopus laevis* and the Surinam toad *Pipa pipa*. Wever (1985) speculated that in *Xenopus* the tympanic disk might have come to serve as a receiver of aquatic sounds and vibration. In fact, the tympanic disk in *Xenopus* is known to contribute significantly as a pathway for sound entering the ear while the animal is underwater; Christensen-Dalsgaard and Elepfandt (1995) found sound-induced vibrations at the disk to be 40 dB greater than those measured in the surrounding tissue. Interestingly, however, in the same study it was found that sound entering through the lung cavity sharply altered the frequency response of the tympanic disk, which highlights the significance of the body wall and lungs as an additional pathway for sound to enter the ear in aquatic frogs. It is also worth noting that frogs in the genera *Ascaphus, Bombina, Xenopus,* and *Pipa* all possess both an AP and a BP exhibiting the standard anuran architecture (Patterson 1948; Lewis 1978; Wever 1985). Therefore, it would appear that neither the absence nor the reduction of the anuran middle ear has driven concomitant changes in the anatomy of the auditory receptor organs, although this may have imposed limits on the functional significance of any further adaptations.

Although it is generally true for amphibians that the reception of airborne sounds requires a tympanic-stapedial sound-transfer pathway, it appears that salamanders and some caecilians have focused their ear to the ground by relying on nontympanic mechanisms of energy transfer to the inner ear. The salamander inner ear is known to be especially responsive to seismic stimulation, and the lungs and body wall are believed to be significantly involved in the transmission of this low-frequency stimuli into the ear (Wever 1979, 1985; Hetherington 1988, 2001; Hetherington and Lindquist 1999). A similar lung-based pathway has been reported for some frogs (Narins ct al. 1988; Ehret et al. 1990, 1994). Adult salamanders are similar to frogs in that they possess both a stapes footplate and an operculum that overlie the oval window (Kingsbury and Reed 1909; Monath 1965; Taylor 1969; Jaslow et al. 1988). Yet salamanders are similar to the gymnophionids in that the tympanum and an air-filled middle ear cavity are absent (Fig. 6.6). Typically either the stapes or an associated ligament extends anteriorly to either the quadrate or squamosal bone, whereas the operculum may provide a pathway for vibrations entering the animal via the shoulder girdle (Kingsbury and Reed 1909). Interestingly, in the caudata the opercular attachment to the oval window does not appear until during metamorphosis, suggesting an association between the function of the opercular system and the animal's emergence into the terrestrial environment (Jaslow et al. 1988). Like salamanders, the opercular apparatus in frogs may have evolved to subserve seismic sensitivity on land (Hetherington 1985, 1988; Hetherington and Lindquist 1999), or it may also enhance tympanic sound transmission by balancing pressure changes caused by breathing (Mason and Narins 2002a).

In the Gymnophiona an exceptionally large stapedial footplate completely covers an equally large oval window; however, no operculum is present (Fig. 6.6). It was previously observed that the middle ear apparatus in the Gymnophiona very much resembles the larval salamander condition (Jaslow et al. 1988), and since the presence of an operculum is considered primitive, the caecilian middle ear may represent an interesting example of *paedomorphosis*, the term applied when there is evidence that evolutionary change in the adult form has occurred due to the premature termination of specific developmental programs.

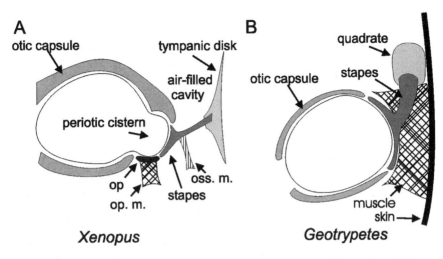

FIGURE 6.6. Two examples of major variations in the middle-ear apparatus of amphibians. (**A**) In *Xenopus* (Anura, Pipidae), the tympanic disk is not externally visible, yet the middle-ear apparatus is otherwise typical of most anurans. (**B**) In the fossorial *Geotrypetes* (Gymnophiona, Caeciliidae), the stapes has been redirected to make a connection with the quadrate bone, and the middle ear cavity has been filled with muscle and connective tissue. Within the amphibian periotic space, a lateral cistern is associated with the presence of a stapes connected to a tympanum, whereas no such specialized space region exists in the gymnophionids. op, operculum; op. m., opercular muscle; oss. m., ossicular muscle.

The absence of an opercular apparatus in caecilians probably correlates with the absence of a pectoral girdle (Hetherington 1992), to which the opercular muscle attaches in other amphibians (Kingsbury and Reed 1909). The stapes itself is, as in some caudates, bent anteriorly to form a loose attachment with the quadrate bone (Sarasin and Sarasin 1890; Wever 1975, 1985). The caecilians are almost exclusively either aquatic or subterranean. Therefore, it is likely that the evolution of the caecilian middle-ear architecture has redirected the earlier sound-conducting pathways to support the flow of vibrations from the animal's skull and body cavity into the inner ear.

Evolution of the vertebrate middle ear may not always have been focused on sound reception. The bullfrog tympanum has also been shown to serve as a sound source by actually broadcasting the high-frequency components of outgoing calls (Purgue 1997). In the bullfrog *Rana catesbeiana* it was found that the sound of the emitted call traveled from the mouth cavity to the middle ear cavity through the eustachian tube, causing the large tympanum to vibrate and effectively trumpet the call from either side of the head. This was consistent with the observation that in these frogs the tympanum diameter of males is greater than that in females, and that only males produce advertisement calls. This example highlights an important principle of the amphibian, and more generally the vertebrate, auditory system, which is that energy flow in nearly every component in the sound transmission and transduction pathway, from the tympanum to the hair cell stereociliary bundle (Manley 2001), can be bidirectional.

4.1 Building a Middle Ear Around New and Old Sense Organs

Concurrent with the appearance of a functional middle ear were significant and necessary modifications of the primitive periotic space. The periotic space in fishes extends from the brain case into the inner ear but not to the lateral edge of the otic capsule. The fish inner ear does not have an oval window. In some fishes such as the ostariophysans, the periotic system serves in conjunction with a specialized apparatus to transmit changes in sound pressure of the swim bladder or similarly adapted structure into displacements within the otic capsule, where the utricle, saccule, or macula neglecta may respond to such stimuli (Popper and Fay 1998). However, for reasons discussed in Manley and Clack, Chapter 1 (see also Lewis and Fay, Chapter 2), a middle ear did not evolve in fish. The fish (and much of the amphibian) inner ear is principally designed to encode near-field stimuli (Ladich and Popper, Chapter 4; Lewis and Narins 1998), essentially vibrations of the whole animal. For early terrestrial vertebrates, reorganization of the periotic canal was necessary for the evolution of sound pressure receptors (van Bergeijk 1967; Fritzsch 1992), which appear to rely heavily on extrapolating movements generated on thin contact membranes (see Figs. 6.2 and 6.8) for the transduction of sound pressure into movement of sensory hair bundles, which in turn initiate electrical activity in the auditory nervous system. It has been proposed (Lombard 1977, 1980; Fritzsch 1992) that the original condition of the periotic labyrinth would have been a simple duct or sac located medially between the otic wall and extending out around the ventromedial aspect of the saccule, as has been described in sarcopterygian fish. Whether a middle-ear sound pathway had to evolve prior to the formation of sense organs capable of transducing sound pressure, or whether at least one sound-pressure–sensitive organ had to be present to drive the evolution of the middle ear, is one of the central issues in discussions of the evolution of vertebrate hearing (Baird 1974; Fritzsch 1998).

Previous speculation on this topic has revolved around two central questions: Is there sufficient evidence that homologous sound pressure receiving organs were already present in the direct ancestors of earliest terrestrial vertebrates (Fritzsch 1987, 1992)? If so, then these organs could have benefited from improvements in the sound transmission pathway, namely the addition of a middle ear. Alternatively, could modifications of the hyomandibular apparatus (or homologous structures), which led to the origination of a functioning middle ear, have been initiated to serve other functions, such as endowing the saccule, lagena, utricle, or macula neglecta with some basic acoustic sensitivity? In anurans, the saccule and lagena both exhibit some sensitivity to airborne sounds

(Frishkopf and Geisler 1966; Lewis 1988; Yu et al. 1991; Christensen-Dalsgaard and Narins 1993; Christensen-Dalsgaard and Jørgensen 1996; Cortopassi and Lewis 1996, 1998), which is surely enhanced by the presence of a middle ear. However, it appears that the evolutionary pressures favoring the formation of a novel sound pressure receiver were so great that it occurred early and perhaps often during the water-to-land transition. Evolution of a terrestrial middle ear apparatus also had to occur in concert with the expansion of the periotic labyrinth. It has been pointed out previously (Burlett 1934; Lewis and Lombard 1988; Fritzsch 1992) that there appears to be a correlation between the emergence of specialized sound pressure receivers and expansion of the periotic labyrinth. Thus, it becomes more likely that the middle ear in amphibians arose in support of either the AP or BP. Since the amphibian macula neglecta is generally not associated with any specialized connections to the periotic canal, it would seem less likely, though not impossible, for it, too, to have promoted the formation of a terrestrial middle ear.

4.2 Evolutionary Implications of the Hearing Apparatus in Tadpoles

An important issue raised by van Bergeijk (1966) was one of ontogeny. Amphibians, especially frogs, may spend a considerable amount of their lives in a fully aquatic larval stage, which concludes with a remarkable metamorphosis of the entire animal, including many features of the ear. The larval tadpole inner ear possesses a complete set of adult inner-ear end organs, including the AP and BP. In anurans, the AP is fully formed by the tadpole stage (Patterson 1948; Hertwig and Schneider 1986; Hertwig 1987) and presumably functions in the aquatic environment for tadpoles in a manner similar to that observed in the adults, despite the absence of an adult middle ear. Consistent with its adult terrestrial vibrational sensitivity, it seems likely that the tadpole AP responds to whole-animal movements generated by underwater near-field stimulation, much like the fish saccule. During metamorphosis the inner ear changes little, whereas the pathway by which sound enters the ear changes dramatically. The tadpole ear possesses a pathway for sound pressure to enter the ear, and it appears notably similar in design to the Weberian apparatus known in ostariophysan fishes (Popper and Fay 1998). Anuran tadpoles possess a "bronchial columella" (or bony connection between a bronchial out-pocketing and the periotic sac), which appears to provide a means for delivering sound pressure into the tadpole inner ear (Witschi 1949, 1955). Upon metamorphosis this pathway is lost during the reconstruction of the middle ear. The Weberian apparatus in teleost fish apparently serves to stimulate the sacculi in those animals, and so too the bronchial columella could just as well be derived by a similar mechanism and for a similar purpose in the presumptive amphibian ancestors, the rhipidistian fishes. If that were the case, then either the AP or BP could have evolved under similar circumstances to enhance sensitivity to sounds entering the ear via the bronchial columella. A terrestrial middle ear that further enhanced the functioning of one or both of these newly established amphibian auditory receptor organs could have been added later.

4.3 Did Sound Pressure Receivers Require a Middle Ear Be Present to Evolve?

Where it has occurred in amphibians, the loss or reduction of the tympanum presumably happened after the appearance of the AP and BP. In some salamanders, the BP is believed to have been lost, presumably after or in parallel with loss of the tympanic sound pathway. An AP remains in all amphibians regardless of the state of the middle ear. This may indicate that the BP is much more dependent than the AP on the presence of a functioning middle ear. For those animals lacking a functional middle ear, it is believed that alternate sound transfer pathways, perhaps those more appropriate to an aquatic or fossorial habitat, evolved to replace the tympanal sound input. In fact, the available evidence would indicate that probably all amphibians rely to some extent on alternate sound transfer pathways (Kingsbury and Reed 1909; Ehret et al. 1983; Wilczynski and Capranica 1984; Wever 1985; Wilczynski et al. 1987; Hetherington 1988; Narins et al. 1988; Ehret et al. 1990, 1994; Hetherington and Lindquist 1999).

Certainly the seismic/vibrational response of the amphibian saccule is not dependent on sound pressure waves arriving via the middle ear. It happens, however, that the frog saccule also exhibits considerable sensitivity to airborne sounds (Moffat and Capranica 1976); in *Rana catesbeiana* the most sensitive saccular auditory afferent fibers are a mere 5 dB less sensitive to airborne sounds than the most sensitive auditory afferent fibers of the AP (Yu et al. 1991). Yet the saccule does not appear to rely on any specialized contact membranes for its acoustic sensitivities. Is it really possible that after adding a middle ear, reorganizing the periotic canal, and evolving a bevy of ultrastructural specializations, the AP has succeeded in providing the animal with only an additional 5 dB of acoustic sensitivity? It must surely be that the acoustic sensitivities of the saccule have benefited from the addition of a middle ear and an extended periotic canal in much the same way as the AP has. It may also be that the acoustic sensitivities of the saccule are mere reflections of its extraordinary seismic sensitivity. As pointed out earlier, the saccular chamber is an important component of the sound transfer pathway to the AP; sound pressure travels through the saccule to get to the AP. So it should not be entirely surprising that the saccule responds to airborne sound. The important point is that emergent acoustic sensitivity of the saccule could have significantly influenced the early reorganization of the terrestrial inner ear and the evolution of a middle ear. Furthermore, there is ample evidence that the AP functions underwater and could have evolved to supplement the bandwidth of the saccule regardless of whether the stimulus is airborne, aquatic, or seismic. Thus one must conclude that the origins of the AP are not necessarily coupled to the evolution of the middle ear, and could just as well have occurred prior to any significant reorganizations of the periotic labyrinth. This is significant because it means that either the AP or the saccule are just as likely as the BP to have been the principal motivator and benefactor of improvements in sound transfer efficiency into the inner ear.

5. Specializations of the Periotic Canal

The condition of the periotic labyrinth prior to amphibians is presumed to have been a simple duct or sac located medially between the otic wall and extending out around the ventromedial aspect of the saccule, as found in the sarcopterygians (Fritzsch 1992). Later, extension of a periotic canal laterally to the area of the oval window might have exploited movements generated near or by the operculum, or other nearby structures, ultimately hastening the evolution of the middle ear. The amphibian periotic labyrinth differs chiefly from that of the amniotes in one peculiar way: in amphibians the periotic canal extends caudally around the endolymphatic space, whereas in the other terrestrial vertebrates it passes rostrally (Lombard 1980; Lewis and Lombard 1988). There is no evidence to suggest that this difference is associated with any significant physiological consequences, yet it has been taken to suggest that the periotic labyrinth evolved independently in amphibians and in the amniotes.

Considering the anatomical diversity of the amphibian middle ear, the considerable morphological variability of the periotic canal comes as no surprise. The periotic canal (POC) illustrated in Fig. 6.2 is a generalized version of perhaps the most derived type of POC found in the advanced frogs. However, since the presence of, collectively, an AP, a BP, and a functional middle ear apparatus such as that in anurans today is considered to be the primitive condition for all amphibians, one must assume that the POC illustrated in Figure 6.2 is not that far removed from the primitive condition as well. Some of the variability described for the periotic duct system in six different amphibians is illustrated in Figure 6.7, including differences in the lateral periotic cistern, medial periotic sac, and the interconnecting periotic canal. The chief difference between a primitive frog such as *Xenopus* and an advanced frog such as *Rana* is the length and origin of the periotic recess extending to the BP. In *Xenopus*, as in other primitive frogs, the BP periotic duct exists as a short branch off the POC, whereas the main course of the POC runs almost directly across the AP contact membrane (there is a short AP recess) (Patterson 1948). Thus, in primitive frogs, the POC is relatively simple: it runs almost directly mediolaterally and makes brief extensions toward the auditory sense organs. In more advanced frogs, the BP is in contact with a separate canal that diverges directly from the periotic sac within the braincase. In *Rana*, the BP periotic extension passes through its own foramen into the braincase, whereas in *Xenopus* a single foramen exists.

Lombard (1977) observed three general categories of POC variability among the caudata, which extend reasonably well to all amphibians so far described.

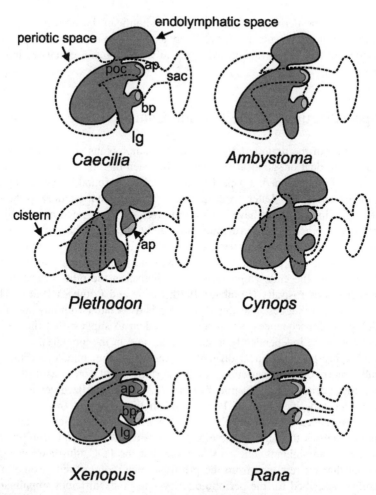

FIGURE 6.7. Anatomy of the periotic space. The periotic space consists of a large region on the lateral side of the otic capsule that sometimes includes a smaller cistern associated with the oval window, a canal wrapping around the endolymphatic space that forms connections with the auditory end organs, and a terminal periotic sac. The anatomy of the periotic canal in particular shows considerable variability in its length, direction, and interactions with the endolymphatic space, and may play a larger role in sound processing than previously understood. Formation of a dedicated periotic extension to the BP such as in *Rana* may correlate with an increased separation in the relative functions of the AP and BP. [Figures based on descriptions in Patterson (1948), Lombard (1977), Wever (1985), and Purgue and Narins (2000a).]

First, there is considerable variability in the point where the POC meets with the periotic cistern, as well as considerable variability in the size of the cistern itself. For frogs, salamanders such as *Ambystoma*, and for those gymnophionids so far studied, the POC usually joins the cistern slightly dorsal and posterior to the oval window, and then passes medioventrally along the utriculosaccular border. In some salamanders, however (for example, *Plethodon*), the POC may join the cistern from a more ventrolateral aspect and/or may reflect a more intimate association with the oval window itself. Second, Lombard recognized three generalized patterns of bending in the periotic canal, which appeared naturally associated with variations in the overall length of the POC. Third, the first two points invariably lead to variations in the nature of the associations between the POC and the AP and BP.

These anatomical variations imply that important physiological differences in the amphibian auditory system that have arisen during evolution can be linked to the architecture of the POC. Purgue and Narins (2000a,b) demonstrated that the shape of the periotic duct can influence the spectrum of acoustic energy flowing to each of the auditory end organs. In much the same way that the shape of a musical horn determines its acoustic resonant properties, the shape and length of the periotic canal and the shape of the recesses leading to the AP and BP make distinct and quantifiable contributions to the acoustic sensitivities of those two auditory sense organs. Thus the anatomical variability of the amphibian periotic system is likely to reflect significant adaptations for the passage of sound waves flowing through the inner ears of select amphibian species.

6. The Amphibian Saccule: A Specialized Terrestrial Seismic Detector

Of all the vertebrate hearing organs, the saccule is the most sensitive to substrate-borne vibrations (Koyama et al. 1982; Narins and Lewis 1984; Christensen-Dalsgaard and Narins 1993). However, low-frequency auditory afferent fibers exiting the AP also exhibit considerable sensitivity to substrate-borne vibrations (Frishkopf and Goldstein 1963; Yu et al. 1991; Christensen-Dalsgaard and Narins 1993). Since the saccule and AP are both sensitive to sounds and vibrations, it was hypothesized that frogs discriminate sounds not so much based on whether or not they are seismic or auditory, but instead rely solely on bandwidth (Yu et al. 1991). Indeed, the seismic and auditory stimuli processed by the saccule and AP project to the same region of the auditory midbrain, separated only by frequency (Koyama et al. 1982). The saccule responds to seismic and auditory sounds below approximately 100 Hz, whereas the AP is sensitive to sounds and vibrations in the range of roughly 100 to 1200 Hz, depending on the species (Frishkopf and Goldstein 1963; Sachs 1964; Frishkopf and Geisler 1966; Liff 1969; Boord et al. 1970; Feng et al. 1975). This difference in frequency ranges appears to derive from differences in the mechanical and electrical frequency tuning mechanisms inherent in each receptor. It appears that the AP evolved at least in part to provide added bandwidth to the seismic sensitivity of the animal. Why not just expand the frequency range of the amphibian saccule? Indeed, some fish have done just that (Fay and Edds-Walton 1997). However, it may be that the anatomical and physiological changes needed to accomplish this might also degraded this organ's sensitivity to terrestrial vibrations, or its ability to encode the fine temporal details of a stimulus, such as signal onset, offset, or phase, all cues that take on added significance for animals localizing sounds traveling through air. Differences between the sensitivity of the saccule and AP extend beyond their respective frequency ranges: individual nerve fibers exiting the saccule are more sharply tuned to seismic than to auditory stimuli over the same bandwidth, whereas the opposite is true for the AP (Yu et al. 1991). Thus, even though these two organs exhibit relatively minor differences in their overall sensitivity to both seismic and auditory stimuli, the saccule has evolved to provide exceptional sensitivity and temporal resolution for the encoding of seismic/vibrational stimulations, whereas the AP has been adapted to provide fine tuning over a broader bandwidth.

Among amphibians, the anatomy of the saccule is generally similar among the three orders, with one notable exception (Wever 1975; Lombard 1977; White and Baird 1982). Both Wever (1975) and White and Baird (1982) noted that in the Gymnophiona the saccular macula is elongated, tapering off laterally (Fig. 6.4). An elongated sensory epithelium in any auditory structure is commonly thought to imply an extra degree of sensory processing, whether it be an extended tonotopy or a sound-localization based on stereovillar bundle orientation patterns. Since no physiology has been reported on the response properties of the gymnophionid auditory system, we can only anticipate the functional significance of the elongated saccular macula. However, one may safely suspect that the caecilian saccule, perhaps the principal hearing organ among the fossorial gymnophionids, may have undergone considerable modifications to enhance its capabilities as a seismic auditory receptor in much the same way as the more derived anurans have extended the sensitivity of the primitive AP.

7. Origins and Evolution of Dedicated Sound Pressure Receptor Organs

Since most amphibians possess two sense organs that detect airborne sounds whereas the other terrestrial vertebrate groups only possess one, it is natural to question whether either the AP or BP is directly homologous to, or in any way contributed to the evolution of, the BP of these other groups. The amphibian AP and BP both share similarities with BPs or cochleae of other terrestrial vertebrates, which also exist as evaginations of the saccule and make contact with the periotic canal. In primitive amphibians such as *Xenopus* or *Cryptobranchus*, a rather amorphous extension of the periotic sac is found closely opposed to the medial wall of the saccule, thus indicating that the AP and BP recesses probably did not have to extend far to reach it during the early stages of evolution. In more derived species, individual, dedicated periotic canals are extended laterally from the periotic sac to meet both of the recesses. But during the earliest stages of terrestrial vertebrate inner-ear evolution, the primitive architecture of the periotic sac probably would have had a strong influence on the initial positioning and orientation of any novel sound pressure receivers. This may explain why the AP and BP are adjacent to each other in amphibians. Therefore, the position and orientation of inner-ear sense organs are unreliable indicators of homology among the terrestrial vertebrates. However, if we consider the origins of the AP and BP separately, we may be able to identify key characters that can strengthen the association between one of them and the reptilian and avian BPs.

7.1 The Macula Neglecta and the Origin of the Amphibian Papilla

In amphibians a macula neglecta has only been observed in caecilians. It was first described by Sarasin and Sarasin (1892), and described in some detail for *Typhlonectes, Geotrypetes, Dermophis*, and *Ichthyophis* (Wever 1975; Wever and Gans 1976; White and Baird 1982). That this structure shares its name with a similar structure in some fishes (Retzius 1881; Corwin 1977, 1981) indicates that previous researchers envisioned a homologous relationship between the two structures. If so, it may be presumed that the earliest amphibians possessed a macula neglecta and that this end organ was lost very early in the anuran and urodele sister groups, and at some later time some of the Gymnophiona also abandoned the macula neglecta. Perhaps those of the Gymnophiona that had soon returned to a predominantly aquatic lifestyle found the structure still beneficial for much the same reasons as had their aquatic ancestors.

Similarities in the microanatomy of the macula neglecta sensory epithelium and its tectorial membrane to those of the AP have led to the hypothesis either that the two organs may be homologous or that the AP may indeed be a macula neglecta that has been moved from the pars superior to the pars inferior (Wever 1975, 1985; Wever and Gans 1976; White and Baird 1982). Indeed, within the Caudata, members of the family Plethodontidae have an otherwise typical-looking AP that extends ventrally as an evagination of the utricular chamber (Lombard 1977; Wever 1985), in much the same way as if a macula neglecta were given its own recess extending downward (vertically) toward and making contact with the periotic canal. Such an analogy may be misplaced, however, since Lombard (1977) concluded that the primitive urodele AP was more likely to have been a horizontal evagination of the pars inferior. Thus the vertical AP in plethodonts may reflect a derived, dorsal migration of the primitive AP into the region once occupied by a macula neglecta. In caecilians the sensory epithelia of the macula neglecta and AP are often positioned near enough to appear as almost mirror images of each other, one (the AP) facing ventrally within its recess and the other facing dorsally within the utricular chamber (Wever 1985). Both organs are always located near the utriculosaccular duct, and in some urodeles the AP actually exists as an out-pocket of the duct itself (Lombard 1977).

Similarly arranged sensory epithelia, similarly constructed tectorial membranes, and the neighboring positions of the two organs within the endolymphatic space suggest a homologous relationship between the macula neglecta and the AP. White and Baird (1982) speculated that for those caecilians possessing both an AP and a macula neglecta, the AP sensory epithelium may have derived from the fission of a more primitive macula neglecta. In fact, in some fishes the macula neglecta has been reported to be composed of two separate sensory patches (Platt 1977). Later, Fritzsch and Wake (1988) found that in the gymnophionids the sensory maculae of the AP and macula neglecta developed from a common patch of sensory primordia, only to be separated from one another by the constriction of the utriculosaccular foramen. For those amphibians possessing only the AP, it may be surmised either that the macula neglecta was lost following the appearance of the AP, or that the entire macula neglecta migrated down into the pars inferior as part of the reorganization of the terrestrial inner ear (Lewis and Narins 1998). Such a move may have facilitated an interaction between the primitive macula neglecta (or future AP) and an expanding role of the periotic canal for terrestrial hearing (Fritzsch and Wake 1988; Fritzsch 1992).

7.2 Origins of the Basilar Papilla

The origins of the amphibian BP have drawn considerable discussion (Baird 1974; Wever 1975, 1985; Lombard 1977; White and Baird 1982; Fritzsch 1987, 1992; Lewis and Lombard 1988; Lewis and Narins 1998) since most researchers favor a homology between the BP of amphibians and amniotes. The early evolutionary events that created the amphibian BP therefore may be considered central to the evolution of hearing in all terrestrial vertebrates. The single most influential piece of evidence supporting a BP homology among all terrestrial vertebrates is that, in those Caudata and Gymnophiona that have a BP, the sensory epithelium is found either within the lagenar recess or within an evagination of the proximal end of the lagenar recess. In amniotes, the BP shares a common recess with the lagena. Thus, a similar association of the BP and lagena in widely disparate taxonomic groups has been taken to imply homology. Unfortunately, the anurans do not fit so easily into this mold. Though the anatomy of the anuran BP is generally consistent with the BP of urodeles and caecilians, Wever (1985) thought the two BPs must have separate origins since the anuran BP always exists as an independent evagination of the saccule. Wever (1985) also cited differences in the microanatomy of the salamander BP tectorial membrane (which he considered similar in design to the anuran AP tectorium) and the absence of a tectorial curtain in the salamander BP as implying different modes of stimulation. These minor differences in BP anatomy, which can be observed when comparing the most primitive frogs and the salamanders, do not necessarily imply separate origins for the two organs, but at least suggest an early separation between the BPs of the two groups.

Among the Caudata the Cryptobranchidae, the Dicamptodontidae, some members of the Salamandridae, and the Ambystomatidae all exhibit prominent BPs (Lombard 1977; Wever 1985; White 1986a). The Amphiumidae and the salamandrids *Triturius, Taricha,* and *Pleurodeles* exhibit reduced BPs, but no BP has been found in *Siren, Necturus, Notophthalmus,* or among the Plethodontidae. Among the Gymnophiona, members of the families Caeciliidae and Typhlonectidae do not appear to possess a BP, but a BP was found in members of the Ichthyophiidae (Sarasin and Sarasin 1890, 1892; Wever 1975, 1985; Wever and Gans 1976; Lombard 1977; White and Baird 1982). Lombard (1977) concluded that the presence of a BP must be the primitive condition for all Caudata since a BP was found lacking in several groups considered otherwise more derived, whereas the BP was common to many animals deemed primitive based on otic and other morphological traits. If indeed the presence of a BP located within, or extending from, the lagenar cavity is the primitive condition for amphibians, then it becomes highly likely that this structure is homologous to the amniote BP. The principal implication is that access to the lagenar cavity was critically important for the evolution of early sound pressure receivers. Perhaps this was because it provided a convenient pathway to the periotic sac, or because positioning of the sensory epithelium within the lagenar cavity effectively isolated it from the mechanical vibrations of the saccule, thereby improving its sensitivity to alternative stimuli. Either way, the lagena clearly influenced the evolution of the BP in terrestrial vertebrates.

In the gymnophionids, caudatans, and in the more primitive anurans, the BP receives its innervation as a twig from the lagenar branch of the VIIIth nerve, which is similar to the BP innervation patterns of amniotes (Fritzsch and Wake 1988; Fritzsch 1992; Fritzsch and Neary 1998). By contrast, the AP receives its innervation from a branch off the posterior ramus of the VIIIth nerve, which is also the innervation pathway for the macula neglecta in gymnophionids (Fritzsch and Wake 1988; Fritzsch 1992; Fritzsch and Neary 1998). Since inner-ear innervation patterns appear to be highly conserved across vertebrate taxa (Fritzsch 1992; Fritzsch and Neary 1998; Fritzsch 2001; Fritzsch et al. 2002), they too indicate that the AP derives from a fission of the more primitive macula neglecta, whereas the amphibian BP is homologous to the amniote BP.

7.3 Specializations of the Auditory Sensory Epithelia

The basic architectural design of the AP is similar for all the amphibians, especially when one considers the relative location of the recess, the arrangement of the sensory macula, the morphology of the tectorial membrane and tectorial curtain, and positioning of the contact membrane. However, there remain some compelling variations in the anatomy and organization of the AP among amphibians, and especially frogs, for which the greatest number of examples has been collected. The primitive AP design as witnessed in the gymnophionids, the Caudata, and in the most primitive anurans consists of a single circular or triangular patch of sensory hair cells located on the dorsomedial wall of the recess (Figs. 6.3 and 6.8). The hair cells are stimulated by the movements of an overlying tectorial membrane. The principal deviation from this primitive condition is the addition of a caudal extension to the sensory epithelium found in the intermediate and advanced anurans (Figs. 6.3 and 6.8), the length of which appears to correlate directly with an upward expansion in the range of frequencies encoded by the AP (Lewis 1977b; Lewis et al. 1982). This range extension appears to be directly associated with the competition for shared acoustic resources, in this case bandwidth being the limited resource.

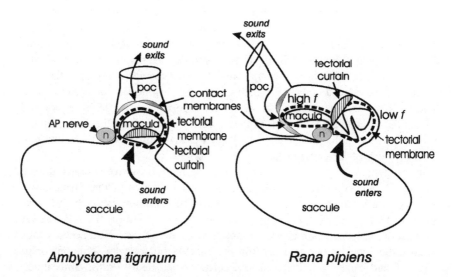

FIGURE 6.8. Functional architecture of the AP. The AP of the salamander *Ambystoma tigrinum* is reflective of the primitive AP architecture seen in the Caudata, Gymnophiona, and some primitive anurans. It consists of a single patch of hair cells located on the dorsal or dorsomedial surface of a recess. Overlying (ventral to) the hair cells is a perforated tectorial membrane. The hair cell stereovillar bundles appear to project into channels in the tectorial membrane (Lewis and Narins 1998). A contact membrane at the distal end of the AP recess serves as an exit for sound pressure waves passing through the AP chamber, and may exhibit complex modes of vibrations that influence the patterns of sound-induced tectorial membrane motion (Purgue and Narins 2000b). In the derived frog *Rana pipiens*, the entire papillar chamber, sensory epithelium, and tectorial membrane have been elongated caudally during the evolution of a broadband tonotopic map within the sensory macula. The AP chamber is bisected by a thin tectorial curtain extending from the tectorial membrane ventrally to the floor of the chamber, and may serve in the transduction of sound energy into movement of the tectorial membrane. f, frequency; poc, periotic canal.

Presumably during the evolution of acoustic communication by frogs, neighboring frogs effectively minimized competition through a progressive upward shift in the fundamental frequencies of their mating calls (Lewis et al. 1992). Frogs have found others ways to achieve acoustic separation in their calling behavior, such as through changes in call structure, adjustments in the temporal parameters of their calling behavior, and by exploiting a variety of alternative acoustic substrates such as water, plants, and the ground (Narins et al. 2000; Garcia-Rutledge and Narins 2001; Lewis et al. 2001). But none of these have yet been correlated with a distinct adaptation of the auditory periphery in the same way as frequency.

The organization of the AP sensory epithelium is in itself something of an enigma; its characteristic S-shaped epithelial outline, a conspicuously bipartite tectorial membrane, a centrally located tectorial curtain, the presence or absence of electrically tuned hair cells, and the puzzling hair cell orientation patterns all exhibit variations and tantalizing trends across taxa. However, until more is known about the functional significance of these and other individual traits, we are left to wonder about their influence on evolution. One trait, electrical tuning of hair cells, is prevalent throughout the nonmammalian vertebrates and so offers an interesting opportunity to study the evolution of a particular cell type (Coffin et al., Chapter 3). Electrical resonances in membrane potential (Corey and Hudspeth 1979) are a means by which the auditory hair cell can maximize the efficient release of neurotransmitter in response to acoustic stimulation of a restricted bandwidth. They have been found to occur in both fish and amphibian saccular hair cells. Electrical resonances probably arose fairly early in the evolution of hair cells (Manley 2000), and they have more than likely provided an important mechanism for frequency filtering well before the first terrestrial vertebrates were in a position to benefit from an increased bandwidth of information (Coffin et al., Chapter 3).

Electrical resonances are present in all amphibian saccular hair cells (Corey and Hudspeth 1979; Hudspeth and Lewis 1988; Holt and Eatock 1995; Smotherman and Narins 1998), in about half of the hair cells of the AP (Smotherman and Narins 1999a), and can be evoked in a small percentage of hair cells from the BP (Smotherman and Narins 1999b). All saccular hair cells exhibit a similar tuning bandwidth. Dissociated hair cells exhibit membrane electrical resonances in the range of 50 to 150 Hz, which correlates reasonably well with the best frequencies of the saccular nerve fibers in situ. All saccular fibers appear to be tuned to these lower frequencies, and it is likely that both electrical tuning of the hair cells and the mechanics of this end organ both contribute to its bandpass characteristics (Egert and Lewis 1995). Within the saccular macula there is no evidence of any frequency-specific organization of hair cells or nerve fiber sensitivities.

Basilar papilla hair cells can exhibit resonances because they generally possess the same complement of ionic currents present in saccular hair cells, though the relative magnitudes of the ionic currents are generally insufficient to support robust electrical resonances such as those in saccular hair cells (Smotherman and Narins 1998, 1999b). When BP hair cells are stimulated to exhibit electrical resonances, they do so over a frequency range nearly an order of magnitude less than the sensitivity range of the organ in situ. The frequency range of the resonances reflects the speed and numbers of the ion channels in the basolateral membrane, which again appear generally similar to the ion channels present in saccular or AP hair cells. Yet BP hair cells would require substantial modifications to support electrical tuning at the frequencies to which the BP responds best (about 1.4 kHz for *Rana pipiens*). Such adaptations are theoretically possible (Wu et al. 1995) but have not been observed in any preparation to date.

The electrically tuned AP hair cells represent a significant step forward in the evolution of auditory sense organs. In the anuran *Rana pipiens*, the hair cell resonant frequencies are carefully organized into a spatial array of frequencies spanning a much broader bandwidth than that observed in the saccule (Pitchford and Ashmore 1987; Hudspeth and Lewis 1988; Smotherman and Narins 1999a). The electrical properties of the saccular and AP hair cells are fundamentally similar; they possess essentially the same set of ionic conductances. However, in the AP, some additional morphogenetic factors must be present during development that designate the frequency of each individual hair cell based on its relative position within the sensory epithelium. In this regard the anuran AP is unique among all the amphibian auditory end organs so far described, but similar to the reptile and bird basilar papillae, which also exhibit a tonotopic arrangement of structural and/or physiological hair cell properties. Within this context, it will be especially interesting to learn whether or not the molecular determinants of tonotopy in the anuran AP are similar to those responsible for generating similar patterns within the auditory sensory epithelia of birds and reptiles.

Hair cell orientation patterns in the anuran AP and BP are as complex and diverse as in any vertebrate auditory end organ, and as yet no hard evidence exists to explain their function. All vertebrate hair cells are morphologically polarized: the hair cell stereovillar bundle must be displaced toward a single kinocilium or the tallest stereovilli for maximum sensitivity of hair cell stimulation (Flock 1965). In fishes the saccular macula displays a variety of different hair cell orientation patterns (Popper and Platt 1983; Popper and Fay 1998) that are believed to contribute to the abilities of those fishes to localize sounds (Fay and Edds-Walton 1997; Lu and Popper 2001). The hair cell orientation patterns of the amphibian saccule differ from the general patterning described for the fish saccule (Popper and Fay 1998). Most BP hair cells are polarized parallel to the axis of the recess, though they appear equally likely to be oriented either toward or away from the opening of the recess into the saccule: Lewis (1978) observed significant intraspecific variability in orientation patterns of the BP macula, though general patterns were discernible. The fairly simple yet ill-defined BP orientation patterns are similar to the simple orientation patterns of the AP sensory epithelia in the Caudata and Gymnophiona

(Lombard 1977; White and Baird 1982). This may reflect similarities in the modes of stimulation utilized by both the primitive AP and the BP. In the more derived frogs, the AP sensory epithelium exhibits a much more complex array of hair cell orientation patterns, but the functional relevance of these patterns is as yet unknown.

7.4 Development of the Inner Ear: Clues to Evolutionary Origins

Early inner-ear development appears to occur similarly in all vertebrates (Patterson 1948; Hertwig 1987; Sokolowski and Popper 1987; Bissonnette and Fekete 1996; Haddon and Lewis 1996; Fritzsch et al. 1998, 2002; Fritzsch 2001; Kil and Collazo 2001), and the basic mechanisms are likely to be highly conserved. The molecular substrate guiding ear morphogenesis from fish to mammals appears to rely upon many homologous genes (Fritzsch 1998, 2001; Fritzsch et al. 2002). For the ear, the phylogenetic distribution of novel sense organs and accessory acoustic structures does to some extent mirror early developmental sequences of the more advanced vertebrates (Fekete 1999). In the frog, the earliest patches of hair cells destined for auditory structures of the pars inferior appear as early as during the invagination of the otic vesicle (Hertwig 1987; Kil and Collazo 2001). This initial group of hair cells, known as the macula communis, rapidly expands dorsally, producing a sheet of hair cells that will ultimately break apart to form the sensory epithelia of the saccule, lagena, BP, and AP, and (in caelicians) the macula neglecta (Fig. 6.9) (Patterson 1948; Hertwig 1987; Fritzsch and Wake 1988). This progressive segregation of the macula communis during morphogenesis may well mirror a similar sequence of events that occurred during the evolution of the terrestrial vertebrate ear (Fritzsch 1998; Fritzsch et al. 1998, 2002; Fritzsch and Neary 1998).

During the earliest stages of vertebrate inner-ear development, the hair cells of the macula communis do not exhibit the adult patterns of stereovillar architecture or orientation patterns (Sokolowski and Popper 1987). Apparently the specialized organization of each sense organ's sensory epithelium arises independently after the breakup of the macula communis. This split occurs during the evagination of their respective recesses (Patterson 1948; Hertwig 1987; Fritzsch and Wake 1988), although further hair cell production occurs after formation of the recesses and apparently continues throughout the life of the adult animal (Li and Lewis 1974). Since some salamanders have lost a BP recess while maintaining a BP sensory epithelium, it would appear that the morphogenesis of sensory epithelia and the formation of papillar recesses, especially the BP recess, are guided by separate molecular cues in amphibians (Fritzsch 1998).

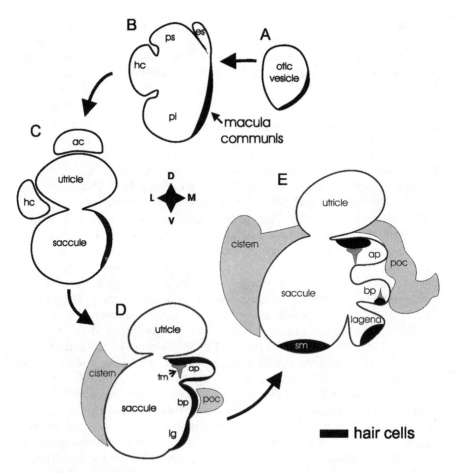

FIGURE 6.9. Morphogenesis of the auditory portion of the *Xenopus* inner ear. During vertebrate embryogenesis, the first cells destined to become inner ear hair cells may appear as early as during invagination of the otic vesicle (**A**) (Sokolowski and Popper 1987). This early patch of hair cells grows rapidly to form the macula communis (**B**), from which all inner ear hair cells eventually derive. In part, the process of sensory structures forming through the evagination of the medial wall of the saccular chamber pulls apart the macula communis during morphogenesis (**C–D**), providing each sense organ with its own patch of hair cells. The molecular cues that define where the macula communis separates are not known, nor is it known what physiological properties these early hair cells possess. Hair cell differentiation continues within the maculae of each sense organ long after development of the inner ear is completed (**E**), and quite possibly throughout an amphibian's life. es, endolymphatic sac; ps, pars superior; pi, pars inferior; hc, horizontal canal; ac, anterior canal; poc, periotic canal; tm, tectorial membrane. Based on Patterson (1948) and Hertwig (1987). Not drawn to scale.

7.5 The Functional Significance of Forming Papillar Recesses

The auditory sense organs of amphibians can generally be characterized as having a sensory macula placed within a recess and possessing a contact membrane. We have already discussed the variations in the sensory epithelia and the role of the contact membrane, but what is the significance of the recess? The presence and position of these evaginations are significant for many reasons. For example, the AP evagination is always located at (and generally ventral to) the junction of the sacculo-utricular foramen, the same position as the origin of the cochlear duct in amniotes (Bever and Fekete 1999). In some species of salamander the BP recess exists without the macula or contact membrane. In others a sensory epithelium is present within the lagenar recess, but a morphologically distinct BP recess is absent. The exact position and length of the BP recess varies among amphibians (Lombard 1977; White and Baird 1982; Wever 1985; White 1986a). Similarly we have discussed how the AP recess can vary considerably in architecture, being considerably longer in derived frogs in which a caudal high-frequency extension of the sensory epithelium is present. Furthermore, in some primitive frogs such as *Cryptobranchus,* the opening of the AP to the saccule is quite large, the entire recess appearing more as a depression than a separate recess (Baird 1974; Lombard 1977), whereas in other instances a large opening is coupled to a deeper recess (as is the case in the salamander family Ambystomatidae). Among the salamanders there is substantial diversity in the size of the saccular opening to the AP recess (Lombard 1977). For auditory sense organs in general, it appears that the recess serves not only as an extension of the sense organ toward the periotic space, but also to differing degrees may serve to isolate the sensory maculae from substantial seismic or vibrational interferences resulting from natural movements of the otoconia within the saccular chamber. The depth of the recess and the size and shape of the opening should influence the amount of sensory input derived from the endolymphatic channels relative to the periotic pathway. The AP is a special case, since it is known to receive and be highly responsive to both airborne and seismic stimulation. In light of the work by Purgue and Narins (2000a,b), we may assume that the AP receives much of its sensory input via the saccular chamber; therefore, anatomical variations in the size and position of the opening to the AP recess influence sound transmission into the AP from the endolymphatic space.

In the case of the BP it is not clear whether or not the presence of a BP recess is the primitive condition in amphibians, although its absence is considered primitive for vertebrates in general (Fritzsch 1987, 1992). At some point early in the evolution of terrestrial vertebrates, the BP sensory epithelium migrated into the lagenar cavity, and then later established its own dedicated recess in amphibians. Moving the BP macula into the lagenar recess probably reduced the influence of seismic and gravistatic stimulation on the BP, allowing it to become relatively more sensitive to sound pressure. Yet typically the lagenar recess also contains some form of otoconial mass. Further enhancements of the BP as a sound pressure receptor within the lagenar recess would have necessitated substantial reorganization of the lagena itself, thereby threatening the functionality of the lagena. One may therefore conclude that, unlike the archosaurs, the success of early amphibians was somehow dependent on conservation of the primitive lagenar architecture. In those amphibians where there was sufficient evolutionary pressure for the enhancement of a (second?) higher-frequency sound pressure receptor, a dedicated BP recess evolved. This may well have been the case for the stem amphibian, consistent with Lombard's (1977) conclusion that the presence of a BP recess is likely to be the primitive condition. In that case the secondary loss of the BP recess could have been fueled by a rededication of the entire lagenar recess to the functioning of the lagena itself, a likely adaptation for those amphibians returning even temporarily (in the evolutionary context) to a fossorial or aquatic habitat.

8. Summary

The amphibian ear continues to offer important insights into the physiology and evolution of hearing for the following reasons: (1) It possesses a relatively simple structure, the BP, which exists in three distinctly different formations—(a) the formation of a sensory epithelium, (b) the formation of a dedicated endolymphatic recess, and (c) the positioning of the epithelium within the recess—making it possible to investigate the genetic underpinnings of some of the major components of all hearing organs. (2) The AP, present in all amphibians, exhibits cellular specializations common to many terrestrial vertebrates, thus offering the opportunity to identify the morphogenetic factors responsible for defining hair cell physiology and the principles of epithelial organization.

A sensory macula within the vertebrate inner ear composed of a simple circular patch of hair cells existed much earlier than the first amphibian. Presumably within this seed epithelium lie the roots of all the vertebrate inner-ear sensory epithelia, such as the lagenar, utricular, neglectar, and saccular maculae. The hallmark of the terrestrial hearing is the addition of novel sense organs dedicated to the neural coding of acoustic stimuli. The addition of new sensory maculae at different locations within the saccular, utricular, and lagenar chambers was undoubtedly attempted multiple times during vertebrate evolution, and one such experiment may have placed a sensory patch near the opening of the lagena in *Latimeria* (Fritzsch 1987). In similar fashion one might imagine that the macula neglecta in fish could have been co-opted by the neighboring region of the pars inferior to create the amphibian AP. Whatever the actual sequence of events was, it remains likely that the sensory maculae for all the different amphibian auditory receptor organs ultimately share many homologous genetic underpinnings (or developmental modules; see Grothe et al., Chapter 10).

The addition of acoustic sensory epithelia required the creation of papillar recesses. It appears that the AP and BP recesses of primitive frogs, salamanders, and caecilians are anatomically and functionally very similar, again indicating homologous origins for the genetic control of creating papillar recesses. The details of positioning the recess, whether it opens directly into the lagenar or saccular chamber, the size of the opening, orientation, depth, and contacts with the POC reflect species-specific derivations arrived at later. Similarities among other supporting structures, such as the otoconial masses or tectorial structures, innervation patterns, etc., also imply a high degree of genetic overlap in the formation of each of the amphibian auditory receptors.

Although the physiology of the amphibian BP is significantly different from the amniote BP, a wealth of evidence suggests that the two are homologous. The AP is, on the other hand, far more similar to the amniote BP in terms of physiology and functions. An interesting question arises when we consider the origins of electrical tuning in the AP. Within the strict context of the amphibian ear, the emergence of a tonotopic arrangement of electrically tuned hair cells appears to represent a secondary specialization of a sensory epithelium that did not possess such an adaptation at the time the AP first appeared. Of course the same must also be true for the amniote BP, since the amphibian BP shows no signs of ever being tonotopically organized. Thus, unless a more primitive auditory end organ is found to be tonotopically organized (i.e., a saccule), it is unlikely that tonotopy within of the frog AP can be considered directly homologous to the tonotopy observed in some amniotes (Crawford and Fettiplace 1981; Fuchs and Evans 1990). On the other hand, given the evidence that electrical tuning is achieved by nearly identical cellular mechanisms in frogs and turtles (Art and Fettiplace 1987; Art and Goodman 1996; Smotherman and Narins 1999a), it would hardly be surprising to find that the molecular determinants of electrical tuning in these groups are achieved by similar genetic constructs. For other evolutionary questions, such as the secondary loss of the BP (or just the BP recess) in some salamanders, or the loss or reduction of the tympanum in the Caudata and Gymnophiona, it may turn out to be easier to understand the developmental mechanisms by which such a change occurred than it will be to infer the biological impetus that originally drove such adaptations.

In conclusion, we would suggest that the evolution of the amphibian auditory system occurred through a series of modular adaptations. Namely, multiple simple acoustic receptors were built from a few similar components (a basic round sensory macula, a tectorial mass, and a dedicated recess), and then each of these components was manipulated independently within each sense organ, while the organs themselves were moved about within the inner ear. For the most part, the amphibian inner ear has not converged on a single generalized architecture. As may also be true for the other classes of terrestrial vertebrates (see Manley, Chapter 7), the amphibian ear appears to have been a very pliable neural substrate that was capable of adapting in concert with each species unique behavioral traits.

References

Art JJ, Fettiplace R (1987) Variation of membrane properties in hair cells isolated from the turtle cochlea. J Physiol (Lond) 385:207–242.

Art JJ, Goodman MB (1996) Ionic conductances and hair cell tuning in the turtle cochlea. Ann N Y Acad Sci 781:103–122.

Baird IL (1974) Some aspects of the comparative anatomy and evolution of the inner ear in submammalian vertebrates. Brain Behav Evol 10:11–36.

Bergeijk WA van (1967) The evolution of vertebrate hearing. In: Neff WD (ed) Contributions to Sensory Physiology. New York: Academic Press, pp. 1–49.

Bergeijk WA van, Witschi W (1957) The basilar papilla of the anuran ear. Acat Anat 30:81–91.

Bever MM, Fekete DM (1999) Ventromedial focus of cell death is absent during development of *Xenopus* and zebrafish inner ears. J Neurocyt 28:781–793.

Bissonnette JP, Fekete DM (1996) Standard atlas of the gross anatomy of the developing inner ear of the chicken. J Comp Neurol 368:620–630.

Bolt JR, Lombard RE (1992) Nature and quality of the fossil evidence for otic evolution in early tetrapods. In: Webster DB, Fay RR, Popper AN (eds) The Evolutionary Biology of Hearing. New York: Springer-Verlag, pp. 377–403.

Boord RL, Grochow LB, Frishkopf LS (1970) Organization of the posterior ramus and ganglion of the VIIIth cranial nerve of the bullfrog Rana catesbeiana. MIT Res Lab Electr Quart Prog Rep 99: 180–182.

Burlett HMd (1934) Vergleichende Anatomie des statoakustishen Organs. In: Bolk L, Göppoert E, Kallius E, Lubosch W (eds) Handbuch der Vergleichenden Anatomie der Wirbeltiere, vol 2. Berlin: Urban und Schwarzenberg, pp. 1293–1492.

Callery EM, Fang H, Elinson RP (2001) Frogs without polliwogs: evolution of anuran direct development. Bioessays 23:233–241.

Capranica RR (1976) Morphology and physiology of the auditory system. In: Llinas R, Precht W (eds) Frog Neurobiology. Berlin: Springer-Verlag, pp. 551–575.

Capranica RR, Moffat AJM (1975) Selectivity of peripheral auditory systems of the spadefoot toad (*Scaphiopus couchi*) for sounds of biological significance. J Comp Physiol 100:231–249.

Christensen-Dalsgaard J, Elepfandt A (1995) Biophysics of underwater hearing in the clawed frog, Xenopus loevis. J. Comp Physiol [A] 176:317–324.

Christensen-Dalsgaard J, Jørgensen MB (1988) The response characteristics of vibration-sensitive saccular fibers in the grassfrog, *Rana temporaria*. J Comp Physiol [A] 162: 633–638.

Christensen-Dalsgaard J, Jørgensen MB (1996) Sound and vibration sensitivity of VIIIth nerve fibers in the grassfrog, *Rana temporaria*. J Comp Physiol [A] 179:437–445.

Christensen-Dalsgaard J, Narins PM (1993) Sound and vibration sensitivity of VIIIth nerve fibers in the frogs *Leptodactylus albilabris* and *Rana pipiens pipiens*. J Comp Physiol [A] 172:653–662.

Clack JA (1997) The evolution of tetrapod ears and the fossil record. Brain Behav Evol 50:198–212.

Clack JA (2002) Patterns and processes in the early evolution of the tetrapod ear. J Neurobiol 53:251–264.

Clack JA, Ahlberg PE, Finney SM, Dominguez A, Robinson J, Ketcham RA (2003) A uniquely specialised ear in a very early tetrapod. Nature 425:65–69.

Coates MI, Clack JA (1991) Fish-like gills and breathing in the earliest known tetrapod. Nature 352:234–236.

Corey DP, Hudspeth AJ (1979) Ionic basis of the receptor potential in a vertebrate hair cell. Nature 281:675–677.

Cortopassi KA, Lewis ER (1996) High-frequency tuning properties of bullfrog lagenar vestibular afferent fibers. J Vestib Res 6:105–119.

Cortopassi KA, Lewis ER (1998) A comparison of the linear tuning properties of two classes of axons in the bullfrog lagena. Brain Behav Evol 51:331–348.

Corwin JT (1977) Morphology of the macula neglecta in sharks of the genus *Carcharhinus*. J Morphol 152:341–362.

Corwin JT (1981) Audition in elasmobranchs. In: Tavolga WN, Popper AN, Fay RR (eds) Hearing and Sound Communication in Fishes. New York: Springer-Verlag, pp. 81–105.

Crawford AC, Fettiplace R (1981) Non-linearities in the responses of turtle hair cells. J Physiol (Lond) 315:317–338.

Duellman WE, Trueb L (1994) The Biology of Amphibians. Baltimore: Johns Hopkins University Press.

Egert D, Lewis ER (1995) Temperature-dependence of saccular nerve fiber response in the North American bullfrog. Hear Res 84:72–80.

Ehret G, Moffat AJ, Capranica RR (1983) Two-tone suppression in auditory nerve fibers of the green treefrog (*Hyla cinerea*). J Acoust Soc Am 73:2093–2095.

Ehret G, Tautz J, Schmitz B, Narins PM (1990) Hearing through the lungs: lung-eardrum transmission of sound in the frog *Eleutherodactylus coqui*. Naturwissenschaften 77: 192–194.

Ehret G, Keilwerth E, Kamada T (1994) The lung-eardrum pathway in three treefrog and four dendrobatid frog species: some properties of sound transmission. J Exp Biol 195: 329–343.

Fay RR, Edds-Walton PL (1997) Directional response properties of saccular afferents of the toadfish, *Opsanus tau*. Hear Res 111:1–21.

Fekete DM (1999) Development of the vertebrate inner ear: insights from knockouts and mutants. Trends Neurosci 22:263–269.

Feng AS, Narins PM, Capranica RR (1975) Three populations of primary auditory fibers in the bullfrog (*Rana catesbeiana*): their peripheral origins and sensitivities. J Comp Phys 100:221–229.

Flock Å (1965) Transducing mechanisms in the lateral line canal organ receptors. In: Frisch L (ed) Cold Spring Harbor Symposium on Quantitative Biology: Sensory Receptors, vol 30. Cold Spring Harbor: CSH Laboratory of Quantitative Biology, pp. 133–145.

Frishkopf LS, Geisler CD (1966) Peripheral origins of auditory responses from the eighth nerve of the bullfrog. J Acoust Soc Am 40:469–472.

Frishkopf LS, Goldstein MH (1963) Responses to acoustic stimuli from single units in the eighth nerve of the bullfrog. J Acoust Soc Am 35:1219–1228.

Fritzsch B (1987) Inner ear of the coelacanth fish *Latimeria* has tetrapod affinities. Nature 327:153–154.

Fritzsch B (1992) The water-to-land transition: evolution of the tetrapod basilar papilla, middle ear, and auditory nuclei. In: Webster DB, Fay RR, Popper AN (eds) The Evolutionary Biology of Hearing. New York: Springer-Verlag, pp. 351–375.

Fritzsch B (1998) Hearing in two worlds: theoretical and actual adaptive changes of the aquatic and terrestrial ear for sound reception. In: Fay RR, Popper AN (eds) Comparative Hearing: Fish and Amphibians. New York: Springer, pp. 15–42.

Fritzsch B (2001) Evolution and develoment of the vertebrate ear. Brain Res Bull 55: 711–721.

Fritzsch B, Neary T (1998) The octavolateralis system of mechanosensory and electrosensory organs. In: Heatwole H, Dawley EM (eds) Amphibian Biology, vol 3. Chipping Norton, England: Surrey Beatty & Sons, pp. 878–922.

Fritzsch B, Wake MH (1988) The inner ear of gymnophione amphibians and its nerve supply: a comparative study of regressive events in a complex sensory system (Amphibia, Gymnophiona). Zoomorph 108:201–217.

Fritzsch B, Barold K, Lomax M (1998) Early embryology of the vertebrate ear. In: Rubel E, Popper AN, Fay RR (eds) Development of the Auditory System. New York: Springer-Verlag, pp. 80–145.

Fritzsch B, Beisel KW, Jones KR, Farinas I, Maklad A, Lee JE, Reichardt LF (2002) Development and evolution of inner ear sensory epithelia and their innervation. J Neurobiol 53:143–156.

Fuchs PA, Evans MG (1990) Potassium currents in hair cells isolated from the cochlea of the chick. J Physiol (Lond) 429:529–551.

Garcia-Rutledge EJ, Narins PM (2001) Shared acoustic resources in an old world frog community. Herpetologica 57:104–116.

Geisler CD, Bergeijk WAv, Frishkopf LS (1964) The inner ear of the bullfrog. J Morphol 114:43–58.

Haddon C, Lewis J (1996) Early ear development in the embryo of the zebrafish, *Danio rerio*. J Comp Neurol 365:113–128.

Hertwig I (1987) Morphogenesis of the inner ear of *Rana temporaria* (Amphibia; Caudata). Zoomorphol 107:103–114.

Hertwig I, Schneider H (1986) Development of the supporting cells and structures derived from them in the inner ear of the grass frog, *Rana temporaria* (Amphibia; Caudata). Zoomorphol 106:137–146.

Hetherington TE (1985) The role of the opercularis muscle in seismic sensitivity in the bullfrog *Rana catesbeiana*. J Exp Zool 235:27–34.

Hetherington TE (1988) Biomechanics of vibration reception in the bullfrog *Rana catesbeiana*. J Comp Physiol 163:43–52.

Hetherington TE (1992) The effects of body size on the evolution of the amphibian middle ear. In: Webster DB, Fay RR, Popper AN (eds) The Evolutionary Biology of Hearing. New York: Springer-Verlag, pp. 421–437.

Hetherington TE (2001) Laser vibrometric studies of sound-induced motion of the body walls and lungs of salamanders and lizards: implications for lung-based hearing. J Comp Physiol [A] 187:499–507.

Hetherington TE, Lindquist ED (1999) Lung-based hearing in an "earless" anuran amphibian. J Comp Physiol [A] 184:395–401.

Holt JR, Eatock RA (1995) Inwardly rectifying currents of saccular hair cells from the leopard frog. J Neurophysiol 73:1484–1502.

Hudspeth AJ, Lewis RS (1988) Kinetic analysis of voltage- and ion-dependent conductances in saccular hair cells of the bull-frog, *Rana catesbeiana*. J Physiol (Lond) 400: 237–274.

Jarvik E (1980) Basic Structure and Evolution of Vertebrates. New York: Academic Press.

Jaslow AP, Hetherington TE, Lombard RE (1988) Structure and function of the amphibian middle ear. In: Fritzsch B, Ryan MJ, Wilczynski W, Hetherington TE, Walkowiak W (eds) The Evolution of the Amphibian Auditory System. New York: Wiley-Interscience, pp. 69–91.

Kil S-H, Collazo A (2001) Origins of inner ear sensory organs revealed by fate map and time-lapse analyses. Dev Biol 233:365–379.

Kingsbury BD, Reed HD (1909) The columella auris in amphibia. J. Morphol 20:549–628.

Koyama H, Lewis ER, Leverenz EL, Baird RA (1982) Acute seismic sensitivity in the bullfrog ear. Brain Res 250:168–172.

Lewis ER (1977a) Comparative scanning electron microscope study of the anuran basilar papilla. Ann Proc Electron Microsc Soc Am 35:632–633.

Lewis ER (1977b) Structural correlates of function in the anuran amphibian papilla. Scan Electron Microsc 2:429–439.

Lewis ER (1978) Comparative studies of the anuran auditory papillae. Scan Electron Microsc 2:633–642.

Lewis ER (1984) On the frog amphibian papilla. Scan Electron Microsc 43:1899–1913.

Lewis ER (1988) Tuning in the bullfrog ear. Biophys J 53:441–447.

Lewis ER, Lombard RE (1988) The amphibian inner ear. In: Fritzsch B, Ryan MJ, Wilczynski W, Hetherington TE, Walkowiak W (eds) The Evolution of the Amphibian Auditory System. New York: Wiley, pp. 93–123.

Lewis ER, Narins PM (1998) The acoustic periphery of amphibians: anatomy and physiology. In: Fay RR, Popper AN (eds) Comparative Hearing: Fish and Amphibians. New York: Springer-Verlag, pp. 101–154.

Lewis ER, Leverenz EL, Koyama H (1982) The tonotopic organization of the bullfrog amphibian papilla, an auditory organ lacking a basilar membrane. J Comp Physiol 145:437–445.

Lewis ER, Hecht EI, Narins PM (1992) Diversity of form in the amphibian papilla of Puerto Rican frogs. J Comp Physiol 171:421–435.

Lewis ER, Narins PM, Cortopassi KA, Yamada WM, Poinar EH, Moore SW, Yu X-L (2001) Do male white-lipped frogs use seismic signals for intraspecific communication? Am Zool 41:1185–1199.

Li CW, Lewis ER (1974) Morphogenesis of auditory receptor epithelia in the bullfrog. In: Johari O, Corvin I (eds) Scanning Electron Microscopy, vol 3. Chicago: IIT Research Institute, pp. 791–798.

Liff H (1969) Responses from single auditory units in the eighth nerve of the Leopard frog. J Acoust Soc Am 45:512–513.

Lombard RE (1977) Comparative morphology of the inner ear in salamanders (Caudata: Amphibia). Cont Vert Evol 2:1–140.

Lombard RE (1980) The structure of the amphibian auditory periphery: a unique experiment in terrestrial hearing. In: Popper AN, Fay RR (eds) Comparative Studies of Hearing in Vertebrates. New York: Springer-Verlag, pp. 121–138.

Lu Z, Popper AN (2001) Neural response directionality correlates of hair cell orientation in a teleost fish. J Comp Physiol [A] 187:453–465.

Manley GA (2000) Cochlear mechanisms from a phylogenetic viewpoint. Proc Natl Acad Sci USA 97:11736–11743.

Manley GA (2001) Evidence for an active process and a cochlear amplifier in nonmammals. J Neurophysiol 86:541–549.

Mason M, Narins PM (2002a) Vibrometric studies of the middle ear of the bullfrog (*Rana catesbeiana*) II: The operculum. J Exp Biol 205:3167–3176.

Mason MJ, Narins PM (2002b) Vibrometric studies of the middle ear of the bullfrog *Rana catesbeiana* I. The extrastapes. J Exp Biol 205:3153–3165.

Mason MJ, Lin CC, Narins PM (2003) Sex differences in the middle ear of the bullfrog (*Rana catesbeiana*). Brain Behav Evol 61:91–101.

Moffat AJM, Capranica RR (1976) Auditory sensitivity of the saccule in the American toad (*Bufo americanus*). J Comp Physiol 105:1–8.

Monath T (1965) The opercular apparatus of salamanders. J Morphol 116:149–170.

Narins PM, Lewis ER (1984) The vertebrate ear as an exquisite seismic sensor. J Acoust Soc Am 76:1384–1387.

Narins PM, Ehret G, Tautz J (1988) Accessory pathway for sound transfer in a neotropical frog. Proc Natl Acad Sci USA 85:1508–1512.

Narins PM, Lewis ER, McClelland BE (2000) Hyperextended call note repertoire of the endemic Madagascar treefrog *Boophis madagascariensis* (Rhacophoridae). J Zool (Lond) 250:283–298.

Noble GK (1931) The Biology of the Amphibian. New York: McGraw-Hill.

Patterson NF (1948) The development of the inner ear of *Xenopus laevis*. Proc R Soc Lond 119:269–291.

Pitchford S, Ashmore JF (1987) An electrical resonance in hair cells of the amphibian papilla of the frog *Rana temporaria*. Hear Res 27:75–83.

Platt C (1977) Hair cell distribution and orientation in goldfish otolithic organs. J Comp Neurol 172:283–287.

Popper AN, Fay RR (1998) The auditory periphery in fishes. In: Fay RR, Popper AN (eds) Comparative Hearing: Fish and Amphibians. New York: Springer, pp. 43–100.

Popper AN, Platt C (1983) Sensory surface of the saccule and lagena in the ears of ostariophysan fishes. J Morphol 176:121–129.

Purgue AP (1997) Tympanic sound radiation in the bullfrog *Rana catesbeiana*. J Comp Physiol [A] 181:438–445.

Purgue AP, Narins PM (2000a) Mechanics of the inner ear of the bullfrog (*Rana catesbeiana*): the contact membranes and the periotic canal. J Comp Physiol [A] 186:481–488.

Purgue AP, Narins PM (2000b) A model for energy flow in the inner ear of the bullfrog (*Rana catesbeiana*). J Comp Physiol [A] 186:489–495.

Retzius G (1881) Das Gehörorgan der Wirbelthiere. I. Gehörorgander Fische und Amphibien. Stockholm: Samson and Wallin.

Sachs MB (1964) Responses to acoustic stimuli from single units in the eighth nerve of the green frog. J Acoust Soc Am 36:1956–1958.

Sarasin P, Sarasin F (1890) Zur Entwicklungsgeschichte und Anatomie der ceylonesischen Blindwuhle *Ichthyophis glutinosis*. In: Das Gehörorgan, vol 2. Wiesbaden: Erg Naturwiss Forsch auf Ceylon, pp. 207–222.

Sarasin P, Sarasin F (1892) Uber das Gehörorgan der Ceaciliiden. Anat Anz 7:812–815.

Smotherman M, Narins P (1998) Effect of temperature on electrical resonance in leopard frog saccular hair cells. J Neurophysiol 79:312–321.

Smotherman M, Narins P (1999a) The electrical properties of auditory hair cells in the frog amphibian papilla. J Neurosci 19:5275–5292.

Smotherman M, Narins P (1999b) Potassium currents in auditory hair cells of the frog basilar papilla. Hear Res 132:117–130.

Smotherman M, Narins P (2000) Hair cell, hearing and hopping: a field guide to hair cell physiology in the frog. J Exp Biol 203:2237–2246.

Sokolowski BHA, Popper AN (1987) Gross and ultrastructural development of the saccule of the toadfish *Opsanus tau*. J Morphol 194:323–348.

Stejneger L (1899) Description of a new genus and species of Discoglossid toad from North America. Proc US Nat Mus 21:899–901.

Taylor EH (1969) Skulls of Gymnophiona and their significance in the taxonomy of the group. Univ Kansas Sci Bull 48:585–687.

Villiers CGS de (1934) Studies of the cranial anatomy of *Ascaphus truei Stejneger*. Bull Mus Comp Zool Harvard Coll 77:1–38.

Wever EG (1975) The caecilian ear. J Exp Zool 191:63–72.

Wever EG (1979) Middle ear muscles of the frog. Proc Natl Acad Sci USA 76:3031–3033.

Wever EG (1985) The Amphibian Ear. Princeton, NJ: Princeton University Press.

Wever EG, Gans C (1976) The caecilian ear: further observations. Proc Natl Acad Sci USA 73:3744–3746.

White JS (1986a) Comparative features of the surface morphology of the basilar papilla in five families of salamanders (Amphibia; Caudata). J Morphol 187:201–217.

White JS (1986b) Morphological and fine structural features of the basilar papilla in ambystomid salamanders (Amphibia; Caudata). J Morphol 187:181–199.

White JS, Baird IL (1982) Comparative morphological features of the caecilian inner ear with comments on the evolution of amphibian auditory structures. Scan Electron Microsc 3:1301–1312.

Wilczynski W, Capranica RR (1984) The auditory system of anuran amphibians. Prog Neurobiol 22:1–38.

Wilczynski W, Resler C, Capranica RR (1987) Tympanic and extratympanic sound transmission in the leopard frog. J Comp Physiol 161:659–669.

Witschi E (1949) The larval ear of the frog and its transformation during metamorphosis. Z Natur 4(b):230–242.

Witschi E (1955) The bronchial columella of the ear of larval Ranidae. J Morphol 96:497–512.

Wu YC, Art JJ, Goodman MB, Fettiplace R (1995) A kinetic description of the calcium-activated potassium channel and its application to electrical tuning of hair cells. Prog Biophys Mol Biol 63:131–158.

Yu XL, Lewis ER, Feld D (1991) Seismic and auditory tuning curves from bullfrog saccular and amphibian papillar axons. J Comp Physiol 169:241–248.

Zardoya R, Meyer A (2000) Mitochondrial evidence on the phylogentic position of caecilians (Amphibia; Gymnophiona). Genetics 155:765–775.

Zardoya R, Meyer A (2001) On the origin of and phylogenetic relationships among living amphibians. Proc Natl Acad Sci USA 98:7380–7383.

7
The Lizard Basilar Papilla and Its Evolution

GEOFFREY A. MANLEY

1. Introduction

Lizards are as structurally diverse as other groups of amniotes, such as mammals and birds. When we look at their hearing organs, however, the structural variety that we see in lizards exceeds that seen in any other amniote group. Its length alone can vary up to a factor of more than 40 times. In fact, this may well be the most variable sense organ seen in vertebrates (Manley 2000c,d).

In general, we associate structural variations with functional specializations. For example, the auditory foveae [a region of the auditory papilla expanded to accommodate a small but very important frequency region (Köppl 2001; Vater 2001)] of the barn owl and of some bat species clearly result from selection pressures for passive or active sound localization. In almost all lizards, however, there is no evidence for specialization to particular, known aspects of the species' lifestyles, in spite of a family-, genus-, and even species-specific structure of the hearing organ (Wever 1978; Miller 1980). This chapter shows that in spite of this enormous structural variety, the hearing organ of lizards demonstrates surprisingly little functional variation, which suggests that many of the structural changes were selectively equivalent. The lizard ear is arguably an excellent case to illustrate the controversial concept that has been termed "neutral" evolution, i.e., in this case changes in structure that are not the result of any particular selective pressure.

This chapter does not consider any changes that have occurred in the evolution of the middle ear in different lizard families. A number of morphological changes are observed in different groups, including loss of the tympanic membrane. Wever (1978) identified three general types of lizard middle ear, based partly on the structure of the extracolumella and the presence or absence of tympanic muscles. Except for similarities in middle-ear structure between related lizard families, however, few data provided insights into the evolution of middle-ear diversity. Wever also studied the efficiency of sound transmission in different species and found some substantial differences. Since then, little work has been carried out on differences between lizard middle ears that is relevant for asking questions about evolutionary trends.

2. The Lepidosaurs

Here is a brief review of some terms used to describe lizards and their relatives. The lizards are only one of several groups descended from a common ancestor in the lineage of the lepidosaurs (Fig. 7.1; see also Clack and Allin, Chapter 5). This group of reptiles evolved separately from the more distantly related archosaurs, a large group that includes the dinosaurs, the crocodilians, and their relatives, the birds. Most modern lepidosaurs are *squamates*, a group that includes both the true lizards and the snakes, but the group lepidosaurs also includes two species of the genus *Sphenodon*, the Tuatara "lizards" (they are not true lizards) of New Zealand.

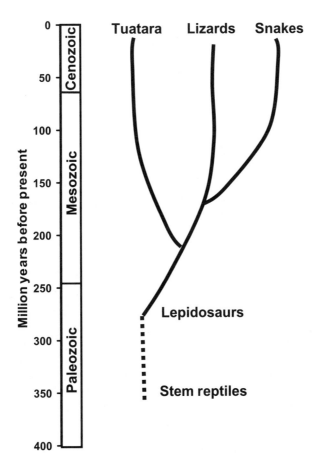

FIGURE 7.1. A schematic diagram showing the times of origin of the three subgroups of the lepidosaurs. (Modified from Manley 2002b. © 2002 Wiley Periodicals. Used with permission.)

The oldest lizard-like animals are known from deposits near the Permian-Triassic boundary [285 million years (MY) ago; Carroll 1988 and personal communication]. Seventy million years later, sphenodontid rhynchocephalians, the ancestors of *Sphenodon*, show up in the upper Triassic and lower Jurassic (215 MY). The lineage leading to snakes apparently separated from a varanoid-like lizard form sometime in the mid-Jurassic (180 MY). Sister taxa representing the modern lizard superfamilies first appeared in the upper Jurassic (145 MY) and the oldest true snakes are only a little more recent, dating from the lower Cretaceous (140 MY).

As shown by Clack and Allin (Chapter 5), the evolution of amniote ears experienced an enormously important boost when, during and after the Permian-Triassic transition (285 MY), middle ears suitable for the transmission of airborne sound developed independently in all amniote lineages. By then, animals that could have been the sister taxon of both sphenodontids and squamates also possessed a structure at least superficially resembling the impedance-matching middle ear of modern lizards (Carroll 1988). Comparative studies suggest that prior to the development of such a middle ear, the hearing component of the inner ear of all the diverging lines of amniotes was a rather simple structure. The newly developed ability to detect faint airborne sounds using the tympanic middle ear initiated a period of strong selection pressures in all amniote groups, but especially in those groups that then began to use such sounds for prey detection and/or interspecific communication. Since the amniote lineages had already diverged in the Paleozoic, their further evolutionary history ran completely independently of each other (Manley 2000a). To suggest what changes occurred early in lepidosaur evolution, we need to understand the structure of the auditory papilla of the stem reptiles in cladistic terms, which will help elucidate the plesiomorphic state.

3. The Hearing Organ of Stem Reptiles

The ears of modern mammals, archosaurs, and lepidosaurs are so diverse that it is not possible to conceive of a common ancestor that already had a complex inner ear. This is unlikely anyway, as the middle-ear systems enabling sensitive hearing of airborne sounds—and thus driving the evolution of more complex inner ears—did not develop until after the divergence of the various amniote groups. The hearing organ of stem reptiles was thus most probably a relatively simple structure. The simple basilar papilla of turtles is the most primitive among the amniotes (Miller 1980) and it has probably hardly changed over a very long period of time, thus giving us an idea as to the structure of the stem-reptile papilla. This idea is supported by information on the inner ear of primitive lepidosaurs of the sphenodontid group (see below).

Miller (1978, 1980) regarded the turtle papilla as being "probably not much different from that possessed by reptilian stem stock." Such papillae are generally less than 1 mm long and contain several hundred to one thousand hair cells (Fig. 7.2A). In a broadly based comparison to lepidosaur groups, Miller identified as follows those features of the turtle papilla that he considered to be primitive:

1. There is a low packing density of hair cells (156–247 hair cells per 10,000 µm^2).
2. The stereovillar bundles of the hair cells are all oriented in the same direction, which Miller calls "unidirectional kinocilial orientation."
3. All hair cells are innervated by both afferent and efferent nerve fibers.
4. All hair cells are covered by a thick, continuous tectorial membrane.
5. There is a general lack of special cytological specialization of hair cells and supporting cells.
6. The hair cell bundles contain a relatively large number of stereovilli (>74).

What Miller could not have known at that time is that turtle papillae are made up of hair cells whose response selectivity—that is, the mechanism determining their selectivity to different frequencies—is mainly determined by the properties of ion channels in the cell membrane. This elegant mechanism was subsequently thoroughly described (reviewed in Hudspeth 1997; Fettiplace and Fuchs 1999) and is often called electrical tuning. In essence, the hair cells receive a mechanical input (their bundle is sheared by relative movement to the tectorial structure over them), and this opens transduction channels in the bundle. The resulting depolarization of the cell membrane opens voltage-sensitive calcium

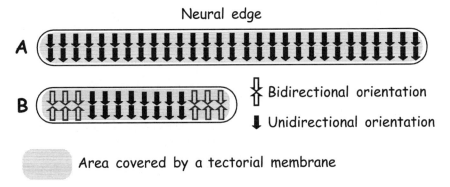

FIGURE 7.2. Primitive and more advanced lizard papillar configurations. **A**: A primitive lepidosaur papilla. It supports many hair cells whose bundles are all oriented in the same direction. Such a papilla would only have responded to low sound frequencies (below about 1 kHz). **B**: An early squamate papilla, which is presumed to have been shorter than that in A, although this is not a necessary assumption. Here, the original, unidirectional hair cell area is confined to the center of the papilla. This area is flanked at both ends by newly developed hair cell areas in which the hair cell bundles face each other and the hair cells respond preferably to sound frequencies above 1 kHz. (From Manley 2002b. © 2002 Wiley Periodicals. Used with permission.)

channels, admitting calcium ions into the cell. These ions can attach to the inside of calcium-sensitive potassium channels in the cell membrane, which are then opened in turn (Fig. 7.3A). This leads finally to potassium outflow through these channels, which pushes the cell's membrane potential back in a hyperpolarizing direction, and this closes the voltage-sensitive calcium channels. The calcium ions are quickly sequestered inside the cell. Thus this process is self-regenerative, but it is restarted at every cycle of the sound wave as long as a sound stimulus is present. The number of channels involved and their kinetics

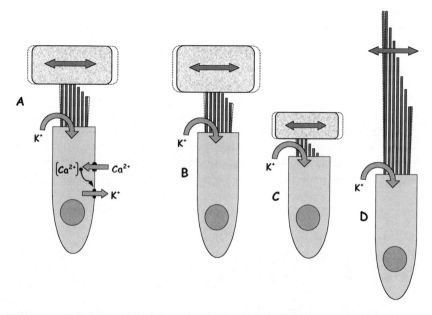

FIGURE 7.3. Two known mechanisms determining hair cell frequency responses. **A**: The biophysical properties of the lateral cell membrane determine the best frequency. In such hair cells, potassium ions that enter through the apical transduction channels of the stereovillar bundle activate voltage-sensitive calcium channels in the lateral membrane. Both the potassium and the calcium ions depolarize the cell membrane. The calcium ions that enter the cell activate calcium-dependent potassium channels of the lateral membrane, leading to an increased loss of potassium ions and a repolarization of the cell membrane. The kinetics of the various channels and their density work together to induce the largest voltage oscillations at a particular sound frequency. **B–D**: The preferred frequency is primarily determined not by cell membrane ion channels, but by the micromechanical properties of the hair cell stereovillar bundle and any tectorial material attached to it. An increase in tectorial mass and a lengthening of the stereovillar bundle will both lead to a lower preferred response frequency. Thus cell B would respond to a substantially lower frequency than cell C. Loss of the tectorial material, as in cell C, can—within limits—be compensated for by a prominent elongation of the stereovillar bundle. (Modified from Manley 2000a, with permission.)

are adjusted for each cell in such a way that the rapidity of the ionic events is such as to operate best (i.e., produce the largest swings of cell potential) at a particular sound frequency—the cell's "best" or "characteristic" frequency. By changing the numbers of channels and their properties systematically along the epithelium, the characteristic frequencies of the hair cells are arranged with low preferred frequencies at the apical end of the papilla and rising steadily to high preferred frequencies at the basal end of the papilla. The papilla is thus tonotopically organized. This kind of frequency-selective mechanism has a temperature-dependent, intrinsic upper frequency limit (Wu et al. 1995) that is partly the result of constraints in the number and kinetics of the channels that can be packed into the cell membrane. In many respects, this primitive papilla resembles some of the vestibular epithelia from which it is derived. It is unique, however, in that it is supported by a freely suspended basilar membrane, whereas all vestibular epithelia are over "solid ground." The highest preferred frequencies found in the turtle papilla are below 1 kHz.

As noted above, this papilla represents the plesiomorphic, or ancestral state, for amniotes, including the early lepidosaurs. An auditory papilla of this kind, responsive only to low frequencies, was presumably adequate to cope with an auditory input that at that stage in evolution was conducted not through a specialized middle ear, but through the tissues of the head. The large bones of this head region in stem reptiles (see Clack and Allin, Chapter 5) would only have been moved by rather substantial, low-frequency forces. During their evolution from this plesiomorphic condition, each of the main lines of amniotes, the Lepidosauria, Archosauria, and Mammalia, independently developed specializations of their auditory papillae, and each shows unique apomorphies (i.e., features newly derived in evolution; see Manley and Clack, Chapter 1). One of the main questions underlying these developments in the various lines diverging from the stem reptiles is: What evolutionary event, or what selective pressures, led to this?

As noted above, the clue seems to lie in the fact that during the late Permian–early Triassic periods (220 to 250 MY ago), each of the groups independently developed a tympanic middle ear (see Clack and Allin, Chapter 5). The term *tympanic* emphasizes developments in the region of the middle ear (that can be observed in fossil remains) involving an eardrum or tympanic membrane. These changes led to the development of middle-ear systems that were dramatically (perhaps 40 to 60 dB) more sensitive to airborne sounds. It was presumably the resultant increased prevalence and salience of sound stimuli to the inner ear that formed a powerful selective pressure for the expansion and specialization of the hearing organs of their descendants. The three characteristic basic patterns unique to mammals, birds, and lizards suggest that, following the arrival of tympanic middle ears, these inner-ear epithelial patterns developed in each lineage quite rapidly.

4. The Early Evolution of Lepidosaur Basilar Papillae

There is a consensus that the small group of rhynchocephalians—the tuatara lizards (genus *Sphenodon*) of New Zealand—represent the most primitive lepidosaurs (Carroll 1988). It is thus not surprising that their auditory papilla strongly resembles that of the turtles. Small extensions of the hair cell region beyond the free basilar membrane and on to the surface of the limbic support at both ends, typical for turtles, are also seen in *Sphenodon* (Wever 1978). Even though we can only see the similarities from the anatomy, it is highly likely that its function is also closely similar. According to Wever (1978), *Sphenodon* only has 225 hair cells, has only low-frequency hearing, and is about 20 to 30 dB less sensitive than good lizard sensitivity (and thus similar to turtles). Since these animals are highly endangered, we may never know more about their inner-ear function than this.

In cladistic terms, the sphenodontid rhynchocephalians can be regarded as an outgroup for the *Squamata*, the lizards and snakes. On this basis, we can also conclude that the turtle-type auditory papilla is the plesiomorphic form for squamates. It was in this condition that the lepidosaur ancestors developed a new middle ear that closely resembles that of the other diapsid descendants, the birds and other archosaur groups. This development made it possible to increase the sensitivity of hearing, but it also made available frequencies higher than 1 kHz. The upper frequency hearing limits of lepidosaurs and archosaurs (including birds) lies today between about 4 and 12 kHz (Manley 1990).

A comparative study (see below) makes it possible to conclude that in squamates, selection pressures to increase the upper frequency response limit of the auditory papilla led very early to the extension and specialization of those hair cell areas found at both ends of the ancestral papilla (Fig. 7.2B). It is very probable that this happened before the various ancestors of the lizard superfamilies diverged in the upper Jurassic (145 MY). In extant lizards, the basilar papilla always has two clearly recognizable types of hair cell areas at different locations along the length of the papilla. Since no other amniotes show this feature, it can be assumed to be synapomorphic to the lizards. This change was accompanied by a rise in hair cell densities, to 320 to 400 hair cells per 10,000 μm^2 (Miller 1978).

Electrophysiological studies (reviewed in Manley 1990) have shown that one of these areas consistently contains hair cells that respond to low frequencies, with an upper frequency limit near 1 kHz. This hair cell area almost certainly derives from and corresponds to the entire papilla of stem reptiles. In the other, new type of hair cell area the hair cells respond best to frequencies above 1 kHz, maximally up to about 8 kHz. The upper limit is species-specific and the selectivity of these hair cells to such frequencies is not based on electrical tuning (see below).

Only in lizards do we find two groups of hair cells that are separated functionally and spatially into a low-frequency and one or two high-frequency groups, based on fundamentally different mechanisms of frequency selectivity. This arrangement arose completely independently of the hair cell specializations seen on the one hand in birds (and crocodilians) and in mammals on the other. Although the evidence is still inconclusive (Manley 1990), we assume that the low-frequency, generally unidirectionally oriented hair cell area in lizards still mainly uses the ancestral electrical tuning mechanism. This is compatible with the lack of strong anatomical gradients in the hair cell bundles of this area. Although there are few data available, the tonotopic axis in this area, that is, the direction in which frequency changes from place to place, can lie either across the papilla's width or along the papilla's length. This tonotopicity follows the orientation of the anatomical gradients that do exist (Manley 1990). Whereas in the Tokay gecko electrophysiological data show that the low-frequency tonotopic axis runs along the papilla (Manley et al. 1999), in the bobtail skink the low frequencies increase from the abneural to the neural side (Köppl and Manley 1992). In the former case, the tonotopic organization of the low-frequency area continues seamlessly into the high-frequency area. In the latter case, this is not so. There is thus obviously no functional necessity for frequency continuity of the ends of the tonotopic distributions of the two hair cell areas. (Fig. 7.4).

The newly developed hair cell area in lizards consisted of hair cells that relied not on electrical tuning but rather on a mechanism of frequency selectivity known as micromechanical tuning. In such cells, the frequency selectivity results from the mechanical properties of the hair cell bundles (such as the bundle stiffness) and of the structures that surround them, such as the mass of the tectorial membrane. Location-specific variations in these properties (e.g., gradients in bundle stiffness) are the basis for the tonotopic organization (Fig. 7.5). Since even in primitive papillae, such as that of the turtle, weak anatomical gradients in hair cell bundle structure are seen, the basis for rudimentary micromechanical tuning was present very early in hair cell evolution. The new hair cell area(s) show much stronger anatomical gradients, however, and modeling work (e.g., Authier and Manley 1995; Manley et al. 1989) has demonstrated that these gradients are consistent with a micromechanical mechanism for frequency tuning and selectivity. The gradient might be such that at one end of this micromechanically tuned hair cell area, the cells look like the cell in Fig. 7.3B, and at the other end like the cell in Fig. 7.3C. Micromechanical tuning was not limited in its upper frequency by the same constraints as those on electrical tuning. This fact played a critical role during the evolution of responses to increasingly high frequencies.

Another important feature of micromechanical tuning also warrants mentioning here, as it strongly influences papillar evolution in lizards. Whereas, in principle, hair cells that are electrically tuned are not influenced by the tuning characteristics of immediately neighboring hair cells, this is not the case for micromechanically tuned hair cells. Since the mechanics of bundle motion are critical in such cells, and they are generally physically coupled through a tectorial membrane, each cell will influence the response of its neighbors to a

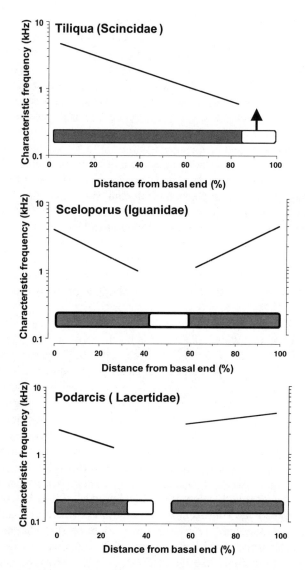

FIGURE 7.4. A graphical presentation of the tonotopic arrangement in three kinds of lizard papillae, in each case with a drawing of the hair cell area arrangement. In the bobtail skink *Tiliqua rugosa*, there are two hair cell areas, a long, basally lying area (gray strip) that responds in a graded fashion to frequencies between about 1 and 5 kHz, and a short, apical area (white strip) whose tonotopic axis is oriented *across* the papilla (arrow). In *Sceloporus orcutti*, the two flanking areas show mirror-image responses to higher frequencies. In *Podarcis* species, there is a physical separation of three hair cell areas. In the basal section, the hair cell areas resond to low (white) and midfrequencies (gray). Apically, the (gray) area responds to the highest frequencies. The axis of tonotopicity of the centrally lying, low-frequency area is uncertain in both *Sceloporus* and *Podarcis*. (The lower two panels from Manley 2002b. © 2002 Wiley Periodicals. Used with permission.)

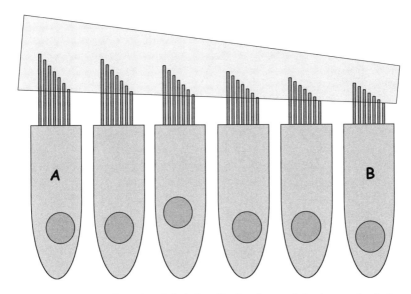

FIGURE 7.5. A sketch of a strip of six hair cells showing graded micromechanical properties of their stereovillar bundles (taller to the left, **A**) and tectorial material (thicker and thus more massive on the left). These morphological trends are complementary, both leading to higher frequency responses in the hair cells on the right (**B**).

greater or lesser degree (Fig. 7.5). To reduce the magnitude of such effects, it is necessary either to widely separate such cells (a solution not realized, because this changes the mechanical relationships of stiffness and mass) or to have many hair cells, with neighbors responding to similar frequencies. The latter is only possible—while maintaining a selective response capability over a wide frequency range—if there is a large number of hair cells in an elongated epithelium. Thus micromechanical tuning, by its very nature, induced a strong selection pressure for an increase in hair cell numbers and, concomitantly, in the total length of the auditory papilla. It is thus not surprising that in all amniote groups that rely on micromechanical tuning, the papillar length increased during evolution. In lizards, however, this increase is only seen in some families (for reasons explained below) and is smaller than that seen in birds and in mammals (Manley 1990). Nonetheless, in lizard papillae in general, the micromechanically tuned area is the longer of the different hair cell areas.

A characteristic feature of micromechanically tuned hair cell areas in lizard papillae is that they are always made up of hair cells that are oppositely oriented, that is, the stereovillar bundles of one group are rotated 180 degrees relative to those of the other group (Figs. 7.2B and 7.6). This arrangement, known as bidirectional orientation, is also found in some vestibular organs and lateral-line neuromasts (see Coffin et al., Chapter 3). In the lizard basilar papilla, the bidirectional organization may be very strict, as for example in geckos and skinks,

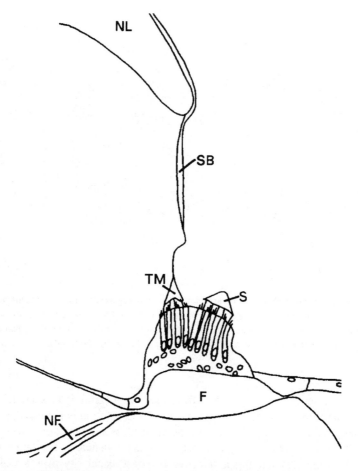

FIGURE 7.6. A diagrammatic cross section of the high-frequency, bidirectionally oriented hair cell area in the apical third of the basilar papilla of the Tokay, *Gekko gecko*, to illustrate the papillary bar or fundus (F) beneath the basilar membrane and the opposite orientation of hair cells. In *Gekko*, the hair cell orientation changes twice across the high-frequency region. Here, there are two parallel-lying hair cell areas with different tectorial coverings, a tectorial membrane (TM) connected by a thin curtain to the neural limbus (NL) on the neural side of the papilla, and sallets (S) over the abneural hair cell area. NF nerve fibers. SB spindle body. (From Köppl and Authier. ©1995 Elsevier. Used with permission.)

or rather loose, as in teiids and varanids (Miller 1980, 1992). We are still not able to say for sure whether any functional change led to this arrangement. It is, however, probably linked to another synapomorphic, or evolutionarily early, feature of lizard papillae, as follows.

In contrast to the situation in turtles, the basilar membrane of lizards is supported by a longitudinal thickening beneath it that Miller (1980) called a papillary bar and Wever (1978) referred to as the fundus (Fig. 7.6). This thickening means that the basilar membrane is not very flexible in the longitudinal direction, which explains why experimental studies have shown that the basilar membrane demonstrates no differential frequency selectivity along its length (Frischkopf and DeRosier 1983; Holton and Hudspeth 1983; Manley et al. 1988). The basilar membrane is thus not thrown into longitudinal waves like those observed in mammals. The in-vitro studies of Frischkopf and DeRosier (1983) and Holton and Hudspeth (1983) showed that instead, the basilar membrane pivots about the neural edge, resulting in a larger up-and-down excursion on the abneural edge. This results in a rotational component to the movement of the abneural edge and the top of the papilla. The hair cell stimulus thus results from the drag of the fluid on the tectorial structure and on the stereovillar bundles. From this movement, a phase difference between the movement of the top and the bottom of the hair cells results, the magnitude of which is determined by the micromechanical properties of the hair cell bundles. It is likely that a bidirectionally oriented hair cell arrangement is better suited to respond in this kind of situation.

Additionally, the presence of two, oppositely oriented groups of hair cells is related to an additional feature of lizard hearing. We have recently shown through in-vivo studies of the Australian bobtail skink *Tiliqua rugosa* that the active process of lizard hair cells is located in the stereovillar bundles (Manley et al. 2001). This active process, responsible for the cochlear amplifier, has also been studied in vitro in the bundles of turtle auditory hair cells and bullfrog saccular hair cells (review in Hudspeth 1997; Manley 2002a). In those preparations, the mechanism driving active bundle motions is the result of calcium binding to open transduction channels, which induces their premature closure (Choe et al. 1998). This exerts a force on the bundle that, if the relative phase is appropriate, can add mechanical energy to the bundle's movement. Such motions also occur spontaneously (Fettiplace and Fuchs 1999; Martin et al. 2003) and result in the presence of sounds that are emitted from the ear, known as spontaneous otoacoustic emissions (reviewed by Köppl 1995; Manley 2000b).

In the context of bidirectionally oriented groups of hair cells, it is likely that spontaneous mechanical activity in one group of hair cells with a particular bundle orientation can actually stimulate a neighboring hair cell group of opposite bundle orientation, inducing that cell group to oscillate further and in turn. This could establish a mutual-stimulation rocking of the papilla even in the absence of a sound stimulus. Theoretical considerations show that such "noise" in fact can lead to an enhanced sensitivity of the organ (Choe et al. 1998). Incidentally, such spontaneous rocking motions could initiate rhythmic activity in hair cells, resulting in the well-known phenomenon of preferred intervals in the spontaneous activity of auditory-nerve afferent fibers (see Manley 1990).

5. The Evolutionary Divergence of Lizard Families and of Their Papillae

The lizards and snakes form the largest groups that arose within the diapsid lepidosaurs, and these two groups diverged at the latest in the early Cretaceous (Fig. 7.1). To discuss the ancestry and relationships of lizard groups in this chapter, I will use a cladogram that is the consensus of extensive comparisons of both soft tissue and osteological features (Fig. 7.7; Lee 1997, 2000). As noted above, sister taxa of the modern superfamilies of lizards are known since the upper Jurassic, 145 million years ago.

The lizards diversified greatly and this diversification was accompanied by a remarkable differentiation of hearing-organ structure, evolving a family-, subfamily-, genus-, and in some cases even species-specific anatomy. The snakes apparently arose from a common ancestor with monitor lizards in the middle to late Jurassic (Caldwell and Lee 1997; Tchernov et al. 2000). This fact suggests that snakes, which have a simple, *Sphenodon*-like papilla, have secondarily lost the more complex lizard type of auditory organ. Presumably

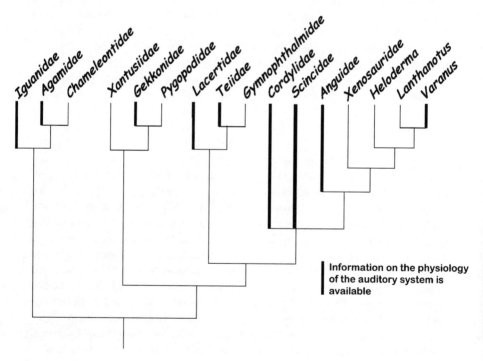

FIGURE 7.7. A cladogram of the evolutionary relationships between lizard families (after Lee 1998). The thicker connecting line indicates that for that family, some physiological data are available (Manley 1990, 2000d). (From Manley 2002b. © 2002 Wiley Periodicals. Used with permission.)

due to the aquatic or burrowing lifestyle of their ancestors (Lee 1997; Caldwell 1999; Clack and Allin, Chapter 5) and changes in eating habits with the consequent reduction of the middle ear (e.g., loss of the eardrum), they have reverted to a simpler papillar structure similar to that of the plesiomorphic condition and responding only to low frequencies (Hartline 1971; Wever 1978). Despite this, snake papillae can be quite large, especially in the more primitive burrowing snakes, which contain up to 1,574 hair cells (Wever 1978). The hair cell density

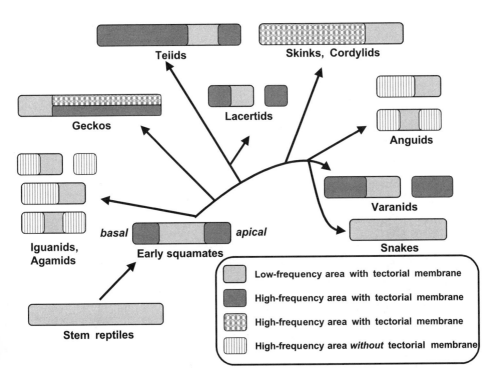

FIGURE 7.8. A schematic of probable evolutionary changes in papillar structure during the evolution of lizards. Some structural features, such as a physical division into subpapillae or a shortened papilla with hair cell areas lacking a tectorial membrane, have arisen more than once. This scheme makes it possible to explain the unusual anatomy and the reversed tonotopic organization of the auditory papilla in geckos. Areas in light gray are low-frequency areas, and their hair cells are always covered by a tectorial membrane and are generally (but not always, see Manley 1990) all oriented abneurally. A darker gray coloration indicates the structure of the tectorial covering of different hair cell areas. A continuous gray indicates a continuous tectorial membrane as in birds and mammals and, indeed, in archosaur and chelonian reptiles. A checkered gray area indicates the presence of tectorial sallets, and a striped gray area completely lacks a tectorial covering. (After Manley 2000c. © 2000 Wiley-VCH Verlag GmbH. Used with permission.)

found in snake papillae [168–250 hair cells per 10,000 μm^2 (Miller 1978)], is low and comparable to that of the turtle, but also to the low-frequency area of lizard papillae. Structural simplification and some loss of function due perhaps to a burrowing lifestyle has also occurred within some lizard families, for example, the amphisbaenids.

Miller (1980, 1992) suggested that within the lizard families, the most primitive structural pattern of the basilar papilla is represented by those families in which the basilar papilla is generally made up of three hair cell areas (some iguanid, agamid, and anguid lizards). The central of these hair cell area responds to low frequencies and corresponds to the plesiomorphic state of stem reptiles (Manley 1990). Here, all the hair cell stereovillar bundles have the same orientation (abneural, unidirectional orientation). The other two hair cell areas, which respond to frequencies above 1 kHz, flank the central area at both ends and are anatomical and functional mirror images of each other (Figs. 7.2B and 7.8). As noted above, these apomorphic, high-frequency areas arose at both ends of the earliest lizard papilla. These two areas likely contained roughly equally large groups of hair cells, and within each group, the stereovillar bundles faced each other (bidirectional orientation). These are also the hair cell areas that respond to frequencies above 1 kHz (Manley 1990).

Miller's (1980) suggestion is supported not only by his and Wever's (1978) extensive comparative anatomical studies, but also by what we know about the function of these papillae. The most parsimonious explanation for the diversity of modern lizard papillae is reached by placing such a tripartite papilla at the base of the lizard cladogram (Fig. 7.8; Manley 2002b). From this basic morphological pattern as a starting point, a remarkable array of evolutionary trends over time is discernible in the different lizard families (Figs. 7.8 and 7.9).

In some lizard families, for example, one of the bidirectional hair cell areas was lost in some or in all species (Figs. 7.8 and 7.9). In some families (e.g., skinks, cordylids), the apical bidirectional area disappeared, and thus the ple-

⎯⎯⎯⎯⎯⎯⎯⎯⎯⎯⎯⎯⎯⎯⎯⎯⎯⎯⎯⎯⎯⎯⎯⎯⎯⎯⎯⎯⎯⎯⎯⎯⎯⎯▶

FIGURE 7.9. A plot showing the apomorphies that arose during the evolution of the lizard basilar papilla. Next to each family name (top) is a small pictogram showing the average length of papillae in each family. The scale is oriented on that of the geckos, which is 2 mm long. Thus it can easily be seen that papillar length varies greatly throughout the lizards. For the purposes of this figure, it is assumed that the stem squamate papilla consisted of three hair cell areas (B, basal area; LF, low-frequency area; A, apical area). During the evolution of the various lineages, loss of either area A or B occurred several times independently; in some groups the papilla was divided into subpapillae. It should be noted that, although there is a continuing discussion as to the placement of the family Xantusiidae, the structure of their papilla most resembles that of skinks and cordylids and certainly not that of geckos. [Modified from Lee's (2000) strict consensus tree.]

7. Evolution of the Lizard Basilar Papilla 215

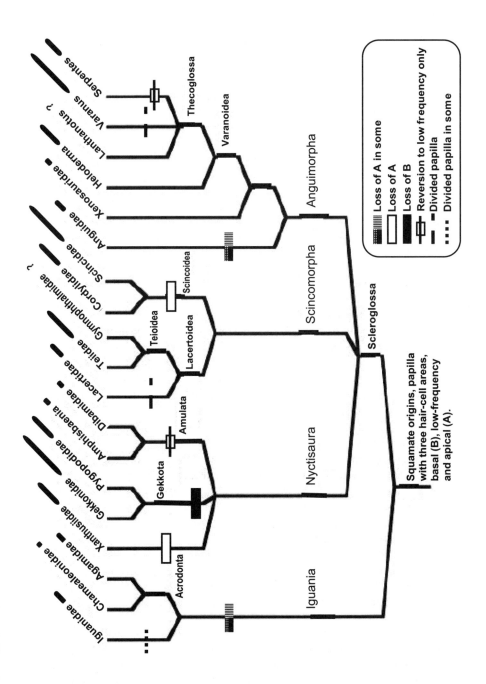

siomorphic, or ancient, low-frequency unidirectional area lies at the apical end of the papilla. The frequency organization of such papillae, in which the lowest frequencies are processed at the apical end of the papilla (Manley 1990) thus (convergently) resembles the organization of all bird and mammal papillae. It has thus been easily accepted as the norm without concern for its independent origin. In contrast to this situation (and uniquely among amniotes), in geckos and pygopods it was the basal bidirectional, high-frequency area that disappeared, placing the low-frequency area at the papillar base. Thus in geckos, the direction of the tonotopic organization is reversed as compared to almost all other amniotes (Manley et al. 1999).

In lacertid, varanid, and basiliscine iguanid lizards, the papilla is even physically divided into two subpapillae (Figs. 7.8 and 7.9). In these species (and presumably linked causally to division of the papilla), the two bidirectional areas have lost the structural mirror imagery, and have specialized and diverged structurally. One consequence of this specialization is that the tonotopic organizations of the two flanking hair cell areas also lose their functional mirror imagery (Manley 1990; Köppl and Manley 1992). The consequence is that each hair cell area is responsible for responses to only part of the high-frequency range, eliminating redundancy and presumably improving the processing of these frequencies by spreading them over more hair cells (Manley 2002b).

In many lizard families, there was also a pronounced tendency over time to elongation of the papilla, an effect that is most pronounced in the high-frequency area(s). The maximal papillar length reached, however, is seldom more than 2 mm (reviews in Wever 1978; Manley 1990). This elongation is accompanied by a roughly proportional increase in the number of hair cells, up to a maximum in lizards of about 2000 (Miller and Beck 1988; Miller 1992). The possible physiological consequences of such variations in structural patterns is discussed below.

As can be seen from Figs. 7.8 and 7.9, the evolution of structural features of the papilla in the various developmental lines of lizards was anything but orderly. Papillar elongation, papillar shortening, loss of hair cell areas, and papillar subdivisions occur in different lineages independently and apparently randomly. There seems to be no rhyme or reason for the changes; even within groups of related families, we observe large differences in papillar size and structure. We can only begin to interpret this apparently rather chaotic situation by understanding the functional consequences, if any, of the structural changes seen.

6. What Functional Changes Accompanied the Structural Ones?

Only those structural changes that result in a change of function at the level of the auditory nerve and auditory brain can be the subject of evolutionary selection pressures. In other words, only those structural changes that influenced the physiological input to the brain were "visible" to selection pressures. The types of structural changes that need to be discussed include papillar elongation, papillar shortening, loss of hair cell areas, and the division into subpapillae.

6.1 Papillar Elongation

Not knowing the size of the papilla of the earliest lizards makes it impossible to know the extent of elongation that has occurred. Assuming that early lizard papillae were about 1 mm long or smaller, as in most turtles, some lizard families have achieved elongations of up to double this length. There is a fairly tight relationship between papillar length and the number of hair cells, so that this change in length implies a doubling of hair cell numbers and, potentially at least, a doubling of the quantity of information passing to the brain about the sound stimulus. However, as Miller and Beck (1988) clearly showed, the number of afferent nerve fibers is not directly proportional to the number of hair cells. Instead, there is a tendency for hair cells in long papillae to be nonexclusively innervated, that is, the afferent nerve fibers contact more than one hair cell. This change in innervation pattern reduces the number of information channels. Thus only in some species, such as geckos, where the hair cell/nerve fiber ratio is nearly one (Miller, 1985) and the papilla is large, did papillar elongation produce large increases in the information input to the brain. Geckos are known to communicate using acoustic signals, and their hearing is perhaps the most sensitive and frequency selective among lizards (Manley 1990). In this family at least, papillar elongation can be correlated with positive evolutionary changes toward a greater importance of hearing in the animal's behavior and, presumably, its survival and reproductive success.

6.2 Papillar Shortening

The basilar papilla in many families has shortened, in some cases down to a size where one may be justified in asking whether the few remaining hair cells (sometimes as few as 50) have an important function at all. However, it should be noted that there is no known case of the total loss of the hearing organ, suggesting that even a small structure plays a useful role. As noted above, papillar shortening mainly manifests itself in a reduction of the high-frequency hair cell areas. The potential problems arising through the coupling of such closely lying hair cells in small papillae (coupling would tend to force the hair cells to respond to similar frequencies) are at least partly counteracted by an accompanying loss of the tectorial membrane. Complete loss of the tectorial membrane occurs only in short papillae. Without a tectorial connection, individual hair cells can respond more independently of their neighbors and thus to different characteristic frequencies. Since in such papilla the number of nerve fibers often greatly exceeds the number of hair cells and the hair cells are generally exclusively innervated, these papillae can also pass a significant amount of information to the brain.

This point is illustrated in Table 7.1. Comparing the iguanid species *Scelo-*

TABLE 7.1. Comparison of the iguanid species.

Feature	Gekko gecko	Sceloporus orcutti	Ophiosaurus/ Gerrhonotus
No. of nerve fibers (NF)	1885	600	799
No. of hair cells (HC)	2100	90	225
NF/HC ratio	0.9	6.7	3.6
Frequency range	4.85	4.05	3.8
Q_{10} max	12	4.5	4.5

porus orcutti and the gekkonid *Gekko gecko*, it can be seen that whereas *G. gecko* has 23 times as many hair cells in its long papilla (papillar length about 2 mm) (Köppl and Authier 1995), the number of nerve fibers in *G. gecko* is only about three times higher than in *S. orcutti* (length 0.3 mm). These structural and innervational measures thus counteract papillar shortening and the impact of shortening on information flow to the brain is not proportional to the change in papillar length. Even tiny papillae are able to communicate relatively sensitive and frequency-selective information to the brain. This information may be perfectly adequate for behavioral tasks such as alerting to imminent danger, etc. Under these conditions, it can be expected that in some species, selection pressures to maintain long papillae were not particularly strong.

Although only limited data are available, the loss of the tectorial membrane that usually accompanied papillar shortening apparently did have physiological consequences. This phenomenon, discussed at some length by Authier and Manley (1995), is the result of the fact that the coupling of oscillators (in this case the tectorial membrane plays the role of coupling the hair cells as oscillators) leads to a frequency-response characteristic that is narrower (i.e., more frequency selective) and has deeper skirts (is more sensitive). The result of the presence or absence of coupling can be seen by comparing sensitivity and frequency selectivity in those papillae with and those without a tectorial membrane (TM) (Fig. 7.10). Here, there is evidence that through abandoning the TM, papillae become less sensitive and there is a loss of frequency selectivity. Thus, the auditory nerve fibers of *Gerrhonotus multicarinatus* are about 10 dB less sensitive and only half as frequency selective as those of *Tiliqua rugosa* (Fig. 7.10). A similar comparison could be made to data from *Sceloporus*, which lacks a tectorial membrane over its higher frequency hair cells and *Gekko*, which, like *Tiliqua*, has tectorial sallets (further details in Manley 1990).

6.3 Loss of One of the Two High-Frequency Hair Cell Areas

Since one or the other high-frequency hair cell area has been lost in a number of families and sporadically within some families, there was probably not a high selection pressure to maintain both of them. This is presumably due to the fact

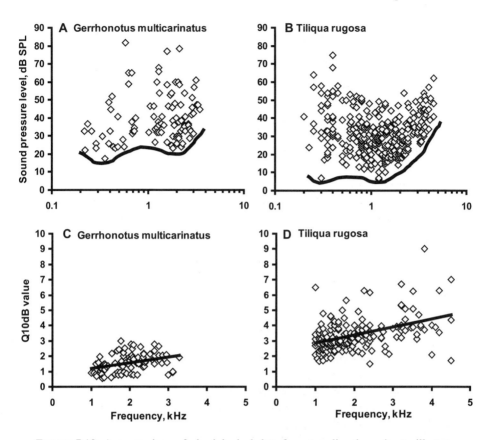

FIGURE 7.10. A comparison of physiological data from two lizard species to illustrate the two consequences of the loss of a tectorial membrane. **A**: Each diamond represents the best sensitivity of a single auditory-nerve fiber at that frequency. The curved line encloses the most sensitive points and indicates the probably audibility curve for the anguid lizard *Gerrhonotus multicarinatus*. **B**: Similar data for the skink *Tiliqua rugosa* indicate that *T. rugosa* is about 10 dB more sensitive. **C**: Similar data show the frequency-tuning sharpness, or selectivity, for single auditory-nerve fibers in *G. multicarinatus*. These can be compared to equivalent data from *T. rugosa* (**D**). The Q10 dB value is a simple mathematical quantity that divides the value of the best frequency by the frequency bandwidth of the selectivity curve for that neuron at 10 dB above the best sensitivity. Thus, larger values indicate higher selectivity. By this measure, auditory-nerve fiber selectivity in *G. multicarinatus* is substantially smaller than it is in *T. rugosa*. Both the loss of sensitivity and the poorer selectivity can be regarded as consequences of the loss of the tectorial membrane. (From Manley 2002b. © 2002 Wiley Periodicals. Used with permission.)

that primitively the two areas code for the same frequency ranges and are thus functionally redundant. If, after loss of one area, the same amount of space were used to expand another hair cell area, this would lead to an expanded space representation of frequency that is presumably advantageous. It would make no difference whether the apical or the basal hair cell area were lost.

6.4 Division into Subpapillae

The mirror-image frequency ranges of the two high-frequency hair cell areas become disparate if a divider were placed across the papilla. In the three families where such dividers occur in all or some species (varanids, lacertids, and the iguanid subfamily of the basilisks), the divider is always placed between the apical high-frequency area and the other two areas (Fig. 7.8). As such divisions in the different families almost certainly arose independently from each other (Figs. 7.8 and 7.9), this consistent fact may indicate some—as yet unknown— underlying functional reason for a separation at this point.

There are no physiological studies of the tonotopic organization in basiliscine iguanids, but two other cases of divided papillae have been examined using electrophysiological frequency-mapping techniques: varanids (Manley 1977) and lacertids (Köppl, cited in Manley 1990). In both cases, the larger basal subpapilla has two hair cell areas that code for the low and midrange frequencies, whereas the smaller apical subpapilla consists of only high-frequency–type hair cells and responds only to the highest frequency range, above the ranges of the other subpapilla (Fig. 7.4). In these species, the effective papillar length devoted to frequencies above 1 kHz is thus longer than it would be if the two flanking hair cell areas were mirror images. This presumably means that the brain receives more detailed spectral information spread over the nerve fibers innervating these areas. Papillar subdivision may thus simply be an alternative to the complete loss of one hair cell area with subsequent elongation of the other. The physiological results are certainly similar and perhaps selectively equivalent.

7. Summary

For all four types of change in structural patterns outlined in this chapter, there are known or presumed functional consequences. In the case of papillar shortening, concomitant changes in innervation patterns may ameliorate or perhaps even fully mask these consequences. In no case are the structural changes accompanied by dramatic changes in physiology; rather, the changes are relatively subtle and perhaps of no great importance to animals whose use of acoustic stimuli is limited. It is therefore of no great surprise that the evolution of the lizard papilla has led to structural "drift" in a number of directions and in a parallel fashion in different families. Structural changes in the periphery can be regarded as having been selectively equivalent, and thus may be discussed under the controversial title of neutral evolution. More modeling work would be valuable to put current assumptions concerning micromechanical tuning and its influence on hair cell spacing onto firmer ground.

Acknowledgments. This chapter is dedicated to the memory of the late Dr. Malcolm Miller, whose extensive electron microscopical work laid the foundation for much of the theoretical and practical work that formed the basis of this chapter. Miller put the amazing variety of lizard ears into a logical, comparative structure and considered which aspects were ancient and which were of more recent origin. All of us who studied his papers and knew him as a friend were awed by the huge amount of work he did and were filled with respect at the profound insights he developed.

I thank Dr. Christine Köppl (Technical University, Munich) for valuable comments on the manuscript and Professors Jenny Clack (Cambridge University, England) and Robert Carroll (McGill University, Montreal, Canada) for helpful consultations regarding interpretations of the fossil story.

References

Authier S, Manley GA (1995) A model of frequency tuning in the basilar papilla of the Tokay gecko, *Gekko gecko.* Hear Res 82:1–13.

Caldwell MW (1999) Squamate phylogeny and the relationships of snakes and mosasauroids. Zoll J Linn Soc 125:115–147.

Caldwell MW, Lee MS (1997) A snake with legs from the marine Cretaceous of the Middle East. Nature 386:705.

Carroll RL (1988) Vertebrate Paleontology and Evolution. New York: Freeman.

Choe Y, Magnasco MO, Hudspeth AJ (1998) A model for amplification of hair-bundle motion by cyclical binding of Ca^{2+} to mechanoelectrical-transduction channels. Proc Natl Acad Sci USA 95:15321–15326.

Fettiplace R, Fuchs PA (1999) Mechanisms of hair cell tuning. Annu Rev Physiol 61: 809–834.

Frishkopf LS, DeRosier DJ (1983) Mechanical tuning of free-standing stereociliary bundles and frequency analysis in the alligator lizard cochlea. Hear Res 12:393–404.

Hartline PH (1971) Physiological basis for detection of sound and vibration in snakes. J Exp Biol 54:349–371.

Hedges SB, Poling LL (1999) A molecular phylogeny of reptiles. Science 238:998–1001.

Holton T, Hudspeth AJ (1983) A micromechanical contribution to cochlear tuning and tonotopic organization. Science 222:508–510.

Hudspeth AJ (1997) Mechanical amplification of stimuli by hair cells. Curr Opin Neurobiol 7:480–486.

Köppl C (1995) Otoacoustic emissions as an indicator for active cochlear mechanics: a primitive property of vertebrate auditory organs. In: Manley GA, Klump GM, Köppl C (eds) Advances in Hearing Research. Singapore: World Scientific Publishing, pp. 207–218.

Köppl C (2001) The auditory fovea of the barn owl. In: Manley GA, Fastl H, Kössl M,

Oeckinghaus H, Klump GM (eds) Auditory Worlds: Sensory Analysis and Perception in Animals and Man. Weinheim: Wiley-VCH, p. 75.

Köppl C, Authier S (1995) Quantitative anatomical basis for a model of micromechanical frequency tuning in the Tokay gecko, *Gekko gecko*. Hear Res 82:14–25.

Köppl C, Manley GA (1992) Functional consequences of morphological trends in the evolution of lizard hearing organs. In: Webster DB, Fay RR, Popper AN (eds) The Evolutionary Biology of Hearing. New York: Springer-Verlag, pp. 489–509.

Lee MS (1997) The phylogeny of varanoid lizards and the affinities of snakes. Philos Trans R Soc Lond [B] 352:53–91.

Lee MS (2000) Soft anatomy, diffuse homoplasy, and the relationships of lizards and snakes. Zool Scripta 29:101–130.

Manley GA (1977) Response patterns and peripheral origin of auditory nerve fibres in the monitor lizard, *Varanus bengalensis*. J Comp Physiol 118:249–260.

Manley GA (1990) Peripheral Hearing Mechanisms in Reptiles and Birds. Heidelberg: Springer-Verlag.

Manley GA (2000a) Cochlear mechanisms from a phylogenetic viewpoint. Proc Natl Acad Sci USA 97:11736–11743.

Manley GA (2000b) Otoacoustic emissions in lizards. In: Manley GA, Fastl H, Kössl M, Oeckinghaus H, Klump GM (eds) Auditory Worlds: Sensory Analysis and Perception in Animals and Man. Weinheim: Wiley-VCH, pp. 93–102.

Manley GA (2000c) Design plasticity in the evolution of the amniote hearing organ. In: Manley GA, Fastl H, Kössl M, Oeckinghaus H, Klump GM (eds) Auditory Worlds: Sensory Analysis and Perception in Animals and Man. Weinheim: Wiley-VCH, pp. 7–17.

Manley GA (2000d) The hearing organs of lizards. In: Dooling R, Popper AN, Fay RR (eds) Comparative Hearing: Birds and Reptiles. New York: Springer-Verlag, pp. 139–196.

Manley GA (2002a) Evidence for an active process and a cochlear amplifier in non-mammals. J Neurophysiol 86:541–549.

Manley GA (2002b) Evolution of structure and function of the hearing organ of lizards. J Neurobiol 53:202–211.

Manley GA, Yates GK, Köppl C (1988) Auditory peripheral tuning: evidence for a simple resonance phenomenon in the lizard *Tiliqua*. Hear Res 33:181–190.

Manley GA, Köppl C, Yates GK (1989) Micromechanical basis of high-frequency tuning in the bobtail lizard. In: Wilson JP, Kemp D (eds) Cochlear Mechanisms—Structure, Function and Models. New York: Plenum Press, pp. 143–150.

Manley GA, Köppl C, Sneary M (1999) Reversed tonotopic map of the basilar papilla in *Gekko gecko*. Hear Res 131:107–116.

Manley GA, Kirk D, Köppl C, Yates GK (2001) In-vivo evidence for a cochlear amplifier in the hair-cell bundle of lizards. Proc Natl Acad Sci USA 98:2826–2831.

Martin P, Bozovic D, Choe Y, Hudspeth AJ (2003) Spontaneous oscillation by hair bundles of the bullfrog's sacculus. J Neurosci 23:4533–4548.

Miller MR (1978) Scanning electron microscope studies of the papilla basilaris of some turtles and snakes. Am J Anat 151:409–435.

Miller MR (1980) The reptilian cochlear duct. In: Popper AN, Fay RR (eds) Comparative Studies of Hearing in Vertebrates. New York, Heidelberg, Berlin: Springer-Verlag, pp. 169–204.

Miller MR (1985) Quantitative studies of auditory hair cells and nerves in lizards. J Comp Neurol 232:1–24.

Miller MR (1992) The evolutionary implications of the structural variations in the auditory papilla of lizards. In: Fay RR, Popper AN, Webster DB (eds) The Evolutionary Biology of Hearing. Heidelberg, New York: Springer-Verlag, pp. 463–487.

Miller MR, Beck J (1988) Auditory hair cell innervational patterns in lizards. J Comp Neurol 271:604–628.

Tchernov E, Rieppel O, Zaher H, Polcyn MJ, Jacobs LL (2000) A fossil snake with limbs. Science 287:2010–2012.

Vater M (2001) Cochlear specializations in bats. In: Manley GA, Fastl H, Kössl M, Oeckinghaus H, Klump GM (eds) Auditory Worlds: Sensory Analysis and Perception in Animals and Man. Weinheim: Wiley-VCH, pp. 61–65.

Wever EG (1978) The Reptile Ear. Princeton, NJ: Princeton University Press.

Wu Y-C, Art JJ, Goodman MB, Fettiplace R (1995) A kinetic description of the calcium-activated potassium channel and its application to electrical tuning of hair cells. Prog Biophys Mol Biol 63:131–158.

8
Hearing Organ Evolution and Specialization: Archosaurs

OTTO GLEICH, FRANZ PETER FISCHER, CHRISTINE KÖPPL, AND GEOFFREY A. MANLEY

1. Introduction

Among amniotes a group named archosaurs includes the crocodilians, extinct dinosaurs, and birds (see Phylogeny, below). Because of these evolutionary relationships, the archosaurs are considered together in this chapter. The available data on inner-ear structure and function from different archosaur species (various birds and *Caiman crocodilus*) is combined and reviewed, in an attempt to identify the putative primitive condition and subsequent specializations that occurred during evolution.

Figure 8.1 serves as an introduction to the avian inner ear and the various anatomical structures. The photograph in the upper left shows a ventrolateral view of the head of a chicken (*Gallus gallus domesticus*) with a dissected left ear, and the sketch on the right identifies relevant structures. The eardrum is slightly bulged outward at the insertion of the middle-ear ossicle, the columella. An opening was dissected into the basal part of the cochlear bony wall (thin dotted line) to enable a view into the inner ear and show the round window and the oval window with the columella footplate. The rim that divides the opening in two compartments is the cartilaginous abneural limbus, which overlies the basilar papilla. The scala tympani is to the left on the side of the round window and the scala vestibuli is to the right on the side of the oval window. In addition, two semicircular canals and the posterior ampulla can be recognized. A schematic cross section through the cochlear duct is illustrated in the lower left of Figure 8.1, and the lower right illustrates the outline of the basilar papilla indicating the basal and apical ends and the neural and abneural sides.

One obvious behavior of archosaurs is acoustic communication. This is most highly developed in songbirds, but most other archosaurs show some type of acoustic interaction. Vocalization is used in many contexts, for example to attract females, to defend territory, to warn about predators, and to identify and interact with other group members. A highly specialized type of vocalization is used by oil birds (*Steatornis caripensis*) and cave swiftlets (*Collocalia linch*) for echolocation (Konishi and Knudsen 1979). Also, crocodilians vocalize for intraspecific communication (Garrick et al. 1978). Since vocalization and acous-

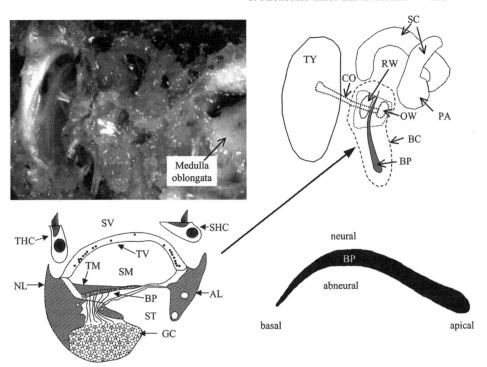

FIGURE 8.1. Top left: A ventrolateral view of a chicken head with the bone dissected away to show the left ear apparatus. Top right: Schematic drawing of relevant structures. Bottom left: Schematic cross section of the cochlear duct at a level indicated by the arrow pointing to the overview. At the top of this sketch typical examples of tall and short hair cells are illustrated at higher magnification. Bottom right: Outline of a basilar papilla, indicating the basal and apical as well as the neural and abneural positions. AL, abneural limbus; BC, bony cochlear wall; BP, basilar papilla; CO, columella; GC, cochlear ganglion; NL, neural limbus; OW, oval window; PA, posterior ampulla; RW, round window; SC, semicircular canal; SHC, short hair cell; SM, scala media; ST, scala tympani; SV, scala vestibuli; THC, tall hair cell; TM, tectorial membrane; TV, tegmentum vasculosum; TY, tympanic membrane.

tic communication are so common among archosaurs, it would be consistent to assume that their inner ears co-evolved while processing these acoustic stimuli during evolution.

It has been shown that many of the basic behavioral measures of hearing are remarkably similar in different bird species (Dooling 1982, 1992) and there are only a few examples that suggest some form of auditory specialization. In the budgerigar (*Melopsittacus undulatus*), behavioral data demonstrate that the critical ratio (a measure of frequency resolution) is unusually low, near 3 kHz, the frequency that is most prominent in budgerigar vocalizations (Okanoya and Dooling 1987a). In pigeons (*Columba livia*) and chickens, specializations for the perception of low frequency "infrasound" have been described (Kreithen and Quine 1979; Lavigne-Rebillard et al. 1985; Warchol and Dallos 1989; Schermuly and Klinke 1990a,b).

Besides using acoustic signals for communication, the barn owl (*Tyto alba*) has a highly specialized auditory system that allows it to catch mice in absolute darkness, based only on acoustic cues produced by the prey (Payne 1971). The barn owl, among the species studied to date, is the one that shows the most obvious signs of auditory specialization. In contrast to other birds, an external ear is formed by a specialized configuration of facial feathers, and this contributes to the barn owl's extraordinary sensitivity. Other specializations include the asymmetrical position of the outer ears, which improves their ability to localize a sound source (Knudsen and Konishi 1979; Moiseff 1989). The basilar papilla of the barn owl is highly elongated, permitting an extended high-frequency limit and the processing of sound produced by prey (Smith et al. 1985; Fischer et al. 1988; Köppl et al. 1993; Köppl 1997). In the barn owl, the auditory nuclei in the ascending auditory pathway also show an over-representation in the frequency range that is used for prey localization (reviewed in Konishi 1993).

2. Phylogeny

The amniotes form a monophyletic group originating from a common ancestor that gave rise to the recent vertebrate classes of "reptiles," birds, and mammals. This typological classification introduced by Linneaus is based on similarities between groups of living animals. It does not, however, appropriately reflect the phylogenetic relationships. A more adequate grouping is based on cladistics, which preserves evolutionary relationships (Carroll 1997). Such a view suggests that the assemblage of mammals is a sister group of the other amniotes (sauropsida or the classes reptiles and birds) and both groups diverged more than 300 million years ago (see Clack and Allin, Chapter 5).

In the Permian ($\approx$250 million years ago), first Chelonia (turtles) and then Lepidosauria (*Sphenodon*, lizards and snakes) diverged among the sauropsida, leaving a monophyletic group of archosauria that includes crocodilians (crocodiles, alligators, caimans, and gavials), extinct dinosaurs, and birds. The lineage of modern birds diverged from dinosaur ancestors around 165 million years ago, and *Archaeopteryx* (150 million years ago) can be regarded as one of the first representative birds. Common features of *Archaeopteryx* and modern birds that distinguish them from dinosaurs are a reduced number of caudal vertebrae, formation of a furcula and a sternum, increased length of forearms, and flight feathers. An important feature in the development of active flight was the reduction of body size (estimated to have been 500 g in *Archaeopteryx*). The fossil record of birds is extremely limited in the Mesozoic, and fossils of ancestral forms of all living bird orders appear at the end of the Mesozoic, approximately 65 million years ago (Carroll 1988; 1997).

Due to the poor fossil record, the relationships of modern bird orders is only poorly resolved, but this is an important issue for any attempt to understand the evolution of the hearing organ by a comparative analysis of inner ear features in representative species. In the following discussion, only those avian groups and species are considered for which data on inner-ear structure are available. Current views on the relationships of modern bird orders that have been developed on the basis of paleontological evidence are briefly summarized. In addition, a different view that was derived from the analysis of the mitochondrial genome and DNA-DNA hybridization is discussed.

Based on paleontological evidence, Feduccia (1995) suggested that living bird orders evolved within a short period of 5 to 10 million years from a group of "transitional shore birds" approximately 65 million years ago. According to Feduccia, song birds arose 20 to 30 million years ago, in a second phase of radiation. However, fossils from Australia suggest that songbirds already existed 55 million years ago (Boles 1995). Based on the characteristics of the palate, two groups of birds have been defined, one with more primitive features (paleognathous birds or ratites) and a more advanced second group including all other birds (neognathous birds). Although this character does not prove the monophyletic origin of ratites, the shared primitive paleognathous palate suggests that they diverged before the origin of more modern birds with a neognathous palate (Carroll 1988; Feduccia 1995). Systematic and quantitative data regarding the inner ear of modern ratites are available for the emu.

Feduccia (1980) and Olson (1985) suggest that the neognathous birds can be grouped into a water-bird and a land-bird assemblage. The water birds separated early from the land birds, and the only representatives of water birds for which sufficiently quantitative data on the inner ear are available are anseriformes (ducks). Among land birds, the galliformes [chicken and quail (*Coturnix coturnix*)], columbiformes (pigeon), and psittaciformes (budgerigar) form a primitive group, whereas strigiformes (barn owl) and passeriformes [starling (*Sturnus vulgaris*), canary (*Serinus canarius*), zebra finch (*Taeniopygia guttata*)] belong to those groups of advanced land birds for which quantitative data on inner-ear structure and function are available.

In contrast to phylogenetic trees based on paleontological data, DNA-DNA hybridization techniques (Sibley and Ahlquist 1990) or the analysis of the mitochondrial genome (Mindell et al. 1999; Cao et al. 2000) have led to alternative trees. These newer trees support the view that crocodiles are a sister group of birds, but the position of turtles and the relationships within bird groups differ in several important respects from the traditional view. The analysis of mitochondrial genes and protein sequences by Cao et al. (2000) suggested that among amniotes, first the mammals then the squamates diverged, followed by birds, and leaving turtles and crocodiles as sister groups.

In the bird groups, Mindell et al. (1999) suggested a basal divergence of songbirds and a more recent divergence of duck and ratites, which were previously considered as basal groups in trees based on the fossil record. However, quite different trees (including one close to the more traditional classification) were also supported by their data with near-equal likelihood. It appears that, as Feduccia (1995) already argued, the divergence of different bird groups within a very short period of time (10 million years) affects the interpretation of both DNA-DNA hybridization and cladistic methods used to determine the higher-level relationships among birds. Thus the identification of primitive and advanced groups based on our current knowledge of amniote evolution is not unequivocally possible. As a guideline, the current classification of birds based on fossil evidence will be used (Feduccia 1980; Olson 1985).

3. Frequency Maps and Papillar Dimensions

An essential feature of any vertebrate hearing organ, which also has a powerful influence on its detailed morphology, is the frequency range covered and its tonotopic organization. Maps derived from single-fiber staining of physiologically characterized auditory nerve fibers are available for the starling (Gleich 1989), chicken (Manley et al. 1987; Chen et al. 1994; Jones and Jones 1995), pigeon (Smolders et al. 1995), and emu [*Dromaius novaehollandiae* (Köppl and Manley 1997)]. A cochlear frequency map is available for the barn owl (Köppl et al. 1993) that is based on small tracer injections in physiologically characterized locations of the cochlear nuclei. The derivation of these maps has been extensively discussed in Gleich and Manley (2000).

To increase the database, maps for additional avian species were derived for which low- and high-frequency limits could be estimated from behavioral audiograms, and at least one additional frequency-mapping point on the papilla was available from some other measure, such as pure-tone sound damage. From polynomial fits of published behavioral audiograms, the most sensitive frequency and the lowest threshold were determined. By comparison of behavioral audiogram data with the detailed results from frequency mapping in the barn owl, starling, and pigeon, the frequency where the threshold was 30 dB above that of the most sensitive frequency was selected as the highest frequency represented on the papilla. The lowest frequency on the papilla was set to one octave below the lowest frequency whose sensitivity lies 30 dB above the best threshold. Using this approach, data from canary (Dooling et al. 1971; Okanoya and Dooling 1985, 1987a,b; Gleich et al. 1997), zebra finch (Okanoya and Dooling 1987a, 1990; Hashino and Okanoya 1989; Ryals et al. 1999), budgerigar (Saunders and Dooling 1974; Dooling and Saunders 1975; Saunders et al. 1979; Saunders and Pallone 1980; Okanoya and Dooling 1987a), and quail (Linzenbold et al. 1993; Niemiec et al. 1994; Dooling and Ryals 1995) became available. For *Caiman*, relevant data were derived from auditory-nerve-fiber recordings (Smolders and Klinke 1986) and measurements of basilar membrane vibration (Wilson et al.

1985). These data on frequency and position were used to calculate best-fit maps according to the method of Greenwood (1990).

By the addition of these estimated maps to those published previously, the question as to which features of place-frequency maps in birds could be ancestral and which could be derived can be addressed. Figure 8.2 shows frequency maps for nine species of birds and for *Caiman*. The plots show frequency both as a function of the normalized distance (Fig. 8.2A,C,E) and as absolute distance (Fig. 8.2B,D,F) from the apical end of the papilla. The top row (Fig. 8.2A,B) shows the map of *Caiman*, as a nonavian representative of the archosaurs, and the map of the emu as a representative of the ratites. The middle row (Fig. 8.2C,D) shows place-frequency maps from four species of primitive land birds. The bottom row (Fig. 8.2E,F) gives examples from the group of advanced land birds. A comparison, especially between advanced and primitive land birds, shows no striking similarities within the groups and no obvious discrepancies between the groups that would reveal prominent evolutionary trends.

These plots show that a simple logarithmic function of frequency versus place is, across different bird orders, a reasonable representation of the data from the emu, pigeon, quail, canary, zebra finch, and *Caiman*. A marked deviation from a logarithmic frequency representation at low frequencies is obvious in the barn owl, starling, chicken, and, to a lesser degree, the budgerigar. This compression has been well documented in neural data from the barn owl (Köppl et al. 1993) and the starling (Gleich 1989). The frequency map derived from data of labeled chicken auditory fibers (Manley et al. 1987; Chen et al. 1994; Jones and Jones 1995) also suggests such a compression, although there are no data points available for the apical quarter of the chicken basilar papilla. Two observations support the view that this compression in the chicken is real. Warchol and Dallos (1989) recorded from very-low-frequency fibers in Nucleus angularis, one of the targets of the central projections of auditory nerve fibers. These fibers responded best to frequencies well below 100 Hz. Based on an analysis of the response latencies derived from phase-versus-frequency plots, they suggested that these responses originate from the most apical portion of the basilar papilla. In addition, Lavigne-Rebillard et al. (1985) reported a specialized area at the most apical tip of the chicken basilar papilla and suggested that this region could be responsible for perception of very low frequencies below 100 Hz. In contrast to the situation in the chicken, single-fiber mapping in the pigeon (Smolders et al. 1995) showed an almost perfect logarithmic frequency representation along the basilar papilla. Schermuly and Klinke (1990b) have shown that infrasound perception in the pigeon originates from abneural hair cells in the apical third of the papilla that is not part of the normal frequency representation along the neural portion.

These data suggest that a deviation from a logarithmic frequency representation should be interpreted as a derived feature. The barn owl deviates from all other species by a dramatically elongated papilla and an extended high-frequency hearing range. In contrast, the compressed frequency representation

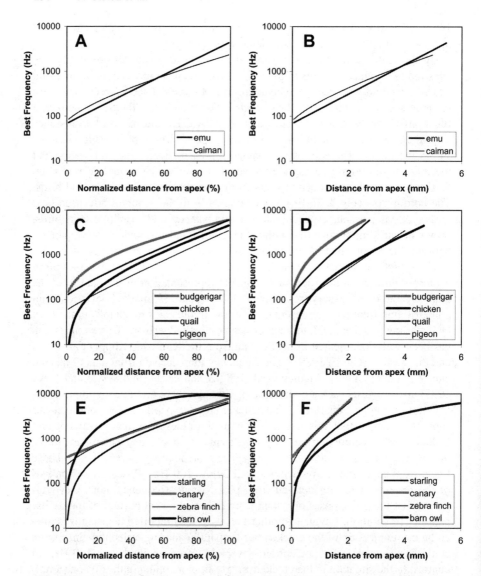

FIGURE 8.2. Place versus frequency maps for different bird species and *Caiman*. The left column (**A, C, E**) shows response frequency as a function of the normalized distance from the apical end. The right column (**B, D, F**) shows response frequency as a function of the absolute distance from the apical end. The top row (A, B) gives the estimated map for *Caiman* and emu. The middle row (C, D) shows maps of primitive land birds (pigeon, chicken, quail, budgerigar), and the bottom row (E, F) shows data from advanced land birds (canary, zebra finch, starling, and barn owl).

at the apex in the starling, chicken, and budgerigar extends their hearing range to lower frequencies. These deviations from a logarithmic frequency representation are not associated with assumed phylogenetic groups and thus presumably developed independently in the diverse groups.

Another feature of frequency representation is evident in the right column of Figure 8.2. In species with short papillae, the frequency range covered is higher and the functions appear steeper than in species with longer papillae. Again, this pattern is observed within separate orders of birds (e.g., passeriformes: canary and starling; galliformes: chicken and quail) and thus results from independent development. A quantitative analysis was performed by calculating the slope of the frequency map for the basal half of the papilla and also for the distance between the 1- and 2-kHz place in each species.

Both graphs of Figure 8.3 demonstrate that the mapping constant is correlated with the length of the papilla, more space per octave being available in species with longer papillae. This suggests that species with shorter papillae do not process proportionately more limited frequency ranges. Both plots demonstrate the grossly deviating pattern of the barn owl from that observed in other species. The basal half of the barn owl is extremely specialized, with an extraordinary mapping constant of more than 5 mm/octave. In the other birds, a linear regression analysis well describes the change in slope (rate of change of frequency) as a function of papilla length. The slope increases from 0.5 mm/octave in the canary (2.1 mm papilla) to 0.9 mm/ octave in the emu (5.5 mm papilla). Also, if an identical frequency range (1–2 kHz, where most maps show a logarithmic frequency representation) is considered, the space devoted to this octave in birds varies between 0.5 and 0.9 mm from canary to emu. Again, the barn owl deviates from the other species. While the barn owl devotes exceptionally much space on the papilla to the frequency range between 5 and 10 kHz, it is much more similar to other birds in the 1- to 2-kHz region (0.9 mm/octave).

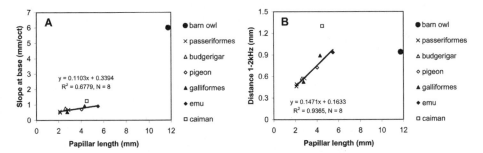

FIGURE 8.3. A: The slope of the place-frequency function calculated for the basal half of the basilar papilla from seven species. B: The distance between the positions on the papilla responding to 1 and 2 kHz. The lines show significant linear regressions that exclude the obviously deviating data point from the barn owl ($p = 0.019$ and 0.002).

This supports the notion stated by Fischer et al. (1988) that the apical half of the barn owl papilla resembles that of "normal" birds, whereas the basal half represents an exceptionally specialized region. The systematic dependence of the represented frequency range on the papillar length and the mapping constant (with the exception of the barn owl) suggests that the length of the papilla is an important factor that determines the mechanics of frequency representation in the avian inner ear.

4. Hearing Range and Papillar Dimensions

In addition to the audiograms of the species mentioned in the section on "maps," audiogram parameters were derived for the starling (Trainer 1946; Kuhn et al. 1982; Dooling et al. 1986), duck (audiogram of *Anas platyrhynchos*, Trainer 1946; compared to papilla of tufted duck, Manley et al. 1996), pigeon (Fay 1988), and barn owl (Konishi 1973; Dyson et al. 1998). For the adult chicken, data published by Saunders and Salvi (1993) allowed only the identification of the best frequency, but no low- and high-frequency limits, because they did not test frequencies at which thresholds were at least 30 dB above the most sensitive threshold.

These data revealed highly significant inverse correlations between the length of the papilla and the best audiogram frequency (Fig. 8.4C, $p = 0.004$) or the upper frequency (30 dB above most sensitive frequency, Fig. 8.4B, $p = 0.002$) for the various bird species—again with the exception of the barn owl. The low-frequency limit (Fig. 8.4A) also showed a trend for an inverse relationship with papillar length ($p = 0.053$), supporting the notion that (again with the exception of the specialized barn owl) longer papillae code a lower frequency range than shorter papillae (see also frequency maps). The lower correlation between papillar length and the lowest frequency represented on the papilla, as compared to either the best frequency or the high-frequency limit, is likely to be due to the compressed frequency mapping at low frequencies observed in several species. There was a trend such that the frequency range in octaves is larger in longer papillae (Fig. 8.4D, $p = 0.088$). Thus, the smaller mapping constant in short papillae is not sufficient to preserve the frequency range as measured in octaves.

The length of the papilla is the main determinant of the highest frequency coded by the ear, and only the barn owl shows a deviating specialization for the perception of high frequencies. The highest frequency that is available for processing in the avian inner ear is limited by the middle ear. The columella type middle ear is not suitable for effective transmission of frequencies above 10 kHz into the inner ear (Manley 1990; Saunders et al., 2000). Together with the observation that papillar length determines the mapping constant (Fig. 8.3), the finding that the frequency range of short papillae is shifted toward higher frequencies, relative to that of longer papillae, supports the view that papillar length

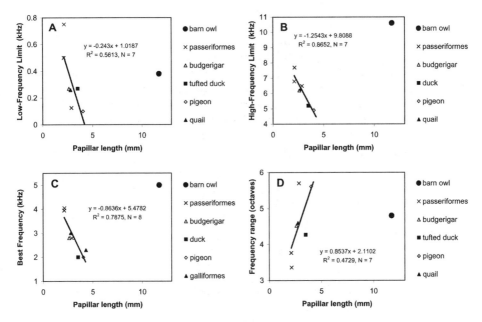

FIGURE 8.4. Shown are the low-frequency (**A**) and high-frequency (**B**) limit, the most sensitive frequency of the audiogram (**C**), and the frequency range in octaves (**D**) as a function of papillar length. The lines show linear regressions through the data points excluding, however, the grossly deviating barn owl.

determines the mechanical response to sound and consequently affects the frequency response.

In parallel, in small species the size of the middle ear is also limited and smaller than in large species. All other things being equal, a small middle ear would be expected to limit the transmission of low frequencies into the inner ear. It is a common observation that smaller structures resonate at higher frequencies than larger structures (e.g., organ pipes or strings on a harp). Thus the extended high-frequency hearing in small species could simply be the consequence that their small heads could not accommodate larger papillae and their middle-ear structures are small, rather than an evolutionary selective pressure to extend high-frequency hearing. Although one would expect that high-frequency hearing is advantageous for the localization of sound sources, only the barn owl shows a real specialization for high-frequency hearing.

In the barn owl, the adaptation for nocturnal hunting created a highly specialized auditory system that allows them to catch prey in complete darkness, using only acoustic cues. As discussed in the section on frequency maps, some species independently developed specializations to extend their low-frequency hearing range (pigeon, chicken, starling). The selection pressure that led to the extension of low-frequency hearing in several species is not as obvious as the pressure that led to the extend high frequency hearing in the barn owl. For the pigeon, Kreithen and Quine (1979) suggested that this species could use infrasound for spatial orientation, the detection of weather fronts, or flight control (altitude determination). The fact that birds from diverse groups have quite different lifestyles (pigeon, chicken, starling) and show signs of specialized low-frequency hearing complicates the identification of selective pressures, especially because it is not yet clear what they gain from perceiving low frequencies. The fact that bigger birds have longer papillae (although they do not maximize papillar length to the degree that head size could accommodate) might be interpreted as a means for extending low-frequency hearing. On the other hand, concomitant with the argument expressed above with regard to high-frequency hearing, physical factors alone would make it easier for larger birds to detect lower frequencies.

5. Relationships Between Body Weight, Papillar Length, and Hair Cell Number

The size of a species clearly affects the size of the inner ear. This is illustrated for the extremes among archosaurs that were included in this study. The left column in Figure 8.5 shows schematically the skeleton of *Brachiosaurus brancai* and illustrations of a chicken and a canary drawn to the same scale. The two birds are illustrated in addition at $20\times$ magnification because the canary is virtually invisible when drawn at the same scale as *Brachiosaurus*. The right column of Figure 8.5 shows the left inner ear including the semicircular canals of these three species. The *Brachiosaurus* specimen is an endocast from an animal that lived approximately 145 million years ago while the chicken and canary specimen show the dissected bony labyrinth. From this illustration it is clear that body size varies to a much higher degree among these species than the size of the inner ear. A more quantitative analysis was performed by comparing the length of the basilar papilla and the body weight.

The estimated lengths of the living basilar papillae and, where available, the number of hair cells, were derived from the literature (Schwartzkopff and Winter 1960; von Düring et al. 1974; Leake 1977; Fischer et al. 1988; Gleich and Manley 1988; Manley et al. 1993; Gleich et al. 1994; Dooling and Ryals 1995; Manley et al. 1996; Köppl et al. 1998). The length of the papilla was also estimated for two dinosaur species. A detailed drawing of the vestibular apparatus of *Allosaurus fragilis* including the lagena was published by Rogers (1998). The length of the total cochlear duct was estimated to be 12 mm from Figure 3 in Rogers (1998). In addition, an inner ear endocast from *Brachiosaurus brancai* from the collection of the Museum of Natural History in Berlin (Germany, see Fig. 8.5) was available for the analysis. The total length of the lagena was approximately 16 mm in this specimen. Assuming that the papilla

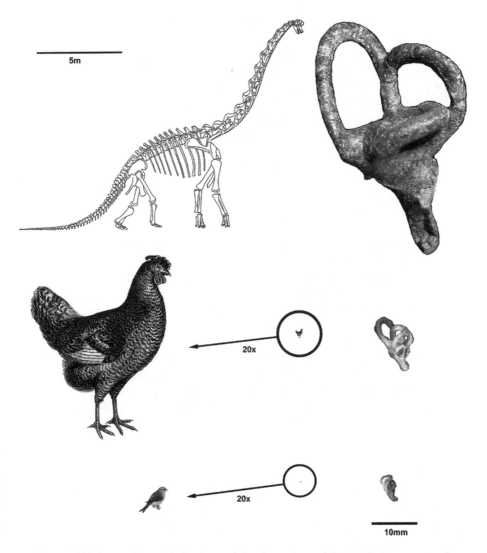

FIGURE 8.5. Comparison of body size and the dimensions of the vestibular apparatus in archosaurs. In the left column body size is illustrated for *Brachiosaurus brancai* (modified from Gunga et al. 1995), chicken, and canary. *Brachiosaurus* is approximately 12 m high and is on display at the Museum of Natural History in Berlin, Germany. The encircled images of the birds are shown at the same scale as *Brachiosaurus*. However, because the canary is virtually not visible at this scale, the birds are also shown at 20× magnification. The right column compares the left labyrinth from these species at the same scale. The image shown for *Brachiosaurus* is from an extremely well-preserved endocast from the Natural History Museum collection (used with permission). The pictures from chicken and canary show the dissected bony labyrinth. Although the size of the labyrinth varies considerably, variation in the size of the animals is much greater.

is approximately two thirds of the total lagena results in 8 mm for *Allosaurus* (body weight 1.4 tons) and 10.7 mm for *Brachiosaurus* (body weight 75 tons; Gunga et al. 1995).

Figure 8.6A shows that there is a strong correlation between papillar length and the size of the body (represented here by body weight; p <0.001), and an exponential fit based on recent species (excluding the specialized barn owl) is a good representation of the data. This fit is shown as continuous line covering the available data from living birds and *Caiman*. There is no clear grouping according to systematic relationships between avian orders, even though all songbirds are much smaller than the Palaeognathae. However, within the songbirds, although they are at the lower end of the distribution, the variation of papillar length with body weight is well represented by the common exponential regression line. This is also true for the chicken and quail (galliformes), species showing a relatively large variation in both parameters. In addition to recent archosaurs, estimates of papillar length for two species of dinosaurs were included. The dotted line is simply an extension of the regression line calculated for living species to show that it provides a fair representation even for these extinct relatives of birds and crocodiles.

Figure 8.6B shows that papillar length and hair cell number are highly correlated. The only exception, which lies far off the distribution of the other species (which includes *Caiman*), is again the barn owl. The causal relationship between basilar papillar and body size is evident only for the smallest species that certainly could not fit papillae that are 5 to 6 mm in length into their small heads. However, in birds that weigh over 100 g, and especially in the Palaeognathae and also in the dinosaurs, the head size could accommodate longer pa-

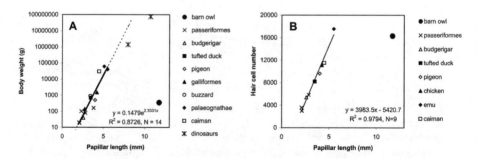

FIGURE 8.6. Body weight (**A**) and hair cell number (**B**) as a function of the length of the basilar papilla. The line in A shows an exponential fit to the data from recent species with a dotted extrapolation, and a linear regression line is shown in B. The papillar length of the 1400 kg *Allosaurus fragilis* was estimated from Figure 3 in Rogers (1998). The papilla length of *Brachiosaurus brancai* was estimated from an inner ear cast from a specimen of the collection of the Museum of Natural History in Berlin, Germany, that is also shown in a photograph in Figure 8.5. Body weight for this specimen of *Brachiosaurus* was previously published (Gunga et al. 1955).

pillae than those found. While body weight varies by a factor of four million between canary and *Brachiosaurus*, papillar length varies by a factor of only 6.7. This clearly demonstrates that with the exception of the barn owl, there was no strong selective pressure to maximize papillar size. The right plot shows that the number of hair cells and the length of the papilla are highly linearly correlated across species ($p < 0.001$).

6. Conclusions on Gross Cochlear Dimensions and Hearing Ranges

Except for the barn owl, the length of the papilla and the number of hair cells are correlated to the body weight of the adult bird. The regression line obtained for recent species including *Caiman* is also a fair representation for data from two dinosaur species. The length of the papilla and/or body size determines the hearing range of the species. Except for the barn owl, the long papillae of large birds code a lower frequency range than the short papillae of small birds. Consequently dinosaurs with their relatively long papillae had probably a very limited high-frequency hearing. The basic frequency representation on the papilla is almost logarithmic. Some species independently developed a compressed representation at the apex of the papilla, presumably to extend their low-frequency hearing range (starling, chicken, budgerigar). The pigeon shows specialized processing of infrasound in the abneural apical half of the papilla that does not affect the logarithmic frequency representation along the neural half. Among the birds studied, the barn owl shows by far the largest deviations from the pattern found in the other bird species. The estimated place-frequency map of the *Caiman* (which belongs to the sister group of birds within the archosaurs) is similar to that of the emu (Fig. 8.2A,B). In addition, when viewed as a function of the normalized distance from the base, the frequency map runs almost parallel to the maps of canary, quail, pigeon, and zebra finch, supporting the view that a logarithmic representation of frequencies is an ancestral feature.

Assuming that the common ancestor of birds was chicken-sized, its basilar papilla was approximately 4 mm long; it had about 10,000 hair cells; its most sensitive hearing was around 1.8 kHz, with a 5 kHz high- and a 0.08 kHz low-frequency limit; and it had a logarithmic representation of frequency on the papilla with a mapping constant of approximately 0.66 mm/octave.

7. A Comparison of Morphological Gradients Along the Papilla in Different Species

Detailed and comparable anatomical data, primarily collected using scanning electron microscopy, are available for emu, representing paleognathous birds; the hooded duck (*Aythya fuligula*), resembling one member of the primitive water birds; chicken, pigeon, and budgerigar, as members of the primitive land birds; and barn owl, starling, canary, and zebra finch, representing advanced land birds (Tilney and Saunders 1983; Fischer et al. 1988; Gleich and Manley 1988; Manley et al. 1993, 1996; Köppl et al. 1998). Comparing these representative species should help identify potential evolutionary trends and facilitate discriminating between primitive and derived features.

The width of the basilar papilla shows a prominent gradient in all species, being narrow at the base and wider toward the apex (Fig. 8.7A,B). This gradient is correlated with the tonotopic frequency representation, such that a narrow and stiffer basilar membrane is tuned to high frequencies whereas the wide and thus more flexible apex is tuned to low frequencies. This pattern is indeed present in all papillae (except for the extreme apical part, where the papilla tapers toward the apical end). However, the papillar width at a given frequency varies noticeably across different species. At 1 kHz, the papillar width is near 140 µm in pigeon and canary, and 190 µm in budgerigar and chicken (Fig. 8.7A). The values from the barn owl are shifted to much higher frequencies as compared to the other species. Plotting the width of the papilla as a function of the distance from the basal end brings the data from different species to a much better alignment (Fig. 8.7B). The basal end of the papilla is quite similar across species and those papillae that are longer get continuously wider. The barn owl again differs from all other species. Very similar patterns were observed for the number of hair cells across the papilla. Compared to other species, the barn owl has many fewer hair cells across the papilla in the highly specialized basal half that is adapted for processing frequencies between 5 and 10 kHz.

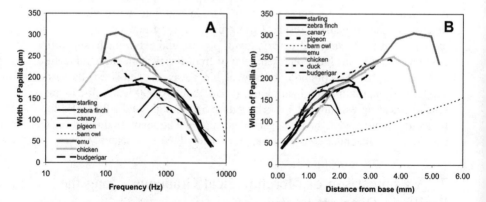

FIGURE 8.7. The width of the basilar papilla as a function of frequency (**A**) and the distance from the basal end of the cochlea (**B**) for different species. Note that the scale for the right column in this and the following figures is limited to a maximum of 6 mm, and consequently the apical half of the barn owl papilla is not shown in these plots. Nevertheless, it is obvious that the barn owl data deviate substantially from those of all other species.

An important measure that contributes to the mechanical resonance properties is the number of stereovilli in the hair cell bundles (Fig. 8.8). Bundles with a large number of stereovilli are stiffer and are predicted to resonate at higher frequencies than bundles with a smaller number. If the number of stereovilli is plotted as a function of frequency, each species shows the correlation between frequency and stereovilli number, but the data from different species vary quite considerably (Fig. 8.8A). A given frequency is not associated with the same number of stereovilli across different species, in part due to other influences on frequency response, such as the bundle height. If the stereovilli number is plotted as a function of the distance from the basal end of the papilla, data from most species lie close to each other (Fig. 8.8B). Only the emu and the barn owl are obviously different from the other species.

Finally, the height of the stereovillar bundle also contributes to the mechanical properties, such that tall bundles are predicted to exhibit lower resonance frequencies than short bundles. Data from the starling, canary, pigeon, and emu were collected from sectioned material, whereas barn owl and chicken data were obtained using scanning electron microscopy. For a direct comparison, chicken and barn owl data were corrected by 25% to compensate for shrinkage during critical point drying. In all species, there was a gradient in stereovillar height from the basal to the apical end of the papilla (Fig. 8.9). The tallest bundles were found at the apex in each species. Bundle height at a given response frequency (Fig. 8.9A) and at a given distance from the basal end of the papilla (Fig. 8.9B), as well as the slope of the height gradient, vary considerably across species.

These examples demonstrate that none of the individual anatomical parameters analyzed determines the response frequency, because data from different species, when plotted as a function of response frequency, do not form a single

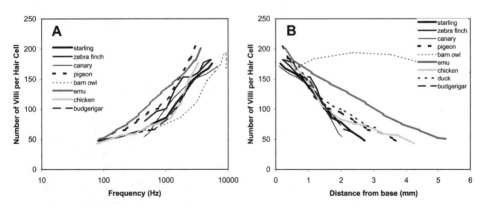

FIGURE 8.8. The number of stereovilli per hair cell for different species as a function of frequency (**A**) and the distance from the basal end of the papilla (**B**).

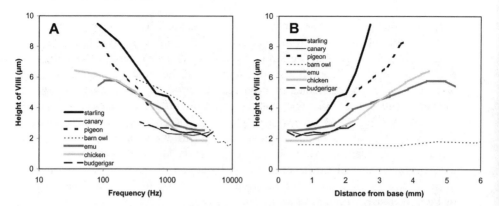

FIGURE 8.9. The height of the tallest stereovilli in a bundle as a function of response frequency (**A**) and of the distance from the basal end of the papilla (**B**), for different species.

overlapping function. The interaction of systematic variations in different morphological gradients (e.g., bundle height, stereovilli number, tectorial membrane mass, basilar membrane properties) results in the response frequency at a given position along the papilla.

Several anatomical parameters are remarkably similar across different species when plotted as a function of the distance from the basal end. In addition, with the exception of the barn owl, the parameters analyzed do not show evidence of specializations within different evolutionary groups. The difference between species appears primarily related to total papillar length, and most variations occur at the apical end. Species with short papillae reduce the apical part as compared to species with longer papillae, while preserving anatomical gradients in the base of the papilla. This trend is also evident in data on the systematic variation of the hair-bundle orientation in the papilla. The maximum rotation of stereovillar bundles shows a basal-to-apical gradient, and the maximum rotation at the apex is smaller in species with a short papilla (e.g., canary) than in species with longer papillae (e.g., emu; summarized in Gleich and Manley 2000).

As emphasized above, some species independently developed an extended low-frequency hearing range that is revealed in place-frequency maps by deviations from a logarithmic frequency representation. The available data show no common mechanism as to how this specialization was achieved. The comparatively tall stereovillar bundles in the apex of the starling basilar papilla (Fig. 8.9) may be one way of extending the hearing range toward low frequencies. Pigeons, a species with a logarithmic frequency representation along the papilla, perceive infrasound in the specialized abneural and apical half of the papilla (Schermuly and Klinke 1990b). The thickening of the basilar and tectorial membranes near the apex could be associated with infrasound perception in the pigeon (Gleich and Manley 1988).

8. Gradients of Hair Cell Dimensions Along and Across the Basilar Papilla

In contrast to birds (see below), there are two distinct types of hair cells in *Caiman* (Baird 1974; von Düring et al. 1974; Leake 1977). Tall hair cells are taller (25–30 µm) than wide (6–14 µm), located on the neural side of the papilla, and their number in any given cross section of the papilla increases from two at the base to 12 near the lagena (apex). Short hair cells are located abneurally over the free basilar membrane; they are much shorter than tall hair cells (6 µm), and their number in a cross section increases from 11 at the base to 25 at the apex. In contrast to birds, in any given cross section there is an abrupt transition from tall to short hair cells, which allows an unequivocal discrimination between these two distinct hair cell types in *Caiman*. The description of the innervation pattern by von Düring et al. (1974) did not go into detail, but suggested that the afferent innervation of the short hair cells is less than that of the tall hair cells.

In birds, hair cell dimensions also show gradients along and across the papilla, and Takasaka and Smith (1971) described them as tall, intermediate, or short. These categories were based on morphological parameters of the hair cells, and in the pigeon they also correlated well with innervation patterns. Tall hair cells were taller than wide and had large afferent and small efferent synapses. Short hair cells were wider than tall with small bouton-like afferent synapses and large efferent synapses. Because the quantitative analysis of innervation patterns using serial ultrathin sections and electron microscopy is extremely demanding, the dimensions of the hair cells, determined using the less demanding technique of light microscopy, have been used also in other species to discriminate between hair cell types. However, as discussed below, hair cell shape and innervation pattern are not always as correlated as in the pigeon. Consequently, Fischer (1994) suggested defining hair cell types on the basis of functionally relevant innervation patterns rather than on criteria based on their height and width ratios. In addition, it has been shown that many other morphological gradients in hair cells vary gradually along and across the papilla (see discussion in Gleich et al. 1994; reviewed in Gleich and Manley 2000) so that it is not possible to define distinct types of tall and short hair cells based on hair cell dimensions across avian species (Fig. 8.10F).

Quantitative systematic measurements of hair cell dimensions are available only for the emu (Köppl et al. 1998), the barn owl (Köppl 1993), the canary (Mandl 1992), the starling (unpublished data), and the pigeon (unpublished data from Olga Ganeshina). For the following comparison, means were calculated for data from hair cells that were within 20% of the papillar width on the neural

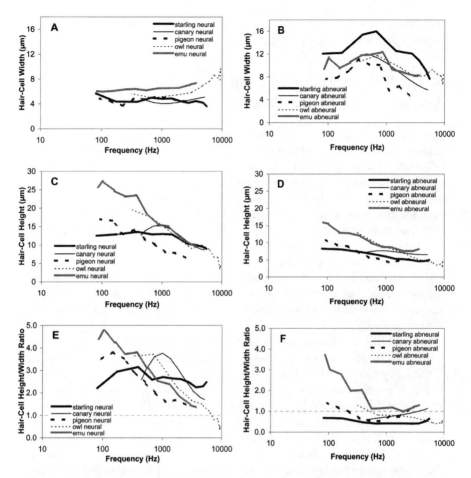

FIGURE 8.10. Hair cell dimensions as a function of response frequency. The left column (**A, C, E**) shows data from neural and the right column (**B, D, F**) data from abneural hair cells. A and B illustrate the width, C and D the height, and E and F the height: width ratio of the hair cells. The horizontal dashed lines in E and F label the value of 1 that in the past was used to discriminate between hair cell types.

or on the abneural sides, to illustrate the extremes. Data are shown as a function of response frequency (Fig. 8.10) and as a function of the distance from the base (Fig. 8.11).

The width of neural hair cells is virtually constant along the papilla in the canary and pigeon, whereas it shows a small systematic increase toward the base and toward higher frequencies in the emu and barn owl and a small decrease in the starling. Gradients of hair cell width along the papilla are much more pronounced in abneural hair cells. Hair cell width increases from the base toward

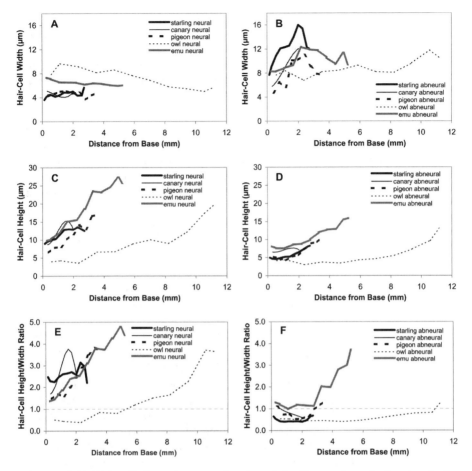

FIGURE 8.11. Hair cell dimensions as a function of the distance from the basal end of the basilar papilla. Organization of the diagrams as in Figure 8.10.

the middle of the papilla and then decreases again toward the apex in the emu, pigeon, and starling. Hair cell width increases monotonically from the base to the apex in the canary, supporting the previous notion that canaries—or perhaps all small passerines—"dropped" the low-frequency portion of their papilla. The width of abneural hair cells in the barn owl shows a continuous increase toward the apex. Plotted as a function of frequency, the data from the barn owl overlap with those from the other species.

The height of the hair cells shows a systematic increase from the base toward the apex in all species, for both neural and abneural hair cells. The slope of this function is steeper for neural than for abneural hair cells. In the canary the height of abneural hair cells is almost constant along the papilla. When plotted as a function of frequency, the height of neural hair cells from the emu, canary, starling, and barn owl are quite similar in the frequency range above 1 kHz, whereas pigeon hair cells are systematically shorter. Below 1 kHz starling and pigeon hair cells are similar in height and shorter than those of emu and barn owl. Looking at hair cell height as a function of the distance from the base shows that starling, canary, and emu are quite similar, whereas pigeon and barn owl have shorter hair cells. Data from abneural hair cells show more variability across species. The frequency representation of abneural hair cells has to be viewed critically because frequency maps based on functionally characterized and labeled neurons (that only innervate neural hair cells) may not adequately represent the response frequency of abneural hair cells. This has been shown directly in the pigeon, where hair cells responsive to infrasound appear to exist abneural to conventional low-frequency responses in more neurally lying hair cells (Schermuly and Klinke 1990b; Smolders et al. 1995). Of course, the frequency response of abneural short hair cells, which lack afferent innervation (see next section), remains speculative.

If one accepts the functional conversion of position into frequency for all locations across the papilla, then two important points become apparent. First, the absolute height and width dimensions of hair cells in equivalent frequency ranges vary considerably within and between the species (Fig. 8.10A–D) and are thus unlikely to have a consistent functional meaning. Second, the ratio of hair cell height to width shows a considerable variability across species for neural hair cells, that is, the traditional tall hair cells (Figs. 8.10E and 8.11E). Height-width ratios of abneural hair cells, that is, the traditional short hair cells, are conspicuously different in the emu as compared to the other species (see Figs. 8.10F and 8.11F; emu data are above the 1:1 line). As pointed out previously (Köppl et al. 1998), this can be interpreted as a primitive condition of the emu, showing less differentiation between tall and short hair cells than more advanced birds. However, this contrasts with the pronounced hair cell differentiation in crocodilians, which would argue for this being an ancient feature of the archosaur papilla. It is of course possible that the condition in crocodilians was arrived at independently. Thus, with the presently available information, it is not possible to identify primitive or advanced features regarding hair cell specialization across the papilla.

9. Hair Cell Types and Innervation Patterns in the Basilar Papilla

Avian hair cell shape, together with the overall innervation pattern, was first used to discriminate between hair cell types. According to this scheme, hair cells that are taller than wide were called tall hair cells, and those that were wider than tall were called short hair cells (Takasaka and Smith 1971), with a continuous transition in between. Tall hair cells are found in the neural part and in proportionally greater numbers in the apex of the papilla, short hair cells are present along the abneural edge and in proportionally greater numbers in the base of the papilla.

Thus, hair cell shape varies systematically along and across the basilar papilla, and this occurs in a species-specific manner. In general, tall hair cells resemble hair cells in more primitive vertebrate groups and in the vestibular system, whereas short hair cells deviate from this basic pattern. Tall hair cells receive more afferent than efferent innervation, whereas short hair cells receive no afferent innervation but have prominent efferent innervation (Takasaka and Smith 1971; Fischer 1994). However, as noted above, the morphology of the hair cells and their innervation pattern is not congruent in different bird species. For example, almost all hair cells in the basal half of the papilla in the barn owl are short hair cells according to the original definition (Takasaka and Smith 1971), but the innervation of the neurally lying cells is not typical of short hair cells (and therefore they presumably have a different function; Fischer 1994). In contrast, the emu has very tall hair cells (see Figs. 8.10C and 8.11C) and based on the older definition virtually no short hair cells would exist (Köppl et al. 1998; Fischer 1998; Figs. 8.10F and 8.11F), although the innervation pattern across the papilla is similar to that in other species in that most abneural hair cells lack an afferent innervation.

In the base of the barn owl's papilla, the more abneural hair cells are very uniform in that they are about 4 μm high and have very little cytoplasm. They are nearly completely filled by a rather condensed nucleus and by the cuticular plate, and they have no afferent innervation. The latter feature was also found in the other species that have been studied by transmission electron microscopy (see below). In view of the above data, a more functional definition that is valid for all investigated species was given by Fischer (1994). He suggested that tall hair cells have afferent-fiber synapses (and generally also efferent synapses), whereas short hair cells have no afferent but efferent innervation. This makes a functional distinction possible in that it is clear that short hair cells cannot fulfill a direct sensory function. Many tall hair cells in birds are also coupled together by a network of direct cell-to-cell contacts, the most interesting ones being true fusions of neighboring hair cells (Fischer et al. 1991).

The innervation pattern in the avian hearing organ is rather uniform among species, with the marked exception, in some quantitative aspects, of the barn owl (see below). Besides the afferent and efferent innervation of the hair cells, efferent fibers also run to the hyaline cells, a supporting-cell type found along the abneural edge of the papilla (Cotanche et al 1992; Oesterle et al. 1992). Myelinated nerve fibers lose their myelin sheath when they enter the basilar papilla. Several authors report that different classes of afferent as well as efferent nerve fibers can be distinguished (e.g., Fermin and Cohen 1984; Zidanic and Fuchs 1996). According to these findings, tall hair cells synapse with thick afferents and thin efferents, whereas abneural hair cells synapse with thin (if any) afferents and thick efferents. Other results argue for a single, rather uniform afferent population (Fischer 1994; Fischer et al. 1994; Köppl et al. 2000) and a less clear morphological separation of efferent fiber and terminal types (Fischer 1994; Köppl 2001).

Most investigators use the chicken as a model bird. It certainly has some primitive features, but it is also a species that has been domesticated for a long time, and, at least as adults, different breeds show vastly different degrees of inner ear pathologies (Durham et al. 2002). The emu is a better model for an unspecialized basilar papilla, and this may be especially true with regard to the innervation pattern. The following summary of innervation patterns is primarily based on transmission electron microscopic studies in the chicken (Fischer 1992), the emu (Fischer 1998), the starling (Fischer et al. 1992), and the barn owl (Smith et al. 1985; Fischer 1994).

The number of afferent nerve terminals is characteristically low in birds (about two to three per hair cell), as compared to mammals. Afferent fibers run directly to innervate one or up to a few hair cells. More primitive birds like the emu and the chicken have a marked number of afferent fibers that innervate hair cells nonexclusively (similar to the situation in *Caiman*; von Düring et al. 1974), but only in the apical, lower-frequency area. In songbirds, the percentage of exclusively innervating afferent fibers increases: in the starling, 67% of the afferents in the apex were exclusive (Fischer et al. 1992), as compared to the emu, where more than 50% of the afferent fibers are nonexclusive (Fischer 1998). The exclusivity is highest in the barn owl (Fischer 1994), where nonexclusive afferents are only, but frequently, found in the most apical part of its papilla. In addition, in this highly specialized bird, up to 20 afferent terminals are present on a given hair cell in the auditory foveal region (Köppl et al. 1993), where the behaviorally most important frequencies are coded.

Generally, there is a gradient in the number of afferent terminals from the neural to the abneural side of the papilla, the densest afferent innervation being found along the neural edge (shown as afferent contact area in Fig. 8.12A–C). Along the abneural edge, however, there is no (starling, barn owl) or only very scarce (chicken, emu) afferent innervation. In the emu, this afferent-fiber–free strip of hair cells is narrow (Fischer 1998), whereas in starling (Fischer 1992) and in the barn owl (Fischer 1994) it is more marked.

The strict separation in abneural hair cells without and medial and neural hair cells with afferent innervation appears as an advanced feature that is more pro-

FIGURE 8.12. The synaptic contact area per hair cell across the width of the papilla shown separately for the afferent (**A, B, C**) and efferent (**D, E, F**) innervation. Data from the 0.6-kHz (A, D) and the 3-kHz (B, E) area of the basilar papilla from emu (open circles, thin lines) and the barn owl (filled triangles, heavy lines) are shown to compare a presumably primitive and advanced species. Data from the 7.3-kHz foveal region of the barn owl show the highly specialized innervation of the most neural hair cells (C) and a relatively large proportion of abneural hair cells with exclusive efferent innervation (F).

8. Archosaur Inner Ear Evolution   247

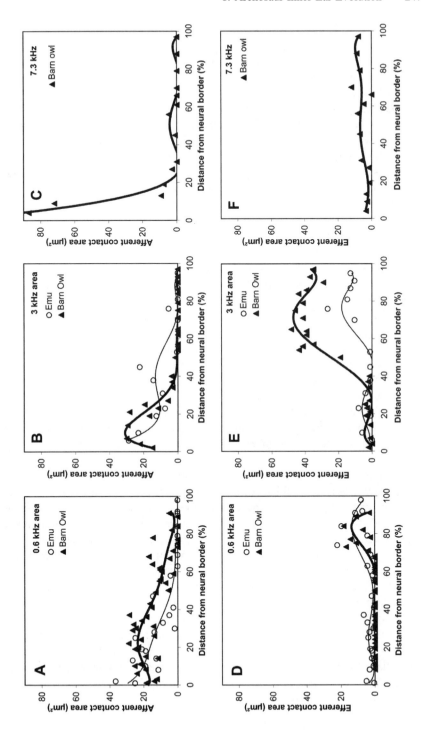

nounced in the basal high-frequency portion of the papilla and correlates with extended high-frequency hearing. As the primitive condition, including the limited data from *Caiman* (von Düring et al. 1974), a papilla with many nonexclusive afferents and with weak afferent innervation toward the abneural edge can be postulated.

Avian hair cells usually have one efferent terminal per hair cell, but some hair cells may have as many as three. Abneural hair cells always have efferent synapses, whereas, especially in the apical third of the papilla, many medial and neural hair cells lack efferent innervation. There is thus an increase in the efferent synaptic area from the neural to the abneural side of the papilla (Fig. 8.12D–F). This situation is probably also the primitive condition. It is apparent that efferent innervation is more important with increasing high-frequency hearing. Extremely specialized and very uniform abneural hair cells that receive only efferent innervation is the dominating feature in the basal two thirds of the papilla in the barn owl, the high-frequency specialist among birds.

In Figure 8.12, the afferent and efferent contact area per hair cell are compared for the emu and the barn owl, for two comparable frequency strips across the papilla that covered the apical, unspecialized portion of the basilar papilla in the owl and in addition for the highly specialized high-frequency region in the barn owl. The afferent and efferent innervation patterns found in the barn owl and emu are qualitatively quite similar and differ only in details. The most obvious difference is the disproportionately high efferent contact area in the 3-kHz region of the barn owl (Fig. 8.12E). In the region of the auditory fovea in the barn owl, afferent synapse area of neural hair cells is disproportionately higher (Fig. 8.12C) than shown here for the "unspecialized" area below 4 kHz (Fischer 1994). Although the efferent contact area appears relatively low in the fovea (Fig. 8.12F) the proportion of abneural hair cells that are innervated only by efferents is high.

The inner ear of crocodilians closely resembles that of birds. The most obvious difference is that tall hair cells and short hair cells are clearly morphologically distinct, instead of showing a continuous gradient. There are few data on their innervation. Von Düring et al. (1974) studied the *Caiman*, and preliminary unpublished data are available for the alligator. *Caiman* tall hair cells are contacted by about four afferents, and the proportion of nonexclusive afferent innervation is high. Short hair cells are also contacted by afferents. One or two efferents synapse with any given hair cell.

In addition to the innervation of hair cells, there is also efferent innervation to the hyaline cells in several bird species (Takasaka and Smith 1971; Fischer et al. 1992; Fischer 1994) but the synaptic structures are not clearly defined. In addition, regular and frequent en-passant synapses between afferent and efferent fibers have been found (von Düring 1985; Fischer and Junker 2000). Both features are also present in the *Caiman* (von Düring 1974), but their function is far from being understood.

10. Implications from the Innervation Pattern

This complex pattern of innervation makes a functional analysis of the avian inner ear, such as the effects of the efferent system, rather difficult. The innervation of the avian inner ear appears to be just as complex as in mammals (Dannhof and Bruns 1993). An essential question concerns the poorly understood function of the short hair cells. Their function is probably not as essential for avian hearing as is the function of the outer hair cells in mammals, since, after acoustic trauma, considerable functional recovery occurs before a complete regeneration of short hair cells (Smolders 1999; Pugliano et al. 1993). Short hair cells are apparently more important at higher frequencies. Comparisons between several avian species have led to the hypothesis that short hair cells co-evolved together with a new subtype of efferent neuron innervating them (Köppl 2001).

11. Comparison of the Archosaur Cochlea with the Inner Ear of Primitive Mammals

The cochlea of primitive mammals and that of the archosaurs is superficially similar (see Vater et al., Chapter 9). Although the auditory epithelium in primitive mammals and archosaurs are more-or-less banana-shaped and have a similar

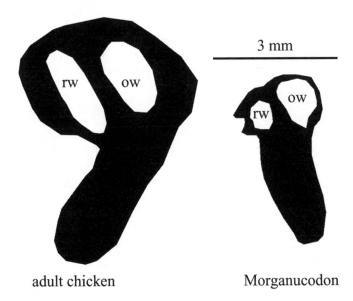

FIGURE 8.13. The ventrolateral view of the right inner ear including round (rw) and oval (ow) windows from an adult chicken and the extinct primitive mammal Morganucodon. (Redrawn from Graybeal et al. 1989.)

size, a closer view shows that the cochleae bend in opposing directions—in birds, the neural side is convex, whereas it is concave in mammals. This difference between the mammalian and archosaur was also observed in a reconstruction of the cochlear cavity of the primitive extinct mammal *Morganucodon*, which lived approximately 200 million years ago (Graybeal et al. 1989; Fig. 8.13). In addition, profound differences in hair cell and supporting cell structure (see Vater et al., Chapter 9) emphasize that despite the superficial similarity with regard to over all dimensions, these two groups of vertebrates have evolved separately for 300 million years.

12. Summary

The analysis of many different anatomical parameters from the basilar papilla, and the characterization of the frequency response along the papilla shows only one species whose patterns deviate greatly from those obtained in all other birds: the barn owl (Figs. 8.2–8.4 and 8.6–8.12). Among the other bird species and crocodilians analyzed, there are no systematic patterns that are associated with specific phylogenetic groups. A remaining enigma is the evolution and, indeed, the definition of hair cell types in birds and crocodiles. Although there are very clearly two different hair cell types (tall and short hair cells), their unambiguous definition across frequency ranges and species remains difficult.

In contrast to the phylogenetic background, the size of the species correlates well with the length of the papilla, which in turn determines the frequency range mapped to the papilla. Contrary to the expectation that there could be a selection pressure to extend high-frequency hearing to the limit imposed by the middle ear, which would improve for example the ability to localize sound sources, several species independently developed specializations for extending the low-frequency hearing range. Thus, low-frequencies are obviously important. At present, however, it is not clear what birds really do with their low-frequency hearing. In small species that weigh only around 20 g, head size obviously puts a limit on the length of the cochlea and consequently on the range of frequencies. The longer papillae in bigger birds could be an adaptation to extend the hearing range to lower frequencies. Alternatively, the size of the inner ear cannot be reduced below a certain minimum in larger archosaurs, which in turn puts limits on their high frequency-hearing.

Acknowledgments. We thank J. Strutz for supporting this work (O.G.). O. Ganeshina provided unpublished data from pigeon hair cell measurements. W.-D. Heinrich from the Museum für Naturkunde of the Humboldt-Universität Berlin provided information on *Brachiosaurus* and made the *Brachiosaurus* specimen accessible for examination. We thank C. Cavannah and R. Nowak for help with photography.

References

Baird IL (1974) Anatomical features of the inner ear in submammalian vertebrates. In: Keidel WD, Neff WD (eds) Handbook of Sensory Physiology, vol V/1. Berlin Heidelberg, New York: Springer, pp. 159–212.

Boles WE (1995) The world's oldest songbird. Nature 374:21–22.

Cao Y, Sorenson MD, Kumazawa Y, Mindell DP, Hasegawa M (2000) Phylogenetic position of turtles among amniotes: evidence from mitochondrial and nuclear genes. Gene 259:139–148.

Carroll RL (1988) Vertebrate Palaeontology and Evolution. New York: Freeman.

Carroll RL (1997) Patterns and process of vertebrate evolution. New York: Cambridge University Press.

Chen L, Salvi R, Shero M (1994) Cochlear frequency-place map in adult chickens: intracellular biocytin labeling. Hear Res 81:130–136.

Cotanche DA, Henson MM, Henson OW Jr (1992) Contractile proteins in the hyaline cells of the chicken cochlea. J Comp Neurol 324:353–364.

Dannhof BJ, Bruns V (1993) The innervation of the organ of Corti in the rat. Hear Res 66:8–22.

Dooling RJ (1982) Auditory perception in birds. In: Kroodsma DE, Miller EH (eds) Acoustic Communication in Birds, vol 1. New York: Academic Press, pp. 95–130.

Dooling RJ (1992) Hearing in birds. In: Fay RR, Popper AN, Webster DB (eds) The Evolutionary Biology of Hearing. Heidelberg, New York: Springer-Verlag, pp. 545–559.

Dooling RJ, Ryals BM (1995) Effects of acoustic overstimulation on four species of birds. In: Manley GA, Klump GM, Köppl C, Fastl H, Oeckinghaus H (eds) Advances in Hearing Research. Singapore: World Scientific, pp. 32–39.

Dooling RJ, Saunders JC (1975) Hearing in the parakeet (*Melopsittacus undulatus*): absolute thresholds, critical ratios, frequency difference limens, and vocalizations. J Comp Physiol Psych 88:1–20.

Dooling RJ, Mulligan JA, Miller JD (1971) Auditory sensitivity and song spectrum of the common canary (*Serinus canarius*). J Acoust Soc Am 50:700–709.

Dooling RJ, Okanoya K, Downing J, Hulse S (1986) Hearing in the starling (*Sturnus vulgaris*): Absolute thresholds and critical ratios. Bull Psychonom Soc 24:462–464.

Durham D, Park DL, Girod DA. (2002) Breed differences in cochlear integrity in adult, commercially raised chickens. Hear Res 166:82–95.

Düring M von, Karduck A, Richter H (1974) The fine structure of the inner ear in *Caiman crocodylus*. Z Anat Entwickl Gesch 145:41–65.

Düring M von, Andres KH, Simon K (1985) The comparative anatomy of the basilar papillae in birds. Fortschr Zool 30:681–685.

Dyson ML, Klump GM, Gauger B (1998) Absolute hearing thresholds and critical masking ratios in the European barn owl: a comparison with other owls. J Comp Physiol [A] 182:695–702.

Fay RR (1988) Hearing in Vertebrates: A Psychophysics Databook. Winnetka, IL: Hill-Fay Associates.

Feduccia A (1980) The Age of Birds. Cambridge, MA: Harvard University Press.

Feduccia A (1995) Explosive evolution in tertiary birds and mammals. Science 267: 637–638.

Fermin CD, Cohen GM (1984) Development of the embryonic chick's statoacoustic ganglion. Acta Otolaryngol (Stockh) 98:42–52.
Fischer FP (1992) Quantitative analysis of the innervation of the chicken basilar papilla. Hear Res 61:167–178.
Fischer FP (1994) Quantitative TEM analysis of the barn owl basilar papilla. Hear Res 73:1–15.
Fischer FP (1998) Hair-cell morphology and innervation in the basilar papilla of the emu (*Dromaius novaehollandiae*). Hear Res 121:112–124.
Fischer FP, Junker M (2000) Complex innervation pattern in the basilar papilla of a bird, the Australian emu. 23rd Midwinter Research Meeting of the Association of Research Otolaryngology, abstract, p. 279.
Fischer FP, Köppl C, Manley GA (1988) The basilar papilla of the barn owl Tyto alba: A quantitative morphological SEM analysis. Hear Res 34:87–102.
Fischer FP, Brix J, Singer I, Miltz C (1991) Contacts between hair cells in the avian cochlea. Hear Res 53:281–292.
Fischer FP, Singer I, Miltz C, Manley GA (1992) Morphological gradients in the starling basilar papilla. J Morphol 213:225–240.
Fischer FP, Eisensamer B, Manley GA (1994) Cochlear and lagenar ganglia of the chicken. J Morphol 220:71–83.
Garrick LD, Lang JW, Herzog HA (1978) Social signals of adult American alligators. Bull Am Mus Nat Hist 160:153–192.
Gleich O (1989) Auditory primary afferents in the starling: correlation of function and morphology. Hear Res 37:255–268.
Gleich O, Manley GA (1988) Quantitative morphological analysis of the sensory epithelium of the starling and pigeon basilar papilla. Hear Res 34:69–86.
Gleich O, Manley GA (2000) The hearing organ of birds and crocodilia. In: Dooling RJ, Fay RR, Popper AN (eds) Comparative Hearing: Birds and Reptiles. New York: Springer, pp. 70–138.
Gleich O, Manley GA, Mandl A, Dooling R (1994) The basilar papilla of the canary and the zebra finch: a quantitative scanning electron microscopic description. J Morphol 221:1–24.
Gleich O, Dooling RJ, Presson JC (1997) Evidence for supporting cell proliferation and hair cell differentiation in the basilar papilla of adult Belgian Waterslager canaries (*Serinus canarius*). J Comp Neurol 377:5–15.
Graybeal A, Rosowski JJ, Ketten DR, Crompton AW (1989) Inner-ear structure in *Morganucodon*, an early jurassic mammal. Zool J Linn Soc 96:107–117.
Greenwood DD (1990) A cochlear frequency-position function for several species—29 years later. J Acoust Soc Am 87:2592–2605.
Gunga H-Chr, Kirsch KA, Baartz F, Röcker L, Heinrich W-D, Lisowski W, Wiedemann A, Albertz J (1995) New data on the dimensions of *Brachiosaurus brancai* and their physiological implications. Naturwiss 82:190–192.
Hashino E, Okanoya K (1989) Auditory sensitivity of the zebra finch (*Poephila guttata castanotis*). J Acoust Soc Jpn (E) 10:51–52.
Jones SM, Jones TA (1995) The tonotopic map in the embryonic chicken cochlea. Hear Res 82:149–157.
Knudsen EI, Konishi M (1979) Mechanisms of sound localization in the barn owl (*Tyto alba*). J Comp Physiol [A] 133:13–21.
Konishi M (1973) How the owl tracks its prey. Am Sci 61:414–424.

Konishi M (1993) Listening with two ears. Sci Am 268:66–73.
Konishi M, Knudsen EI (1979) The oilbird: hearing and echolocation. Science 204:425–427.
Köppl C (1993) Hair-cell specializations and the auditory fovea in the barn owl cochlea. In: Duifhuis H, Horst JW, van Dijk P, van Netten SM (eds) Biophysics of Hair Cell Sensory Systems. Singapore: World Scientific Publishing, pp. 216–222.
Köppl C (1997) Number and axon calibres of cochlear afferents in the barn owl. Auditory Neurosci 3:313–334.
Köppl C (2001) Efferent axons in the avian auditory nerve. Eur J Neurosci 13:1889–1901.
Köppl C, Manley GA (1997) Frequency representation in the emu basilar papilla. J Acoust Soc Am 101:1574–1584.
Köppl C, Gleich O, Manley GA (1993) An auditory fovea in the barn owl cochlea. J Comp Physiol [A] 171:695–704.
Köppl C, Gleich O, Schwabedissen G, Siegl E, Manley GA (1998) Fine structure of the basilar papilla of the emu: implications for the evolution of avian hair-cell types. Hear Res 126:99–112.
Köppl C, Wegscheider A, Gleich O, Manley GA (2000) A quantitative study of cochlear afferent axons in birds. Hear Res 139:123–143.
Kreithen ML, Quine DB (1979) Infrasound detection by the homing pigeon: a behavioural audiogram. J Comp Physiol [A] 129:1–4.
Kuhn A, Müller CM, Leppelsack H-J, Schwartzkopff J (1982) Heart rate conditioning used for determination of auditory thresholds in the starling. Naturwiss 69:245–246.
Lavigne-Rebillard M, Cousillas H, Pujol R (1985) The very distal part of the basilar papilla in the chicken: a morphological approach. J Comp Neurol 238:340–347.
Leake PA (1977) SEM observations of the cochlear duct in *Caiman crocodilus*. Scan Electron Micosc 2:437–444.
Linzenbold A, Dooling RJ, Ryals BM (1993) A behavioral audibility curve for the Japanese quail (*Coturnix coturnix japonica*). 16th Midwinter Research Meeting of the Association of Research Otolaryngology, abstract, p. 211.
Mandl A (1992) Eine quantitative, morphologische Untersuchung der Papilla basilaris des Kanarienvogels (*Serinus canarius*). Diplom-Thesis at the Department of Zoology of the Technical University of Munich.
Manley GA (1990) Peripheral Mechanisms in Reptiles and Birds. Berlin: Springer-Verlag.
Manley GA, Brix J, Kaiser A (1987) Developmental stability of the tonotopic organization of the chick's basilar papilla. Science 237:655–656.
Manley GA, Schwabedissen G, Gleich O (1993) Morphology of the basilar papilla of the budgerigar *Melopsittacus undulatus*. J Morphol 218:153–165.
Manley GA, Meyer B, Fischer FP, Schwabedissen G, Gleich O (1996) Surface morphology of the basilar papilla of the tufted duck *Aythya fuligula* and the domestic chicken Gallus gallus domesticus. J Morphol 227:197–212.
Mindell DP, Sorenson MD, Dimcheff DE, Hasegawa M, Ast JC, Yuri T (1999) Interordinal relationships of birds and other reptiles based on whole mitochondrial genomes. Syst Biol 48:138–152.
Moiseff A (1989) Binaural disparity cues available to the barn owl for sound localization. J Comp Physiol [A] 164:629–636.
Niemiec AJ, Raphael Y, Moody DB (1994) Return of auditory function following struc-

tural regeneration after acoustic trauma: behavioral measures from quail. Hear Res 79:1–16.

Oesterle EC, Cunningham DE, Rubel EW (1992) Ultrastructure of hyaline, border, and vacuole cells in chick inner ear. J Comp Neurol 318:64–82.

Okanoya K, Dooling RJ (1985) Colony differences in auditory thresholds in the canary (*Serinus canarius*). J Acoust Soc Am 78:1170–1176.

Okanoya K, Dooling RJ (1987a) Hearing in passerine and psittacine birds: a comparative study of masked and absolute auditory thresholds. J Comp Psychol 101:7–15.

Okanoya K, Dooling RJ (1987b) Strain differences in auditory thresholds in the canary (*Serinus canarius*). J Comp Psychol 101:213–215.

Okanoya K, Dooling RJ (1990) Detection of gaps in noise by budgerigars (*Melopsittacus undulatus*) and zebra finches (*Poephila guttata*). Hear Res 50:185–192.

Olson SL (1985) The fossil record of birds. In: Farner D, King J, Parkes K (eds) Avian Biology, vol 8. New York: Academic Press, pp. 79–238.

Payne RS (1971) Acoustic location of prey by barn owls (*Tyto alba*). J Exp Biol 54: 535–573.

Pugliano FA, Wilcox TO, Rossiter J, Saunders, JC (1993) Recovery of auditory structure and function in neonatal chicks exposed to intense sound for 8 days. Neurosci Lett 151:214–218.

Rogers SW (1998) Exploring Dinosaur neuropalaeobiology: computed tomography scanning analysis of an *Allosaurus fragilis* endocasts. Neuron 21:673–679.

Ryals BM, Dooling RJ, Westbrook E, Dent ML, MacKenzie A, Larsen ON (1999) Avian species differences in susceptibility to noise exposure. Hear Res 131:71–88.

Saunders J, Dooling RJ (1974) Noise-induced threshold shift in the parakeet (*Melopsittacus undulatus*). Proc Natl Acad Sci USA 71:1962–1965.

Saunders J, Pallone R (1980) Frequency selectivity in the parakeet studied by isointensity masking contours. J Exp Biol 87:331–342.

Saunders J, Rintelmann W, Bock G (1979) Frequency selectivity in bird and man: a comparison among critical ratios, critical bands and psychophysical tuning curves. Hear Res 1:303–323.

Saunders JC, Duncan RK, Doan DE, Werner YL (2000) The middle ear of reptiles and birds. In: Dooling RJ, Fay RR, Popper AN (eds) Comparative Hearing: Birds and Reptiles. New York: Springer, pp. 13–69.

Saunders SS, Salvi RJ (1993) Psychoacoustics of normal adult chickens: thresholds and temporal integration. J Acoust Soc Am 94:83–90.

Schermuly L, Klinke R (1990a) Infrasound sensitive neurones in the pigeon's cochlear ganglion. J Comp Physiol [A] 166:355–363.

Schermuly L, Klinke R (1990b) Origin of infrasound sensitive neurones in the papilla basilaris of the pigeon: a HRP study. Hear Res 48:69–78.

Schwartzkopff J, Winter P (1960) Zur Anatomie der Vogel-Cochlea unter natürlichen Bedingungen. Biol Zentralblatt 79:607–625.

Sibley CG, Ahlquist JE (1990) Phylogeny and Classification of Birds: A Study in Molecular Evolution. New Haven, CT: Yale University Press.

Smith CA, Konishi M, Schull N (1985) Structure of the barn owl's (*Tyto alba*) inner ear. Hear Res 17:237–247.

Smolders JWT (1999) Functional recovery in the avian ear after hair cell regeneration. Audiol Neurootol 4:286–302.

Smolders JWT, Klinke R (1986) Synchronized responses of primary auditory fibre populations in *Caiman crocodilus* (L.) to single tones and clicks. Hear Res 24:89–103.

Smolders JWT, Ding-Pfennigdorff D, Klinke R (1995) A functional map of the pigeon basilar papilla: correlation of the properties of single auditory nerve fibres and their peripheral origin. Hear Res 92:151–169.

Takasaka T, Smith CA (1971) The structure and innervation of the pigeon's basilar papilla. J Ultrastruct Res 35:20–65.

Tilney LG, Saunders JC (1983) Actin filaments, stereocilia, and hair cells of the bird cochlea. I. Length, number, width, and distribution of stereocilia of each hair cell are related to the position of the hair cell on the cochlea. J Cell Biol 96:807–821.

Trainer JE (1946) The Auditory Acuity of Certain Birds. PhD Thesis, Cornell University, Ithaca, NY.

Warchol ME, Dallos P (1989) Neural response to very low-frequency sound in the avian cochlear nucleus. J Comp Physiol [A] 166:83–95.

Wilson JP, Smolders JWT, Klinke R (1985) Mechanics of the basilar membrane in *Caiman* crocodilus. Hear Res 18:1–24.

Zidanic M, Fuchs PA (1996) Synapsin-like immunoreactivity in the chick-cochlea: specific labeling of efferent nerve terminals. Auditory Neurosci 2:347–362.

9
Hearing Organ Evolution and Specialization: Early and Later Mammals

MARIANNE VATER, JIN MENG, AND RICHARD C. FOX

1. Introduction

An enormous amount of data is available on the structure and function of the mammalian cochlea and mammalian hearing characteristics. Most of this knowledge is derived from studies of placental mammals, which is only one of the three groups of mammals that exist today. There are only a few studies on the ears and hearing characteristics of the other two living mammalian groups, the monotremes and the marsupials, and only recently have fossil specimens with sufficiently preserved inner ears become available that document important steps in evolution of the mammalian ear from the early Cretaceous up to the present. Mammals are an ancient group among terrestrial vertebrates that arose from mammal-like reptiles probably during the Triassic. The diagnostic features that clearly separate a mammal from its reptilian ancestors and other tetrapods include a property of the hearing apparatus, namely the presence of the three-ossicle middle ear. This key innovation appears to be one prerequisite for the capability of high-frequency hearing that is unique to mammals among tetrapods. The fossil record demonstrates that a middle ear with three ear ossicles precedes the evolution of an elongated and, in the more derived condition, coiled cochlea capable of hearing high frequencies well above 10 kHz.

By combining data from different streams of research, namely comparative anatomy and physiology of living (extant) mammals, and paleontological evidence, this chapter addresses the questions of structural and functional changes in the auditory receptor organ that ultimately led to a cochlea design that is characterized by active mechanical amplification of traveling waves via fast outer hair cell motility and the widest frequency range of hearing among terrestrial vertebrates.

Throughout the chapter, we use the term *cochlea* for the hearing organ, although this term is not perfectly appropriate to denote the uncoiled structure in some mammals and their ancestors.

2. Mammalian Systematics and Phylogeny

Several general questions are important for understanding the evolution of the mammalian cochlea: (1) How are mammals defined and taxonomically grouped? (2) What is their phylogenetic line of descent and the geological time frame of their evolution? (3) What is their geological and geographical distribution and which ecological niches are occupied? (4) Which salient morphological modifications occurred through time? Identification of a living being as a mammal is easy: they are distinct from other vertebrates in many features, such as hair, milk glands, etc. Extant mammals are placed in three groups—monotremes, marsupials, and placentals (Fig. 9.1)—with the latter two being grouped as theria.

Monotremes survive with only three recognized species isolated in New Guinea and Australia (Nowak 1995). These are the platypus, *Ornithorhynchus anatinus*, and the echidnas or spiny anteaters, which are represented by *Tachyglossus aculeatus* and *Zaoglossus bruijnii*. Monotremes are considered the most primitive extant mammals and resemble the reptilian ancestors of mammals in several anatomical details (Griffith 1968, 1978). Most noteworthy, they are the only mammals that lay shell covered eggs that are hatched outside the female reproductive tract. The body size of monotremes is small to medium (platypus: 30–40 cm body length; weight of the male, 1.7 kg, echidna: weight of the male 1.4 kg) and the range of ecological niches is very restricted. Echidna are highly specialized crepuscular anteaters, whereas the platypus is a nocturnal, semiaquatic creature that feeds on small invertebrate prey. Prey capture in platypus depends on a specialized electrosensory system, an exceptional feature among amniotes (Proske et al. 1998). Little is known about the acoustic behavior of monotremes. The platypus is known to vocalize in its burrow and to emit rumbling growls and squeaks, but there are no available sonagrams of vocalizations (Pettigrew et al. 1998). The echidna is known to react strongly to rustling noises but the vocal repertoire has not been characterized. Though differing in details of organization (see below), their auditory apparatus is characterized by a three-ossicle middle ear attached to an only partly curved cochlea. The cochlea shape thus resembles that of birds and crocodilians more than that of marsupials and placentals.

Living marsupials consist of about 250 species and are nowadays restricted to Australia and the New World (Nowak 1995). They differ from placental mammals in their very short gestation periods and in raising their young in a pouch, but display numerous examples of convergence in lifestyle and body plan with the placentals. The vocal repertoire and acoustic behavior is known for some species (Aitkin 1995, 1998). However, there are no terrestrial and aquatic echolocators among marsupials and the range of body size is much more restricted. As in placental mammals, the cochlea is elongated and coiled (Fig. 9.1).

Living eutherians (placentals) are the largest and most diverse mammalian

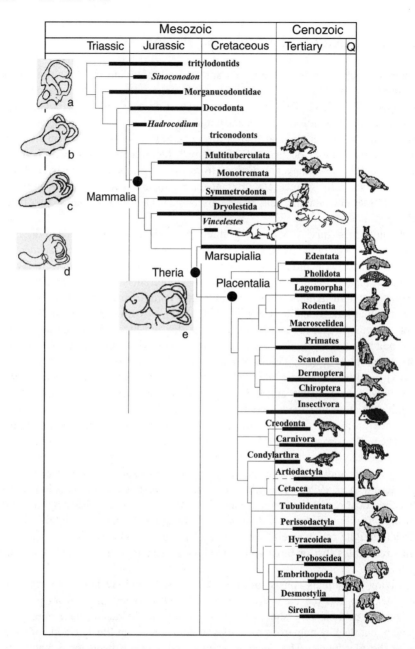

FIGURE 9.1. Simplified phylogenetic tree of mammals (modified from Novacek 1992 and Wang et al. 2001) and examples of inner-ear shape in mammal-like reptiles, mammaliformes, nontherian, and therian mammals. (**A**) *Probainognathus*, based on Allin 1986; (**B**) *Sinoconodon*, based on Luo 2001; (**C**) *Morganucodon*, based on Graybeal et al. 1989 and Luo and Ketten 1991; (**D**) a multituberculate, based on Fox and Meng 1997; (**E**) a therian mammal, based on Meng and Fox 1995a.

group, consisting of about 3800 species that are classified into 19 orders and have a cosmopolitan distribution (Nowak 1995). They adapted to a variety of ecological niches, and differ considerably in body size, ranging from the smallest, a bat weighing about 2 g (*Craseonycteris thonglongyai*), to the largest living mammal, the blue whale (120 tons).

Further back in time, additional mammalian clades existed (McKenna and Bell 1997), some of which are only documented by isolated fragmented fossils. Traditionally, extant and extinct mammals are defined and recognized by derived features of the skull that are not found in other vertebrates, such as three ear ossicles and the dentary-squamosal jaw articulation (Olson 1959; Simpson 1960; Kermack et al. 1973; Crompton and Jenkins 1979; Hopson 1994; Crompton 1995; Cifelli 2001). Though this appears straightforward, in the history of the study on mammalian evolution, there has never been a full consensus on what is a mammal, and the phylogenetic relationships among mammalian taxa are hotly disputed. Consequently, the nomenclature of the major mammalian lineages and their phylogenetic relationships, which are introduced in Figure 9.1, represent only one of several views of mammalian phylogeny.

The common ancestor of mammals probably existed in the early Jurassic (Fig. 9.1). Monotremes are considered living fossils and thought to retain the most primitive characters (Simpson 1945; Cifelli 2001). Marsupials and placentals are thought to be more closely related to each other than to the monotremes and thus are placed as sister groups, although a monotreme-marsupial relationship is also considered (Gregory 1947; Kühne 1973; Janke et al. 1997).

The earliest known, unequivocal monotreme is *Steropodon* from the early Cretaceous of Australia (Archer et al. 1985), which, however, is represented by only a fragmentary jaw and teeth. The earliest known ear region of monotremes is from the Miocene *Obdurodon* (Archer et al. 1993), which is quite similar to that of the living platypus.

The earliest metatherian (marsupials and their stem taxa) is from the Albian-Cenomanian of North America (Cifelli and Muizon 1997), and the best record of early metatherians is in North America (Cifelli 2001). Metatherians are also known from the late Cretaceous of Asia, and the tertiary from South America, Australia, and Antarctica (Cifelli 2001). Deltatheroida, an Asiatic group, is probably related to marsupials (Rougier et al. 1998).

The Cenozoic is the "age of placentals," with an explosive radiation into 22 orders abundantly documented in the fossil record. Cretaceous eutherians (placental and stem taxa) are known from the late early Cretaceous of Asia, and North America (Cifelli 2001). *Ausktribosphenos* from the early Cretaceous of Australia was considered an eutherian, probably an erinacid (Rich et al. 1997, 2001; Flynn et al. 1999) but the taxonomic identification of this taxon is questionable (Kielan-Jaworowska et al. 1998; Archer et al. 1999).

For several Jurassic and Cretaceous groups, there is evidence on the hearing apparatus. These include *Vincelestes* from the early Cretaceous of South America (Rougier 1990; Rougier et al. 1992) and the Multituberculata, Triconodonta (Austrotriconodontidae, Amphilestidae, and Triconodontidae), and Symmetrodonta (Fig. 9.1). Of these, the multituberculates enjoyed the longest evolutionary history. They were a successful and diverse group represented by small to intermediate-sized herbivore animals that are thought to exhibit convergence in lifestyle to rodents. The placement of triconodonts and multituberculates with respect to the other mammalian groups is disputed (for a detailed discussion see Rowe 1988; Wible et al. 1995; Hu et al. 1997; Luo et al. 2001; Wang et al. 2001). Significantly, for all sufficiently documented forms of extinct nontherian mammals, the middle ear is of the mammalian type but the cochlea resembles the hearing organ of reptiles and birds in that it is either straight or only slightly curved (Fig. 9.1). Interestingly, if a curvature is present as in some multituberculates it goes in the opposite direction to that seen in birds (see Gleich et al., Chapter 8, Figure 8.13).

Several further groups that are represented by sporadic fossil evidence from the Triassic/Jurassic period were traditionally considered as mammals but are now placed in the mammaliaformes (McKenna and Bell 1997). These include the Sinoconodontidae, Morganucodonta, Docodonta, and Haramiyoidea. Morganucodon, one of the earliest mammaliaformes had a three-ossicle middle ear attached to a bird/reptile-shaped cochlea (Allin and Hopson 1992; Rosowski 1992).

The separation of the mammalian line from an advanced group of mammal-like reptiles, the cynodonts, is thought to have occurred in the mid-Trias. The likely sister group of the mammaliaformes is represented by the tritylodontids (Allin and Hopson 1992). Though the cynodonts are also relevant to the evolution of the mammalian inner ear, our review and discussion focuses on the above-mentioned groups.

3. Hearing Range and Sensitivity in Extant Mammals

One major evolutionary achievement of mammals as compared to other tetrapods is the capacity to process sounds well above 10 kHz (Fay 1988; Manley 1990, 2000). This extension of the upper frequency limit can be attributed both to differences in middle-ear transmission (i.e., a one-ossicle middle ear as compared to a three-ossicle middle ear) (Manley 1973; Saunders et al. 2000) and cochlear processing of sounds (Manley 2000; Ruggero and Temchin 2002).

3.1 Placentals

Placental mammals differ widely in the frequency range of hearing and this variation is larger than in any other vertebrate taxon (Masterton et al. 1969; Heffner and Masterton 1980; Fay 1988; Echteler et al. 1994) (Fig. 9.2A). Many species, such as cats and several rodents, are sensitive to a wide range of frequencies ranging from well below 5 kHz up to above 60 kHz. In some species, such as elephants, mole rats, and humans, the range of hearing is restricted to frequencies below 20 kHz. Most echolocating bats and cetaceans can hear well

9. Hearing Organ Evolution in Mammals 261

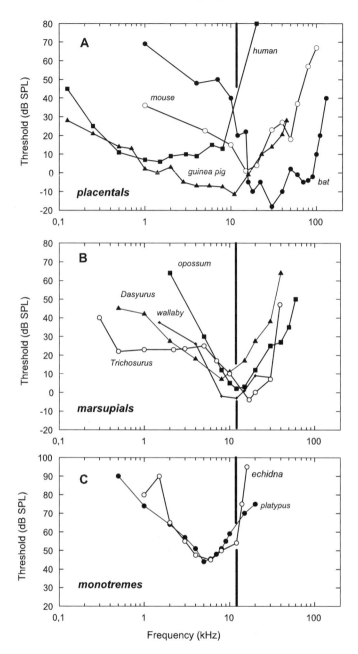

FIGURE 9.2. Comparison of audiograms in three mammalian taxa. **A**: Placentals (guinea pig, mouse, and human after Fay 1988; bat after Schmidt et al. 1984). **B**: Marsupials (*Trichosurus* and *Dasyurus* after Aitkin 1995; opossum *Monodelphis* after Reimer 1996; wallaby after Cone-Wesson et al. 1997). **C**: Monotremes (platypus after Gates et al. 1974; echidna ABR audiogram after Mills and Shepherd 2001). The vertical lines indicate the maximum upper limit of the audiogram in nonmammalian tetrapods (barn owl; Konishi 1973).

above 100 kHz. Among placentals, horseshoe bats and mustached bats have evolved the most highly tuned cochleae known (Kössl and Vater 1995). The differences in hearing capabilities among placentals depend on several factors such as body size, middle-ear functions, and cochlear specializations and can in many cases be interpreted as adaptations to the species' characteristic "acoustic biotope" (Heffner and Heffner 1992; Echteler et al. 1994; Kössl and Vater 1995; Ketten 2000). The acoustic biotope (Aertsen et al. 1979, cited in Aitkin 1995) includes the species' own vocalizations, calls or noises produced by putative predators or prey, and the sound-producing and sound-conducting properties of the physical environment (see Lewis and Fay, Chapter 2). Aquatic, terrestrial or arboreal habitats each pose specific demands on the hearing system.

3.2 Marsupials

Audiograms obtained with a variety of different techniques are available for six marsupial species: the opossums *Didelphis virginiana* (conditioned suppression: Ravizza et al. 1969), and *Monodelphis domestica* [conditioned suppression: Frost and Masterton 1994; acoustic distortion products: Faulstich et al. 1996; auditory brainstem responses (ABR): Reimer 1996], *Marmosa elegans* (Frost and Masterton 1994), the brush-tailed possum *Trichosurus vulpecula* (thresholds of auditory cortical neurons; Gates and Aitkin 1982), the Northern quoll *Dasyurus hallucatus* (ABR: Aitkin et al. 1994), and the tammar wallaby *Macropus eugenii* (ABR: Cone-Wesson et al. 1997). The audiograms reveal differences among species in the best frequency of hearing and the absolute sensitivity (Fig. 9.2B). For example, *Didelphis* has a shallow and insensitive hearing characteristic with the best hearing between 16 and 30 kHz at 20 dB sound pressure level (SPL) (Frost and Masterton 1994). The brush-tailed possum *Trichosurus* is most sensitive at 18 kHz and −2 dB SPL, and *Dasyurus* is most sensitive at 8 kHz and 8 dB SPL. ABR-threshold curves of the opossum *Monodelphis* and the wallaby are similar, with the best hearing between about 8 and 16 kHz around 0 to −5 dB SPL (Reimer 1996; Cone-Wesson et al. 1997). Sensitivity is poor at and below 1.5 kHz. Considering interspecific differences and methodological aspects, the audiograms of marsupials exhibit properties that fall well within the adaptive range of placentals. In some species at least, the sensitivity matches that of placental mammals. In *Monodelphis*, the high thresholds below 8 kHz combined with a wide ultrasonic hearing range were interpreted as a primitive feature for mammals (Faulstich et al. 1996; Heffner and Masterton 1980). Since only a handful of marsupial species has been studied, it is not known whether among marsupials there exist similar correlations between hearing and habitat as among placental mammals.

3.3 Monotremes

Cochlear microphonic recordings in the platypus (Gates et al. 1974) revealed a V-shaped audiogram with the best sensitivity at 5 kHz and a sharp roll-off at both high and low frequencies (approx. 20 dB/octave). At 40 dB SPL, the best sensitivity is poor compared with placental mammals (Fig. 9.2C).

The ABR audiogram of *Tachyglossus* (Mills and Shepherd 2001) is remarkably similar in absolute sensitivity and frequency range to the hearing curve of Platypus. Aitkin and Johnstone (1972) found that the middle ear of echidna is considerably less efficient in transmitting airborne vibrations than that of placental mammals, and suggested an adaptation to hearing of substrate-borne sound. Thus the inferior hearing sensitivity of monotremes may reflect differences to placentals at both the level of middle-ear transmission and the cochlea. Significantly, hearing abilities and middle-ear transmission at frequencies below 6 kHz are also inferior to those of lizards (Aitkin and Johnstone 1972) and birds (Manley 1990). In terms of hearing evolution, the monotremes occupy an intermediate stage between the hearing systems of therapsid reptiles and therian mammals (Gates et al. 1974) since the high-frequency slope of the audiogram extends to higher frequencies (about 15 kHz) than in all nonmammalian tetrapods including the barn owl, which possesses the highest upper limit of hearing among nonmammalian tetrapods at 11 to 12 kHz (Konishi 1973). However, it is lower than that of most therian mammals.

Recent measurements of distortion-product otoacoustic emissions (DPOAEs) in *Tachyglossus* (Mills and Shepherd 2001) yielded audiograms with a mean low-frequency limit of 2.3 kHz and an upper limit of 18.4 kHz (criterion: 30 dB SPL above the best threshold) with the best sensitivity between 4 and 8 kHz. The effective hearing range thus amounts to only three octaves, in contrast to six to eight octaves in generalized placental mammals of comparable body size (Fay 1988; Echteler et al. 1994). However, absolute sensitivity revealed by DPOAE audiograms for monotremes was significantly better than expected from electrophysiological data, which leads to an estimate of best behavioral sensitivity of about 10 dB SPL, close to that of many therian mammals. Clearly, further physiological and behavioral measurements are needed to clarify this important issue.

4. Cochlear Function and Cochlear Amplification Mechanisms in Extant Mammals

In placental mammals, sensitive cochlear tuning is achieved by active nonlinear amplification of the traveling waves (Robles and Ruggero 2001). The prevailing view is that fast motility of the outer hair cell (OHC) body, first described by Brownell et al. (1985), is the likely basis of the "mammalian cochlear amplifier" (Holley 1996; Ashmore et al. 2000; Zheng et al. 2002). In vitro, fast OHC-motility has been demonstrated up to frequencies of about 80 kHz (Frank et al. 1999). In vivo, the motile response is assumed to provide a local feedback into the passive mechanics on a cycle-by-cycle basis (Holley 1996; Ashmore et al. 2000), but how this is achieved at the highest frequencies of hearing is an unresolved issue. The capacity to exert fast length changes in the kilohertz range is unique to OHCs and based on specialized motor proteins in the lateral cell membrane (Kalinec et al. 1992). The responsible motor molecule was characterized as prestin (Zheng et al. 2000a), although other membrane integral proteins are also considered part of the motor (Ashmore et al. 2000). The motor action combined with the intricate macro- and micromechanics of the cochlear duct thus appears to be a salient prerequisite for placental mammals to hear frequencies higher than in nonmammals. In nonmammalian tetrapods, phylogenetically older cochlear amplification mechanisms are implemented that depend on a stereovilli based-mechanism (Manley 2000; Manley et al. 2001).

Despite the fact that monotremes and marsupials could provide salient information for understanding evolution of the mammalian cochlea, our knowledge of basic cochlear function in these taxa is still rudimentary. Fernandez and Smith (1963) reported that, similar to placentals, the endocochlear potential (EP) in marsupials is between 75 and 100 mV and sensitive to anoxia. This contrasts with a much smaller (<20 mV) EP in birds and reptiles (Schmidt and Fernandez 1962; Runhaar et al. 1991). These EPs are anoxia sensitive in birds but not in reptiles. Unfortunately, there are no comparative data on echidna yet.

The principles of tonotopic representation in the monotreme cochlea are unknown, but frequency mapping of the echidna auditory cortex showed an orderly cochleotopic representation of frequencies between 1 and 16 kHz (Krubitzer 1998).

There are only three studies of cochlear functional properties in nonplacental mammals (*Monodelphis*: Müller et al. 1993; Faulstich et al. 1996; echidna: Mills and Shepard 2001). Frequency mapping of the *Monodelphis* cochlea (Müller et al. 1993) shows a baso-apical gradient of high to low frequencies that qualitatively follows the placental mammalian scheme. However, the maximum slope of the frequency-place map in *Monodelphis* amounts to 1.8 mm/octave, which is smaller than in rat (2.1 mm/octave) and cat (3.5 mm/octave). There are no direct measurements of the hydromechanical properties of the cochlear duct or of fast OHC electromotility in any marsupial. Faulstich et al. (1996), measured DPOAEs in *Monodelphis*. These noninvasive measurements provide important clues on nonlinear active hydromechanical cochlear processing (Brown and Kemp 1984; Kössl 1997). The DPOAEs of this marsupial are similar to those of placental mammals, indicating common principles in cochlear function. The best ratios of the primary frequencies ranged between 1.1 and 1.3, which is similar to placental mammals including humans. The gross similarity of distortion product measurements and suppression tuning curves between *Monodelphis* and placental mammals indicates the use of common frequency analysis mechanisms. One putatively different functional feature of the *Monodelphis* cochlea is suggested by the finding that ratio curves of distortion products exhibited multiple peaks at frequency intervals of 4.97 Hz within the frequency range between 20 and 30 kHz (Faulstich et al. 1996). This could argue for a different design in tectorial membrane filter mechanisms as part of the cochlear frequency analyzer (Brown et al. 1992; Allen and Fahey 1993) but clearly more basic data on cochlear function need to be obtained to further highlight possible subtle differences or evolutionary transitions.

Significantly, DPOAE measurements in *Tachyglossus* (Mills and Shepherd 2001) revealed many similarities to those of placental mammals. This suggests the presence of a cochlear amplification mechanism that boosts sensitivity to low-level sound in the monotreme cochlea. However, the nature of the mechanism involved in the echidna is not known to date.

The relative lack of data on cochlear function in monotremes and marsupials clearly leaves a prominent gap in knowledge that could give clues on the evolution of hearing mechanisms in tetrapods. There are, however, several comparative anatomical studies of the hearing organ that allow at least some educated guesses on the nature of transitional changes between cynodont reptiles and mammals.

5. Comparative Cochlear Anatomy in Extant Mammalian Groups

The cochlea of several species of placental mammals has been studied extensively both at the light-microscopic and ultrastructural levels (Lim 1986; Echteler et al. 1994; Kössl and Vater 1995; Slepecky 1996; Ketten 2000). We will point out the salient similarities and differences in cochlear design among extant mammalian groups.

5.1 Marsupials

The very limited comparative anatomical database nevertheless points out remarkable similarities of the cochlear "bauplan" in marsupials and placental mammals (Fernandez and Schmidt 1963) (Fig. 9.3A). At both evolutionary levels, the cochlear duct is coiled and the organ of Corti houses specialized supporting cells (e.g., pillar and Deiters' cells) and two sets of sensory cells: one row of inner hair cells (IHC) and typically three rows of OHCs (Fernandez and Schmidt 1963). At the qualitative light-microscopic level, there are no obvious differences from the placental mammals, although opossums are "living fossils" that have changed little since the cretaceous and evolved in parallel to the placentals. The basilar membrane (BM) of *Monodelphis domestica* is 6.4 mm long (Müller et al. 1993), which is similar to the BM length of the mouse (summarized in Echteler et al. 1994). The qualitative evidence on cochlear structure, together with data on the central auditory pathway (Aitkin 1995, 1998), suggests that as in placentals, the IHCs act as the main information relays to the central auditory system and the OHCs act as local active mechanical elements that enhance sensitivity and tuning to faint signals. Still, it would be highly interesting to obtain measurements of hair cell numbers and densities,

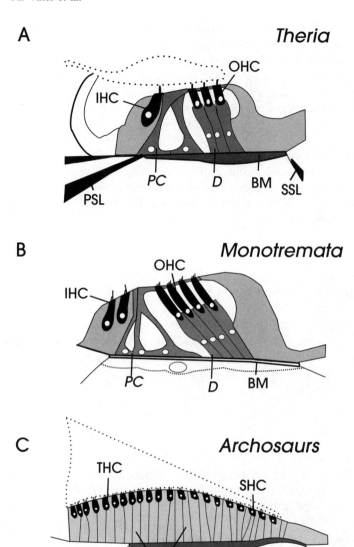

FIGURE 9.3. Schematic drawings of cross-sectioned organs of Corti in eutheria (**A**: Bat: *Tadarida brasiliensis* after Vater and Siefer 1995); **B**: monotreme (combined after Smith and Takasaka 1971 and a transmission electron microscope (TEM) photomicrograph kindly provided by Ladhams and Pickles); and **C**: bird (starling after Fischer et al. 1992). The natural position of the tectorial membrane is unknown in monotremes. BM, basilar membrane; D, Deiters' cells; IHC, inner hair cell; OHC, outer hair cell; PC, pillar cells; PSL, primary osseous spiral lamina; SHC, short hair cells; SSL, secondary osseous spiral lamina; THC, tall hair cells.

afferent and efferent innervation patterns, and structural dimensions of the organ of Corti in marsupials to provide a quantitative basis of the similarities and possible differences between the two therian lines.

5.2 Monotremes

The first anatomical description of the monotreme cochlea by Pritchard (1881) demonstrated that, contrary to expections from its avian or reptilian-like external anatomy, the internal structure bears typical mammalian features. As in placental mammals, there is a clear distinction between IHCs and OHCs, although there are two rows of IHCs instead of a single row (Fig. 9.3B). It needs to be emphasized that the original criterion for the distinction of two hair cell populations and thus the implicit homologies with the placental types of hair cells was the presence of a tunnel of Corti formed by specialized supporting cells. Such a geographical landmark is lacking in all nonmammalian tetrapods (Fig. 9.3C), but differences in afferent and efferent innervation patterns and size and shape of hair cells across the radial extent of the cochlear duct also suggest a division of labor among hair cells in the archosaurs (crocodiles and birds; Manley and Köppl 1998).

Monotremes have a tripartite cochlear duct with scala vestibuli, scala media, and scala tympani, and, similar to the marsupial and placental condition, the lateral wall of the cochlear duct has a well-formed stria vascularis. Reminiscent of the bird and reptile status, the monotreme cochlea houses a vestibular organ, the macula lagena at the apical end.

This pioneering work was followed by three brief reports (Hinojosa in Smith and Takasaka 1971; Chen and Anderson 1985; Pickles 1992), and the most thorough description of the monotreme cochlea is by Ladhams and Pickles (1996) using scanning and transmission electron microscopy. Basilar membrane length in the platypus (4.43 mm) and the echidna (7.6 mm) (Ladhams and Pickles 1996) is comparable to small placentals (mouse: 6.8 mm, see Echteler et al. 1994) but distinctly shorter than in placentals of comparable body size such as the cat (22.5 mm; see Echteler et al. 1994). The BM is clearly tapered in the echidna with maximum width (374 µm) found at 1.9 mm distance from the apex, narrowing down to 180 µm at 5 mm distance from the apex and then remaining constant in width toward the base. Minimal tapering of BM was observed in the platypus (max. width of 185 µm at 50% length; minimal 180 µm at apex). A BM width of 180 µm at the basal cochlear end is greater than in placentals, even including those with a hearing range restricted to below 20 kHz (human: 100 µm; mole rat: 120 µm) (Echteler et al. 1994). There are no further data on dimensions or fine structure of the BM in monotremes, and thus we lack important estimates for stiffness and vibration properties.

Ultrastructural evidence (Ladhams and Pickles 1996) supports the homology of IHCs and OHCs of monotremes with those of placentals. Besides the criterion of location relative to the tunnel of Corti, the criteria for homology include shape and arrangement of stereovilli bundles and internal organization of the hair cell, but unfortunately, quantitative data on the innervation patterns are lacking. Ladhams and Pickles (1996) showed that there are three to five rows of IHCs throughout most regions of the monotreme cochlea (Fig. 9.4) with a reduction to one irregularly spaced row at the extreme base and apex. IHCs throughout the cochlea of monotremes have linearly arranged stereovillar bundles like those of placentals (Fig. 9.4) but typically possess only two rows of stereovilli. In placental mammals this feature is seen in high-frequency IHCs of the cochlear base in bats (Vater and Lenoir 1992), whereas more apically located IHCs or those of most other placental mammals commonly possess multiple graduated rows of stereovilli (Slepecky 1996).

Stereovillar bundles of monotreme OHCs are either straight or shaped like shallow Vs or Ws. Throughout most of the cochlea duct, monotremes establish six rows of OHCs, whereas eutherians typically have three rows of OHCs (Fig. 9.4). Only at the extreme cochlear apex are more than three rows of OHCs found in primates, rats, and mice (Lim 1986; Echteler et al. 1994; Slepecky 1996). Monotreme OHC stereovilli are arranged in five to seven rows of graduated height (Ladhams and Pickles 1996), reminiscent of the condition seen in primates but different from most other placentals where there are mostly three rows (Slepecky 1996).

The shape and internal organization of monotreme OHCs agree with the placental plan (Ladhams and Pickles 1996): the cell bodies are cylindrical, the lateral walls appear surrounded by spaces of Nuel, there is one layer of laminated

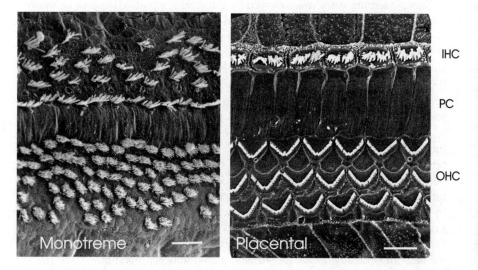

FIGURE 9.4. Scanning electron microscopy of the surface morphology of the organ of Corti in echidna (from Ladhams and Pickles 1996, with permission) and a bat (*Tadarida brasiliensis*). IHC, inner hair cell; OHC, outer hair cell; PC, pillar cells. Scale bars: left: 10 μm, right: 5 μm.

cisternae along the lateral wall, and the mitochondria are distributed in infracuticular and subnuclear regions and along the cisternae. The arrangements of the stereovillar bundles of monotreme hair cells appear more irregular than in placentals (Fig. 9.4). They are oriented radially in IHCs and OHCs that are located closest to the tunnel of Corti, whereas those of hair cells furthest away from the tunnel have an irregular orientation. A tectorial membrane-mediated shearing of stereovilli bundles is indicated by the insertions of the tallest rows of stereovilli of OHCs in the subsurface of the tectorial membrane (Ladhams and Pickles 1996). Similar to most placental mammals, there is no sign of tectorial membrane contact with IHC stereovilli, and there is no indication of a Hensen's stripe, which is present in high-frequency regions of cochlea in placental mammals (Vater and Lenoir 1992). Unfortunately, there are no descriptions or illustrations of the monotreme tectorial membrane in its natural position above the organ of Corti and no investigations of its fine structure.

Kinocilia, a typical feature of embryonic hair cells in placental mammals and of hair cells in adult reptiles and some birds (Manley 1990) occur occasionally in OHCs of the adult echidna but not in the platypus (Ladhams and Pickles 1996).

Interestingly, the number of IHCs in monotremes compares closely with data in the cat, mole, rat, and human despite gross differences in BM length (Table 9.1). This highlights the different strategy in arranging the presumed afferent messenger system in montremes as compared to therians: several rows versus a single row distributed along the length of a spiral. In contrast, the number of OHCs is smaller than in the cat and human, but comparable to the mole rat. Consequently, the ratio of IHC/OHC is lower in monotremes than in placentals (Table 9.1).

One further key feature of the monotreme cochleae concerns the development of specialized supporting cells: pillar cells and Deiters' cells. Such highly specialized nonsensory cells that play a central role in mechanical motion of the organ of Corti in placentals (Slepecky 1996) are not found in nonmammalian vertebrates (Manley 1990). The implementation of micromechanically specialized supporting cells in mammals could have sacrificed their function as a source

TABLE 9.1. Hair cell numbers and basilar membrane (BM) length in monotremes and placentals.

	No. of IHCs	No. of OHCs	Length of BM	Ratio IHC/OHC
Echidna	2710	5550	7.6	1/2
Platypus	1600	3750	4.43	1/2.3
Cat	2600	10,000	22.5	1/3.8
Human	3000	12,000	35	1/4
Mole rat	1179	4881	12.6	1/4.1

IHC, inner hair cell; OHC, outer hair cell.
Echidna and platypus: Ladhams and Pickles 1996; cat, human: Nadol 1988; mole rat: Burda et al. 1989.

for hair cell regeneration. Unique among mammals, the monotreme cochlea features three to four rows of pillar cells whose stalks form the tunnel of Corti and whose complexly interlocked heads form the roof of the tunnel (Ladhams and Pickles 1996) (Fig. 9.3). Thus all three rows of pillar cells contribute to the strength of the arch forming the tunnel of Corti. The base of each OHC of the monotreme cochlea is contained in a Deiters' cup as in placentals; however, different from the situation in placentals, the processes of the Deiters' cells that ultimately form the reticular lamina are joined by cross-bridges.

With distinct classes of IHCs and OHCs and specialized supporting cells, the monotreme organ of Corti has a fundamentally mammalian configuration. The differences between monotremes and therians concern hair cell and supporting cell number and patterning. Our understanding of the developmental mechanisms as the salient motor of cochlear evolution toward a design characterized by one row of IHCs and three rows of OHCs separated by a tunnel formed by two pillar cells is still limited. A number of genes coding for transcription factors, secreted factors, receptor tyrosine kinases, cyclin-dependent kinase inhibitors, and membrane-bound signaling proteins have been implicated in inner-ear morphogenesis and hair cell differentiation (Fekete 1999; Löwenheim et al. 1999; Zheng et al. 2000b). A balance of *Math1*, a candidate gene for control of hair cell differention, and *Hes1*, a negative regulator of neurogenesis, is crucial for the production of an appropriate number of IHCs in the placental ear (Zheng et al. 2000a). Multiple rows of IHCs can be evoked in placentals by application of retinoic acid in the young organ of Corti in vitro (Kelley et al. 1993). Supernumery rows of IHCs, OHCs, and Deiters' cells combined with dysmorphic pillar cells are found in mice lacking $p27^{Kip1}$, a cell-selective regulator of cell proliferation (Löwenheim et al. 1999).

6. Evolution of the Mammalian Cochlear Amplifier

The cochlea of extant monotremes can serve, to a certain degree, as a model of the ancestral mammalian cochlea. However, because of its fairly young age on a geological time scale combined with a significant time of independent parallel evolution, it is clearly not the true archetype nor is it a true "missing link" for the transition between therapsid reptiles and mammaliforms (see Fig. 9.1). It does, however, represent the only known hearing organ outside the theria where OHCs defined by a basic set of common criteria can be identified, and there is evidence of active nonlinear cochlear amplification (Mills and Shepherd 2001). The frequency range of the monotreme cochlear amplifier as judged from DPOAE data appears intermediate between birds/reptiles and placental mammals. It goes up to frequencies of at least 20 kHz (Mills and Shepherd 2001), that is, about an octave higher than in nonmammalian tetrapods (Manley 2000), but its range appears about two octaves lower than the estimated limit of the cochlear amplifier in placental mammals (Kössl 1997; Frank et al. 1999). Addressing the nature of the cochlear amplification mechanism in monotremes is a matter of pure speculation at the present state of knowledge. Three alternatives need to be considered (Mills and Shepherd 2001): (1) monotremes use fast OHC motility as do placentals and probably marsupials; (2) a stereovilli-based mechanism similar to that proposed for nonmammalian tetrapods; or (3) a still-unknown amplification mechanism. Each of these hypotheses leads to a different evolutionary scenario.

Are monotreme OHCs capable of fast motility? There is no experimental evidence to answer this question, but if the status "outer hair cell" defined by morphological data coincides with fast electromotility, then this key invention could have been made by the most recent common ancestor of all mammalian taxa, which would date its origin back to about 200 million years ago.

Equally interesting would be the possibility that monotremes use other mechanisms for active cochlear amplification. In birds and reptiles, there is evidence of a stereovilli-based amplification mechanism (Manley et al. 2001), and the possibility was raised that it could be operative in mammals as well. Like the hypothesis discussed above, this needs more experimental evidence not only in monotremes but also within the Eutheria. If the monotreme amplifier indeed resided at the level of the sensory transduction apparatus, only then at least two different cochlear amplifiers would be used within the taxon mammalia. Furthermore, the invention of OHC motility could either replace the phylogenetically older mechanism or be superimposed on the ancient trait. Given the remarkably convergent cochlear design and the evidence for excellent high-frequency hearing in extant marsupials, the most parsimonous hypothesis is that they employ OHC motility as do placentals. This would date OHC motility back at least 130 million years in early Cretaceous in the common ancestors of the theria.

Postulating an unknown cochlear amplification mechanism for the monotreme cochlea (Mills and Shepherd 2001) is a legitimate approach and emphasizes the fact that our general understanding of active processes in any hearing organ and their exact mode of operation within the complex passive mechanical system is still limited. This unknown mechanism (1) could have evolved independently within the monotremes, thus representing a derived trait of this group only; (2) could represent an ancient trait within the mammalian line and thus be present in theria as well, forming an integral part of the "mammalian cochlear amplifier"; or (3) could also be present in birds and reptiles, and thus unify cochlear amplification to a common mechanism in all amniotes with the consequence that it dates further back in time to the stem reptiles.

Measurements of otoacoustic emissions have documented clear differences in upper frequency limits of cochlear amplification mechanisms among birds, reptiles, monotremes, and placentals. But do these measurements reveal the functional limits of the specific cellular mechanisms employed? Stated differently, would it be sufficient to create sensitive hearing at frequencies beyond 20 kHz by integrating a prestin-like motor protein in the lateral wall of the monotreme OHCs while holding the passive mechanical design of the cochlea and middle ear constant? According to Aitkin and Johnstone (1972), the middle-ear transmission characteristic in monotremes is poor for airborne sounds. Furthermore, with respect to macro- and micromechanical features of anatomical design, the cochlea of monotremes does not qualify as a high-frequency receiver and transducer.

A basic anatomical criterion for a high-frequency specialization is the presence of a narrow and thickened BM at the basal end of the cochlea (Echteler et al. 1994) combined with the presence of a secondary bony osseous spiral lamina (Bruns 1980; Fox and Meng 1997; Ketten 2000). These features are lacking in monotremes. Other high-frequency traits of the placental cochlea are the short cell bodies of basal OHCs and decreased stereovilli height at the basal end of the cochlea (Pujol et al. 1992; reviewed in Echteler et al. 1994). These features are thought to define local passive mechanical resonances. Monotreme OHCs are described as very tall and their height matches that of IHCs in all available illustrations, but there clearly is a need to increase the database. The multiple rows of supporting and sensory cells of the monotreme organ of Corti may also affect local resonant properties by representing a mass load of the BM and thus limit its high-frequency motion. Significantly, in the barn owl, hair cell numbers across the radial extent of the BM are greatly reduced at the basal end of the cochlea (Manley and Köppl 1998). Furthermore, the ratio of IHC/OHC in monotremes is much lower than in placentals, and a synchronized concerted action among multiple rows of OHCs with appropriate phase relationships between BM motion and active reverse transduction is even harder to envisage than in placentals.

The arrangement of stereovillar bundles in the monotreme cochlea reveals irregularities; thus, the force vectors during shear motion appear nonuniformly distributed, a feature that may lead to differences in thresholds among hair cells at a given position of the cochlear duct and also affect putative reverse transduction from motile OHCs into BM mechanics.

The optimizing strategy seen both in placentals and marsupials is a redistribution of sensory cells on a considerably lengthened BM, combined with an increase in the OHC/IHC ratio. There is no living species that bridges the morphological gap between the cochlea of monotremes and those of placentals/marsupials. There is, however, abundant evidence from x-ray investigations, computed tomography, and scanning electron microscopy of fossil ears that allows the identification of phylogenetic relationships and hints at the evolution of hearing function.

7. Cochlea and Hearing in Extinct Mammals

7.1 Early Forms

The inner ear of early mammals appears more conservative than the middle ear because the cochlea is either straight or slightly curved, with minor changes over a long time span (Luo 2001) (Fig. 9.1). For instance, multituberculates have the longest history among mammals, but the bony structures of the cochleae have undergone very little modification from late Jurassic forms (Lillegraven and Hahn 1993) to early Tertiary ones (Miao 1988; Meng and Wyss 1995). An endocast of the cochlea of *Lambdobsalis* revealing an almost straight cochlear duct is shown in Figure 9.5. In some multituberculates the cochlea has a curvature, bending laterally in a direction that is the same as that of therians, but opposite to that of birds (Fig. 9.1). The curvature is certainly less than 180 degrees or a half circle (Fox and Meng 1997). This observation resembles that of Simpson (1937) on *Ptilodus* but differs from that of Sloan (1979) on *Ectypodus* in which a hook-like cochlea, similar to that of the platypus, has been reported. The multituberculate inner ear lacks inner and outer bony laminae for attachment of the basilar membrane (Kielan-Jaworowska et al. 1986; Luo and Ketten 1991; Meng and Fox 1993; Fox and Meng 1997). In this respect it differs from the therians (MacIntyre 1972; Fox and Meng 1997) and resembles primitively all other early forms such as Morganucodon (Kermack et al. 1981; Graybeal et al. 1989). Moreover, the cochlear nerve that innervated the sensory cells passed through the petrosal as a single bundle, as indicated by a single cochlear foramen. Judging from the size of the foramen, it is probable that the number of nerve fibers, therefore perhaps also of the sensory cells, were fewer than those of therians.

The perilymphatic recess in multituberculates marked the dorsal roof of the "round" window but did not merge with the scala tympani of the cochlea. A bony aqueductus cochlea is not developed and a true fenestra cochlea is not formed (Wible 1990; Wible and Hopson 1993; for discussion of round window homologies see Fox and Meng 1997). Furthermore, the cochlea of multituberculates probably possessed a lagena, similar to that of monotremes, as indicated by an apical inflation of the cochlea in some well-preserved multituberculates (Meng and Fox 1993). Some multituberculates show an inflation of the vestibule of the inner ear (Luo 1989; Simmons 1993; but see Lillegraven and Hahn 1993; Meng and Wyss 1995), the only known advanced feature in comparison to the inner ear of mammaliforms. The absolute size of the vestibule in *Lambdopsalis bulla*, a late Paleocene multituberculate (Fig. 9.5), is about two orders of magnitude greater than in *Morganucodon*, 30 times greater than in extant monotremes, and more than 10 times greater than in opossum and human (Luo and Ketten 1991). Based on external morphology of the promontorium of the petrosal, an inner ear similar to that of multituberculates may well be present in triconodonts, such as *Priacodon* (Rougier et al. 1996) and *Repenomamus* (Wang et al. 2001), in symmetrodonts, such as *Zhangheotherium* (Hu et al. 1997), and in dryolestoids, such as *Henkelotherium*, although in the latter, a single-layered primary osseous spiral lamina is developed within the cochlear canal (Hu and Krebs 1999) (for phylogenetic relationships see Fig. 9.1).

7.2 Mesozoic Monotremes

Cranial morphologies of Mesozoic monotremes are unknown (Archer et al. 1985), but their living relatives retain an ear akin to that of other Mesozoic

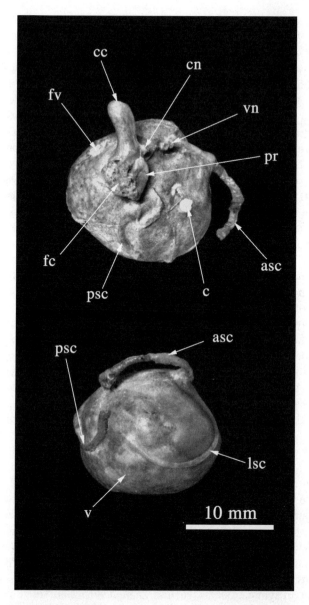

FIGURE 9.5. The inner ear of the multituberculate *Lambdopsalis bulla*, ventral view (paleocene-eocene; from Meng and Wyss 1995, with permission). Note the straight cochlear canal (cc) and the large vestibule (v). Acs, anterior semicircular canal; c, crus commune; cn, cochlear nerve; fc, fenestra cochleae (perilymphatic foramen); fv, fenestra vestibuli; lcs, lateral semicircular canal; pr, perilymphatic recess; psc, posterior semicircular canal; vn, vestibular nerve.

mammals. Radiographic studies (Fox and Meng 1997) show that living monotremes (platypus and echidna) have a cochlear canal with roughly a 180-degree turn. In particular, the cochlear canal of the adult platypus has a relatively straight basal part and an abruptly bent apical part, and is certainly less than 270 degrees and even less than that of echidna that has a nearly semicircular canal. This observation is different from the conventional view that the cochlea of platypus turns through 270 degrees, whereas that of echidna turns through 180 degrees (Kermack et al. 1981; Kermack and Mussett 1983; Luo and Ketten 1991). Radiographs of the inside morphology of the cochlea of the adult platypus (Fox and Meng 1997) did not reveal any bony structures (i.e., inner and outer bony laminae) comparable to those of therians. The cochleae thus resemble those of the early mammals and mammaliformes mentioned above. However, the cochlear nerve passes through a cribriform plate in the internal acoustic meatus, similar to that of therians, although it is not spiral (Simpson 1938; Kermack et al. 1981). Therefore, the pathway for the cochlear nerve in monotremes seems to be more similar to that of the therians than to that of multituberculates. Monotremes are the only extant mammals that retain the lagena at the apical region of the cochlea, resembling that of birds and reptiles (e.g., Ladhams and Pickles 1996). Luo and Ketten (1991) found a peculiar feature in the monotreme ear: "The bony cochlear cavity does not coil in correspondence to the coiled membranous labyrinth in monotremes." This is quite different from the condition in therians, where the membranous labyrinth takes the shape of the bony labyrinth.

7.3 Mesozoic Therians

In Cretaceous therian mammals, the inner-ear structures are dramatically different from those of the non-therians described above. The following characters represent major evolutionary steps that differentiate tribosphenic therians, if not all therians, from non-therians. These synapomorphies include:

1. A cochlea that is fully coiled, that is, one that spirals through at least 360 degrees (Figs. 9.1 and 9.6). The earliest mammalian ear having such a cochlea is a therian petrosal from the Milk River Formation of Alberta, Canada, about 80 million years ago (Meng and Fox 1993). The radiograph of the specimen reveals that the cochlear canal has approximately one and a quarter turns, coiling rather loosely and exhibiting a constant diameter from base to apex. In later therians, such as the marsupials and placentals from the Bug Creek Locality of Montana, the cochlear canal has one and a half turns and coils tightly, as in extant primitive mammals such as the marsupial mole and the hedgehog (Gray 1908a,b; Lewis et al. 1985). A fully coiled cochlea is recorded by at least middle-late Cretaceous time and it may well date from the early Cretaceous or even an earlier time, as anticipated by Fernández and Schmidt (1963).
2. Development of primary and secondary osseous spiral laminae (Fig. 9.6A, D).

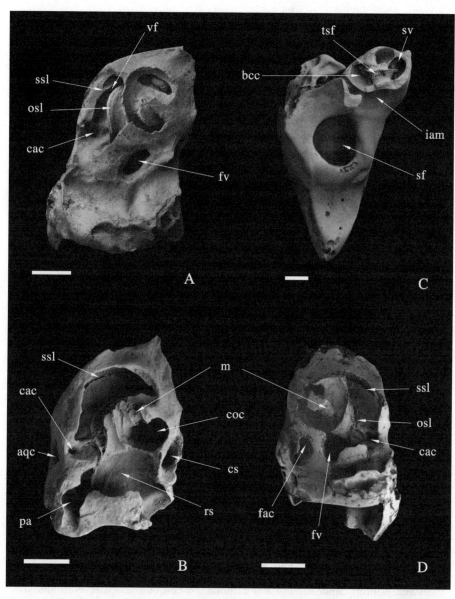

FIGURE 9.6. **A, B**: Ventral views of placental petrosals (left side) from the late Cretaceous Hell Creek formation at Bug Hill Anthills, Montana. **C**: A petrosal of a late Cretaceous placental mammal (right side). **D**: Ventral view of a marsupial petrosal (right side) from the late Cretaceous Hell Creek formation at Bug Hill Anthills, Montana. Scale bars: 1 mm. Aqc, aqueductus cochleae; bcc, basal cochlear canal; cac, cochlear orifice of the aqueductus cochleae; coc, cochlear canal; cs, cavum supracochleae; fac, facial canal; iam, internal acoustic meatus; m, modiolus; osl, primary osseous spiral lamina; pa, posterior ampulla; rs, recessus sphericus; sf, subarcuate fossa; ssl, secondary osseous spiral lamina; sv, scala vestibuli; tsf, tractus spiralis foraminosus; fv, vestibular fissure.

The earliest specimens known having these structures are petrosals from the Judithian Oldman Formation of Alberta, Canada, about 80 million years ago (Meng and Fox 1993), but the structures are best known in specimens of later age (MacIntyre 1972; Meng and Fox 1995a, b). Development of the primary osseous spiral lamina probably resulted from the coiling of the cochlear canal. It provides a rigid support for a narrow and elongate basilar membrane and serves as a pathway for the cochlear nerve fibers that innervate the organ of Corti. Development of the secondary osseous spiral lamina is a feature of the basal cochlear regions in many extant placentals, and its presence has been linked to high-frequency hearing capabilities (Bruns 1980; Ketten 2000).

3. A cochlear nerve that was arranged in a radial pattern. In this fashion, the cochlear nerve passes through the wall of the cochlear canal as numerous branches; thus a spongy belt in the internal acoustic meatus (the tractus spiralis foraminosus in the modiolus; Fig. 9.6) is formed. The radial pattern of the cochlear nerve permits a greater number of nerve fibers to pass through the petrosal and minimizes length differences among the fibers extending to the basal and apical regions of the basilar membrane (West 1985; Meng and Fox 1995b).

4. A narrow BM that proximally extends to the region between the fenestra cochlea and fenestra vestibuli. It has been known that coiling of a cochlea provides an economical means of housing an elongate BM in the skull (Meyer 1907; Békésy 1960). However, the studies of Fox and Meng (1997) show that elongation of the BM is not one-directional, but two-directional, both apically and basally. The presence of the secondary spiral lamina on the internal surface of the bony wall between the oval and round windows suggests that the BM in the ear of Cretaceous therians had already extended basally. In non-therians, there is no evidence for the basal extension of the BM and the bony bridge between the "round" and oval windows is usually narrow.

5. A perilymphatic recess that merges with the scala tympani of the cochlea and is entirely enclosed in the cochlea.

6. Development of a bony aqueductus cochlea and formation of a true fenestra cochlea (also see Wible 1990; Wible and Hopson 1993).

The fossil record documents a sequence of critical morphological changes in the cochlea of therian mammals. In the non-tribosphenic therian *Vincelestes* from the early Cretaceous, the cochlear canal has only a 270-degree turn (Rougier et al. 1992). In the earliest known tribosphenic therian (Meng and Fox 1993, 1995a,b), the cochlea has only one and a quarter turns. In the therian ears from Bug Creek locality, the cochlea has one and a half turns and coils more tightly. In later therians, the number of cochlear coils is usually more than two, with greater tapering (Lewis et al. 1985). This suggests that a coiled cochlea in therians was probably gradually achieved. By contrast, other inner-ear features such as the six listed above display a distinct morphological gap between non-therians and therians. This gap may be bridged as additional morphology becomes available in these features from transitional taxa such as *Vincelestes*.

8. Implications of Monotreme and Multituberculate Cochleae for Mammalian Phylogeny

As presently known, inner-ear structures offer little phylogenetic signal for unraveling the relationships among early mammals because they are mostly primitive in those groups. Two features, however, may have relevance for hypotheses on the relationships among the major mammalian clades: the lagena and the curvature of the cochlea. Among extant mammals, only monotremes primitively possess the lagena, as in reptiles and birds (Pritchard 1881; Denker 1901; Alexander 1904; Gray 1908a,b; Fernández and Schmidt 1963; Griffiths 1968, 1978). Manley (1973) speculated that the cochlear canal of *Triconodon* probably also contained a lagena macula, although the canal is very short, e.g., 3 to 4 mm (Kermack 1963). Meng and Fox (1993, 1995a,b) show that the lagena was probably present in multituberculates as well, as suggested by an apical expansion of the cochlea in at least one well-preserved specimen.

The curvature of the cochlea has been used in the past to differentiate monotremes plus therians from other mammals, including multituberculates (Luo and Ketten 1991), or to group monotremes, therians, and multituberculates (Rowe 1988). According to the observations of Fox and Meng (1997), however, this is a problematic feature. The problem is rooted in the confusion of two anatomically related structures: the osseous cochlear canal and the soft cochlear duct. The cochlear duct, part of the membranous labyrinth, is contained in the osseous cochlear canal. It was observed that the osseous cochlear canal does not coil along with the soft cochlear duct in monotremes (Luo and Ketten 1991; also see Zeller 1993, Fig. 8.6). The coiled soft cochlear duct was then compared with the osseous cochlear canal of other early mammals to reach the conclusion that a cochlea with a curvature greater than 180 degrees supports the monotreme-therian grouping (Luo and Ketten 1991). However, the soft cochlear duct of multituberculates and other early mammals could have coiled within the cochlear canal the same way as in monotremes. Obviously, what should be compared are the osseous cochlear canals in monotremes and other early mammals. In doing so, Fox and Meng (1997) found that the cochlear canals in monotremes and multituberculates show less difference than previously believed and therefore they regard the cochlear curvature as a questionable character for phylogenetic relationships concerning therians, monotremes, and multituberculates.

Moreover, the peculiar coiling condition of the monotreme cochlea poses the question whether it is homologous to that of therians (Zeller 1989), because the osseous cochlear canal always coils along with the cochlear duct in therians (Gray 1908a,b; Fernández and Schmidt 1963; Lewis et al. 1985). Fox and Meng (1997) showed that the cochlear canal in both platypus and echidna does not show substantial difference in its diameter throughout its entire length; therefore, a 270-degree bending of the cochlear duct within such a narrow cochlear canal must be quite abrupt, which further raises question about how the basilar membrane vibrates appropriately.

According to Meng and Fox (1993), tribosphenic therians share the following synapomorphies of inner-ear structures: a fully coiled cochlea, primary and secondary osseous spiral laminae, the perilymphatic recess merging with the scala tympani of the cochlea, an aqueductus cochlea, a true fenestra cochlea, a radial pattern of the cochlear nerve, and an elongate BM extending basally to the region between the fenestra vestibuli and fenestra cochlea. The inner-ear structures of living therians differ from those of their late Cretaceous relatives mainly in having a greater number of spiral turns of the cochlea and a longer BM.

9. Hearing in Early Mammals

The most marked changes in the evolution of the mammalian ear are two: acquisition of a triossicular middle ear and development of a coiled cochlea that is coupled with the above-mentioned inner-ear features. Both the inner and middle ears are important in the evolution of mammalian hearing, although it has been previously stated that the evolution of three middle-ear ossicles provides an explanation for the radical difference between mammals and nonmammals in the upper limit of hearing (Masterton et al. 1969). Evidence has accumulated for cochlear structural correlates for hearing in particular frequency ranges among extant and extinct mammals (Echteler et al. 1994; Meng and Fox 1997).

Concerning the "archaic" frequency range of hearing in mammals, there have been two conflicting views: (1) The earliest mammals had a low-frequency hearing range similar to reptiles and birds. (2) The earliest mammals already had a high-frequency hearing range (Masterton et al. 1969; Rosowski 1992). A further possibility is derived from comparison of the audiograms and cochlea anatomy of extant mammals, showing that the monotremes are intermediate between nonmammalian tetrapods and therians and can thus serve as a model for primitive hearing capacities of the mammalian clade (Gates et al. 1974; Ladhams and Pickles 1996; Mills and Shepherd 2001). It is important, however, to appreciate that no living species perfectly represents the ancestral form.

Based primarily on middle-ear anatomy, some specialists (Rosowski and Graybeal 1991; Rosowski 1992) have come to the conclusion that high-frequency hearing predated *Morganucodon*, a mammaliaform in which the quadrate and articular bones were not yet suspended as purely auditory ear ossicles in the tympanic cavity. For *Morganucodon*, Rosowski (1992) calculated an effective hearing range of 6 to 30 kHz at 30 dB above minimum threshold from middle-ear dimensions. The inner-ear structures of *Morganucodon*, however, are far from being a high-frequency device that is comparable to that of therians; instead, they are bird-like but curve in the opposite direction (Kermack et al. 1981; Graybeal et al. 1989; Rosowski and Graybeal 1991). A similar inner ear is seen in triconodonts (Kermack 1963) and multituberculates (Kielan-Jaworowska et al. 1986; Luo and Ketten 1991; Meng and Fox 1993), although multituberculates are now known to have a triossicular system (Miao and Lillegraven 1986; Miao 1988; Meng and Wyss 1995; Rougier et al. 1996). In all of these forms and in extant monotremes, the inner ear lacks several structures, such as the osseous spiral laminae, that characterize the high-frequency ear of therians. Without these structures, the inner ears of these forms may not have been able to cope with high frequencies as those of living therians, even though their middle ears could have been physically capable of transmitting high-frequency vibrations. In support of this notion, comparisons of audiograms and transfer functions for stapes or columella velocity in extant tetrapods show that the middle ear by itself does not alone limit the high-frequency hearing range and suggest that the tonotopic organization of the cochlea crucially determines the frequency limits of sensitivity (Manley 1972; Ruggero and Temchin 2002).

A diversity of hearing adaptations already in early mammals is indicated by the finding that an inflated vestibule is present in some multituberculate petrosals. This structure may play a role in low-frequency and bone-conducted hearing (Miao 1988; Luo and Ketten 1991; Fox and Meng 1997). This is consistent with the assumed fossorial life of some of these species (Miao 1988; Kielan-Jaworowska and Qi 1990). Speaking for this hypothesis are case reports on the function of comparable specializations in extant amphibia (for discussion see Luo and Ketten 1991) and experiments and models of Békésy (1960) as discussed by Fox and Meng (1997). In early mammals in general, the BM did not seem to extend basally to the area between the oval and round windows, as mentioned above. Consequently, the distance from the fenestra vestibuli, and hence the footplate of the stapes, to the BM is relatively far so that high-frequency vibrations could easily have dissipated before they were able to activate the BM. In multituberculates in particular, the distance between the fenestra vestibuli and the basal BM is further increased as a result of vestibule inflation. An exaggerated example of this pattern is provided by the inner ear of *Lambdopsalis* (Miao 1988), in which the fenestra vestibuli is positioned quite far from the cochlea; in fact, its distance matches the total length of the cochlear canal itself. Furthermore, according to Békésy (1960), vibrations produced by bone conduction of the skull can roughly be resolved into parallel, compressional, and rotational vibrations of the cochlea and middle ear. If an asymmetry is present between the scala tympani and scala vestibuli, that is, the space of the scala vestibuli is larger than that of the scala tympani, compressional vibrations can produce a significant displacement of fluid near the stapes, which in turn causes a disturbance of the BM (Békésy 1960). Inflation of the vestibule in some multituberculates greatly increases the space near the scala vestibuli, making bone-conducted hearing mechanically feasible.

According to Hurum (1998), however, some late Cretaceous multituberculates, such as *Chulsanbaatar*, may have had the capability to hear high-frequency sounds based on "the short cochlea and the speculations about the areas of the tympanic membrane and the footplate of the stapes compared to extant mammals."

Early therians were probably capable of sensitive hearing of high-frequency airborne sound, that is, well above 20 kHz. The basis for this inference is that the inner-ear structures of the early therians are nearly identical to those of living therians (Meng and Fox 1995a,b; Fox and Meng 1997). At least some of the inner-ear structures in therians must have played a role in high-frequency hearing. For instance, development of the primary and secondary spiral laminae provides a stiff framework for the attachment of a narrow BM, and basal extension of the osseous spiral laminae decreases the distance between the stapes and the basal part of the BM.

Furthermore, comparisons across the major tetrapod taxa suggest that an elongation of the cochlea is associated with an increase in the hearing range (Manley 1971, 1973; Wever 1974) and enhanced discrimination of frequencies (Manley 1973; Allin and Hopson 1992). Along with lengthening of the BM, there is an increase in the number of hair cells from amphibians to mammals (Wever 1974), with those vertebrates having a larger number of hair cells being more sensitive to sound (Manley 1973). Elongation of the BM by coiling of the cochlea and development of the radial pattern of the cochlear nerve system suggest an increase of the number of the sensory cells in early therians (Meng and Fox 1995b); thus, late Cretaceous therians probably have developed a sensitive ear good for high-frequency hearing of airborne sounds that had a broad frequency range. Still, it needs to be kept in mind that among extant placentals there is no convenient way to exactly predict the hearing range from gross morphology of the cochlea without independent knowledge of specializations of the soft cochlear anatomy (Echteler et al. 1994) and data on the ecological niche of the species in question.

What is the significance and evolutionary advantage of high-frequency hearing as a trait apparently unique to therian mammals? Masterton et al. (1969) argued that "the reception of high frequencies is tantamount to an expansion of the range of binaural spectral disparities because even little heads induce a large disparity at high-frequencies." A significant correlation between interaural distance and high-frequency limit of hearing is found only in therians and has been confirmed in a large sample of species (Heffner and Heffner 1992).

10. Summary

The evidence collected in this chapter leads to the following conclusions: (1) The evolution of a three-ossicle middle ear and the establishment of a specialized hair cell type, the OHC, as a likely basis of the cochlea amplifier predated the evolution of an elongated coiled cochlea capable of processing frequencies well above 20 kHz. (2) In contrast to therians, early mammals were probably able to hear only relatively lower frequency airborne sounds in a narrower frequency range possibly modeled by the hearing characteristic of extant monotremes, and some multituberculates were specialized for bone-conducted hearing. (3) The ability to hear frequencies well above 20 kHz is a specialized trait of marsupials and placentals and appears linked to a refinement of passive micromechanical structures and possibly active processes together with an elongation of the sensory surface and a redistribution of sensory cells.

Acknowledgments. We wish to thank Jim Pickles, Charles Steele, Ian Russell, and Manfred Kössl for fruitful discussions, valuable comments, and suggestions. Jim Pickles kindly provided photographs of the monotreme cochlea. Lorraine Meeker and Chester Tarka took the photographs of the inner ear of multituberculates, and G.D. Braybook provided the scanning electron microscope images of the petrosals.

References

Aitkin LM (1995) The auditory neurobiology of marsupials: a review. Hear Res 82:257–266.
Aitkin LM (1998) Hearing, the Brain and Auditory Communication in Marsupials. New York: Springer-Verlag.
Aitkin LM, Johnstone BM (1972) Middle-ear function in a monotreme: the echidna (*Tachyglossus aculeatus*). J Exp Zool 180:245–250.
Aitkin LM, Nelson GE, Shepherd RK (1994) Hearing, vocalization and the external ear of a marsupial, the northern quoll, *Dasyurus hallucatus*. J Comp Neurol 349:377–388.
Alexander G (1904) Entwicklung und Bau des inneren Gehörorgans von *Echidna aculeata*. Semonds Zoologische Forschungsreisen in Australien 3:1–118.
Allen JB, Fahey PF (1993) A second cochlear frequency map that correlates distortion product and neural tuning measurements. J Acoust Soc Am 94:809–816.
Allin EF (1986) Auditory apparatus of advanced mammal-like reptiles and early mammals. In: Hotton N, MacLean PD, Roth JJ, Roth EC (eds) The Ecology and Biology of Mammal-Like Reptiles. New York and London: Smithsonian Press, pp. 283–294.
Allin EF, Hopson JA (1992) Evolution of the auditory system in Synapsida ("mammal-like reptiles" and primitive mammals) as seen in the fossil record. In: Webster DB, Fay RR, Popper AN (eds) The Evolutionary Biology of Hearing. New York: Springer-Verlag, pp. 587–614.
Archer M, Flannery TF, Ritchie A, Molnar R (1985) First Mesozoic mammal from Australia—an Early Cretaceous monotreme. Nature 318:363–366.
Archer M, Murray P, Hand SJ, Godthelp H (1993) Reconsideration of monotreme relationships based on the skull and dentition of the Miocene *Obdurodon dicksoni*. In: Szalay FS, Novacek MJ, McKenna MC (eds) Mammal Phylogeny: Mesozoic Differentiation, Multituberculates, Monotremes, early Therians, and Marsupials. New York: Springer-Verlag, pp. 75–94.
Archer M, Arena R, Bassarova M, Black K, Brammall J, Cooke B, Creaser P, Crosby K, Gillespie A, Godthelp H, Gott M, Hand SJ, Kear B, Krikmann A, Mackness B, Muirhear J, Musser A, Myers T, Pledge N, Wang Y, Wroe S (1999) The evolutionary history and diversity of Australian mammals. Australian Mammalogy 21:1–45.
Ashmore F, Geleoc GSG, Harbott L (2000). Molecular mechanisms of sound amplification in the mammalian cochlea. Proc Natl Acad Sci USA 97:11759–11764.
Békésy G von (1960) Experiments in Hearing. New York: McGraw-Hill.
Brown AM, Kemp DT (1984) Suppressibility of the 2f1-f2 stimulated acoustic emission in gerbil and man. Hear Res 42:143–156.
Brown AM, Gaskill SA, Williams DM (1992) Mechanical filtering of sound in the inner ear. Proc R Soc B 250:29–34.
Brownell WE, Bader CR, Bertrand D, de Ribeaupierre Y (1985) Evoked mechanical responses of isolated cochlear outer hair cells. Science 227:194–196.

Bruns V (1980) Basilar membrane and its anchoring system in the cochlea of the greater horseshoe bat. Anat Embryol 161:29–50.
Burda H, Bruns V, Nevo E (1989) Middle ear and cochlear receptors in the subterranean mole-rat, *Spalax ehrenbergi*. Hear Res 39:225–230.
Chen CS, Anderson LM (1985) The inner ear structures of the echidna—an SEM study. Experientia 41:1324–1326.
Cifelli RL (2001) Early mammalian radiations. J Paleontol 75:1214–1226.
Cifelli RL, de Muizon C (1997). Dentition and jaw of *Kokopellia juddi*, a primitive marsupial or near marsupial from the medial Cretaceous of Utah. J Mammalian Evol 4:241–258.
Cone-Wesson BK, Hill KG, Liu G-B (1997) Auditory brainstem response in tammar wallaby (*Macropus eugenii*). Hear Res 105:119–129.
Crompton AW (1995) Masticatory function in nonmammalian cynodonts and early mammals. In: Thomason JJ (ed) Functional Morphology in Vertebrate Paleontology. Cambridge: Cambridge University Press, pp. 55–75.
Crompton AW, Jenkins FA Jr (1979) Origin of mammals. In: Lillegraven JA, Kielan-Jaworowska Z, Clemens WA (eds) Mesozoic Mammals—The First Two-Thirds of Mammalian History. Berkeley: University of California Press, pp. 59–73.
Denker A (1901) Zur Anatomie des Gehörorganes der Monotremata. Denkschriften der Medizinisch-Naturwissenschaftlichen Gesellschaft zu Jena 6:635–662.
Echteler SM, Fay RR, Popper AN (1994) Structure of the mammalian cochlea. In: Fay RR, Popper AN (eds) Comparative Hearing, Mammals. Springer Handbook of Auditory Research, vol 4. New York: Springer-Verlag, pp. 134–172.
Faulstich M, Kössl M, Reimer K (1996) Analysis of non-linear cochlear mechanics in the marsupial *Monodelphis domestica*: ancestral and modern mammalian features. Hear Res 94:47–53.
Fay RR (1988) Hearing in Vertebrates: A Psychophysics Data Book. Winnetka, IL: Hill-Fay Associates.
Fekete DM (1999) Development of the vertebrate ear: insights from knockouts and mutants. TINS 22:263–269.
Fernandez C, Schmidt RS (1963) The opossum ear and evolution of the coiled cochlea. J Comp Neurol 121:151–159.
Fischer FP, Miltz C, Singer I, Manley GA (1992) Morphological gradients in the starling basilar papilla. J Morphol 213:225–240.
Flynn JJ, Parrish M, Rakotosamimanana B, Simpson WF, Wyss AR (1999) A middle Jurassic mammals from Madagascar. Nature 401:57–60.
Fox RC, Meng J (1997) An X-radiographic and SEM study of the osseous inner ear of multituberculates and monotremes (Mammalia): implications for mammalian phylogeny and evolution of hearing. Zool J Linnean Soc 121:249–291.
Frank G, Hemmert W, Gummer AW (1999) Limiting dynamics of high-frequency electromechanical transduction of outer hair cells. Proc Natl Acad Sci USA 96:4420–4425.
Frost SB, Masterton BR (1994) Hearing in primitive mammals: *Monodelphis domestica* and *Marmosa elegans*. Hear Res 76:67–72.
Gates GR, Aitkin LM (1982) Auditory cortex in the marsupial opossum *Trichosurus vulpecula*. Hear Res 7:1–11.
Gates GR, Saunders J, Bock GR (1974) Peripheral auditory function in the platypus, *Ornithorhynchus anatinus*. J Acoust Soc Am 56:152–156.
Gray AA (1908a) An investigation on the anatomical structure and relationships of the labyrinth in the reptile, the bird, and the mammal. Proc R Soc B 80:507–528.

Gray AA (1908b) The Labyrinth of Animals, vol 2. London: J. & A. Churchill.

Graybeal A, Rosowski J, Ketten DR, Crompton AW (1989) Inner ear structure in *Morganucodon*, an early Jurassic mammal. Zool J Linnean Soc 96:107–117.

Gregory WK (1947) The monotremes and the palimpsest theory. Bull American Museum of Natural History 88:1–52.

Griffiths M (1968) Echidnas. Oxford: Pergamon Press.

Griffiths M (1978) The Biology of the Monotremes. New York: Academic Press.

Heffner HE, Masterton B (1980) Hearing in Glires: domestic rabbits, cotton rat, feral house mouse, and kangaroo rat. J Audit Res 9:12–18.

Heffner RS, Heffner HE (1992) Evolution of sound localization in mammals. In: Webster DB, Fay RR, Popper AN (eds) The Evolutionary Biology of Hearing. New York: Springer-Verlag, pp. 691–711.

Holley M (1996) Outer hair cell motility. In: Dallos P, Popper AN, Fay RR (eds) The Cochlea. Springer Handbook of Auditory Research, vol 8. New York: Springer-Verlag, pp. 386–435.

Hopson JA (1994) Synapsid evolution and the radiation of non-eutherian mammals. In: Spencer RS (ed) Major Features of Vertebrate Evolution. Knoxville: Paleontological Society, pp. 190–219.

Hu Y-M, Krebs B (1999) Discovery of the *Henkelotherium* petrosal and its implication for the evolution of mammalian hearing. J Vertebrate Paleontol 19 (3 suppl): 53A.

Hu Y-M, Wang Y-Q, Luo Z-X, Li C-K (1997) A new symmetrodont mammal from China and its implications for mammalian evolution. Nature 390:137–142.

Hurum JH (1998) The inner ear of two late Cretaceous multituberculate mammals, and its implications for multituberculate hearing. J Mammal Evol 5:65–94.

Janke A, Xu X, Arnason U (1997) The complete mitochondrial genome of the wallaroo (*Macropus robustus*) and the phylogenetic relationship among Monotremata, Marsupialia, and Eutheria. Proc Natl Acad Sci USA 94:276–281.

Kalinec F, Holley MC, Iwasa KH, Lim DJ, Kachar B (1992) A membrane based force generation mechanism in auditory sensory cells. Proc Natl Acad Sci USA 89:8671–8675.

Kelley M, Xu X-M, Wagner MA, Warchol ME, Corwin CT (1993) The developing organ of Corti contains retinoic acid and forms supernumery hair cells in response to exogenous retinoic acid in culture. Development 119:1041–1053.

Kermack KA (1963) The cranial structure of the triconodonts. Philos Trans R Soc Lond [B] 246:83–103.

Kermack KA, Mussett F (1983) The ear in mammal-like reptiles and early mammals. Acta Palaeontol Polonica 28:147–158.

Kermack K, Mussett AF, Rigney HW (1973) The lower jaw of *Morganucodon*. Zool J Linnean Soc 53:87–175.

Kermack KA, Mussett F, Rigney HW (1981) The skull of *Morganucodon*. Zool J Linnean Soc 71:1–158.

Ketten DR (2000) Cetacean ears. In: Au WL, Popper AN, Fay RR et al. (eds) Hearing by Whales and Dolphins. Springer Handbook of Auditory Research, vol 12. New York: Springer-Verlag, pp. 43–109.

Kielan-Jaworowska Z, Qi T (1990) Fossorial adaptations of a taeniolabidoid multituberculate mammal from the Eocene of China. Vertebrata PalAsiatica 28:81–94.

Kielan-Jaworowska Z, Presley R, Poplin C (1986) The cranial vascular system in taeniolabidoid multituberculate mammals. Philos Trans R Soc Lond [B] 313: 525–602.

Kielan-Jaworowska Z, Cifelli RL, Luo Z-X (1998) Alleged cretaceous placentals from down under. Lethaia 31:267–268.

Konishi M (1973) How the owl tracks its prey. Am Sci 61:414–424.

Kössl M (1997). Sound emission from cochlear filters and foveae—Does the auditory sense organ make sense? Naturwissenschaften 84:9–16.

Kössl M, Vater M (1995) Cochlear structure and function in bats. In: Popper AN, Fay RR (eds) Hearing by Bats. Springer Handbook of Auditory Research, vol 5. New York: Springer-Verlag, pp. 191–235.

Krubitzer L (1998) What can monotremes tell us about brain evolution? Philos Trans R Soc Lond [B] 353:1127–1146.

Kühne QG (1973) The systematic position of monotremes reconsidered. Z Morpho Tiere 75:59–64.

Ladhams A, Pickles JO (1996) Morphology of the monotreme organ of Corti and macula lagena. J Comp Neurol 366:335–347.

Lewis ER, Leverenz EL, Bialek WS (1985) The Vertebrate Inner Ear. Boca Raton, FL: CRC Press.

Lillegraven JA, Hahn G (1993) Evolutionary analysis of the middle and inner ear of late Jurassic multituberculates. J Mammal Evol 1:47–74.

Lim DJ (1986) Functional structure of the organ of Corti. A review. Hear Res 22:117–146.

Löwenheim H, Furness DN, Kil J, Zinn C, Gültig K, Fero ML, Frost D, Gummer AW, Roberts JM, Rubel EW, Hackney CM, Zenner HP (1999) Gen disruption of p27^{Kip1} allows cell proliferation in the postnatal and adult organ of Corti. Proc Natl Acad Sci USA 96:4084–4088.

Luo Z (1989) Structure of the petrosal of Multituberculata (Mammalia) and the morphology of the molars of early arctocyonids. PhD thesis, University of California at Berkeley.

Luo Z-X (2001) Inner ear and its bony housing in tritylodonts and implications for evolution of mammalian ear. Bull Museum Comparative Zool 155:621–637.

Luo Z, Ketten DR (1991) CT scanning and computerized reconstructions of the inner ear of multituberculate mammals. J Vertebrate Paleontol 11:220–228.

Luo Z-X, Crompton AW, Sun A-l (2001) A new mammaliaform from early Jurassic and evolution of mammalian characteristics. Science 292:1535–1540.

MacIntyre GT (1972) The trisulcate petrosal pattern of mammals. In: Dobzhansky T, Hecht MK, Steere WC (eds) Evolutionary Biology, vol 6. New York: Appleton-Century-Crofts, pp. 275–302.

Manley GA (1971) Some aspects of the evolution of hearing in vertebrates. Nature 230: 506–509.

Manley GA (1972) Frequency response of the middle ear of geckos. J Comp Physiol [A] 81:251–258.

Manley GA (1973) A review of some current concepts of the functional evolution of the ear in terrestrial vertebrates. Evolution 26:608–621.

Manley GA (1990) Peripheral Hearing Mechanisms in Reptiles and Birds. New York: Springer-Verlag.

Manley GA (2000) Cochlear mechanisms from a phylogenetic viewpoint. Proc Natl Acad Sci USA 97:11736–11743.

Manley GA, Köppl C (1998) Phylogenetic development of the cochlea and its innervation. Curr Opinion Neurobiol 8:468–474.
Manley GA, Kirk DL, Köppl C, Yates GK (2001) In vivo evidence for a cochlear amplifier in the hair-cell bundle of lizards. Proc Natl Acad Sci USA 98:2826–2831.
Masterton RB, Heffner H, Ravizza R (1969) The evolution of human hearing. J Acoust Soc Am 45:966–985.
McKenna MC, Bell SK (1997) Classification of Mammals Above the Species Level. New York: Columbia University Press.
Meng J, Fox RC (1993) Inner ear structures from late Cretaceous mammals and their systematic and functional implications. J Vertebrate Paleontol 13 (Suppl 3):50A.
Meng J, Fox RC (1995a) Therian petrosals from the Oldman and Milk River formations (Late Cretaceous), Alberta, Canada. J Vertebrate Paleontol 15:122–130.
Meng J, Fox RC (1995b) Osseous inner ear structures and hearing in early marsupials and placentals. Zool J Linn Soc 115:47–71.
Meng J, Wyss AR (1995) Monotreme affinities and low-frequency hearing suggested by multituberculate ear. Nature 377:141–144.
Meyer M (1907) An introduction to the mechanics of the inner ear. In: Brown WG (ed) Science Series, University Missouri Studies, vol 2. Columbia, MI: EW Stephens, pp. 1–140.
Miao D (1988) Skull morphology of *Lambdopsalis bulla* (Mammalia, Multituberculata) and its implications to mammalian evolution. Contributions to Geology, University of Wyoming, Special Paper 4:1–104.
Miao D, Lillegraven JA (1986) Discovery of three ear ossicles in a multituberculate mammal. National Geographic Res 2:500–507.
Mills DM, Shepherd RK (2001) Distortion product otoacoustic emission and auditory brainstem responses in the echidna (*Tachyglossus aculeatus*). J Assoc Res Otolaryngol 2:130–146.
Müller M, Wess FP, Bruns V (1993) Cochlear place-frequency map in the marsupial *Monodelphis domestica*. Hear Res 67:198–202.
Nadol JB (1988) Comparative anatomy of the cochlea and the auditory nerve in mammals. Hear Res 34:253–266.
Novacek MJ (1992) Mammalian phylogeny: shaking the tree. Nature 356:121–125.
Nowak RM (1995) Walker's Mammals of the world. Baltimore and London: Johns Hopkins University Press.
Olson EC (1959) The evolution of mammalian characters. Evolution 13:344–353.
Pettigrew JD, Manger PR, Fine SLB (1998) The sensory world of the platypus. Philos Trans R Soc Lond [B] 353:1199–1210.
Pickles JO (1992) Scanning electron microscopy of the echidna: morphology of a primitive mammalian cochlea. In: Cazals Y, Demany L, Horner K (eds) Auditory Physiology and Perception. Oxford: Pergamon, pp. 101–107.
Pritchard U (1881) The cochlea of the *Ornithorhynchus platypus* compared with that of ordinary mammals and birds. Philos Trans R Soc Lond 172:267–282.
Proske U, Gregory JE, Iggo A (1998) Sensory receptors in monotremes. Philos Trans R Soc Lond [B] 353:1187–1198.
Pujol R, Lenoir M, Ladrech S, Tribillac F, Rebillard G (1992) Correlation between the length of outer hair cells and the frequency coding of the cochlea. In: Cazals Y, Demany L, Horner K (eds) Auditory Physiology and Perception, Advances in Biosciences. New York: Pergamon Press, pp. 45–52.

Ravizza RJ, Heffner HE, Masterton B (1969) Hearing in primitive mammals. I. Opossum (*Didelphis virginiana*). J Audiol Res 9:1–7.

Reimer K (1996) Ontogeny of hearing in the marsupial, *Monodelphis domestica*, as revealed by brainstem auditory evoked potentials. Hear Res 92:143–150.

Rich TH, Vickers-Rich P, Constantine A, Flannery TF, Kool L, van Klaveren N (1997) A tribosphenic mammal from the Mesozoic of Australia. Science 278:1431–1438.

Rich TH, Flannery TF, Trusler P, Kool AL, van Klaveren N, Vickers-Rich P (2001) Early Cretaceous mammals from Flat Rocks, Victoria, Australia. Records of Queen Victoria Museum 106:1–30.

Robles L, Ruggero MA (2001) Mechanics of the mammalian cochlea. Physiol Rev 81: 1306–1352.

Rosowski JJ (1992) Hearing in transitional mammals: predictions from the middle-ear anatomy and hearing capabilities of extant mammals. In: Webster DB, Fay RR, Popper AN (eds) The Evolutionary Biology of Hearing. New York: Springer-Verlag, 615–631.

Rosowski JJ, Graybeal A (1991) What did *Morganucodon* hear? Zool J Linn Soc 101: 131–168.

Rougier GW (1990) Primeras evidencias sobre la morfología del oído interno en un terio no tribosfénico. Resumenes VII Jornadas Argentinas de Paleontología de Vertebrados, Ameghiniana 26:249.

Rougier GW, Wible JR, Hopson JA (1992) Reconstruction of the cranial vessels in the Early Cretaceous mammal *Vincelestes neuquenianus*: implications for the evolution of the mammalian cranial vascular system. J Vertebrate Paleontol 12:188–216.

Rougier GW, Wible JR, Novacek MJ (1996) Middle-ear ossicles of the multituberculate *Kryptobaatar* from the Mongolian late Cretaceous: implications for mammaliamorph relationships and the evolution of the auditory apparatus. American Museum Novitates 3187:1–43.

Rougier GW, Wible JR, Novacek MJ (1998) Implications of *Deltatheridium* specimens for early marsupial history. Nature 396:459–463.

Rowe T (1988) Definition, diagnosis, and origin of Mammalia. J Vertebrate Paleontol 8:241–264.

Ruggero MA, Temchin AN (2002) The roles of external, middle, and inner ears in determining the bandwidth of hearing. Proc Natl Acad Sci USA 99:13206–13210.

Runhaar G, Schedler J, Manley GA (1991) The potassium concentration of cochlear fluids of the embryonic and posthatching chick. Hear Res 56:227–238.

Saunders JC, Duncan RK, Doan DE, Werner YL (2000) The middle ear of reptiles and birds. In: Dooling RJ, Fay RR, Popper AN (eds) Comparative Hearing: Birds and reptiles. Springer Handbook of Auditory Research, vol 13. New York: Springer-Verlag, pp. 13–69.

Schmidt RS, Fernandez C (1962) Labyrinthine DC potentials in representative vertebrates. J Cell Comp Physiol 59:311–322.

Schmidt S, Türke B, Vogler B (1984) Behavioural audiogram from the bat, *Megaderma lyra* (Geoffroy, 1810; Microchiroptera). Myotis 22:62–69.

Simmons NB (1993) Phylogeny of Multituberculata. In: Szalay FS, Novacek MJ, McKenna MC (eds) Mammal Phylogeny, vol 1: Mesozoic Differentiation, Multituberculates, Monotremes, Early Therians, and Marsupials. New York: Springer-Verlag, pp. 146–164.

Simpson GG (1937) Skull structure of the Multituberculata. Bull American Museum of Natural History 73:727–763.

Simpson GG (1938) Osteography of the ear region in monotremes. American Museum Novitates 978:1–15.
Simpson GG (1945) The principles of classification and a classification of mammals. Bull American Museum of Natural History 85:1–350.
Simpson GG (1960) Diagnosis of the classes Reptilia and Mammalia. Evolution 14:388–391.
Slepecky NB (1996) Structure of the mammalian cochlea. In: Dallos P, Popper AN, Fay RR (eds) The Cochlea. Springer Handbook of Auditory Research, vol 8. New York: Springer-Verlag, pp. 44–130.
Sloan RE (1979) Multituberculata. In: Fairbridge RW, Jablonski D (eds) The Encyclopedia of Paleontology. Stroudsberg: Dowden, Hutchinson & Ross, pp. 492–498.
Smith CA, Takasaka T (1971) Auditory receptor organs of reptiles, birds and mammals. Contrib Sens Physiol 5:129–178.
Vater M, Lenoir M (1992) Ultrastructure of the horseshoe bat's organ of Corti. I. Scanning electron microscopy. J Comp Neurol 318:367–379.
Vater M, Siefer W (1995) The cochlea of *Tadarida brasiliensis*: specialized functional organization in a generalized bat. Hear Res 91:178–195.
Wang Y, Hu Y, Meng J, Li C (2001) An ossified Meckel's cartilage in two Cretaceous mammals and origin of the mammalian middle ear. Science 296:357–361.
West CD (1985) The relationships of the spiral turns of the cochlea and the length of the basilar membrane to the range of audible frequencies in ground dwelling mammals. J Acoust Soc Am 77:1091–1101.
Wever EG (1974) The evolution of vertebrate hearing. In: Keidel WD, Neff WD (eds) Handbook of Sensory Physiology, vol V-1: Auditory System, Anatomy, Physiology (Ear). New York: Springer-Verlag, pp. 423–454.
Wible JR (1990) Petrosals of late Cretaceous marsupials from North America, and a cladistic analysis of the petrosal in therian mammals. J Vertebrate Paleontol 10:183–205.
Wible JR, Hopson JA (1993) Basicranial evidence for early mammal phylogeny. In: Szalay FS, Novacek MJ, McKenna MC (eds) Mammal Phylogeny—Mesozoic Differentiation, Multituberculates, Monotremes, Early Therians, and Marsupials. New York: Springer-Verlag, pp. 45–62.
Wible JR, Rougier GW, Novacek MJ, McKenna MC, Dashzeveg D (1995) A mammalian petrosal from the early Cretaceous of Mongolia: implications for the evolution of the ear region and mammaliamorph interrelationships. American Museum Novitates 3149:1–19.
Zeller U (1989) Die Entwicklung und Morphologie des Schädels von *Ornithorhynchus anatinus* (Mammalia: Prototheria: Monotremata). Abhandlungen der Senckenbergischen Naturforschenden Gesellschaft 545:1–188.
Zeller U (1993) Ontogenetic evidence for cranial homologies in monotremes and therians, with special reference to *Ornithorhynchus*. In: Szalay FS, Novacek MJ, McKenna MC (eds) Mammal Phylogeny, vol 1: Mesozoic Differentiation, Multituberculates, Monotremes, Early Therians, and Marsupials. New York: Springer-Verlag, pp. 95–107.
Zheng J, Shen W, He DZ, Long KB, Madison LD, Dallos P (2000a) Prestin is the motor molecule of cochlear outer hair cells. Nature 405:149–155.
Zheng JL, Shou J, Guillemot F, Kageyama R, Gao W-Q (2000b) Hes1 is a negative regulator of inner ear hair cell differentiation. Development 127:4551–4560.
Zheng J, Madison LD, Oliver D, Fakler B, Dallos P (2002) Prestin, the motor protein of outer hair cells. Audiol Neuro-Otol 7:9–12.

10
The Evolution of Central Pathways and Their Neural Processing Patterns

BENEDIKT GROTHE, CATHERINE E. CARR, JOHN H. CASSEDAY,
BERND FRITZSCH, AND CHRISTINE KÖPPL

1. Introduction

A comprehensive and conclusive description of the evolution of the central auditory system in vertebrates is a difficult, if not impossible, task. We simply lack important basic information. For instance, we do not know how and what the common ancestors of all the terrestrial vertebrates could hear (certainly not airborne sound, because they had no tympanic middle ear) and how they might have processed basic sounds (such as substrate vibrations). However, a comparative approach allows us to define some principles of auditory processing that we find in all hearing vertebrates and a basic outline of its neural substrate. There is a striking similarity among all vertebrates concerning the principal design of the central auditory system. It seems to result from the fact that all vertebrate central auditory systems are based on similar basic neural building blocks that work with similar underlying principles. These building blocks were then shaped by evolutionary constraints that were similar for all hearing vertebrates, simply because the acoustic cues that can be used for sound recognition or sound localizations are limited. However, an important issue in this chapter is the increasing evidence that the elaborated central auditory systems in the different clades of recent vertebrates are to a large extent a result of parallel, independent evolution.

This chapter reviews the current knowledge about the origin of the vertebrate central auditory pathway, which is mainly based on the analysis of its ontogeny, and compares the basic stages of the ascending system in different clades of vertebrates to exemplify the independent evolution of its substructures based on specific evolutionary constraints, focusing on a few specific examples.

2. The Evolutionary Origin of the Central Auditory System

In 1848 Owen coined the word *homology* to characterize identical organs, no matter what variation in form and function they may have in different species. Owen (1848) wanted to distinguish those organs from organs that look alike because functional requirements have shaped them that way (analogy). The middle ear featured extensively in the early specification of the use of the term *homology* when Reichert (1837) showed that indeed two of the middle-ear ossicles were derived from jaw bones. In recent years, the definition of homology has shifted from classical morphology to a more evolutionary definition that relies heavily on transgenerational information flow, that is, the genetic material that governs development of a given organ. Classic examples are the roles played by the gene *Pax6* in eye formation across phyla and the gene *distalless* in limb development of insects and mammals (Pichaud and Desplan 2002). These examples show that the function of crucial transcription factors can be conserved across phyla. Such transcription factors not only are homologous based on their sequence similarity, but also guide homologous developmental processes in otherwise fairly dissimilar organs. If we accept such homology based on genetic identity of developmental modules, we will have to appreciate that apparently dissimilar organs within and across phyla might be derived from common primordia and thus can be homologous. Of course, we must keep in mind that much like in the case of functional similarity (analogy or homoplasy) such developmental modules can be used in different organs and do not necessarily prove homology of entire organs but only of genes and their developmental modules. Other genes, which likely are not homologous across phyla, are needed in addition, to form unique, phyla-specific organs. For example, the fibroblast growth factor FGF10 is essential for the formation of limbs and lungs and is involved in the formation of the semicircular canals of the ear (Pauley et al. 2003). Moreover, together with a second factor, FGF3, FGF10 is needed to generate an ear placode and transform it into an otocyst in mammals (Wright and Mansour 2003).

In the central nervous system the distinction between *homology* and *analogy* is not yet established with the same precision as in peripheral organs, and it is bound to show numerous analogies necessitated by the functional constraints imposed by the auditory signal and its processing. To illustrate that point, we will briefly outline what genetic analysis has revealed in terms of previously unlikely homologies in the ear and the efferent fibers to the ear that were not predicted by morphological and functional analysis.

The last few years have revealed a cascade of gene expression that specifies cell lineage, induces and governs morphogenesis, and regulates cell proliferation and the survival of sensory neurons (Brigande et al. 2000; Farinas et al. 2001; Fritzsch and Beisel 2001; Fekete and Wu 2002). Those molecular data have shown, much like the molecular evolution of eyes outlined above, a deep homology of sensory cells across phyla (Fritzsch and Beisel 2003). Beyond the limited rescue of eye formation in vertebrate with the fly *Pax6* homolog, the fly atonal gene can also substitute for *Math1* and can induce the formation of hair cells (Wang et al. 2002). In addition to these cell specific features, evolution of molecular networks apparently generates maps of transcription factors that specify areas irrespective of their morphological details. For example, the interaction of the transcription factor orthodenticel (otx) and the gastrulation homeobox gene (*gbx*) specifies a specific area apparently in all deuterostome animals. However, only chordates have developed a complex midbrain/hindbrain boundary that apparently uses this conserved transcription factor expression domain common to all triploblastic metazoans to specify a rather different morphology (Lowe et al. 2003). Most interestingly, otx expression, together with other genes, apparently specifies the first gill slit, thus indicating that the area that will eventually become the auditory canal is one of the oldest conserved parts of the deuterostome body. In contrast to this information on brain evolution, the map of transcription factors that specifies developmental domains in the ear is rudimentary and therefore cannot be used as yet to establish identities of auditory receptors across taxa (Fekete and Wu 2002.) However, this example highlights the importance of topological information for the establishment of conserved developmental domains across phyla, no matter what their actual morphology will be like.

While the mechanosensory transduction system and its development appear to have evolved early and are conserved across phyla, the vertebrate ear has evolved a rather unique structure around this conserved neurosensory element. This apparently was achieved by importing existing genes from other developmental modules of the body into the ear to govern formation of semicircular canals, formation of the auditory and vestibular endorgans, and formation of sensory neurons that connect the hair cells to the brainstem (Cantos et al. 2000; Fritzsch and Beisel 2001). Based on our current insights, the vertebrate ear can be viewed as the product of continuous alteration of an existing program to govern sensory hair cell development by importing novel or existing genes to generate a novel context that alters function of those genes (Venter et al. 2001). Such a context dependent action has recently been reported for neuronal fate specifying genes (Cau et al. 2002) and likely will play an even greater role in the evolution of the brain.

The reasons behind the difficulties to establish or refute homologies of tetrapod auditory nuclei that will be apparent in this chapter can best be exemplified with the inner ear efferents, a unique fiber projection system that contacts inner ear hair cells in all vertebrates. Comparative analysis has shown a rich diversity in position of these cells (Roberts and Meredith 1992) allowing only one criterion to suggest homology across taxa, the common hair cell target (Fritzsch 1999b). Nevertheless, developmental data have shown that all these diversely positioned cells derive from a single primordium, which also gives rise to the facial branchial motoneurons (Karis et al. 2001). Developmental segregation of these two populations requires a number of genes that govern the distinct projection of the axon as well as migration differences in the two neuronal populations (Karis et al. 2001). This example suggests that adult cellular distribution is not a good criterion to exclude or establish homology in the central nervous system (CNS). It also shows that the target cannot be used to exclude homology as efferents to the ear, and facial branchial motoneurons clearly have distinct targets but share a common origin in rhombomere 4 of the hindbrain. In addition, efferents can reroute into the facial nerve if they cannot access the ear (Ma et al. 2000). Obviously, exclusion and/or establishment of homology in central auditory pathways requires molecular and developmental data well beyond what is currently known. Although this cell-specific evolution of gene expression domains is not yet understood for the development of the central auditory pathways, a brief outline of certain important genes can be given. We will also revisit ideas about the functional replacement of aquatic sensory systems (lateral line, electroreception) by the central auditory system.

2.1 The Evolutionary Relationship to Other Sensory Pathways

The ancestral vertebrate ear included both vestibular and auditory endorgans sensitive to particle motion (see Chapters 3 and 4). These projected to a ventral octaval column (McCormick 1999). This system was transformed by the evolution of sound pressure–sensitive receivers, and by the evolution of a dorsal acoustic octaval column and central auditory targets. The task of relating the evolution of sound pressure receivers to the evolution of the central auditory system requires resolving the evolution of three interlocked developmental steps:

1. Generate a redundant endorgan that is not required for vestibular responses and therefore can be modified to extract sound pressure from the various stimuli impinging on the ear.
2. Generate sensory neurons that interconnect this new endorgan to an area of the brainstem dedicated to sound pressure reception.
3. Evolve auditory nuclei de novo, or from some undifferentiated precursors, or respecify existing octaval targets throughout the CNS to form targets specifically dedicated to process sound pressure signals.

It comes as no surprise that ideas of transformational identity, similar to the evolutionary transformation of the hyomandibular bone into the stapes of the terrestrial ear, featured in the initial quest to resolve this issue. Larsell (1967) and others proposed that auditory nuclei derive from mechanosensory lateral line areas, a suggestion rooted in the limited resolution of early neuroanatomical tracing studies. It was further suggested that auditory input from the ear replaced, evolutionarily as well as during metamorphosis of frogs, the disappearing lateral line input. Although this idea is interesting, the main problem is that not all frogs lose the lateral line system during metamorphosis, and even those that retain the lateral line system have auditory nuclei (Fritzsch 1988a). Specialized auditory nuclei are found even in some bony fish, which fully retain the mechanosensory lateral line system (McCormick 1999). In addition, there is a direct spinal output in the lateral line nuclei but not the auditory nuclei (Coombs et al. 1989). Thus, neither the apparent coincidence of loss of one and appearance of another nucleus nor the detailed connections support this idea of transformational identity. The only factors supporting it are more generalized connections to the midbrain.

More recent data suggest yet another transformation, related to the discovery that an electroreceptive sense is primitive for all jawed vertebrates (Bullock and Heiligenberg 1986). Loss of this primitive sense and the formation of specific auditory nuclei correlate among amphibians and amniotes (Fritzsch 1990a; McCormick 1999). Moreover, among animals of a given class of vertebrates, such as amphibians, some have electroreception associated with the specialized CNS nuclei but no specialized auditory nuclei (salamanders, caecilians). Others have lost the sense of electroreception but have gained central auditory nuclei with or without loosing the mechanosensory lateral line (Fritzsch 1990a). In contrast to the mechanosensory lateral line nuclei, electroreceptive and auditory nuclei show topological similarity in their rostrocaudal extent (from trigeminal to glossopharyngeal roots or from rhombomere 2 to 6), whereas lateral line and vestibular nuclei extend from rhombomere 1 to rhombomere 7/8. Thus, at least a spatial correlation exists between loss of the ancient sense of passive electroreception and gain of auditory nuclei in the brainstem with the overall topology of the tetrapod auditory nuclei. However, connectional details show that the auditory system has brainstem nuclei (superior olive, lemniscal nuclei) that are not recognized in the electroreceptive or mechanosensory lateral line system (McCormick 1999). More complicated variations of this theme, therefore, have to consider different areas of the brain independently. Thus, midbrain or forebrain areas do not necessarily follow brainstem evolution. It needs to be pointed out here that the novel electroreceptive sense of teleost fish seems to have evolved anew in a pattern grossly similar but not identical to the ancestral passive electroreceptive sense after this sense was apparently lost in ancestral teleost fish (Bullock and Heiligenberg 1986). If the auditory nuclei of tetrapods represent a comparable transformation of a dormant developmental program, refuting such scenarios on topological and connectional grounds will be almost impossible. Other information needs to be considered to establish what the likely evolutionary history of the auditory pathway might be.

Such replacement suggestions assume at their core an independent loss of peripheral sensory organs and their afferents, as well as the central nuclei. Such loss would denervate those central nuclei, which would then either disappear or receive a different input. Such functional reconnections have been experimentally shown for forebrain connections (von Melchner et al. 2000) and certainly are a possibility worth exploring. However, attempts to experimentally reorganize the lateral line or auditory input by transplanting the ear or extirpating the ear in frogs showed little alteration in the remaining projections (Fritzsch 1990b; Fritzsch 1999a). It also needs to be stressed that this entire idea is not yet tested on the molecular level: no specific lateral line electroreceptive or auditory markers are known that highlight all sensory nuclei of each of these systems across taxa independently. More recent molecular data, which suggest that functional systems may be developmentally connected by shared activation of transcription factors (Bermingham et al. 2001; Qian et al. 2001), suggest that such molecules may exist. Moreover, auditory nuclei appear to express unique markers in some songbirds (Akutagawa and Konishi 2001).

It is therefore possible that identical transcription factors govern development of both the peripheral receptors (i.e., sensory neurons, sensory cells) and central nuclei (i.e., lateral line, electroreceptive, and auditory nuclei in the brainstem). If this can be demonstrated, it would appear more plausible that all parts of these systems are lost concomitantly during evolution, presumably owing to a single or few mutations. If proven, this would eliminate the core of the replacement hypothesis and both the auditory periphery and the functionally related auditory nuclei would then need to be considered as evolutionary novelties that nevertheless might represent transformations of a general column of cells extending throughout the hindbrain and spinal cord (Maklad and Fritzsch 2003).

In the following discussion we treat the formation of pressure-sensitive organs in the auditory periphery and central nuclei as an evolutionary novelty. Logic requires that formation of specific endorgans predates that of specific auditory nuclei (one needs to extract auditory information before it can be processed), that is, the evolution of the auditory system needs to be resolved in an ear-to-brain progression (McCormick 1999). We will briefly outline the evolution of the inner ear first (for details see Ladich and Popper, Chapter 4, and Clack and Allin, Chapter 5), followed by the limited insights into the formation of endorgan specific connections and the development of auditory nuclei.

2.2 Evolution of a Sound Pressure–Sensitive Ear

Specialized auditory sound pressure receptors of the ear are uniformly characterized by otoconia-free tectorial membranes and do not respond to gravistatic or angular acceleration stimuli (Fritzsch 1999a). A second, unifying feature of sound pressure receivers is the association of the sensory epithelium with a periotic system that allows sound pressure to induce fluid motion only in the vicinity of this sound pressure–receiving organ (de Burlet 1934). In contrast to the multitude of apparently parallel transformations of gravistatic receptors into sound pressure–receiving auditory endorgans among teleost fish (see Coffin et al., Chapter 3, and Ladich and Popper, Chapter 4) and the unique amphibian papilla of amphibians (see Smotherman and Narins, Chapter 6), there is apparent uniformity in topology and innervation of the main sound pressure–receiving endorgan of tetrapods, the basilar papilla (Fritzsch 1999a). Development in amphibians and mammals suggest that saccule and cochlea become progressively segregated from a common precursor (Fritzsch et al. 2002), suggesting that the basilar papilla evolved only once, among the aquatic ancestors of tetrapods.

Irrespective of their peripheral distribution or their embryonic origin, all sound pressure receivers appear to project to discrete nuclei of the brainstem that receive only limited input from other inner-ear endorgans (McCormick 1999). This requires that sound pressure receiver afferents carry specific markers that allow them to project differently from the nearby vestibular afferents. In mammals, such a differential projection develops prior to the formation of sensory hair cells (Fritzsch 2003) and thus cannot be mediated by hair cell activity. Indeed, topologically correct projections of the auditory system, including the tonotopic organization of afferents, develops before birth and even in the absence of hair cells (Leake et al. 2002; Xiang et al. 2003). Most recent data have identified a gene that is uniquely expressed in cochlear sensory neurons and is, in other systems, involved in pathway selection (Karis et al. 2001). If this gene, *GATA3*, or others can be demonstrated in the auditory sensory neurons of other vertebrates, but is absent in electroreceptive and lateral line afferents as in vestibular afferents, it could provide a unique molecular identifier of auditory sensory neurons. In addition, although hardly anything is known about the molecular guidance of auditory neuron projection to auditory nuclei (Rubel and Fritzsch 2002), the apparent involvement of *GATA3* in pathfinding (Karis et al. 2001) suggests that *GATA3* may mediate selective projections of auditory sensory neurons to auditory nuclei.

This brings us back to the question of the origin of auditory nuclei in the brainstem. Beyond the temporal coincidences in the appearance and loss of various sensory systems, recent experimental work suggests that specific genes, such as basic helix-loop-helix (bHLH) genes and homeotic selector genes, may be uniquely associated with the formation of specific components of the auditory nuclei. For example, like the cerebellar granule cells, the granule cells of the dorsal cochlear nuclei of mice require the bHLH gene *Math1* for their development (Bermingham et al. 2001). This similarity in the regulating network connects evolution of those cells to the evolution of other brainstem novelties such as the inferior olive and the pontine nuclei, for which similar regulation mechanisms are realized. Descriptive analyses of gene expression domains suggest that the vestibular and auditory nuclei may derive from a neurogenin 1 expressing domain, among which is the gene that also regulates sensory neuron formation in the ear (Ma et al. 2000; Gowan et al. 2001; Maklad and Fritzsch 2003). However, experimental verification of this topologically compelling idea needs to be conducted in neurogenin 1, neurogenin 2, and *Math1*-null mutant mice. In summary, although recent data have rendered earlier ideas about the transformation of lateral line nuclei into auditory nuclei doubtful, there is not yet enough molecular developmental data to demonstrate that indeed those nuclei derive from different embryological sources. One possibility is that the entire anlage of lateral line and electroreceptive nuclei has been transformed by the expression of another set of genes into auditory nuclei in amniotic vertebrates. Such ideas would also be the likely explanation for the formation of novel endorgans in the auditory periphery, where existing octaval developmental programs may have been modified by implementing co-opted or novel genes (Fritzsch and Beisel 2003). Moreover, such ideas are analogous to the emerging overall concept of a stepwise transformation of the ear by sequential implementation of novel genes. Such a process is apparent in the formation of the novel angular receptor of jawed vertebrates, the horizontal canal (Cantos et al. 2000; Fritzsch et al. 2001). Understanding how novel auditory nuclei arise in a network of highly conserved transcription factors that predates evolution of the vertebrate brain (Lowe et al. 2003) and that provides a blueprint for topographically specific neuronal development is the most challenging and yet least understood aspect of brain evolution (Fritzsch 2002). The example provided with the efferent system stresses the need for more detailed molecular and developmental analysis of more tetrapods to untangle the evolutionary history of auditory nuclei.

3. The Basic Design of the Vertebrate Central Auditory Pathway

The overall design of the central auditory system exhibits strong similarities among all vertebrates that possess functional ears, able to extract sound pressure. This is often taken as evidence for common ancestry of all auditory systems in early fish. However, there is no proof for such an assumption, but it cannot be excluded that the evolution of the central auditory system simply followed the overall organizational patterns of the octavo-lateral system. Therefore, we will only outline the basic design of the central auditory system without strong assumptions about homology or nonhomology of the principal structures.

The auditory nerve is part of the octavolateral system, sharing hair cells as receptors. It is important to note that the lateral line nerves as they exist in modern extant fish (the only cranial nerves not numbered in the classical nomenclature) are ontogenetically and phylogenetically distinct nerves, and are not associated with the eighth nerve (Northcutt 1990). The latter, therefore, is sometimes referred to as "octaval nerve" (Liem et al. 2001) and transmits information from all parts of the inner ear, whether related to the sense of balance or of hearing. The octaval nerve fibers specialized into auditory nerve fibers in direct association with the evolution of the epithelia that perceive acoustic information. Therefore, we have no proper basis for speculating on the homology of auditory nerve fibers across vertebrates. In all vertebrates to which we can attribute an auditory system with some certainty, auditory nerve fibers enter the hindbrain and synaptically contact first-order auditory neurons in the dorsolateral regions of the medulla (the octaval column), but in some groups they also contact neurons in the reticular formation, including the Mauthner cells in fish. Additionally, some auditory fibers at least in some teleost fish seem to enter the cerebellum (review: McCormick 1999). Hence, in modern fish there is a rather direct connection of auditory nerve input to motor control and fast motor actions such as escape behavior. In tetrapods such connections seem to be missing. Whether they were lost during evolution or whether they never existed in ancient fish clades remains open.

The auditory nerve connects to the central auditory system, which is characterized by several synaptic stages (Fig. 10.1; Table 10.1). This ascending system can be divided into different ascending streams or functionally segregated pathways. For instance, we can divide the lower auditory system into monaural and binaural pathways that share similar synaptic levels, but incorporate different subgroups of neurons at these levels. The structural and functional segregation of these pathways already starts at the level of the auditory nerve; after entering

The vertebrate ascending auditory pathway

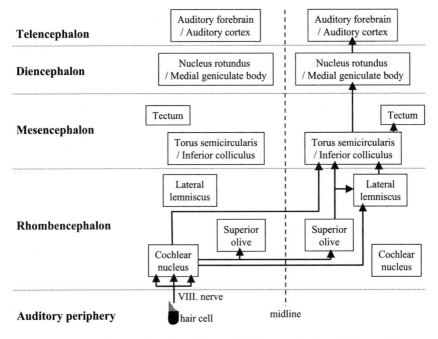

FIGURE 10.1. Simplified schematic of the principle organization of the vertebrate ascending auditory pathway. Each box represents an area that may contain multiple subareas.

TABLE 10.1. Auditory structures and their function.

Brain area	Auditory structures	Function in auditory processing
Telencephalon	6. Different areas	Auditory object formation
Diencephalon	5b. Auditory thalamus	Interface for midbrain-forebrain interactions
Mesencephalon	5a. Tectum	Multimodal integration for motor output
	4. TS/IC	Integration of inputs leads to complex representation of acoustic cues
Rhombencephalon	3. Lateral lemniscus	Temporal analysis of acoustic cues
	2. Superior olive	Temporal and spatial analysis of acoustic cues
	1. Cochlear nucleus	Segregation into different auditory pathways; temporal and spectral analysis

TS, torus semicircularis; IC, inferior colliculus.

the hindbrain it bifurcates and contacts different subdivisions of the cochlear nucleus (Ryugo and Parks 2003). The latter give rise to different connections within the lower auditory pathways that by and large includes groups of neurons in the superior olive and the lateral lemniscus before these pathways reunite at the level of the auditory midbrain. Despite this basic bauplan, the auditory brainstem shows substantial anatomical diversity between the major vertebrate taxa (reviews: Cant 1992; McCormick 1999; Carr and Code 2000). For instance, the cochlear nuclei, the superior olive, and the lateral lemniscus show considerable variations in the number of subnuclei and neuron types, which are interconnected forming a maze of circuits and which are specialized for the processing of different aspects of auditory information. Currently, there is hardly a single auditory nucleus below the midbrain (including the cochlear nucleus) that is agreed to be homologous across all vertebrates. For some nuclei that perform similar functions, a good argument can be made that they are definitely not homologous. We will use some well-studied examples of auditory brainstem processing to illustrate this in more detail below (see sections 4.2 and 4.3). As we will also discuss in more detail below (see section 6), the organization of the vertebrate auditory system above the level of the midbrain is difficult to compare, which is partly due to the very different routes taken by the overall organization of the forebrain in the different clades.

The basic synaptic levels are the following:

1. The auditory nerve.
2. The first, obligatory synaptic level, which is in the rhombencephalon, where the auditory nerve bifurcates and contacts different groups of neurons in the dorsolateral region, the so-called octaval column [in terrestrial vertebrates commonly referred to as the cochlear nucleus (CN)]. Here, all ascending pathways have their origin.
3. The second rhombencephalic structures, such as the nucleus laminaris (NL) in archosaurs (crocodiles and birds) and the superior olive in all tetrapods, which are not an obligatory station and are bypassed by many ascending CN outputs. The superior olive can be a single nucleus (superior olivary nucleus, SON) or a complex structure composed of many nuclei (superior olivary complex, SOC). In many vertebrates these structures are the first stage of binaural interactions.
4. The third and most rostral rhombencephalic auditory structure, the lateral lemniscus (LL). Like the SOC it is not an obligatory station and is bypassed by many ascending fibers from CN and SON/SOC.
5. The auditory midbrain (torus semicircularis or inferior colliculus), which is, by and large, an obligatory synaptic station for all ascending pathways. It is characterized by a high degree of convergence and by rather complex response properties compared to those found at lower auditory stations.
6. The optic tectum or superior colliculus, where one stream of information from the auditory midbrain converges with inputs from other sensory systems, resulting in the control of orienting behavior such as eye or pinna movements. Its structural organization is dominated by the visual system.

7. The auditory thalamus (medial geniculate), which is an obligatory synaptic station for the pathway from the auditory midbrain to the high auditory stations in the telencephalon.
8. The highest more-or-less purely auditory centers in the telencephalon. Because of the high diversity of the telencephalon and its principal organization (see below), these centers are the most diverse in structure and function within the ascending auditory system.

In addition to the ascending pathways, the central auditory system is characterized by multiple descending projections which have been highly integrated into the ascending auditory system and interact with it at almost all levels.

4. The Auditory Brainstem—Specific Adaptations

This section describes the evolution of the different stages of auditory processing and relates its structural and functional characteristics to evolutionary adaptations. This section also addresses the concept of parallel evolution due to similar evolutionary constraints, exemplifying the relationship of structure and function and discussing a few specific examples.

4.1 The Cochlear Nucleus

The termination sites of auditory afferents coming from the inner ear define the first-order nucleus of the brainstem. The term *cochlear nucleus* is used here. However, it should be emphasized that homologies are far from certain and the question of the origin(s) of an exclusively auditory cochlear nucleus in vertebrate evolution is currently unanswered (Fritzsch 1988b; McCormick 1999). Four basic patterns of cochlear-nucleus organization can be defined that correlate with different configurations of the auditory periphery. We first summarize the characteristics of each basic pattern to emphasize the different morphological substrates. We then highlight the functional convergence between them.

There are four basic patterns of cochlear-nucleus organization among vertebrates:

1. Ray-finned and cartilaginous fish do hear, but different inner-ear epithelia serve as sound receptors in different groups and these could in principle also convey vestibular information (see Ladich and Popper, Chapter 4; Popper and Fay 1999). Accordingly, it is currently unclear whether there consistently is a cochlear nucleus, that is, a first-order nucleus with exclusively auditory input. At least two of the four first-order octaval nuclei typically seen are involved in auditory processing; the descending nucleus and the anterior nucleus. Auditory processing clusters in the dorsal regions of both nuclei (McCormick 1999). There is no evidence for functional specialization within these regions (Edds-Walton 1998; Edds-Walton et al. 1999) or for differential projection patterns (McCormick 1999) except for the catfish (Bleckmann et

al. 1991), but future studies may well reveal such specializations in other species as well.

2. Amphibians show great variety in the degree of specialization of the auditory periphery (see Smotherman and Narins, Chapter 6). Substrate-borne vibrations are an important part of the sensory input for all amphibians, and these are processed by otolithic organs, the amphibian papilla, and (if present) the basilar papilla. In addition, most anurans (frogs and toads) possess a tympanic middle ear sensitive to airborne sound that is processed by the amphibian and basilar papillae (see Smotherman and Narins, Chapter 6; Lewis and Narins 1999). Correlated with the perception of airborne sound, a cochlear nucleus, also called the dorsolateral nucleus (DLN), appears in anurans only. Its origins and possible homology to the amniote cochlear nuclei are still unclear (Fritzsch 1988b; McCormick 1999). The DLN receives tonotopic projections from the amphibian and basilar papillae, as well as inputs from the lagenar macula and possibly the saccular macula (review: McCormick 1999). Interestingly, it also receives extensive commissural projections from the contralateral DLN, indicating significant binaural interaction at this early stage (review: Wilczynski 1988). At least six morphological cell types were defined, some of which are clustered within subregions of the DLN (Feng and Lin 1996). Physiologically, a variety of temporal discharge patterns is found that does not exist in the primary afferents (review: Feng and Schellart 1999). Importantly, both morphological and physiological cell types bear many similarities to well-known cell types in the mammalian and avian cochlear nucleus. It is currently not known whether morphological and physiological types in the DLN correlate and whether they may differ in their projection patterns.

3. Most sauropsids (turtles, lizards, crocodiles, and birds) perceive airborne sound via a tympanic middle ear and the basilar papilla, which is unanimously considered homologous to the mammalian organ of Corti (see Manley, Chapter 7, and Gleich et al., Chapter 8). They possess a cochlear nucleus that receives tonotopic, primary input exclusively from the basilar papilla and consistently shows two main subdivisions, nucleus magnocellularis (NM) and nucleus angularis (NA) (review: Carr and Code 2000). Interestingly, there is a lot of confusion and disagreement regarding the distinction between the NM and the second-order nucleus laminaris (NL) in turtles and lizards (Glatt 1975a,b; Miller 1975; Barbas-Henry and Lohman 1988). Some authors even reported primary projections from the auditory nerve to the NL (DeFina and Webster 1974; Barbas-Henry and Lohman 1988; Soares 2002). As first suggested by Glatt (1975b), this may indicate that the NL gradually evolved from the NM (see section 4.3.4, below).

The NM is the much better studied of the two cochlear-nucleus subdivisions. Comparative data suggest that this nucleus is primitively a low-frequency nucleus. It is already well developed in turtles (Glatt 1975b; Miller and Kasahara 1979), whose sensitive hearing range is limited to frequencies below about 1 kHz. In lizards, whose sensitive hearing range typically extends to 4 to 5 kHz (review: Köppl and Manley 1992), the NM receives its input almost exclusively from low-frequency (<1 kHz) auditory nerve fibers (Manley 1981; Szpir et al. 1990). Only in archosaurs (birds and Crocodilia) does the frequency range of the NM extend to higher frequencies of several kHz (review: Carr and Code 2000). Correlated with the frequency extension, well-known specializations for high-frequency temporal processing appear, that is, large end-bulb of Held primary synapses and large cell bodies with short, thick dendrites (review: Carr and Code 2000). The low-frequency region of the NM (below approx. 1 kHz) retains more primitive characteristics (Köppl 1994; Köppl and Carr 1997); end-bulb–like primary synapses are only rarely seen there, if at all (Szpir et al. 1990). NM neurons have a single projection target, the NL (reviews: in Carr and Code 2000; Carr and Soares 2002).

The NA is a very small nucleus in turtles and snakes (Glatt 1975b; Miller and Kasahara 1979; Miller 1980; Defina and Kennedy 1983). It appears to gain in size with an extension of the high-frequency hearing range in all other sauropsids (DeFina and Webster 1974; Glatt 1975b; Szpir et al. 1995; Carr and Code 2000). All authors agree that the NA contains a heterogeneous mix of morphological cell types. Detailed characterizations, however, are available only for birds, where four to five distinct types are seen, some of which resemble cell types in the mammalian cochlear nucleus (Soares and Carr 2001). Physiologically, five different types appear to be typical for birds (Köppl and Carr 2003). These show numerous specific similarities to well-known types in the mammalian cochlear nucleus. It remains to be investigated whether the different response types correlate with morphological types and whether they establish different processing streams in the auditory brainstem.

4. Mammals show the most complex type of cochlear nucleus. There are several excellent reviews (Cant 1992; Rhode and Greenberg 1992; Smith and Spirou 2002; Young and Davis 2002; Cant and Benson 2003; Ryugo and Parks 2003), on which the following summary is based. Three major divisions are generally recognized: the anteroventral (AVCN), posteroventral (PVCN), and dorsal (DCN) cochlear nucleus. They also represent the major output regions of the cochlear nucleus. In addition, several groups of small cell and granular cell regions are found in between the major divisions. These are believed to represent only interneurons. The auditory nerve fibers project into the different divisions of the CN complex via two principal axon collaterals. An ascending branch travels into the AVCN, and a descending branch runs into the PVCN and continues to the DCN. The tonotopic organization of the receptor surface is maintained in the auditory nerve input and is also conserved in all CN subdivisions, but subgroups of physiologically different auditory nerve fibers distribute their terminals differentially among the various types of CN neurons. AVCN, PVCN, and DCN each contain several different neuron types that also form the basis for the definition of further subdivisions.

A hallmark of the mammalian CN complex is a multitude of intrinsic connections and nonprimary inputs that appear to have no equivalent in nonmammals. Together with the differential input pattern from the auditory nerve, this indicates an unparalleled complexity of processing at this early level of the auditory pathway. This particularly holds for the DCN, with its cerebellar-like organization, where auditory and somatosensory inputs (associated with the pinna) are integrated. Also, the granular cell domains appear to contain a subsystem integrating type II primary afferent input (from the outer hair cells in the cochlea) and collateral input from the cochlear efferent system. Besides the multitude of interneurons within each CN complex, all major subdivisions are known to receive, in addition, reciprocal connections from the contralateral CN complex, but not necessarily from the same subdivision.

The increased complexity of the mammalian cochlear nucleus may be associated with the significant high-frequency extension of the mammalian hearing range (see Vater et al., Chapter 9). This enabled mammals to make increased use of level cues in sound localization, both interaural difference cues and, through the addition of pinnae, monaural cues of level differences across frequencies. Mobility of the pinnae, which is typical for many mammals, would have further raised the complexity of sound processing, necessitating the incorporation of information about pinna positions. Thus, the uniquely mammalian pinnae probably had a major influence on the evolution of cochlear nucleus processing.

4.2 Functional Convergence in the Cochlear Nucleus

One of the common principles of the vertebrate auditory system is that, in the CN, a number of distinctly different neuron types is created, which establish several separate auditory brainstem processing streams. Here we will argue that the patterns of cochlear nuclei in land vertebrates (CN patterns 2 to 4 in the above list) evolved independently, converging on basically equivalent solutions

FIGURE 10.2. Schematic representation of neuron types in the cochlear nucleus of frogs. Six morphological cell types have been described for the dorsolateral nucleus (DLN) of the frog *Rana pipiens* (Feng and Lin 1996; this material is used by permission of Wiley-Liss, Inc., a subsidiary of John Wiley & Sons, Inc.), drawings of five of those are shown in the first column. Hall and Feng (1990; with permission from The American Physiological Society) defined six physiological response types in the same species, based on the poststimulus time histogram (PSTH) at 10 dB above threshold. Examples of these are shown in the second column, scaled identically; note that the time axis covers 300 ms. In the third column, the PSTH for some of the same neurons are shown, for stimulation at 30 dB above threshold and covering only the initial 75 ms of the response. Note that often a regular, chopper-like response appears at these higher levels. It is unknown whether a correlation exists between morphology and physiology.

10. Evolution of the Central Auditory System   303

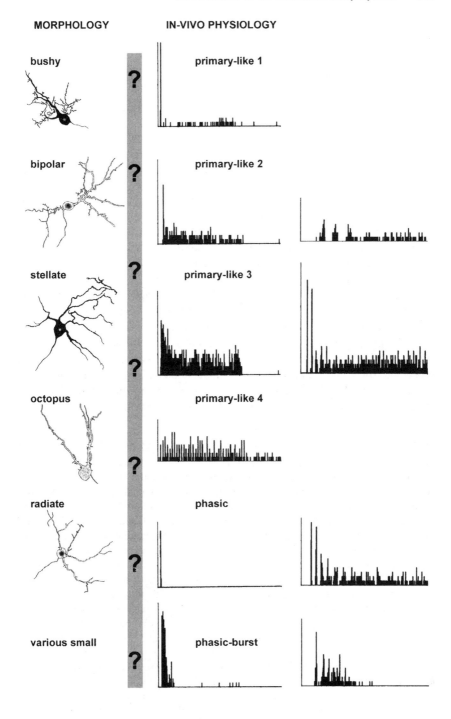

304   B. Grothe et al.

in response to the same selective pressures for the processing of airborne sound. Our conclusion of a parallel evolution of neuron types in the cochlear nuclei—in contrast to those neuron types being the plesiomorphic condition inherited from a common ancestor—is based on two arguments: (1) The likely absence of a dedicated auditory CN in ray-finned and cartilaginous fish and, by inference, its absence in the sarcopterygian ancestors of land vertebrates. The appearance of a CN is correlated with the evolution of sensitivity to airborne sound, which happened late and independently in the major taxa of land vertebrates (see Manley and Clack, Chapter 1, and Clack and Allin, Chapter 5). (2) Despite striking functional similarities between many CN neuron types across vertebrates, significant differences are seen in the underlying neural morphologies and circuits, which suggests that we are looking at differently derived solutions to similar problems. In the following discussion we compare some well-studied examples of cell types, mainly from the cochlear nuclei of birds and mammals, and highlight salient similarities and differences supporting this argument. Figures 10.2 to 10.4 summarize and contrast graphically the major neuron types in the CN of frogs, birds, and mammals.

Surely the most striking and best documented example of functional convergence is that between the spherical bushy cell population of the mammalian AVCN and the neurons of the archosaur NM. Their detailed similarities in both anatomy and physiology have often been pointed out (review: Carr and Soares 2002) and the specialized role of that particular cell type in temporal auditory processing for sound localization is well understood and highlighted again below (see section 4.3.1). These neurons preserve the temporal information provided by the auditory nerve through phase-locking and convey it to binaural compar-

———————————————————————————————▶

FIGURE 10.3. Schematic representation of neuron types in the cochlear nucleus of birds. Five morphological neuron types have been described for the nucleus angularis (NA) and two for the avian nucleus magnacellularis (NM). Column 1 shows examples from the barn owl, taken from Soares and Carr (2001) for NA and from Köppl and Carr (1997; both used by permission of Wiley-Liss, Inc., a subsidiary of John Wiley & Sons, Inc.) for NM; all drawings are scaled identically. In the adjacent column, the characteristic voltage responses of those neuron types to step de- and hyperpolarization (of about +0.15 and −0.5 nA), recorded *in vitro* from chicken brain slices are shown (from Carr and Soares [2002, with permission from S. Karger AG, Basel]; all traces are scaled identically). Finally, two columns of panels on the right-hand side illustrate the response types seen with pure-tone stimulation *in vivo*. Column 3 shows individual examples of PSTH to 50-ms stimuli at characteristic frequency (CF), 20 to 35 dB above threshold, recorded in the barn owl; all PSTHs are scaled identically. In column 4, schematic rate-intensity functions illustrate the typical features for each type, as summarized in Köppl and Carr (2003, with permission from The American Physiological Society); all panels are scaled identically and the spontaneous rate is indicated by a dashed line, where applicable. In NM, only primary-like responses are found. The NA neurons show a variety of response types, but correlations with *in-vitro* response types and morphology have not yet been established.

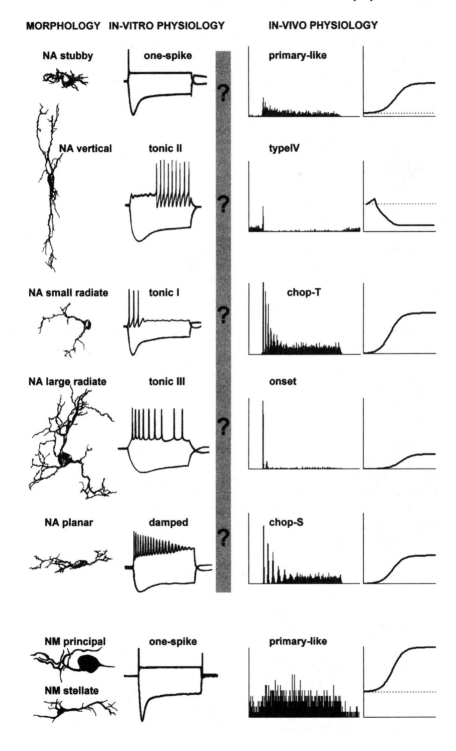

ison circuits at the next higher brainstem level, where interaural time differences are encoded. An interesting difference is that all spherical bushy cells of mammals receive large, calyx-type terminals from the auditory nerve (Fekete et al. 1982; Ryugo and Rouiller 1988; Liberman 1991), whereas in sauropsids, this is only typical for the higher-frequency regions of NM (see pattern 3 in the above list). This is of no known functional consequence, and calyx-type terminals are thus believed to be an adaptation for secure synaptic transmission with minimal losses of temporal accuracy at high frequencies. It is currently unclear whether amphibia have an equivalent of the spherical bushy cell population. Although a morphologically bushy cell type has been identified (Feng and Lin 1996), possible correlations with physiological responses and the projection patterns of the different cell types have not been worked out.

A further intriguing example of functional convergence in the CN is the type IV response type found in both the mammalian DCN and the avian NA (reviews: Young and Davis 2002; Köppl and Carr 2003). In this case, we are faced with detailed similarities in the physiological characteristics and thus presumably the function of these cells, but an apparently very different morphological substrate. Type IV units are characterized by a complex spectral response, displaying interleaving excitatory and inhibitory areas when stimulated with pure tones. In mammals, they are the principal neurons of the DCN and an elaborate circuit has been shown to underlie this response. This circuit includes direct input from the auditory nerve, as well as a multitude of excitatory and inhibitory connections from other CN neurons and proprioceptive input concerning pinna position. There are no anatomical equivalents to the mammalian DCN in birds, although

FIGURE 10.4. Schematic representation of the major types of output neurons in the mammalian cochlear nucleus. Five neuron types are summarized for the ventral cochlear nucleus (VCN) and two for the dorsal cochlear nucleus (DCN). Column 1 shows examples of the cellular morphology (spherical bushy, globular bushy, octopus, fusiform, and giant cells, from Cant 1992; T-stellate and D-stellate cells from Doucet and Ryugo (1997; used by permission of Wiley-Liss, Inc., a subsidiary of John Wiley & Sons, Inc.) all drawings are scaled similarly). In column 2, the characteristic voltage responses of those neuron types to step de- and hyperpolarization (of about $+0.1$ nA and about -0.5 nA), recorded in brain slices are shown (bushy cell taken from Trussell 2002; T-stellate and D-stellate cells from Fujino and Oertel 2001; octopus cell from Oertel et al. 2000; fusiform cell from Zhang and Oertel [1994, with permission from The American Physiological Society] giant cell from Zhang and Oertel 1993; all traces are scaled similarly). On the right-hand side, the response types seen with pure-tone stimulation in vivo are illustrated. Column 3 shows individual examples of PSTH to 25 ms stimuli at CF, approximately 30 to 40 dB above threshold (all PSTH are from Rhode and Greenberg 1992, and are scaled identically). In column 4, schematic rate-intensity functions illustrate the typical features for each type; all panels are scaled identically and the spontaneous rate is indicated by a dashed line, where applicable. Note that for DCN giant cells, definite correlations with *in-vivo* responses have not yet been established.

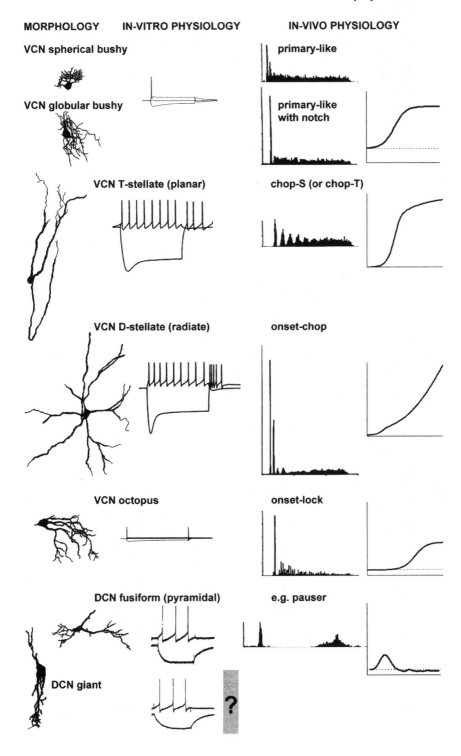

a homology between the DCN and the avian NA had been suggested in early studies (review: Carr and Soares 2002). More recent work has shown that the NA is a heterogeneous nucleus with no prominent subdivisions and with several morphological cell types distributed all across the tonotopic gradient (Soares and Carr 2001). All cell types project outside the nucleus, so there is no evidence for interneurons in the NA. Moreover, there is no morphological cell type in the NA that resembles the principal DCN cells. It therefore appears that birds and mammals have implemented a particular physiological response, the type IV, through very different means. What is it used for? What were the selective pressures driving its parallel evolution? In mammals, it is thought that type IV DCN neurons are involved in the detection of spectral notches, characteristic nulls in the spectrum that are caused by the acoustic filtering properties of the pinna and that provide reliable cues to sound direction in elevation (May 2000; Young and Davis 2002). For most birds, such cues do not exist because of the lack of comparable outer-ear structures. It is thus unlikely that type IV neurons are used in sound localization (Köppl and Carr 2003). They may instead serve the more general function of detecting sharp spectral features (notches or peaks) in communication or other environmental sounds.

Neurons with chopper responses to pure tones have been found across a wide variety of cochlear nuclei, in frogs (review: Feng and Schellart 1999), birds (Sullivan 1985; Warchol and Dallos 1990; Köppl and Carr 2003), and mammals (reviews: Rhode and Greenberg 1992; Young and Davis 2002). Chopper responses are characterized by a relatively regular discharge (independent of stimulus frequency) with low variability over many repetitions of a stimulus, which has led to the prediction that they are involved in the coding of sound level (Shofner and Dye 1989). This seems to be a universal requirement for auditory systems. However, the precise role of most CN chopper neurons is currently not known. According to the most detailed studies in mammals, there are several subtypes that have distinct projection patterns and thus serve quite different functions. It is currently unclear whether and where most nonmammalian chopper units would fit into such a classification scheme. Recent *in-vivo* data from the barn owl suggest that chopper units in the NA of birds may be of two types, a common one similar to the transient chopper of mammals and a rare sustained chopper (Köppl and Carr 2003). However, *in-vitro* electrical properties of chicken NA neurons revealed no close match to any of the chopper types of mammals (Carr and Soares 2002; Soares et al. 2002). Again, this suggests independent evolution in mammals and birds, achieving functionally similar chopper neurons through different means. There was no evidence for onset choppers in birds, a type known to serve as an inhibitory interneuron in the mammalian CN (Young and Davis 2002).

Neurons with pronounced onset responses to pure tones are also a consistent feature of cochlear nuclei, and are found in frogs (review: Feng and Schellart 1999), birds (Sachs and Sinnott 1978; Sullivan 1985; Warchol and Dallos 1990; Köppl and Carr 2003), and mammals (review: Rhode and Greenberg 1992). The most prominent onset type in the mammalian CN, the onset locker, identified morphologically as the octopus cells of the PVCN, is assumed to be involved in temporal processing by detecting broadband transient sounds (Oertel et al. 2000). None of the nonmammalian *in-vivo* studies currently provide sufficient detail to judge whether their onset responses are comparable beyond a superficial similarity of the post-stimulus-time histogram. Characterization of the *in-vitro* electrical properties of chicken NA neurons, however, failed to produce a match to mammalian octopus cells (Soares et al. 2002) and there are also no morphological equivalents (Soares and Carr 2001).

Finally, a significant proportion of neurons in all cochlear nuclei belongs to the primary-like category, that is, neurons that differ little from their auditory nerve inputs in their physiology. In birds and mammals, a subset of those are the spherical bushy cells and NM cells, respectively, discussed above. In mammals, the remaining primary-like units are attributed to globular bushy cells of the AVCN, which feed into interaural comparison circuits of the brainstem (see also below). In birds, the remaining primary-like units make up the majority of cells in the NA (Köppl and Carr 2003), whose function is currently unknown. There is no obvious morphological equivalent to the mammalian globular bushy cell. Instead, primary-like NA units probably correspond to stubby cells, the most common morphological type in the avian NA (Soares and Carr 2001), which also shows matching in vitro electrical properties (Carr and Soares 2002; Soares et al. 2002).

In summary, there are basic physiological types of CN neurons that are widespread and recognizably similar across vertebrate groups. However, the underlying cellular morphologies, the neural circuits, and electrical membrane properties rarely match between the best-studied examples of birds and mammals. We interpret this as evidence for functionally parallel and independent evolution.

4.3 Parallel Evolution of Circuits Involved in Sound Localization

Processing of the cues used in sound localization is an important part of auditory brainstem processing that has been extensively studied in birds and mammals. Importantly, it is the predominant context of much of the nonmammalian data on the auditory brainstem and is therefore a key topic for our evolutionary discussion. Because acoustic space is not mapped by the auditory receptors themselves, sound localization is a computational problem that represents a basic requirement for any auditory system. The basic bauplan of the ascending auditory system, with multiple parallel pathways, allows direct linkage between specific structures (nuclei or subpopulations of neurons within the nuclei) and particular functions in sound localization. However, the cues that can be used are highly dependent on physical dimensions, such that small animals have very different prerequisites than large animals and localization of high-frequency sound differs significantly from that of low-frequency sound. In addition, parameters of the outer, middle, and inner ear partly determine which cues can be used for the central computation of sound location, e.g., the presence or absence of pinna-like structures, middle-ear pressure-gradient receivers, and the overall frequency range of hearing.

We focus on the two binaural cues used for sound localization: the difference in the time of arrival of a sound at the two hearing epithelia (interaural time difference, ITD) and the difference in the sound pressure level at the two ears (interaural intensity difference, IID). Both birds and mammals use these cues, and we can directly compare the neuronal circuits that evolved in these two groups of vertebrates. We start with the ITD processing structures, since the mechanisms we have to deal with are directly incorporated into the structures specialized for temporal processing discussed above, that is, mainly bushy cells in the CN and nucleus magnocellularis. Additionally, using ITDs for sound localization requires a temporal resolution in the microsecond range which is far beyond that normally found in tetrapod brain circuits. Hence, we deal with one of the most unique adaptations in sensory processing that is reflected in particular structural specializations.

4.3.1 ITD Processing Circuits in Birds and Mammals

Although minute, ITDs are the chief cue for localizing low-frequency sounds in the horizontal plane. They depend on the position of a sound source, the interear distance, the size of the skull, and the ways the sound pressure waves reach the eardrum. The combination of these physical constraints yields ITDs in the range of only a few up to some hundred microseconds (μs). In all classes of tetrapods at least some orders adapted to use ITDs. Examples of this adaptation can be found in frogs (amphibians), archosaurs (crocodiles and birds), and different mammalian orders including primates, hoofed animals, carnivores, and some rodents. Although birds (and probably crocodiles) and many mammals achieved a neuronal resolution in the microsecond range, directional hearing based on ITDs can be achieved without such extraordinary neuronal specializations. In frogs, sound waves reach the eardrum in three ways: (1) directly from outside; (2) through the open mouth and the eustachian tube, from inside the middle ear cavity; and (3) via the lungs. This allows the frog tympanic membrane to act as a pressure-gradient receiver (Rheinländer et al. 1979, 1981) that creates directional sensitivity already at the level of the receptor organ and the auditory nerve (review: Lewis and Narins 1999). This directional sensitivity requires the two ears to be intact, but simply because the two ears are acoustically coupled through the eustachian tubes (Feng et al. 1976). Because of this, binaural neurons in the frog auditory brainstem, e.g., the superior olivary nucleus, can act with a comparably moderate ITD resolution (Feng and Capranica 1976, 1978). A different situation can be found in archosaurs, where acoustic coupling of the two middle ear cavities through the eustachian tubes causes a temporal extension of ITDs beyond that created by the pure inter-ear distance (Rosowski and Saunders 1980). The resulting interference expands ITDs particularly at low frequencies so that even chickens experience ITDs of up to 180 µs (Hyson et al. 1994), which is basically double the amount estimated from the pure inter-ear distance (Kuhn 1977; Woodworth 1962). In the case of birds, this mechanism does not compensate for a lack of precise neuronal processing as in the case of frogs, but is rather appended to the neuronal mechanisms for localizing sounds based on ITDs beyond that possible without the acoustic coupling of the middle ears. Mammals lack effective coupling between the middle ears. Therefore, the ITDs they experience are purely based on the shape of the head and the inter-ear distance. Mammals and birds evolved highly specialized areas in the brainstem that allow the processing of ITDs. In birds, this structure is the nucleus laminaris (NL); the mammalian equivalent (analog) is the medial superior olive (MSO).

Normally, the temporal threshold for detection by vertebrate neural circuits is in the range of several milliseconds (Viemeister and Plack 1993). Note that the time course of an action potential is in the range of a millisecond. Hence, during phylogeny the temporal resolution of the neuronal circuits underlying ITD processing in birds and mammals had to be improved by more than an order of magnitude. For instance, neurons in the barn owl NL show a temporal resolution of only a few microseconds (Carr and Konishi 1988, 1990). The evolution of such an exquisite precision required several structural adaptations that reach from the level of single synapses through special membrane properties, through the arrangement of inputs on compartments of single neurons, up to the level of the connection patterns of time processing circuits. Despite the identical physical constraints, the neuronal system encoding ITDs in birds shows profound differences from that in mammals, again indicating parallel, independent evolution of complex auditory processing in archosaurs and mammals.

4.3.2 ITD Processing in the Archosaurs

The most consistent knowledge about processing of auditory space comes from the ITD processing circuit in birds. The bird ITD pathway starts at NM bushy cells, which reliably convey the temporal information inherent in the acoustic stimulus to their projection target (see above; Köppl 1997), the neurons in NL. The NL neurons perform a kind of cross-correlation of the arrival time of action potentials coming in from NM cells from both sides. Adult-like NL neurons are bipolar with principal dendrites arising from the dorsal and the ventral poles of the cell bodies (Parks and Rubel 1975). In crocodiles and many birds, the higher frequency part of the NL (characteristic frequencies [CF] above approximately 1 kHz) consists of a monolayer of neurons arranged in an almost perfect horizontal plane (Parks and Rubel 1975; Young and Rubel 1983; Fig. 10.5A). In the barn owl, however, the sheet of cells is extended in the dorsoventral plane, forming several layers of a modified, nearly adendritic cell type (Fig. 10.5B;

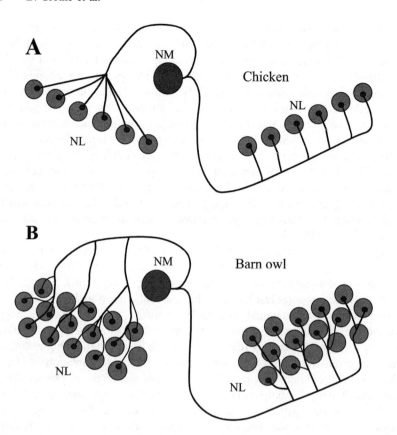

FIGURE 10.5. The excitatory projections from nucleus magnocellularis (NM) to the bird interaural time difference (ITD) detector, nucleus laminaris (NL), as it appears in chicken and the barn owl. **A**: In the chicken, the projections from NM to NL are asymmetric in that only the projection to the contralateral NL takes the form of delay lines. **B**: In contrast, the NM projections in the barn owl are symmetric, creating delay lines to both the ipsilateral and the contralateral NL. Note that the axon parts that run within the NL orthogonal to the mediolateral axis significantly contribute to the absolute delay. (From Carr et al. 2001, Curr Opin Neurobiol; with permission from Elsevier.)

Carr and Boudreau 1993). The former represents the original, plesiomorphic, condition, the latter an apomorphic condition that may be related to processing ITDs of high-frequency tones, which seems unique among birds to owls. There is also evidence that the circuits seen in the chicken and the barn owl represent two extremes of a spectrum of NL morphology within birds (see section 4.3.4). Like almost all brainstem nuclei, NL is tonotopically organized, with the frequency axis running from caudolateral (low-frequency) to rostromedial (high-frequency) (Rubel and Parks 1975; Takahashi and Konishi 1988a).

Neurons in NL receive two groups of inputs. The first is a precise bilateral excitation from NM bushy cells, the second is a tonic inhibition from the SON, not to be confused with the mammalian superior olive. Let us consider only one mediolateral frequency contour in a chick-like NL (Figs. 10.5A and 10.6A). Axons from ipsilateral NM bushy cells project to the dorsal aspect of NL and contact only the dorsal dendrites of NL neurons. Each axon branches, sending collaterals to NL neurons throughout the entire frequency contour. The branching pattern is arranged in such a way that the overall axonal length to each NL neuron is roughly equal. The excitatory inputs from the contralateral NM project to the ventral aspect of NL. They also branch, giving rise to collaterals that cover the entire frequency contour. Their arrangement, however, is significantly different from the ipsilateral input, in that the axonal length systematically increases from lateral to medial (Parks and Rubel 1975; Young and Rubel 1983; Fig. 10.5A). The significantly longer axonal connections to medial NL neurons, compared to that of lateral neurons, represent "delay lines" that create a difference in the arrival time of action potentials (Overholt et al. 1992). The NL neurons act as coincidence detectors that fire maximally when their binaural excitatory inputs arrive simultaneously (Carr and Konishi 1990; Overholt et al. 1992; Joseph and Hyson 1993; Reyes et al. 1996). The anatomical arrangement of bushy cell inputs via the delay lines creates a topographic map of neurons with different preferred ITDs (Overholt et al. 1992). Hence, the place of maximal activity within this map corresponds to a particular location in the horizontal plane outside of the animal's head. This arrangement for encoding azimuthal space via delay lines creating a space map was proposed by Jeffress in 1948.

A variation of this systematic arrangement with delay lines coming from both sides is found in the barn owl (Fig. 10.5B). The overall input pattern is the same. However, the monolayer is expanded to a multilayer due to a tenfold increase of the number of NL neurons (Carr and Friedman 1999). The axons entering from dorsal and ventral interdigitate across the nucleus and make connections to neurons throughout its dorsoventral extent. The conduction delay increases the deeper an axon goes. This increase in conduction delay from the two opposite inputs creates multiple dorsoventrally oriented ITD maps in the barn owl NL (Carr and Konishi 1988, 1990), again creating a topographic arrangement of NL neurons tuned to different "best" ITDs. As shown for the barn owl in vivo, the azimuthal space map created in NL is preserved at higher stations of the auditory system and is finally aligned with the visual space map in the auditory midbrain and the optic tectum (Knudsen and Konishi 1978; Gold and Knudsen 2000).

Only recently did the putative role of inhibitory input to NL neurons gain attention. SON neurons are driven by NA and NL neurons and provide a γ-aminobutyric acid (GABA)ergic feedback projection to NM, NA, and NL (Takahashi and Konishi 1988a; Lachica et al. 1994). This SON input is most likely tonic, not following the temporal pattern of its own phase-locked inputs (Yang et al. 1999; Monsivais et al. 2000), a remarkable contrast to the inhibitory projections in the mammalian ITD encoding system. Interestingly, this tonic

A The avian ITD-detection system

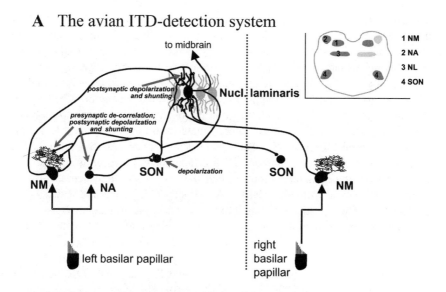

B The mammalian ITD-detection system

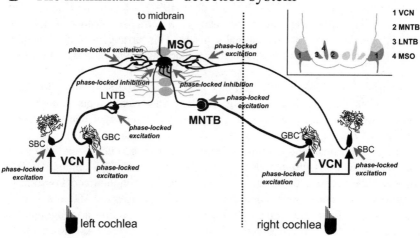

inhibition from SON acts via depolarizing the cell, causing opening of voltage-gated potassium channels. The physiological effect is a shunt and a decrease of the membrane resistance, which increases the spike threshold (Hyson et al. 1995; Monsivais et al. 2000; Monsivais and Rubel 2001). Therefore, the GABAergic circuit prevents unwanted temporal summation of excitatory potentials, ensuring high fidelity of temporal processing across a wide range of sound amplitudes in NM and NL. Additionally, it helps to maintain the coincidence detector in the optimal range of operation (Funabiki et al. 1998) and might also compensate for interaural intensity differences (note the crossed SON projection

FIGURE 10.6. The ITD encoding systems in mammals and birds. **A:** The avian ITD detector is the nucleus laminaris (NL; here shown for the chick; position in the brain stem shown in the inset). The bipolar NL neurons are arranged in one horizontal plane and receive exquisitely phase-locked excitation from nucleus magnocellularis (NM). Additionally, NL neurons receive γ-aminobutyric acid (GABA)ergic inhibition from the superior olivary nucleus (SON). The SON receives inputs from the NL and another source, the nucleus angularis (NA), and provides inhibitory feedback to all of their input sources and to the contralateral SON. The inhibition by SON is decorrelated, thereby eliminating phase-locking, and acts via depolarization and shunting of their target cells. This inhibition provides a differential gain control for the ITD detector mechanism. **B:** The mammalian ITD encoder is the medial superior olive (MSO; position in the brainstem shown in the inset). The bipolar MSO neurons are arranged in one parasagittal plane. Excitatory projections originate from the left and right cochlea via spherical bushy cells (SBC). Globular bushy cells (GBC) project to the medial nucleus of the trapezoid body (MNTB) on the contralateral side and the lateral nucleus of the trapezoid body (LNTB) on the ipsilateral side. The inputs to the MNTB and LNTB are highly specialized for preserving temporal information. The MNTB and LNTB neurons provide phase-locked, hyperpolarizing inhibition to the ipsilateral MSO. The well-timed inhibition is crucial for adjusting the ITD sensitivity of MSO cells to the physiological relevant range of ITDs. (From Grothe 2003; with permission from Nature Publishing Group.)

to NL in Fig. 10.6A; Yang et al. 1999), which could interfere with ITD processing. In vivo recording from the barn owl are consistent with this interpretation (Pena et al. 1996; Takahashi and Konishi 2002). As we will see below, inhibitory inputs to the mammalian ITD detector are of fundamentally different anatomy, physiology, and function, indicating a completely independent evolution.

Hearing of airborne sound via a columella-equipped middle ear evolved in all tetrapod groups in the Triassic (compare Clack and Allin, Chapter 5). Sauropsids started hearing airborne sounds in the low-frequency domain and remained within that range, most likely due to mechanical limitations of their middle ear (Manley and Clack, Chapter 1). Hence, it is safe to assume that in this group of tetrapods, ITD processing was crucial for localizing low-frequency sounds and evolved early in the evolutionary history of hearing. Therefore, it can be assumed to be a plesiomorphic feature of the sauropsid auditory system. However, we need more comparative in vivo data showing how the auditory system in a variety of bird taxa processes and represents ITDs to close the case.

4.3.3 ITD Processing in Mammals

In mammals, the structure that encodes ITDs is the medial superior olive (reviews: Yin 2002; Grothe 2003). At first glance, the functional similarity between the bird NL and the MSO is striking—and led many authors to consider them as homologs. In addition, MSO neurons show anatomical parallels to NL neurons. MSO principal cells are, as NL neurons, bipolar (Ramon y Cajal 1907) and receive bilateral inputs from CN spherical bushy cells (Warr 1966). Superficially, this arrangement seems almost identical to that described above for the bird NL. However, there are distinct and important differences between NL and the MSO concerning the anatomy, the immunocytochemistry, the mechanism of ITD tuning, and the neuronal representation of azimuthal space. First, in contrast to the bird nucleus laminaris, the MSO is located in the ventral brainstem and is embedded in the SOC. The mammalian SOC is composed of numerous nuclei (review: Schwartz 1992), of which most if not all have no anatomical equivalent in the nonmammalian auditory system. Although the overall anatomy is rather similar among mammalian species, there are significant modifications that can be related to different behavioral constraints (Grothe 2000). Since it is not known from which rhombomeres MSO forming cells derive, it remains unclear whether these cells have the same phylogenetic origin as neurons in the bird NL. But given the high diversity of cell types and nuclei in the mammalian SOC, which clearly represent a uniquely mammalian evolution, the question of the exact origin is of rather minor concern. Second, there is some debate about whether the MSO inputs take the course of delay lines, as shown for NL in birds. Although some data from the cat suggest that the contralateral, but not the ipsilateral, inputs take the form of delay lines, the situation is far from being straightforward as in birds (Smith et al. 1993; Beckius et al. 1999). Moreover, the ipsilateral inputs contact only a very limited part of one isofrequency contour (Smith et al. 1993; Beckius et al. 1999), which makes it impossible to judge what the timing relationship of ipsi- and contralateral inputs might be. Third, MSO neurons receive inhibitory inputs that are considerably different from NL inputs, both in cytochemistry and in function. SON inputs to NL are GABAergic (Lachica et al. 1994; Hyson et al. 1995), not phase-locked (Yang et al. 1999), and might provide a gain control for the bird ITD detector (see above). The inhibitory MSO inputs, however, are glycinergic, phase-locked, and arise mainly from the medial and, to some lesser extent, from the lateral nucleus of the trapezoid body (MNTB and LNTB, respectively; review: Grothe 2003). Therefore, the MSO and related structures have no equivalent in nonmammalian species.

The MNTB is most interesting in terms of structure and its role in timing functions. Its neurons have an exceptionally high concentration of the inhibitory transmitter glycine (Wenthold et al. 1987) and receive their excitatory input via a gigantic synapse, the calyx of Held (Morest 1968). It allows for extraordinarily fast and most secure synaptic transmission (review: von Gersdorff and Borst 2002) maintaining the phase-locking of inputs in their inhibitory output (Joris and Yin 1995; Smith et al. 1998). The LNTB neurons projecting to the MSO also show structural adaptations that indicate a temporally very precise transmission (Spirou and Berrebi 1996), and their output to the MSO is also glycinergic (Grothe and Sanes 1993).

Obviously, these phase-locked, glycinergic inhibitory MSO inputs differ significantly from the GABAergic, non–phase-locked inhibitory input to NL.

Moreover, the elaborate mechanism that has evolved in the mammalian auditory system to maintain the timing of the inhibitory input implies a different role in ITD detection than pure gain control. In fact, a recent study shows that the glycinergic inhibition actively takes part in shaping the ITD sensitivity to the physiologically relevant range of ITDs (Brand et al. 2002). Hence, the ITD detector requires, not two, but four exactly timed inputs, a fact that complicates the problem of the existence of simple excitatory delay lines.

Interestingly, in the gerbil, the inhibitory input undergoes a prominent experience-dependent refinement in the first days after hearing onset. Initially, the glycinergic inputs are diffusely distributed over dendrites and MSO cell somata, but during a critical period become refined to the somata. In animals that do not use ITDs, this refinement is absent (Kapfer et al. 2002).

Yet the differences in how birds and mammals encode ITDs even goes beyond the simple encoding mechanism. It also concerns the neuronal representation of ITDs. Recent evidence from guina pigs and gerbils indicates that in mammals, ITDs are not represented by maximal discharges of a small subpopulation of neurons within one frequency contour, but rather by the relative activity of the frequency contours of both MSOs. This activity is regulated by inhibition, not delay lines (reviews: Grothe 2003; McAlpine and Grothe 2003). Therefore, there seems to be no topographic representation of azimuthal space in the MSO itself. Together with the lack of convincing evidence for any ITD map at any higher stations of the mammalian auditory system, one has to consider the possibility that neuronal representation of auditory space fundamentally differs in sauropsids and mammals, indicating again a parallel, independent evolution of spatial hearing. Mammals evolved tympanic ears and hence, almost certainly, the use of ITDs, independently from sauropsids. Therefore, it should not be surprising when the ITD processing structure, and maybe also the neural representation of azimuthal space, took different roads during some 200 million years of the evolution of the processing of airborne sounds.

4.3.4 The Evolution of Nucleus Laminaris and the Medial Superior Olive

Based on the similarities in basic circuitry and function outlined above, many authors have labeled the avian NL and the mammalian MSO as homologs. However, this is clearly a misuse of the term *homolog* in the true phylogenetic sense (for a discussion of homology in the CNS see section 2). The mammalian MSO is universally regarded as part of the olivary complex. The position of the mature NL in the dorsal half of the brainstem, in contrast, suggests no association with the ventrally located olivary complex, and there is no evidence for NL precursors originating and/or migrating from the olivary region. Although the rhombomeric origin is only known for NL but not for MSO neurons, there is evidence that at least some mammalian auditory nuclei derive from rhombomeres that do not contribute to the avian auditory nuclei (review: Maklad and Fritzsch 2003). This, again, shows how problematic it is to homologize NL and MSO.

Instead, Glatt (1975a,b) repeatedly emphasized that the NL is very difficult to demarcate at its caudal end, apparently grading into the NM, and suggested that the NL gradually developed from the NM. Indeed, recent studies suggest that NM and NL are derived from overlapping sources (rhombomeres; review: in Maklad and Fritzsch 2003). Such a development may also explain why several authors found additional primary input to the NL from the auditory nerve (Barbas-Henry and Lohman 1988; Soares 2002). It appears that the NL is present in turtles, which may represent the plesiomorphic state of the sauropsid auditory system (Carr and Code 2000). However, its cells are not arranged in a monolayer (Glatt 1975b). In lizards, Miller (1975) could not identify the NL, whereas other authors identified it in some species that also have a relatively large NM (Barbas-Henry and Lohman 1988). More information on the connections of the NL in turtles and lizards would be highly valuable. In the archosaurs, the NL is consistently identified, mostly by its distinct monolayer structure, although difficulties in defining its caudal end against the NM were also reported (Rubel and Parks 1975).

In summary, the NL in its primitive, all low-frequency condition, as seen in turtles and lizards, is poorly developed and does not show a monolayer arrangement of its cells. In the independent evolutionary line leading to the archosaurs, the NL appears to enlarge and attain the typical monolayer structure, probably correlated with an extension of the NM's and thus also the NL's frequency range. Some observations on crocodilians (Glatt 1975a; Soares 2002) as well as birds (Winter and Schwartzkopff 1961; Köppl and Carr 1997) indicate that the caudolateral low-frequency end of the NL may retain more primitive characteristics in showing no clear monolayer structure and a close association with the NM. The barn owl's medial, high-frequency NL also deviates from the monolayer arrangement, but is clearly hypertrophied as a secondary specialization for unusually accurate sound localization (review: Carr and Code 2000). Anatomically, several other bird species also show this secondary specialization and even intermediate forms of NL were found (Kubke et al. 2002). It remains to be determined whether a hypertrophied NL morphology generally correlates with highly redundant ITD coding and improved behavioral performance, as seen in the barn owl.

Whereas NL seems to be well developed and easily recognizable at least in all archosaurs, the situation is less clear concerning the mammalian MSO. In fact, for some time the existence of an MSO had been attributed to low-frequency hearing mammals only (Harrison and Irving 1966), hence, to mammals that heavily depend on ITDs (Masterton and Diamond 1967). Later studies, however, show the MSO also exists in only or predominantly high-frequency hearing mammals like bats, rats, or mice (review: Grothe 2000). Interestingly, only in mammals that hear low frequencies are the somata of the bipolar MSO principal cells arranged in only one parasagittal plane, whereas in mammals that do not hear low frequencies they are distributed throughout the width of the nucleus, suggesting that the alignment is a structural correlate to the functional specialization related to ITD coding (Fig. 10.7). In fact, the MSO

10. Evolution of the Central Auditory System   319

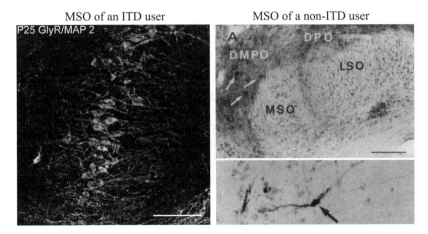

FIGURE 10.7. MSO in an ITD versus a non-ITD user. In the gerbil, an animal with well-developed low-frequency hearing that uses ITDs for sound localization, the bipolar principal MSO neurons are arranged in one parasagittal plane with their dendrites oriented medially and laterally, respectively. The bipolar cells in the MSO of the free-tailed bat, a non-ITD user that does not hear low frequencies, are also bipolar and also point medially and laterally (labeled neuron in the lower panel). However, the somata are not arranged in one parasagittal plane [Nissl staining, counterstained against acetylcholinesterase (ACHE)]. From Kapfer et al. (2002; with permission from Nature Publishing Group) and Grothe 1994 (with permission from Wiley-Liss Inc., a subsidiary of John Wiley & Sons, Inc.).

in bats that do not show this specialization shows a lower ITD resolution that can only be interpreted as epiphenomenal (review: Grothe 2000).

The first mammals were tiny animals, and remained so almost throughout the Mesozoic (Hopson 1973; Crompton and Jenkins 1979). Therefore, in mammals ITD processing might be a very late adaptation that is reflected in the different morphology and ITD resolutions found. For instance, the early mammal *Morganucodon* had an inter-ear distance of about 6 mm (Kermack et al. 1981) resulting in ITDs of probably less than 25 µs (cf. Kuhn 1977). Even the oldest known placental mammal (125 million years old) was so small (Ji et al. 2002) that it would have experienced ITDs in the range of 50 µs at most. Hence the less organized MSO might reflect the plesiomorphic state, and the MSO with aligned somata and a microsecond resolution the derived, apomorphic state.

Because of a lack of physiological studies, the function of the putative MSO in small mammals that show poor thresholds for low frequencies is unclear. Inbody and Feng (1981) showed that rat MSO cells are not sensitive to ITDs, unlike MSO cells in low-frequency specialists (review: Yin 2002). The function of the rat MSO remains unknown. Only from studies on bats comes the observation that even though MSO cells are not involved in ITD processing, they are nevertheless involved in encoding temporal information. For instance, MSO cells might be involved in detection of stimulus pauses or amplitude modulations (Grothe et al. 2001; review: Grothe 2000). A large proportion of bat MSO cells performs such temporal processing in a spatial context, either due to interaural intensity difference sensitivity or because of a rough ITD sensitivity (review: Grothe 2000). The latter would not be suitable for direct sound localization, but such a crude ITD sensitivity could be involved in processing spatial reverberations (cf. Fitzpatrick et al. 2000; Grothe and Neuweiler 2000). Therefore, one hypothesis is that a low-resolution ITD processor found in small mammals evolved into a high-resolution ITD encoder when low-frequency hearing became important (Grothe 2000). This functional evolution is most likely reflected in the structural differences found in the MSO of ITD using and non-ITD using animals (Fig. 10.7).

4.3.5 IID Processing Circuits in Birds and Mammals

A careful analysis was necessary to appreciate the evolutionary convergence of the ITD processing systems in birds and mammals. Much more obvious is the likelihood of parallel evolution of IID processing systems in these groups. IIDs are particularly prominent for frequencies with a wavelength shorter than the inter-ear distance (Thompson 1882; Rayleigh 1907), but in the near field, even lower frequencies create prominent IIDs (Shinn-Cunningham et al. 2000). Because of the generality of good high-frequency hearing in mammals, IIDs are the most common cue for sound localization in terrestrial mammals. Because of the generality of good low-frequency hearing in sauropsids and the low-frequency nature of their communication signals, IIDs are, compared to ITDs, of rather minor importance for these animals (for a detailed discussion see Klump 2000).

Accordingly, our knowledge about IID processing in birds is limited. Most of what we know comes from studies in barn owls. However, these animals are unique in that they use IIDs for localization in the vertical plane, unlike other vertebrates that use IIDs for localizing in the horizontal plane (Norberg 1978; Moiseff 1989). The specialization of the barn owl and some other owls is due to a morphological asymmetry of the outer ear, the facial ruff, and the preaural flap (review: Klump 2000). Volman and Konishi (1990) discussed the asymmetry in some owls as a special adaptation for high frequency (>5 kHz) stimuli that cause prominent IIDs.

Despite the specialization in asymmetrical owls, the neuronal substrate for encoding IID in birds is, most likely, similar in all birds. The bird IID pathway, as described for the barn owl, originates in the cochlear NA; NA neurons project to, among other targets, the contralateral posterior part of the dorsal nucleus of the lateral lemniscus [LLDp (formerly called VLVp); Takahashi and Konishi 1988a,b]. This projection is assumed to provide the contralateral excitatory input to LLDp neurons. The LLDp neurons themselves project to the LLDp on the opposite side (Takahashi and Keller 1992) as well as to the inferior colliculus on both sides. Many LLDp neurons are GABAergic (Carr et al. 1989), and therefore are possibly responsible for the inhibition evoked by stimulation of the ipsilateral ear as found in the LLDp and the inferior colliculus (IC) (Manley et al. 1988; Takahashi and Keller 1992). This inhibition is graded along the dorsoventral axis of the LLDp (Takahashi et al. 1995) and correlates with increasing inhibitory effects along this axis, creating a systematic map of IIDs in the LLDp (Manley et al. 1988).

The mammalian IID processing system is much more elaborate and includes at least three different stages within the ascending pathway. The earliest and by far best known stage at which IID sensitivity is observed is the lateral superior olive (LSO), a nucleus without equivalent in the bird auditory system (cf. Carr and Code 2000). The LSO is excited by stimulation of the ipsilateral and inhibited by stimulation of the contralateral ear (review: Yin 2002). A systematic map of IIDs, comparable to that found in the bird LLDp, has not been found in the LSO. The IID sensitivity created in the LSO is relayed to higher auditory centers (lateral lemniscus and IC) via contralateral excitatory and ipsilateral inhibitory connections (review: Oliver and Huerta 1992). Additional IID processing takes place in the dorsal nucleus of the lateral lemniscus (DNLL) and later in the IC (review: Pollak et al. 2003). Some of their neurons show binaural properties that reflect the direct LSO inputs. In many cases, the IID sensitivity, however, is significantly modified by additional binaural interactions. Additionally, IID sensitivity can be created de novo at the level of the IC itself (review: Pollak et al. 2003).

Despite these differences between birds and mammals, there is an interesting similarity: one connection in the much more complex IID processing system in mammals resembles one found in birds. This is the presumably GABAergic connection from LLDp to the contralateral LLDp. Neurons in the mammalian DNLL are also GABAergic and project to their contralateral counterpart (review: Wu 1999). However, the way these nuclei operate within the localization system is functionally quite different. The reciprocal LLDp connection in birds creates the initial IID sensitivity, whereas its mammalian equivalent adds a temporal component to the IID sensitivity created earlier. Its neurons do not affect the first wave of spatial information going from LSO to IC, but do affect the second wave (Pollak et al. 2003). The DNLL, therefore, might play a crucial role in echo suppression. Since this function of the DNLL requires the IID sensitive inputs from LSO, it is unlikely that LLDp has a similar function in birds. Therefore, in birds and mammals different circuits perform the task of localizing high-frequency sounds by analyzing IIDs.

Taken together, the auditory brainstem circuits underlying sound localization in birds and mammals share many similarities. But the similarities seem to have different reasons. On the one hand, all these circuits are based on the basic bauplan of the ascending auditory system outlined above (see section 3). On the other hand, sauropsids and mammals came up with different variations of the same theme, based on a similar origin and following similar constraints, but adapting their auditory system in quite different fashions.

5. The Auditory Midbrain

The auditory midbrain is thought to be homologous among all vertebrates in terms of structure and connections and embryonic origin (Wilczynski 1988; McCormick 1999). To obtain clues to how it has evolved, we examine common themes in structure and function in the auditory midbrain. We shall briefly describe the connections to show that, although there are species-specific differences, there is a general plan that is common among vertebrates. The similarity in the organization of the ascending pathways to the midbrain (Wilczynski 1988) is remarkable considering the diverse origins of the hearing endorgans and the primary acoustic nuclei of the medulla among vertebrates (see sections 3 and 4). Then we shall focus on response properties that emerge in the midbrain, specifically selectivity for time-varying sounds. This function is likely to be one of the earliest biological functions for hearing. The earliest hearing vertebrates probably had no structural basis for frequency discrimination, somewhat like present-day fish (Feng and Schellart 1999). In fish, the absence of tonotopy leaves temporal processing as the principal basis for discriminating auditory signals (Crawford 1997).

5.1 Internal Organization of the Auditory Midbrain

The auditory midbrain (Figs. 10.8–10.10) lies within a region containing input from several other sensory systems. In nonmammals it lies within the torus semicircularis (TS), and in mammals it is the IC. Note that often the bird TS is referred to as IC (a problematic term if it is used to imply homology of the internal and functional organization of the auditory midbrain in mammals and birds) or as nucleus mesencephalicus lateralis pars dorsalis (MLD). Adjacent to the auditory midbrain are other structures that are important for auditory function such as the intercollicular area and the deep superior colliculus.

A useful example is the TS of the tiger salamander (*Ambystoma tigrinum*). As with all uredeles, this animal does not have a tympanic ear, and like most urodeles, it does not vocalize. It therefore can be expected to have a relatively simple system for processing sounds. In the TS of the tiger salamander, Herrick (1948) found indeed an area that was cytoarchitecturally undifferentiated, and he proposed that it may represent a primitive form of the auditory midbrain. In the European fire salamander (*Salamandra salamandra*), Manteuffel and Naujoks-Manteuffel (1990) confirmed that this area was auditory. They suggested that the pathway was a representative of a plesiomorphic acoustic pathway because its input arises from the intermediate nucleus of the octavolateral complex as it does in cartilaginous fishes (Boord and McCormick 1984).

Vertebrates that have more elaborate hearing than the salamander have a central area of the TS that is purely auditory and generally located medially in the TS. For example, Knudsen (1977) reported a functional segregation in the fish TS, with auditory units located in the most medial part, which he called nucleus

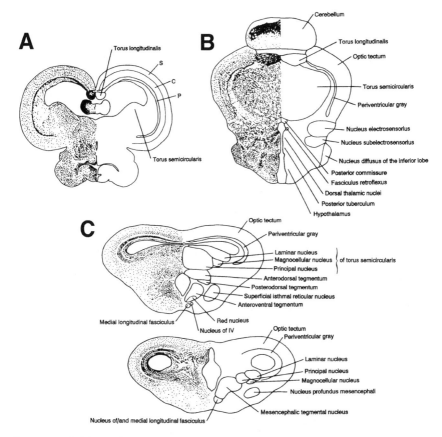

FIGURE 10.8. The auditory midbrain in fish and amphibians. **A**: Drawing of transverse view of midbrain of a teleost fish, *Lepomis gibbosus*, to show cytoarchitecture (left hemisection) and position of torus semicircularis. **B**: Drawing of transverse view of midbrain of a weakly electric fish *(Apteronotus leptorhynchus)*, to show cytoarchitecture (left hemisection) and position of torus semicircularis. Note greatly expanded size of torus. In another species of weakly electric fish the auditory area is in the dorsal medial part of the TS (see text). **C**: Drawing of transverse view of midbrain of an anuran amphibian *(Rana catesbeiana)*, to show cytoarchitecture of the torus semicircularis (TS) (left hemisection). The position of the TS in amphibians is similar to that in fishes. From Butler and Hodos (1996, ISBN: 0471888893; this material is used by permission of Wiley-Liss, Inc., a subsidiary of John Wiley & Sons, Inc.).

centralis. Electroreceptive units were found in the most lateral part of TS and mechanoreceptive units in the area between the auditory and electroreceptive areas. He demonstrated that nucleus centralis was cytoarchitecturally distinct from the electroreceptive and mechanoreceptive areas. Connectional and physiological studies of the TS of fish have confirmed this general plan in three species of vocalizing fish *(Pollimyrus adspersus*: Kozloski and Crawford 1998;

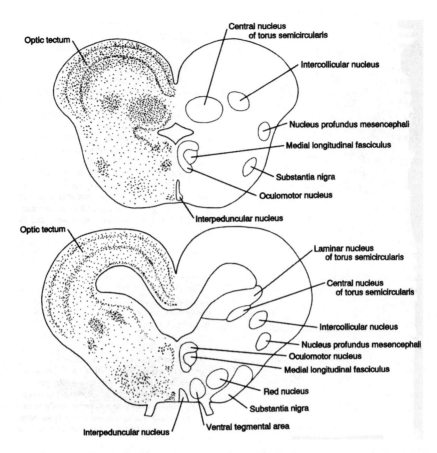

FIGURE 10.9. The auditory midbrain in a reptile. Drawings of two transverse sections of midbrain of a lizard (*Tupinambis nigropunctatus*), to show cytoarchitecture (left hemisection) and position of torus semicircularis. From Butler and Hodos (1996, ISBN: 0471888893; this material is used by permission of Wiley-Liss, Inc., a subsidiary of John Wiley & Sons, Inc.).

Porichthys notatus: Bass et al. 2000; *Opsanus beta*: Bass et al. 2001). In amphibians, there is a similar lateral to medial organization of sensory inputs (Wilczynski 1988). Physiological evidence indicates that response properties from lateral to medial are tactile, tactile and auditory, and auditory only (e.g., *Rana pipiens*: Comer and Grobstein 1981).

Likewise, sauropsids (including birds) and mammals have an ascending somatosensory target, the "intercollicular" nucleus that is adjacent to the auditory part of the midbrain. For example, in the turtle the spinal cord and dorsal column nuclei project to the caudal and lateral part of the TS (*Pseudemys scripta elegans*: Künzle and Woodson 1982). A number of other studies show somatosensory projections to the lateral part of the TS, to the laminar TS, or to the

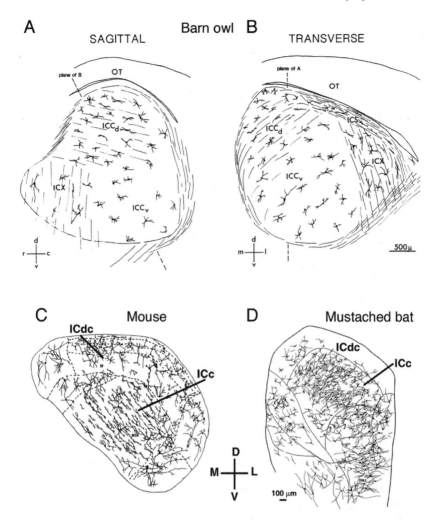

FIGURE 10.10. Lamination in the core of auditory midbrain of a barn owl, a mouse, and a bat. **A, B**: Camera lucida drawings of Golgi-stained cells to show subdivisions of the auditory midbrain (IC) in the barn owl and the orientation of cells as seen in parasagittal (A) and transverse (B) planes. The thin lines show the orientation of fibers. Note the orientation of dendrites and laminated appearance in ICCd. From Knudsen 1983, used by permission of Wiley-Liss, Inc., a subsidiary of John Wiley & Sons, Inc. **C, D**: Drawings of Golgi impregnated neurons to illustrate main subdivisions and the orientation of dendrites in the IC of two mammals. C: Mouse (from Meininger et al. 1986). D: Mustached bat (from Zook et al. 1985; used by permission of Wiley-Liss, Inc., a subsidiary of John Wiley & Sons, Inc.). IC, inferior colliculus; ICc, central nucleus of IC; ICC_d, dorsal division of ICc; ICC_v, ventral division of ICc; ICS, superficial nucleus of IC; ICX, external nucleus of IC; OT optic tectum; c, caudal; d, dorsal; l, lateral; m, medial.

intercollicular area (review: Pritz and Stritzel 1989), whereas medial parts of the torus remain mostly free of nonauditory input.

5.1.1 Core and Belt Auditory Organization

At least for most vertebrates the auditory midbrain can be divided into a central auditory "core" around which is a "belt" area (Wilczynski 1988). The exception might be bony fish, where a core-belt organization has not yet been reported. In all other vertebrates, the core is the main target of the ascending inputs from the auditory structures in the lower brainstem. In contrast, the belt areas are not the main target of ascending auditory pathways but receive auditory inputs from the core or from higher auditory structures. For example, the amphibian TS has a core ventral area that receives ascending auditory input and a shell that includes the laminar nucleus, not to be mistaken with the nucleus laminaris in the medulla of archosaurs (Fig. 10.8A). The laminar nucleus (Fig. 10.8) is seen in amphibians (Feng 1983; Wilczynski 1988; Luksch and Walkowiak 1998), some nonavian sauropsids (Fig. 10.9) (Browner and Rubinson 1977; Belekhova et al. 1985; Künzle 1986), and some fishes, but not birds and mammals (Fig. 10.10). In reptiles, as in amphibians, the laminar nucleus receives little or no ascending auditory input (Künzle 1986).

In all vertebrates studied, the auditory midbrain is the main output to auditory thalamus. However, there are some important differences in the details of the connections. For example in amphibians, the auditory projections from the TS to the thalamus arise mainly from the laminar nucleus and the magnocullular nucleus (Feng and Lin 1991; Luksch and Walkowiak 1998). The principal nucleus, which is the core area, does not project to the thalamus. In mammals, the target of the ascending auditory pathways, the central nucleus of the inferior colliculus, is also the major source of projections to the auditory thalamus. This difference illustrates a difficulty in determining what subdivisions of the amphibian or reptile midbrain are analogous to the subdivisions in the auditory midbrain in birds and mammals, indicating an independent evolution of the recent internal organization of the TS/IC.

An interesting evolutionary variation in the organization of the amphibian torus is seen in *Xenopus*, which retains a lateral line system after metamorphosis. The ascending input from the lateral line nuclei in the medulla terminate in the magnocellular nucleus and in the principal nucleus, whereas the ascending auditory input terminates in the laminar nucleus (Edwards and Kelley 2001). Edwards and Kelley (2001) propose that the magnocellular nucleus and the principal nucleus adapted to process near field water pressure changes, whereas the laminar nucleus processes sound in water.

Although the core region is the main input processing center, there are differences in the internal organization that are related to the range of frequencies to which the animal is sensitive. In animals that have an elongated papilla or basilar membrane, there is a tendency for the central nucleus to have a laminar organization. This organization is most prominent in the mammalian IC, where the orientation of disk-shaped cells and the course of input fibers and terminals produce a laminar appearance in the central nucleus (Fig. 10.10C,D) (reviews: Oliver and Huerta 1992; Casseday et al. 2002). The laminar structure corresponds to the tonotopic organization of the central nucleus of the IC.

In mammals, there are species differences in the range of frequencies represented in the central nucleus of the IC. In some echolocating bats, the IC contains an expanded representation of frequencies that are biologically important in the bats' behavior. In some cases, this expanded representation is related to an expanded receptor surface (as in *Pteronotus parnellii*; Kössl and Vater 1985), in other cases (*Eptesicus fuscus*) the expansion appears to be independent of the representation at the receptor surface (Casseday et al. 2002). This neural magnification of a frequency range suggests that the IC has adapted to represent biologically important sounds during evolution.

Sauropsids also have a tonotopic organization that follows the orientation of dendrites of some cells, although the laminar pattern is less distinct than in mammals (e.g., *Tupinambis nigropunctatus*: Browner and Rubinson 1977; Manley 1971; Fig. 10.10).

In amphibians, there is little evidence of a laminar pattern in the core (Feng 1983), and the tonotopy is difficult to interpret because some neurons have more than one peak of frequency sensitivity (review: Feng and Schellart 1999).

The relationship between core and belt areas is hierarchical in that the pathways from core to motor pathways usually travels via belt regions. For example, in the barn owl each of these parts plays a specific role in the neural formation of a map of auditory space. In the core area information about the binaural cues are combined (Pena and Konishi 2001). The core areas project to the external nucleus, a belt region, such that a two-dimensional map of auditory space is formed (Knudsen and Konishi 1978; Gold and Knudsen 2000; review: Carr and Code 2000). Although a clear functional distinction between core and belt remains to be shown for all vertebrates, this structural organization is nevertheless generally present.

5.1.2 Parallel Pathways and Delay Lines

All vertebrates so far examined have multiple auditory pathways that ascend to the auditory core of the midbrain. Although there are differences in the number of pathways, their organization follows the same general plan. There is direct input from the cochlear nucleus or its analog, and indirect input from one or more nuclei that receive their inputs from the cochlear nucleus. For example, in fish there is one direct pathway as well as indirect input from one or possibly two nuclei (McCormick 1999). In birds and mammals (reviews: Carr and Code 2000; Casseday et al. 2002) the auditory midbrain receives a complex set of convergent inputs from a large number of quite separate auditory centers in the lower brainstem.

The auditory pathways to the mammalian midbrain have an especially elaborate system to create delay lines and other transformations of auditory information (Casseday et al. 2002). The multiple pathways to the IC supply it with different kinds of signal transformation. Some of the intervening synaptic connections transform the input from excitation to inhibition. The temporal pattern of response seen in the auditory nerve, onset followed by activity sustained for the sound duration, is transformed to other patterns such as onset, offset, chopping or pauser patterns and variations of these. A further temporal change is simply to increase latency in some pathways by intervening synapses so that the range of latencies seen at the auditory nerve is markedly expanded at the input to the IC. Thus, the IC can be seen as the target of a complex pattern of delay lines with different discharge characteristics that can be excitatory or inhibitory, monaural or binaural. This patterning of input seems to create temporal templates that make some neurons in the IC highly selective for time-varying sounds. Physiological evidence of these multiple inputs and the possibility of the creation delay lines is seen in the very broad range of response latencies at the IC of mammals (Haplea et al. 1994; Klug et al. 2000) as well as in the TS of fish (Feng and Schellart 1999).

5.1.3 Relation of the Midbrain to Motor Systems

In all vertebrates that have been examined, the auditory midbrain projects to motor systems. The main source of these motor projections is the belt region. For instance, the external nucleus of the IC in birds and mammals projects to the deep superior colliculus. The laminar nucleus in reptiles and amphibians projects to the spinal cord. These belt areas receive input from the core. Thus, another feature of the auditory midbrain is that it can supply the motor system with processed auditory information. These connections may be a basis for responding to biologically important sounds (Casseday and Covey 1996).

The most extensive analysis of the relationship of the auditory midbrain to motor systems comes from work on the barn owl. There the pathway for orientation to sound in space includes a projection from the external nucleus to the deep layers of the optic tectum, and from the tectum to the pontine tegmentum and to spinal cord (Knudsen and Knudsen 1983; Masino and Knudsen 1992). A similar pathway is also seen in mammals and serves much the same function as in birds. Analogous pathways may exist in amphibians and other sauropsids if one considers the projection from laminar nucleus to the spinal cord, but the function of this connection is not known.

Evolution of vocal behavior is one of the most distinctive evolutionary developments related to hearing. Most vocal functions require auditory feedback, necessitating an auditory-motor connection. The connection is in part from the auditory midbrain. For example, Goodson and Bass (2002) suggest that there is a vocal-acoustic circuitry common to vertebrates. In all vertebrate groups, there is some evidence of connections from the auditory midbrain to circuits involving vocalization. In vocal fish the TS projects to vocal areas in the periaqueductal gray and paralemniscal tegmentum (Goodson and Bass 2002). Similar connections in mammals are from the IC to the paralemniscal area (Metzner 1996), and in birds from the auditory midbrain to the nucleus intercollicularis (ring dove, *Streptopelia risoria*: Akesson et al. 1987). Goodson and Bass (2002) suggest that connections from auditory nuclei, including the auditory midbrain, to vocal centers are a result of evolutionary convergence.

In birds, the intercollicular nucleus (ICo) plays a role in the interface between auditory and motor systems. The auditory input to ICo arises from TS (ring dove, *S. risoria*: Akesson et al. 1987). The ICo is part of the descending pathway from the telencephalic structure, the archistriatum, for the control of vocalization in birds (Wild et al. 1993; Mello et al. 1998). The archistriatum (arcopallium in the new nomenclature; see section 6) projects to the central nucleus of TS (Mello et al 1998). Thus, the primary auditory center in the midbrain of many birds has extensive interaction with motor systems via its reciprocal connections with the ICo as well as direct projections from the motor system. All of these motor-related pathways concern the control of a specific motor behavior, vocalization. In mammals, the IC receives input from motor systems, athough the function is unclear (review: Casseday et al. 2002).

5.1.4 Descending Input to the Auditory Midbrain

Descending connections from the forebrain obviously must be related to the evolution of size and complexity of the forebrain itself. Thus, it is not surprising that the most extensive descending connections to the auditory midbrain are found in birds and mammals. Nevertheless, it might well be a general rule in all vertebrates that there is the possibility for the forebrain to modulate operations of the auditory midbrain.

5.2 Processing Temporal Patterns

The ability to make temporal discriminations of sounds is a common thread among vertebrates. Therefore, it is reasonable to speculate that the common ancestor of vertebrates had temporal processing ability, whether or not it was capable of frequency or spatial discrimination. The connectional evidence (see section 5.1.2) indicates that all vertebrates have some form of synaptic delay lines in the ascending pathways to the auditory midbrain. Delay lines of this sort combined with interactions of excitatory and inhibitory inputs can serve as the structural base for selectivity to temporal features of sound. Temporal selectivity that is created in the auditory midbrain is an important clue in assessing what functions have evolved there.

In vocalizing fish, the TS plays an important role in selective detection of the temporal structure of the sounds they produce. The mormyrid fish *Pollimyrus adspersus* uses species-specific acoustic displays during courting (Crawford

1997). The vocalizations are clicks repeated at different rates comprising three different elements of the acoustic display, grunt, groan, and growl (Crawford et al. 1997). Crawford (1997) found that approximately one third of the auditory neurons in the TS were selective for repeated clicks and responded preferentially to specific click rates. The range of interval selectivity was the same range of click rates the fish use for vocal communication. Crawford points out that interval selective neurons are not found at earlier stages of auditory processing, and so it is likely an emergent property in the TS (although not investigated in nonvocalizing fish).

Bass and colleagues have studied midshipman fish (*Porichthys notatus*), a species in which the males produce long lasting (~60 min) courting sounds or "hums" to attract females. The hums have a fundamental frequency of about 100 Hz, differing slightly among individuals. When two males are in the same vicinity the hums from one male will physically interact with those of a nearby male to produce beats at a frequency equal to the difference between the frequencies produced by the two males (Fig. 10.11A). Midbrain neurons in the midshipman respond synchronously to the beat frequencies of 10 Hz or less (Fig. 10.11B), which corresponds to the beats from two concurrent hums (Bodnar and Bass 1997). In contrast, peripheral auditory afferents synchronize well

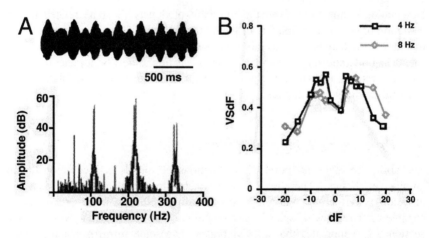

FIGURE 10.11. Processing of concurrent vocal signals by neurons in the midbrain of a vocalizing fish, the midshipman (*Porichthys notatus*). **A**: Beat waveform with amplitude modulations produced by concurrent hums of male midshipman fish (top) and amplitude spectrum showing frequencies of beats produced by the difference frequency (dF) between the fundamentals and higher harmonics of two signals. **B**: Examples of vector strength of sychronization to dF (VSdF) at different dFs for two neurons. The dF that produced peak synchronization is indicated (4 and 8 Hz). Both neurons showed symmetrical tuning, peak synchronization at positive and negative dFs. (Adapted from Bodnar et al. 2001 with permission from Springer Verlag.)

to the waveform of the individual components of the sound but poorly to the beats (McKibben and Bass 1999). On this and other evidence, Bodnar and Bass (1997) suggest that there is a transformation in the neural code from periodicity code of the frequency waveform at the primary afferents to a periodicity code of the difference frequency at the midbrain. Anatomical evidence of direct and indirect auditory pathways that converge at the midbrain might account for the transformation in the code according to Bass et al. (2000).

From an evolutionary aspect it is significant that this sort of change in code from the periphery to the midbrain appears in fish. Analogous roles for direct and indirect pathways have been proposed in other orders, so either this is a primitive feature of the midbrain, or it has evolved independently several times.

For anuran amphibians, discrimination of temporal patterns of sound is essential to the discrimination of conspecific vocalizations such as the mating or advertisement calls of males. The simplest temporal parameter is how long a sound lasts. Neurons that are selective for specific sound durations were first discovered in the auditory midbrain of amphibians (*Rana catesbeiana*: Potter 1965; *Eleutherodactylus coqui*: Narins and Capranica 1980). This discovery was important because it established that there is a neural code for sound duration such that different neurons can represent different durations. The TS is the first stage in the auditory system where neurons are tuned to a specific narrow range of sound duration (Gooler and Feng 1992; review: Feng and Schellart 1999).

Many of the advertisement calls of anuran amphibians are amplitude modulated (AM), consisting of trains of pulses with very regular durations and interpulse intervals (IPIs). Extensive work on selectivity for AM rate in the TS shows that the temporal selectivity in a given species is related to the temporal properties of the species vocalizations (review: Feng and Schellart 1999). An important recent example (Fig. 10.12) is the finding that some torus semicircularis neurons require a specific number of pulses at specific IPIs to respond (Alder and Rose 2000; Edwards et al. 2002). The relative absence of this kind of selectivity below the TS suggests that discrimination of biologically important time-varying sounds emerges in the TS. Because it is a source of motor pathways, the TS could play an important role in the response to biologically important sounds.

The observation that processing in the TS is related to a given species' vocalization pattern has significance for questions of evolution. Specifically, it raises the question, To what extent has the evolution of processing in the auditory midbrain been driven by the need to discriminate vocalizations? In this regard, it would be of considerable comparative interest to examine the auditory processing in the torus of nonvocalizing urodele amphibians. Such information would contribute to theoretical considerations about the role of temporal processing in the auditory midbrain (Casseday and Covey 1996).

Other than tonotopy there is nothing, to our knowledge, on the response properties of auditory neurons in the TS of sauropsids, except birds. Although very

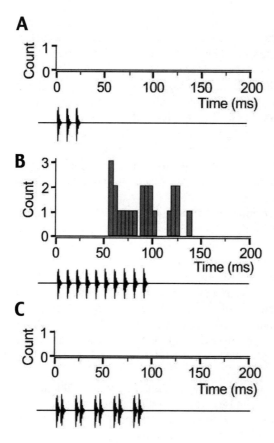

FIGURE 10.12. Poststimulus time histograms (top graphs in **A**, **B**, and **C**) of spike counts of a single anuran midbrain neuron's response to sound pulses (traces below graphs) that differ in number of pulses or interpulse intervals (IPI). **A**: The neuron did not respond to only three pulses (10 ms IPI), but responded well to 10 pulses, represented at a similar IPI (**B**). **C**: The neuron also did not respond to ten pulses, when the IPI was alternated between 5 and 15 ms. From Edwards et al. 2003 (with permission from Nature Publishing Group).

little is known about processing of time-varying sounds in the auditory midbrain of birds, some evidence suggests that it also selects temporal features. In the auditory midbrain of the guinea fowl (*Numida meliagris*), Scheich et al. (1977) showed that some cells, called "complex neurons" responded best to complex sounds such as conspecific calls, multiharmonic sounds, or other sounds with multiple frequency components. Most of the sounds were characterized by a high degree of temporal structure, such as sound pulses repeated at regular intervals.

Surprisingly little is known about the response properties of neurons in the auditory midbrain of songbirds. Recent abstracts (Woolley and Casseday 2001a,b) suggest that the auditory midbrain of zebra finches is an important stage in transforming the representation of auditory signals that eventually reach the specialized song areas of the forebrain. Some (~16%) of the cells have two peaks of sensitivity to pure tones; the response to sinusoidal frequency or sinusoidal amplitude modulation was synchronized only within particular ranges of modulation. Furthermore, many cells respond better to conspecific songs than to heterospecific songs.

As in the frog auditory midbrain, some IC neurons are selective to specific ranges of sound duration in the IC of bats (*Eptesicus fuscus*: Casseday et al. 2000; Faure et al. 2003; *Antrozous pallidus*: Fuzessery and Hall 1999). In the big brown, *Eptesicus fuscus*, most duration selective neurons respond to short durations, typically 1 to 10 ms, which corresponds with the duration of most of the echolocation calls used by the animal (Ehrlich et al. 1997; Faure et al. 2003). Duration selective cells have also been found in the IC of several species, although to different ranges of sound duration (chinchillas: Chen 1998; mice: Brand et al. 2000), showing that duration selective cells are not specific adaptations to echolocation. The pallid bat, *Antrozous pallidus*, is a good example that duration selectivity is most likely an adaptation to the general listening requirements of an animal (Fuzessery and Hall 1999). This bat uses echolocation for navigation but also uses passive hearing for detecting the sounds made by the small animals and insects on which it preys. This dual lifestyle means that the IC is arranged by its tonotopy into a high-frequency (30 to 80 kHz) echolocation part, which is ventromedial in the IC, and a low-frequency (3 to 30 kHz) prey-detector part, which is lateral and dorsal. Neurons selective for short sound durations (<7 ms) are found in both the lateral IC and the ventral IC. However, in the dorsal IC most duration selective neurons responded only to long duration sounds. Duration selectivity requires delay lines to adjust the temporal sequence of inhibitory and excitatory inputs, all of which have approximately the same sound frequency.

Delay tuning is another good example of processing of biologically important sounds in the IC. The time difference between a broadcast sound and its echo from a reflective target is thought to be the cue that bats use to detect target distance, and in fact neurons that are tuned to the delay of two signals are found in the auditory midbrain (Feng et al. 1978). These combination-sensitive, delay-tuned neurons are clear examples of specific neural adaptation for echolocation. These neurons were first discovered in the auditory cortex (Suga et al. 1978), and were later also found in the IC (Mittmann and Wenstrup 1995; Portfors and Wenstrup 1999). These neurons are particularly interesting because they require specific frequency combinations and temporal delays to respond. In the mustached bat, *Pteronotus parnellii*, the frequencies and delays are similar to those in the echolocation calls of the bat. The echolocation pulse of the mustached bat has multiple harmonics and frequency modulated (FM) components. Combination-sensitive delay-tuned neurons responded poorly if at all to individual FM components. However, they responded well to combinations of the fundamental FM (FM1) component and one of the higher harmonic FM (FMn) components, but only if there was a specific time delay between FM1 and FMn (Fig. 10.13). These neurons require convergence of different frequency inputs in which there is a temporal delay in one input relative to the other. The convergence of multiple ascending pathways to the IC

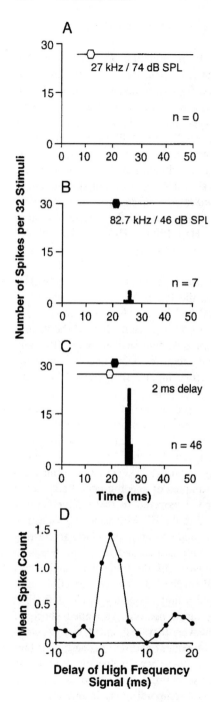

FIGURE 10.13. Delay tuning of a neuron in the inferior colliculus of an echolocating bat (*Pteronotus parnellii*). **A**: This neuron failed to respond to a 27-kHz tone presented alone. **B**: The neuron responded with only a few spikes to an 82.7-kHz tone. **C**: The neuron gave a vigorous response when the 27-kHz tone followed the onset of the 82.7-kHz tone by 2 ms. **D**: A delay tuning curve shows the spike count as a function of the delay between the high tone relative to the low tone. A delay of zero indicates that the onsets of the tones were simultaneous, and negative delays mean the onset of the high-frequency tone preceded the onset of the low-frequency tone. From Portfors and Wenstrup 1999 (with permission from the American Physiological Society).

seems particularly well suited to provide these spectral and temporal requirements. It should be mentioned that evidence for convergence of different frequency input is also found in the auditory midbrain of amphibians (Scheich et al. 1977; Fuzessery 1988; Feng and Schellart 1999; Hall 1999) and fish (Lu and Fay 1966).

5.3 Conclusions: Salient Features and Similarities in the Auditory Midbrain

We propose that the following features are common in the auditory midbrain of vertebrates: (1) core and belt organization, (2) output to motor systems, (3) somatosensory/lateral line input, (4) descending pathways from the forebrain, and (5) parallel pathways and delay lines. Most of these ideas for organization and function of the auditory midbrain have been pointed out previously (cf. Wilczynski 1988). The fact that these similarities are found in all vertebrates examined so far suggests either that they are primitive conditions, or that they represent parallel evolution serving a common role. The common role may be processing temporal features of biologically important sounds (Casseday and Covey 1996) making the midbrain a most important station in the context of evolutionary adaptations in sound analysis.

6. The Auditory Forebrain

Wilczynski (1984) has pointed out that the greatest changes in the auditory system are found in the forebrain. He suggested that these changes accompany the increased use of sound for communication in amphibians, archosaurs, and mammals. The other significant change in the forebrain occurs with mammals, which are distinguished from other vertebrates by the evolution of the cortex in place of the more usual nuclear organization (Karten and Shimizu 1989; Aboitiz et al. 2004). At present the evolutionary origins and homologies of the different forebrains areas are a subject of active research and controversy.

We summarize the basic organization of the forebrain (Fig. 10.14) here and refer the reader to reviews on forebrain evolution (Northcutt and Kaas 1995; Striedter 1997; Aboitiz et al. 2003). All forebrains are divided into a dorsal pallium and a ventral subpallium. In agnathans, the olfactory bulbs project to the whole pallial surface (Northcutt 1996). In vertebrates with jaws, however, the olfactory input is usually restricted to the lateral pallium and other sensory projections expand into the forebrain via multiple relays from the diencephalon. The only group to which this organization does not apply is the amphibians, which are considered to have a secondarily simplified brain (Northcutt 1981).

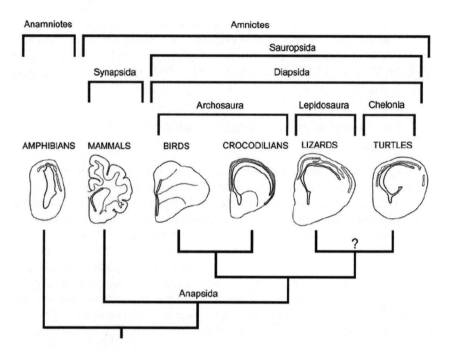

FIGURE 10.14. Tetrapod forebrains. Cladogram indicates phylogenetic relationships in tetrapods. In each crown taxon, a figure of a coronal section of one cerebral hemisphere (only one hemisphere is shown; lateral is to the right and medial is to the left) of a representative species is included. From Aboitiz et al. 2003 (reprinted with the permission of Cambridge University Press).

6.1 Basic Patterns of Auditory Forebrain Organization Among Vertebrates

The basic vertebrate forebrain pattern has projections from the auditory midbrain to the dorsal thalamus to the dorsal pallium. We will identify some of the common features and significant differences between each major vertebrate group.

In bony fish, the central nucleus of the auditory midbrain has major ascending projections to both dorsal and ventral thalamus, the posterior tuberculum, and the hypothalamus (McCormick 2002). A cladistic analysis of the connections between the midbrain and forebrain in catfish revealed that the pathway through the dorsal thalamus is probably a primitive character for all jawed vertebrates, whereas the pathway through the posterior tuberculum is probably a derived character for actinopterygian fishes. The auditory pathways through the hypothalamus appear to have evolved independently in catfishes and frogs (Striedter 1991). The telencephalon in bony fish is divided into dorsal and ventral areas, with the dorsal area proposed to be homologous to the pallium of other vertebrates, and the ventral area to the subpallium. Dorsal areas respond to sound, but little is known about auditory processing rostral to the midbrain (Echteler 1985; Lu and Fay 1995).

In frogs, the auditory midbrain torus projects to the central and posterior nuclei of the thalamus and to the striatum. Recordings from the posterior nucleus show sensitivity to the frequency combination present in the frog advertisement calls, whereas many neurons in the central thalamus are broadly tuned and sensitive to specific temporal features of the call (Feng and Schellart 1999). The central thalamus projects to the striatum, anterior preoptic area, and the ventral hypothalamus. Ascending auditory pathways terminate mainly in the striatum, which is clearly not homologous to either the cortex or the dorsal ventricular ridge (DVR) (Ten Donkelaar 1998). These connections may mediate control of reproductive and social behavior in frogs (Wilczynski et al. 1993). The anterior thalamic nucleus supplies ascending information to the medial pallium, although little is known about pallial auditory processing in this group.

The sauropsids all have very similar ascending auditory projections, and will be described first as group (review: Carr 1992). In turtles, lizards, crocodilians, and birds, the cells of the central nucleus of the auditory midbrain (torus) project bilaterally to a dorsal thalamus region made up of a central nucleus surrounded by a shell. The core and shell receive ascending input from the central nucleus of the auditory midbrain and from the lemniscal nuclei, respectively. In both birds and crocodilians, the core and shell thalamic regions project to separate, nonoverlapping regions of forebrain, suggesting a basic division of auditory input to the forebrain into two separate streams (Wild et al. 1993). Names are different, with dorsal thalamus called nucleus reuniens in turtles, nucleus medialis in lizards and crocodilians, and nucleus ovoidalis in birds. Nevertheless, the consistent pattern of connections and dorsal thalamic location throughout amniote phylogeny suggests these structures are homologous.

The telencephalic terms are also different, but again the organization is similar among all sauropsids. The sauropsid pallium is made up of a three-layered cortex and a prominent periventricular structure, the DVR, which is the largest telencephalic region in reptiles, and its medial region contains neurons that are sensitive to sound stimuli (Pritz 1974). Neurons in the shell area surrounding the thalamic nucleus reuniens project to the nearby caudocentral portion of the DVR (Pritz and Stritzel 1994), which suggests the presence of a second auditory region in the DVR. The auditory region of the medial DVR projects mainly to the subpallial corpus striatum and to the posterior DVR. The posterior DVR corresponds to the archistriatum (now termed arcopallium) in birds (Lanuza et al. 1999). The bird equivalent to the auditory region of the DVR is Field L. Most avian forebrain nomenclature has been changed through a consensus-based discussion in 2002.

Field L is the first telencephalic stage of the ascending auditory pathway in birds (Cohen et al. 1998; Scheich 1990). It is a tonotopically or cochleotopically organized structure comparable in many respects with the mammalian primary auditory cortex (AI) (Scheich 1990) in that it is the principal target of ascending input from the thalamic nucleus ovoidalis. Field L is divided into three parallel layers, L1, L2, and L3, with L2 being further divided. Auditory units in L2 generally have narrow tuning curves with inhibitory sidebands, which might be expected from their direct input from thalamus, whereas the cells of L1 and L3 exhibit more complex responses in the guinea fowl (Scheich et al. 1979). The general avian pattern is that Field L projects to the adjacent hyperstriatum (now termed mesopallium) and to other nuclei of the caudal neostriatum (now termed nidopallium), including the higher vocal center (HVc) in songbirds. These regions contain neurons that are sensitive to communication signals such as the bird's own song. In barn owls, areas of the auditory archistriatum contain a high proportion of space-tuned neurons (Cohen and Knudsen 1999). In birds, pallial auditory nuclei project to the auditory areas of the archistriatum (now arcopallium), which project back down to the auditory thalamus and midbrain (review: Carr and Code 2000). Projections from the archistriatum to the intercollicular area (and directly to the hypoglossal nucleus in songbirds) appear to mediate vocalization.

For mammals, the ascending auditory projections share many features with other vertebrates, but the origins and homologies of mammalian cortex are presently controversial (Aboitiz et al. 2004). Comparative studies of the embryological origins of the forebrain nuclei in all vertebrates are therefore an area of extensive investigation (Puelles et al. 1999; Striedter and Keefer 2000). Present theories suggest that the mammalian cortex originates at least in large part from the dorsal pallium (Northcutt and Kaas 1995; Aboitiz et al. 2004), but that the mammalian isocortex and the DVR in sauropsids have separate origins. Thus at present it appears that sauropsid and mammalian auditory systems evolved in parallel, although they share many features.

Three sets of connections characterize the mammalian auditory forebrain (Winer 1992; de Rebaupierre 1997; Kaas et al. 1999). First, auditory information is relayed from the ventral nucleus of the medial geniculate complex to three core primary or primary-like areas of auditory cortex that are cochleotopically organized and highly responsive to pure tones. Second, auditory information is then distributed from the core areas to a surrounding belt of about seven areas that are less precisely cochleotopic and generally more responsive to complex stimuli. Current thinking (Kaas et al. 1999) suggests that belt and parabelt regions may be concerned with integrative and associative functions involved in pattern perception and object recognition. These regions are characterized by ascending projections from midbrain tegmentum, dorsal medial geniculate body, and nonprimary auditory cortex. Third, an even more broadly distributed set of connections/affiliations link auditory forebrain with cortical and subcortical components of the limbic forebrain and associated autonomic areas, and elements of motor system that organize behavioral responses to biologically significant sounds (Kaas et al. 1999). The above view largely characterizes studies on primates (Fig. 10.15). At least one tonotopic representation has been found in all species studied (review: Merzenich and Schreiner 1991).

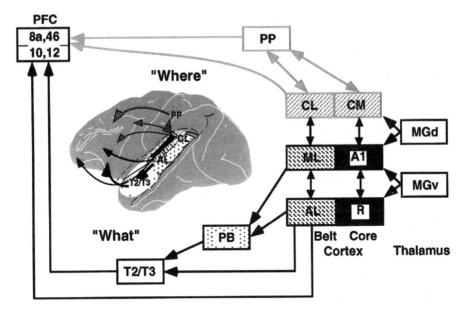

FIGURE 10.15. Mammal auditory forebrain. Schematic flow diagram of "what" and "where" streams in the auditory cortical system of primates. The ventral "what" stream is shown in black, the dorsal "where" stream in gray. PFC, prefrontal connections, PP, posterior parietal cortex; PB, parabelt cortex; MGd and MGv, dorsal and ventral parts of the MGN. From Rauschecker and Tian (2000; Copyright National Academy of Sciences, U.S.A.).

Primitive mammals like tenrecs have few auditory areas (Krubitzer et al. 1997), but there are at least seven tonotopic maps in the cat and mustached bat (Winer 1992).

After sound frequency, the stimulus dimension that has received the most attention in terms of cortical representation is the binaural response type. A common feature in all species studied is a clustering of neurons according to aural type. In the cat (Imig and Adrian 1977; Middlebrooks et al. 1980) and ferret (Kelly and Judge 1994), these clusters appear as binaural bands that run orthogonal to isofrequency contours, suggesting that each isofrequency band has access to different binaural types.

The nature of spatial representation in the mammalian auditory cortex has proven a difficult issue. The clustering of aural types in several species, and their relationship to frequency representation, suggests a modular organization, with one function being the extraction of spatial information. Free-field stimulation studies have reported that neurons are also clustered with respect to spatial sensitivity. However, unlike the superior colliculus, no topographic organization of binaural cues or spatial sensitivity that might provide an ordered representation of space has been found in the mammalian auditory cortex (review: Cohen and Knudsen 1999). Large receptive fields and effects of intensity level, which tend to degrade spatial tuning observed near intensity thresholds, have led to suggestions that auditory space may be represented in the cortex through nontopographically organized population codes, using such information as the temporal pattern of neural spikes, mean spike latencies, or coordinated firing among multiple neurons. For example, using artificial neural network modeling based on cat cortical data, the temporal spike patterns of single neurons were found to carry more spatial information than spike counts (review: Middlebrooks et al. 2002). Most of these neurons appeared to encode sound locations over a wide range of azimuths, suggesting a code based on information extracted by a distributed population of neurons. Other modeling data suggest that relative spike counts across a population of neurons may also contribute to sound localization. The common theme underlying these proposed codes is a representation of stimulus features that does not require a systematic anatomical spatial correlate.

The many similarities in the physiology of avian and mammalian auditory systems may have evolved in parallel when the ancestors of modern birds and mammals developed sensitivity to high-frequency airborne sounds. An understanding of these parallel developments may reveal common computational principles in the auditory system. We next discuss the organization and shared features of forebrain auditory structures devoted to communication in birds and mammals.

6.2 Birdsong and Human Speech

Many animals make communication sounds, but few learn them. The exceptions are humans, and the songbirds, parrots, and hummingbirds. Bird song and speech learning share common features reviewed in Doupe and Kuhl (1999) and summarized here. First, humans and songbirds learn their vocal motor behavior early in life, with a strong dependence on hearing the adults they will imitate as well as hearing themselves when they practice (Konishi 1985). Second, vocal perception and production are tightly linked in the vocal learning process. Finally, both groups exhibit sensitive periods for song and speech learning (Doupe and Kuhl 1999).

Despite the differences in forebrain organization in birds and mammals, the pathways for vocal production and processing are similar (Fig. 10.16). The vocal control system includes the brainstem motorneurons involved in control of vocal tract structures, and medullary structures that integrate respiratory and vocal tract control. These structures receive input from midbrain regions involved in the production of complex species-specific calls, including the caudal periaqueductal gray and adjacent parabrachial tegmentum in mammals. In birds, the midbrain dorsomedial intercollicular nucleus connects directly to the vocal motor neurons for the muscles of the syrinx and to the medullary nuclei involved with respiratory control (for review see Wild 1997). Stimulation of the archi-

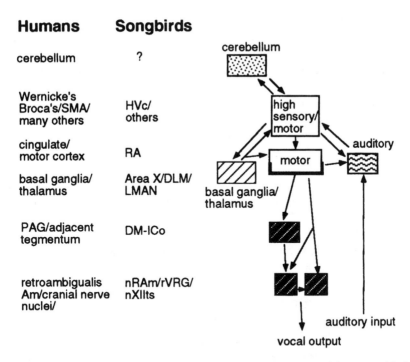

FIGURE 10.16. Neural substrates for vocal learning in songbirds and humans with the names of approximately equivalent areas at left. The two lowest levels (brain stem and midbrain controllers and integrators of vocalization, *striped black*) are shared with all nonlearners. White boxes indicate unique structures with specific roles in vocalization. Both vocal circuits involve a forebrain/basal ganglia/thalamus loop. Finally, both circuits have strong auditory inputs and overlapping auditory/motor centers, as well as feedback from motor centers to auditory areas. SMA, supplementary motor area; PAG, periaqueductal gray; DM-Ico, dorsomedial nucleus of the intercollicularis; nRAm, retroambigualis; Am, ambiguous; rVRG, rostral ventral respiratory group. From Doupe and Kuhl 1999 (with permission).

striatum (now arcopallium) or anterior cingulate cortex also results in calls in many vertebrates. Vocal control in oscine songbirds and parrots differs from that in other sauropsids, however, because the preexisting projection from the archistriatum to the intercollicular region appears to have been modified. In these birds the robust nucleus of the archistriatum (RA) projects directly to the motor nucleus of the hypoglossal nerve in the medulla, in addition to the intercollicular region.

Two major steps in the evolution of learned vocalizations in both primates and songbirds appear to have taken place in the high-level circuits for control of vocal production (Doupe and Kuhl 1999). First, songbirds and humans have evolved forebrain areas to allow for control of the preexisting vocal control pathways described above (Fig. 10.16). In humans, lesions of Broca's area in the posterior frontal inferior cortex cause disruption of speech production. In songbirds, lesions of the posterior forebrain song pathway disrupt song production. This posterior pathway consists of nucleus HVc and RA (Figs. 10.16 and 10.17). The HVc projects to the RA, which then connects directly to the nuclei involved with vocal motor and respiratory control (Wild 1997). In the second major evolutionary change, songbirds and humans have evolved forebrain areas to learn song and speech, and tie auditory feedback to sound production. Studies of songbirds have had a profound effect on our understanding of how speech and song are learned.

The song system contains a forebrain–basal ganglia circuit that plays a critical role in song learning, in addition to the "motor" posterior pathway. This circuit is termed the anterior forebrain pathway and is made up of a projection from the HVc to nucleus X to DLM to lateral magnocellular nucleus of the anterior neostriatum (LMAN) to RA (Fig 10.17). The current assumption is that the motor programs underlying the bird's unique song are stored in the posterior

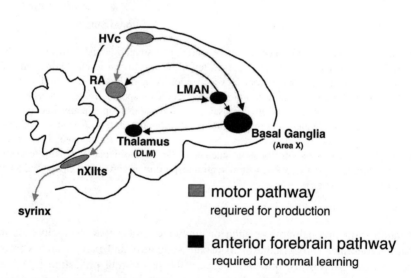

FIGURE 10.17. Anatomy of the song system, which consists of two pathways. Motor pathway nuclei are shown in gray, and the anterior forebrain pathway (AFP) nuclei are in black. The motor pathway, necessary for normal song production throughout life, includes the higher vocal center (HVc), the robust nucleus of the archistriatum (RA), and the tracheosyringeal portion of the hypoglossal nucleus (nXIIts). The AFP comprises Area X (X), the medial nucleus of the dorsolateral thalamus (DLM), and lateral magnocellular nucleus of the anterior neostriatum (LMAN). The Field L complex and related areas provide auditory input to the song system. From Brainard and Doupe 2000 (with permission from Nature Publishing Group).

pathway, whereas the anterior pathway contains neurons that respond to song stimuli, consistent with the idea that this pathway is a possible site of song evaluation (review: see Brenowitz et al. 1997). The anterior pathway projects to the posterior pathway, so it is well positioned to provide a guiding influence on the developing motor program. It is also homologous to cortical basal-ganglia circuits in mammals (Bottjer and Johnson 1997). When song is learned, it appears that the anterior pathway provides or transmits an instructive signal, indicating a mismatch between auditory feedback and the internal song model (Brainard and Doupe 2000). Birdsong studies also show how interrelated auditory and motor processes are in vocal learning, as is true for brain areas involved in speech. Less is known about the neurobiology of speech learning.

The parallels between birdsong and speech are striking, but there are important differences. Doupe and Kuhl (1999) state that appearance of meaning in human speech is the most obvious difference with birdsong. Human speech must be built on preexisting brain structures in other primates (see below), but also must have made a large evolutionary step to create language. By contrast, smaller modifications of the pathways in suboscine birds could lead to the song system. The critical step shared by avian vocal learners with humans appears to have been to involve the auditory system, both for learning the sounds of others and for allowing the flexibility to change their own vocal output (Doupe and Kuhl 1999).

Variation in the connections of song system to auditory system are illustrated in the independent evolution of birdsong in oscine songbirds, parrots, and hummingbirds. Comparisons of oscine and parrot song systems provide an example of the parallel evolution of forebrain auditory pathways for perception and production of song. In the songbirds, Field L provides auditory inputs to the song system, but it is not the main source of auditory information in budgerigars. Instead, auditory inputs to the vocal control system appear to derive from nucleus basalis and the frontal neostriatum (Striedter 1994). These nuclei receive auditory input via ascending projections from the lemniscal nuclei (Arends and Zeigler 1986). The auditory nucleus basalis projects upon the overlying neostriatum (now nidopallium), which in turn projects upon more caudal and lateral pallial regions more directly involved in motor control, as well as to Field L (Wild 1994). A phylogenetic analysis suggests that the midbrain and medullary vocal control pathways are homologous across all birds, but that most of the vocal control circuits in the forebrain have evolved independently in parrots and songbirds (Striedter 1997).

6.3 What Sets the Mammalian Forebrain Apart?

Although only humans use speech, most mammals use sound for communication, and a vocal pathway must exist for processing vocal communication sounds in addition to "where" and "what" pathways proposed for auditory cortex (Kaas et al. 1999). Recordings from marmoset auditory cortex reveal very similar preferences for species-specific vocalizations to those found in songbirds (Wang and Kadia 2001). The majority of neurons in marmoset A1 respond more strongly to natural marmoset vocalizations than to the time-reversed vocalizations. In a clever experiment to determine whether the differences in neural responses were simply due to the difference between the acoustic structures of natural and time-reversed vocalizations, Wang and Kadia (2001) compared neural responses to natural and time-reversed marmoset twitter calls in A1 of cats with those obtained from A1 of marmosets using identical stimuli. The preference for natural marmoset twitter calls in marmoset A1 was absent in cat A1. Although both cortices responded approximately equally to time-reversed twitter calls, marmoset A1 responded much more strongly to natural twitter calls than did cat A1. This differential representation of marmoset vocalizations in two cortices suggests that experience-dependent and possibly species-specific mechanisms are involved in cortical processing of communication sounds.

Mechanisms underlying the emergence of responses to complex sounds also have been addressed in the most vocal mammals, the bats, which use sound for both echolocation and communication. Bats determine target distance from the time delay between their sonar vocalizations and the returning echoes and use the three-dimensional information about target location to guide the features of their sonar vocalizations and to position themselves to grasp the insect with their wing or tail membrane (review: Moss and Schnitzler 1995). A great deal is known about the organization of the auditory forebrain in the mustached bat, *Pteronotus parnellii*. *Pteronotus* auditory cortex is differentiated into at least seven areas, many of which are related to identification of echolocated prey. A1 systematically represents frequency with an enlarged representation of the Doppler shift compensation region (pulse frequency range), mapping not just frequency, but amplitude. There are several maps of echo delay, for delays that represent near, midrange, and far targets. There are also maps of the contralateral azimuth and a second Doppler shift region. Suga (1988) has used these data to construct a parallel-hierarchical scheme for signal processing in which different cortical areas encode different aspects of the auditory scene.

In addition to the appearance of multiple cortical areas, bat cortex is also marked by neurons sensitive to combinations of sounds. *Pternotus'* echolocation call is composed of both a constant frequency and frequency-modulated components, and these different components are processed in parallel channels in the auditory system by virtue of their tonotopy. In the cortex and also the inferior colliculus, combination sensitive neurons are created by comparing across frequency channels (Suga 1988). Combination sensitive neurons have also been found in the songbird auditory system (Margoliash and Fortune 1992), but what may be unique to mammals is the appearance of a precisely tuned corticofugal system. Suga and his colleagues (2000) propose that this corticofugal system mediates a highly focused positive feedback to physiologically "matched" subcortical neurons, and widespread lateral inhibition to physiologically "unmatched" subcortical neurons, to adjust and improve information processing of complex sounds. It remains to be seen whether similar feedback systems characterize the forebrain circuits of other vertebrates, or whether these developments are unique to mammals and a consequence of corticothalamic organization.

6.4 Conclusions: Salient Features and Similarities in the Auditory Forebrain

The following features characterize the auditory forebrain of vertebrates: (1) evolution of cortex in mammals; (2) large changes associated with the increased use of sound for communication in amphibians, archosaurs, and mammals; (3) multiple tonotopic representations; (4) core and shelf organization; (5) specialized communication systems that have evolved in parallel in humans, songbirds, parrots, and hummingbirds; and (6) precisely tuned corticofugal systems in mammals. The similarities observed in all vertebrates examined suggests that they represent parallel evolution serving a common role. The common role may be the encoding of the auditory scene, and the detection of biologically important sounds.

7. Summary

The central auditory system of all vertebrates is based on similar building blocks with similar developmental principles resulting in similar basic functions. However, the detailed neuronal substrates and mechanisms by which these functions are achieved were shaped to a significant extent by independent evolution. The predominant reason was the late and independent evolution of tympanic ears in land vertebrates with its associated sensitivity gain to airborne sound and higher frequencies. This opened up a whole new sensory world and probably provided the major evolutionary pressures for specializations and refinements in central auditory processing—at a time when the major lines of vertebrates had already separated from each other. Thus, the central auditory systems observed in modern vertebrates show many examples of parallel solutions for similar problems. Great caution should be exercised in homologizing individual substructures in the auditory system between clades. Anatomical or functional similarities alone are not sufficient grounds. This is perhaps most impressively demonstrated in the parallel evolution of the auditory forebrain centers.

References

Aboitiz F, Morales D, Montiel J (2004) The evolutionary origin of the mammalian isocortex: towards an integrated developmental and functional approach. Behav Brain Sci 27 (in press).

Akesson TR, De Lanerolle NC, Cheng MF (1987) Ascending vocalization pathways in the female ring dove: projections of the nucleus intercollicularis. Exp Neurol 95: 34–43.

Akutagawa E, Konishi M (2001). A monoclonal antibody specific to a song system nuclear antigen in estrildine finches. Neuron 31:545–556.

Alder TB, Rose GJ (2000) Integration and recovery processes contribute to the temporal selectivity of neurons in the midbrain of the northern leopard frog, Rana pipiens. J Comp Physiol 186:923–937.

Arends J, Zeigler HP (1986) Anatomical identification of an auditory pathway from a nucleus of the lateral lemniscal system to the frontal telencephalon (nucleus basalis) of the pigeon. Brain Res 398:375–381.

Barbas-Henry HA, Lohman AHM (1988) Primary projections and efferent cells of the VIIIth cranial nerve in the monitor lizard, *Varanus exanthematicus*. J Comp Neurol 277:234–249.

Bass AH, Bodnar DA, Marchaterre MA (2000) Midbrain acoustic circuitry in a vocalizing fish. J Comp Neurol 419:505–531.

Bass AH, Bodnar DA, Marchaterre MA (2001) Acoustic nuclei in the medulla and midbrain of the vocalizing Gulf toadfish (Opsanus beta). Brain Behav Evol 57:63–79.

Beckius GE, Batra R, Oliver DL (1999) Axons from anteroventral cochlear nucleus that terminate in medial superior olive of cat: observations related to delay lines. J Neurosci 19:3146–3161.

Belekhova MG, Zharskaja VD, Khachunts AS, Gaidaenko GV, Tumanova NL (1985) Connections of the mesencephalic, thalamic and telencephalic auditory centers in turtles. Some structural bases for audiosomatic interrelations. J Hirnforsch 26:127–152.

Bermingham NA, Hassan BA, Wang VY, Fernandez M, Banfi S, Bellen HJ, Fritzsch B, Zoghbi HY (2001) Proprioceptor pathway development is dependent on Math1. Neuron 30:411–422.

Bleckmann H, Niemann U, Fritzsch B (1991) Peripheral and central aspects of the acoustic and lateral line system of a bottom dwelling catfish, Ancistrus sp. J Comp Neurol 314:452–466.

Bodnar DA, Bass AH (1997) Temporal coding of concurrent acoustic signals in auditory midbrain. J Neurosci 17:7553–7564.

Bodnar DA, Holub AD, Land BR, Skovira J, Bass AH (2001) Temporal population code of concurrent vocal signals in the auditory midbrain. J Comp Physiol 187:865–873.

Boord RL, McCormick CA (1984) Central lateral line and auditory pathways: a phylogenetic perspective. Am Zool 24:765–774.

Bottjer SW, Johnson F (1997) Circuits, hormones, and learning: vocal behavior in songbirds. J Neurobiol 33:602–618.

Brainard MS, Doupe AJ (2000) Auditory feedback in learning and maintenance of vocal behaviour. Nat Rev Neurosci 1:31–40.

Brand A, Urban R, Grothe B (2000) Duration tuning in the mouse auditory midbrain. J Neurophysiol 84:1790–1799.

Brand A, Behrend O, Marquardt T, McAlpine D, Grothe B (2002) Precise inhibition is essential for microsecond interaural time difference coding. Nature 417:543–547.

Brenowitz EA, Margoliash D, Nordeen KW (1997) An introduction to birdsong and the avian song system. J Neurobiol 33:495–500.

Brigande JV, Kiernan AE, Gao X, Iten LE, Fekete DM (2000) Molecular genetics of pattern formation in the inner ear: do compartment boundaries play a role? Proc Natl Acad Sci USA 97:11700–11706.

Browner RH, Rubinson K (1977) The cytoarchitecture of the torus semicircularis in the Tegu lizard, Tupinambis nigropunctatus. J Comp Neurol 176:539–557.

Bullock TH, Heiligenberg W (1986) Electroreception. New York: Wiley and Sons, pp. 637.

Butler AB, Hodos W (1996) Comparative Vertebrate Neuroanatomy: Evolution and Adaptation. New York: Wiley-Liss.

Cant NB (1992) The cochlear nucleus: Neuronal types and their synaptic organization. In: Webster DB, Popper AN, Fay RR (eds) The Mammalian Auditory Pathway: Neuroanatomy. New York: Springer-Verlag, pp. 66–116.

Cant NB, Benson CG (2003) Parallel auditory pathways: projection patterns of the different neuronal populations in the dorsal and ventral cochlear nuclei. Brain Res Bull 60:457–474.

Cantos R, Cole LK, Acampora D, Simeone A, Wu DK (2000) Patterning of the mammalian cochlea. Proc Natl Acad Sci USA 97:11707–11713.

Carr CE (1992) The evolution of the central auditory system in reptiles and birds. In: Webster DB, Fay RR, Popper AN (eds) The Evolutionary Biology of Hearing. New York: Springer-Verlag, pp. 511–544.

Carr CE, Boudreau RE (1993) Organization of the nucleus magnocellularis and the nucleus laminaris in the barn owl: encoding and measuring interaural time differences. J Comp Neurol 334:337–355.

Carr CE, Code RA (2000) The central auditory system of reptiles and birds. In: Dooling RJ, Fay RR, Popper AN (eds) Comparative Hearing: Birds and Reptiles. New York: Springer-Verlag, pp. 197–248.

Carr CE, Friedman MA (1999) Evolution of time coding systems. Neural Comput 11: 1–20.

Carr CE, Konishi M (1988) Axonal delay lines for time measurement in the owl's brainstem. Proc Natl Acad Sci USA 85:8311–8315.

Carr CE, Konishi M (1990) A circuit for detection of interaural time differences in the brain stem of the barn owl. J Neurosci 10:3227–3246.

Carr CE, Soares D (2002) Evolutionary convergence and shared computational principles in the auditory system. Brain Behav Evol 59:294–311.

Carr CE, Fujita I, Konishi M (1989) Distribution of GABAergic neurons and terminals in the auditory system of the barn owl. J Comp Neurol 286:190–207.

Carr CE, Soares D, Parameshwaran S, Perney T (2001) Evolution and development of time coding systems. Curr Opin Neurobiol 11:727–733.

Casseday JH, Covey E (1996). A neuroethological theory of the operation of the inferior colliculus. Brain Behav Evol 47:311–336.

Casseday JH, Ehrlich D, Covey E (2000) Neural measurement of sound duration: control by excitatory-inhibitory interactions in the inferior colliculus. J Neurophysiol 84: 1475–1487.

Casseday JH, Fremouw T, Covey E (2002) The inferior colliculus: a hub for the central auditory system. In: Oertel D, Fay RR, Popper AN (eds) Springer Handbook of Auditory Research: Integrative Functions in the Mamalian Auditory Pathway. New York: Springer-Verlag, pp. 238–318.

Cau ES, Casarosa S, Guillemot F (2002) Mash1 and Ngn1 control distinct steps of determination and differentiation in the olfactory sensory neuron lineage. Develop 129:1871–1880.

Chen GD (1998) Effects of stimulus duration on responses of neurons in the chinchilla inferior colliculus. Hear Res 122:142–150.

Cohen YE, Knudsen EI (1999) Maps versus clusters: different representations of auditory space in the midbrain and forebrain. Trends Neurosci 22:128–135.

Cohen YE, Miller GL, Knudsen EI (1998) Forebrain pathway for auditory space processing in the barn owl. J Neurophysiol 79:891–902.
Comer C, Grobstein P (1981) Organization of sensory inputs to the midbrain of the frog, Rana pipiens. J Comp Physiol 142:161–168.
Coombs S, Görner P, Münz H (1989). The Mechanosensory Lateral Line: Neurobiology and Evolution. New York: Springer-Verlag, pp. 724.
Crawford JD (1997) Feature-detecting auditory neurons in the brain of a sound-producing fish. J Comp Physiol 180:439–450.
Crawford JD, Cook AP, Heberlein AS (1997) Bioacoustic behavior of African fishes (Mormyridae): potential cues for species and individual recognition in Pollimyrus. J Acoust Soc Am 102:1200–1212.
Crompton AW, Jenkins AF Jr (1979) Origin of Mammals. In: Lillegraven JA, Kielan-Jaworowska Z, Clemens WA (eds) Mesozoic Mammals. Berkeley: University of California Press, pp. 55–73.
de Burlet HM (1934) Vergleichende Anatomie des statoakustischen Organs. a) Die innere Ohrsphäre. In: Bolkl, Göppert E, Kallius E, Lubosch W (eds) Handbuch der Vergleichenden Anatomie der Wirbeltiere, vol 2. Berlin: Urban and Schwarzenberg, pp. 1293–1432.
DeFina AV, Kennedy MC (1983) The cochlear nuclei in colubrid and boid snakes: a qualitative and quantitative study. J Morphol 178:285–301.
DeFina AV, Webster DB (1974) Projections of the intraotic ganglion to the medullary nuclei in the tegu lizard, *Tupinambis nigropunctatus*. Brain Behav Evol 10:197–211.
de Rebaupierre F (1997) Acoustical information processing in the auditory thalamus and cerebral cortex. In: Ehret G, Romand R (eds) The Central Auditory System. New York: Oxford University Press, pp. 317–398.
Doucet JR, Ryugo DK (1997) Projections from the ventral cochlear nucleus to the dorsal cochlear nucleus in rats. J Comp Neurol 385:245–264.
Doupe AJ, Kuhl PK (1999) Birdsong and human speech: common themes and mechanisms. Annu Rev Neurosci 22:567–631.
Echteler SM (1985) Organization of central auditory pathways in a teleost fish, Cyprinus carpio. J Comp Physiol 156:267–280.
Edds-Walton PL (1998) Projections of primary afferents from regions of the saccule in toadfish (*Opsanus tau*). Hear Res 115:45–60.
Edds-Walton PL, Fay RR, Highstein SM (1999) Dendritic arbors and central projections of physiologically characterized auditory fibers from the saccule of the toadfish, *Opsanus tau*. J Comp Neurol 411:212–238.
Edwards CJ, Kelley DB (2001) Auditory and lateral line inputs to the midbrain of an aquatic anuran: neuroanatomic studies in Xenopus laevis. J Comp Neurol 438:148–162.
Edwards CJ, Alder TB, Rose GJ (2002) Auditory midbrain neurons that count. Nat Neurosci 5:934–936.
Ehrlich D, Casseday JH, Covey E (1997) Neural tuning to sound duration in the inferior colliculus of the big brown bat, Eptesicus fuscus. J Neurophysiol 77:2360–2372.
Farinas I, Jones KR, Tessarollo L, Vigers AJ, Huang E, Kirstein M, de Caprona DC, Coppola V, Backus C, Reichardt LF, Fritzsch B (2001) Spatial shaping of cochlear innervation by temporally regulated neurotrophin expression. J Neurosci 21:6170–6180.
Faure PA, Fremouw T, Casseday JH, Covey E (2003) Temporal masking reveals prop-

erties of sound-evoked inhibition in duration-tuned neurons of the inferior colliculus. J Neurosci 23:3052–3065.

Fekete DM, Wu DK (2002) Revisiting cell fate specification in the inner ear. Curr Opin Neurobiol 12:35–42.

Fekete DM, Rouiller EM, Liberman MC, Ryugo DK (1982) The central projections of intracellularly labeled auditory nerve fibers in cats. J Comp Neurol 229:432–450.

Feng AS (1983) Morphology of neurons in the torus semicircularis of the northern leopard frog, Rana pipiens pipiens. J Morphol 175:253–269.

Feng AS, Capranica RR (1976) Sound localization in anurans. I. Evidence of binaural interaction in dorsal medullary nucleus of bullfrogs (Rana catesbeiana). J Neurophysiol 39:871–881.

Feng AS, Capranica RR (1978) Sound localization in anurans. II. Binaural interaction in superior olivary nucleus of the green tree frog (Hyla cinerea). J Neurophysiol 41:43–54.

Feng AS, Lin WY (1991) Differential innervation patterns of three divisions of frog auditory midbrain (torus semicircularis). J Comp Neurol 306:613–630.

Feng AS, Lin WY (1996) Neuronal architecture of the dorsal nucleus (cochlear nucleus) of the frog, *Rana pipiens pipiens*. J Comp Neurol 366:320–334.

Feng AS, Schellart NAM (1999) Central auditory processing in fish and amphibians. In: Fay RR, Popper AN (eds) Comparative Hearing: Fish and Amphibians. New York: Springer-Verlag, pp. 218–268.

Feng AS, Gerhardt HC, Capranica RR (1976) Sound localization behavior of green treefrog (Hyla-Cinerea) and barking treefrog (Hyla-Gratiosa). J Comp Physiol 107:241–252.

Feng AS, Simmons JA, Kick SA (1978) Echo detection and target-ranging neurons in the auditory system of the bat Eptesicus fuscus. Science 202:645–648.

Fitzpatrick DC, Kuwada S, Batra R (2000) Neural sensitivity to interaural time differences: beyond the Jeffress model. J Neurosci 20:1605–1615.

Fritzsch B (1988a) The amphibian octavo-lateralis system and its regressive and progressive evolution. Acta Biol Hung 39:305–322.

Fritzsch B (1988b) Phylogenetic and ontogenetic origin of the dorsolateral auditory nucleus of anurans. In: Fritzsch B, Ryan MJ, Wilczynski W, Hetherington TE, Walkowiak W (eds) The Evolution of the Amphibian Auditory System. New York: John Wiley, pp. 561–585.

Fritzsch B (1990a) On the coincidence of loss of electroreception and reorganization of brain stem nuclei. In: Scheich H (ed) The Neocortex: Ontogeny and Phylogeny. New York: Plenum Press, pp. 103–111.

Fritzsch B (1990b) Experimental reorganization in the alar plate of the clawed toad, Xenopus laevis. I. Quantitative and qualitative effects of embryonic otocyst extirpation. Brain Res Dev Brain Res 51:113–122.

Fritzsch B (1999a) Hearing in two worlds: Theoretical and realistic adaptive changes of the aquatic and terrestrial ear for sound reception. In: Fay RR, Popper AN (eds) Comparative Hearing: Fish and Amphibians. New York: Springer-Verlag, pp. 15–42.

Fritzsch B (1999b) Ontogenetic and evolutionary evidence for the motoneuron nature of vestibular and cochlear efferents. In: Berlin CI (ed) The Efferent Auditory System: Basic Science and Clinical Applications. San Diego: Singular Publishing, pp. 31–59.

Fritzsch B (2002) Evolution of the ancestral vertebrate brain. In: Arbib MA (ed) The Handbook of Brain Theory and Neural Networks. Cambridge: MIT Press, pp. 373–376.

Fritzsch B (2003) Development of inner ear afferent connections: forming primary neurons and connecting them to the developing sensory epithelia. Brain Res Bull 60: 423–433.

Fritzsch B, Beisel KW (2001) Evolution and development of the vertebrate ear. Brain Res Bull 55:711–721.

Fritzsch B, Beisel KW (2003) Molecular conservation and novelties in vertebrate ear development. Cur Top Develop Biol 57:1–44.

Fritzsch B, Signore M, Simeone A (2001) Otx1 null mutant mice show partial segregation of sensory epithelia comparable to lamprey ears. Dev Genes Evol 211:388–396.

Fritzsch B, Beisel KW, Jones K, Farinas I, Maklad A, Lee J, Reichardt LF (2002) Development and evolution of inner ear sensory epithelia and their innervation. J Neurobiol 53:143–156.

Fujino K, Oertel D (2001) Cholinergic modulation of stellate cells in the mammalian ventral cochlear nucleus. J Neurosci 21:7372–7383.

Funabiki K, Koyano K, Ohmori H (1998) The role of GABAergic inputs for coincidence detection in the neurones of nucleus laminaris of the chick. J Physiol 508:851–869.

Fuzessery ZM (1988) Frequency tuning in the anuran central auditory system. In: Fritzsch B, Wilczynski W, Ryan MJ, Hetherington TE, Walkowiak W (eds) The Evolution of the Amphibian Auditory System. New York: Wiley, pp. 253–273.

Fuzessery ZM, Hall JC (1999) Sound duration selectivity in the pallid bat inferior colliculus. Hear Res 137:137–154.

Glatt AF (1975a) Vergleichend morphologische Untersuchungen am akustischen System einiger ausgewählter Reptilien. A: *Caiman crocodilus*. Rev Suisse Zool 82:257–281.

Glatt AF (1975b) Vergleichend morphologische Untersuchungen am akustischen System einiger ausgewählter Reptilien. B: Sauria, Testudines. Rev Suisse Zool 82:469–494.

Gold JI, Knudsen EI (2000) A site of auditory experience-dependent plasticity in the neural representation of auditory space in the barn owl's inferior colliculus. J Neurosci 20:3469–3486.

Goodson JL, Bass AH (2002) Vocal-acoustic circuitry and descending vocal pathways in teleost fish: convergence with terrestrial vertebrates reveals conserved traits. J Comp Neurol 448:298–322.

Gooler DM, Feng AS (1992) Temporal coding in the frog auditory midbrain: the influence of duration and rise-fall time on the processing of complex amplitude-modulated stimuli. J Neurophysiol 67:1–22.

Gowan K, Helms AW, Hunsaker TL, Collisson T, Ebert PJ, Odom R, Johnson JE (2001) Crossinhibitory activities of Ngn1 and Math1 allow specification of distinct dorsal interneurons. Neuron 31:219–232.

Grothe B (2000) The evolution of temporal processing in the medial superior olive, an auditory brainstem structure. Prog Neurobiol 61:581–610.

Grothe B (2003) Sensory systems: New roles for synaptic inhibition in sound localization. Nat Rev Neurosci 4:540–550.

Grothe B, Neuweiler G (2000) The function of the medial superior olive in small mammals: temporal receptive fields in auditory analysis. J Comp Physiol [A] 186:413–423.

Grothe B, Sanes DH (1993) Bilateral inhibition by glycinergic afferents in the medial superior olive. J Neurophysiol 69:1192–1196.

Grothe B, Schweizer H, Pollak GD, Schuller G, Rosemann C (1994) Anatomy and projection patterns of the superior olivary complex in the Mexican free-tailed bat, Tadarida brasiliensis mexicana. J Comp Neurol 343:630–646.

Grothe B, Covey E, Casseday JH (2001) The medial superior olive in the big brown bat: neuronal response to pure tones, amplitude modulations, and pulse trains. J Neurophysiol 86:2219–2230.

Hall JC (1999) GABAergic inhibition shapes frequency tuning and modifies response properties in the auditory midbrain of the leopard frog. J Comp Physiol [A] 85:479–491.

Hall JC, Feng AS (1990) Classification of the temporal discharge patterns of single auditory neurons in the dorsal medullary nucleus of the northern leopard frog. J Neurophysiol 64:1460–1473.

Haplea S, Covey E, Casseday JH (1994) Frequency tuning and response latencies at three levels in the brainstem of the echolocating bat, Eptesicus fuscus. J Comp Physiol 174:671–683.

Harrison JM, Irving R (1966) Visual and nonvisual auditory systems in mammals. Anatomical evidence indicates two kinds of auditory pathways and suggests two kinds of hearing in mammals. Science 154:738–743.

Herrick CJ (1948) The Brain of the Tiger Salamander, Ambystoma tigrinum. Chicago: University of Chicago Press.

Hopson JA (1973) Endothermy, small size, and the origin of mammalian reproduction. Am Nature 107:446–452.

Hyson RL, Overholt EM, Lippe WR (1994) Cochlear microphonic measurements of interaural time differences in the chick. Hear Res 81:109–118.

Hyson RL, Reyes AD, Rubel EW (1995) A depolarizing inhibitory response to GABA in brainstem auditory neurons of the chick. Brain Res 677:117–126.

Imig TJ, Adrian HO (1977) Binaural columns in the primary field (A1) of cat auditory cortex. Brain Res 138:241–257.

Inbody SB, Feng AS (1981) Binaural response characteristics of single neurons in the medial superior olivary nucleus of the albino rat. Brain Res 210:361–366.

Jeffress LA (1948) A place theory of sound localization. J Comp Physiol Psychol 41:35–39.

Ji Q, Luo ZX, Yuan CX, Wible JR, Zhang JP, Georgi JA (2002) The earliest known eutherian mammal. Nature 416:816–822.

Joris PX, Yin TC (1995) Envelope coding in the lateral superior olive. I. sensitivity to interaural time differences. J Neurophysiol 73:1043–1062.

Joseph AW, Hyson RL (1993) Coincidence detection by binaural neurons in the chick brainstem. J Neurophysiol 69:1197–1211.

Kaas JH, Hackett TA, Tramo MJ (1999) Auditory processing in primate cerebral cortex. Curr Opin Neurobiol 9:164–170.

Kapfer C, Seidl AH, Schweizer H, Grothe B (2002) Experience-dependent refinement of inhibitory inputs to auditory coincidence-detector neurons. Nat Neurosci 5:247–253.

Karis A, Pata I, van Doorninck JH, Grosveld F, de Zeeuw CI, de Caprona D, Fritzsch B (2001) Transcription factor GATA-3 alters pathway selection of olivocochlear neurons and affects morphogenesis of the ear. J Comp Neurol 429:615–630.

Karten HJ, Shimizu T (1989) The origins of neocortex: connections and lamination as distinct events in evolution. J Cognit Neurosci 1:291–301.

Kelly JB, Judge PW (1994) Binaural organization of primary auditory cortex in the ferret (Mustela putorius). J Neurophysiol 71:904–913.

Kermack KA, Mussett F, Rigney HW (1981) The skull of Morganucodon. Zool J Linn Soc 71:1–185.

Klug A, Khan A, Burger RM, Bauer EE, Hurley LM, Yang L, Grothe B, Halvorsen MB,

Park TJ (2000) Latency as a function of intensity in auditory neurons: influences of central processing. Hear Res 148:107–123.

Klump GM (2000) Sound localization in birds. In: Dooling RJ, Fay RR, Popper AN (eds) Comparative Hearing: Birds and Reptiles. New York: Springer-Verlag, pp. 249–307.

Knudsen EI (1977) Distinct auditory and lateral line nuclei in the midbrain catfishes. J Comp Neurol 173:417–431.

Knudsen EI (1983) Subdivisions of the Inferior Collieulus in the Barn Owl (*Tyto alba*). J Comp Neurol 218:174–186.

Knudsen EI, Knudsen PF (1983) Space-mapped auditory projections from the inferior colliculus to the optic tectum in the barn owl (Tyto alba). J Comp Neurol 218:187–196.

Knudsen EI, Konishi M (1978) A neural map of auditory space in the owl. Science 200: 795–797.

Konishi M (1985) Birdsong: from behavior to neuron. Annu Rev Neurosci 8:125–170.

Köppl C (1994) Auditory nerve terminals in the cochlear nucleus magnocellularis: differences between low and high frequencies. J Comp Neurol 339:438–446.

Köppl C (1997) Phase locking to high frequencies in the auditory nerve and cochlear nucleus magnocellularis of the barn owl, Tyto alba. J Neurosci 17:3312–3321.

Köppl C, Carr CE (1997) Low-frequency pathway in the barn owl's auditory brainstem. J Comp Neurol 378:265–282.

Köppl C, Carr CE (2003) Computational diversity in the cochlear nucleus angularis of the barn owl. J Neurophysiol 89:2313–2329.

Köppl C, Manley GA (1992) Functional consequences of morphological trends in the evolution of lizard hearing organs. In: Webster DB, Fay RR, Popper AN (eds) The Evolutionary Biology of Hearing. New York: Springer-Verlag, pp. 489–510.

Kössl M, Vater M (1985) The cochlear frequency map of the mustached bat, Pterouotus parnellii. J Coup Physiol [A] 157:687–987.

Kozloski J, Crawford JD (1998) Functional neuroanatomy of auditory pathways in the sound-producing fish Pollimyrus. J Comp Neurol 401:227–252.

Krubitzer L, Künzle H, Kaas J (1997) Organization of sensory cortex in a Madagascan insectivore, the tenrec (Echinops telfairi). J Comp Neurol 379:399–414.

Kubke MF, Carr CE, Dooling RJ (2002) Organization of nucleus laminaris in different species of birds. Assoc Res Otollaryngol Abstr 426.

Kuhn GF (1977) Model for interaural time differences in azimuthal plane. J Acoust Soc Am 62:157–167.

Künzle H (1986) Projections from the cochlear nuclear complex to rhombencephalic auditory centers and torus semicircularis in the turtle. Brain Res 379:307–319.

Künzle H, Woodson W (1982) Mesodiencephalic and other target regions of ascending spinal projections in the turtle, Pseudemys scripta elegans. J Comp Neurol 212:349–364.

Lachica EA, Rubsamen R, Rubel EW (1994) GABAergic terminals in nucleus magnocellularis and laminaris originate from the superior olivary nucleus. J Comp Neurol 348:403–418.

Lanuza E, Martinez-Marcos A, Martinez-Garcia F (1999) What is the amygdala? A comparative approach. Trends Neurosci 22:207–208.

Larsell O (1967) The Comparative Anatomy and Histology of the Cerebellum from Myxinoids Through Birds. Hansen J (ed). Minneapolis: University of Minnesota Press, p. 291.

Leake PA, Snyder RL, Hradek GT (2002) Postnatal refinement of auditory nerve projections to the cochlear nucleus in cats. J Comp Neurol 448:6–27.

Lewis ER, Narins PM (1999) The acoustic periphery of amphibians: anatomy and physiology. In: Fay RR, Popper AN (eds) Comparative Hearing: Fish and Amphibians. New York: Springer-Verlag, pp. 101–154.

Liberman MC (1991) Central projections of auditory-nerve fibers of differing spontaneous rate. I. Anteroventral cochlear nucleus. J Comp Neurol 313:240–258.

Liem KF, Bemis WE, Walker WF, Grande L (2001) Functional Anatomy of the Vertebrates—An Evolutionary Perspective. Orlando, FL: Hartcourt College.

Lowe CJ, Wu M, Salic A, Evans L, Lander H, Stange-Thomann N, Gruber CE, Gerhart J, Kirschner M (2003) Anteroposterior patterning in hemichordates and the origins of the chordate nervous system. Cell 113:853–865.

Lu Z, Fay RR (1966) Tow-tone interaction in primary afferents and midbrain neurons of the goldfish, *Carassius Auratus*. Audit Neurosci 2:257–273.

Lu Z, Fay RR (1995) Acoustic response properties of single neurons in the central posterior nucleus of the thalamus of the goldfish, Carassius auratus. J Comp Physiol [A] 176:747–760.

Luksch H, Walkowiak W (1998) Morphology and axonal projection patterns of auditory neurons in the midbrain of the painted frog, Discoglossus pictus. Hear Res 122:1–17.

Ma Q, Anderson DJ, Fritzsch B (2000) Neurogenin 1 null mutant ears develop fewer, morphologically normal hair cells in smaller sensory epithelia devoid of innervation. J Assoc Res Otolaryngol 1:129–143.

Maklad A, Fritzsch B (2003) Development of inner ear projections and their central targets. Brain Res Bull 60:407–510.

Manley GA (1981) A review of the auditory physiology of the reptiles. In: Ottoson D (ed) Progress in Sensory Physiology. Berlin, Heidelberg, New York: Springer-Verlag, pp. 49–134.

Manley GA, Koppl C, Konishi M (1988) A neural map of interaural intensity differences in the brain stem of the barn owl. J Neurosci 8:2665–2676.

Manley JA (1971) Single unit studies in the midbrain auditory area of caiman. Z Vergl Physiol 71:255–261.

Manteuffel G, Naujoks-Manteuffel C (1990) Anatomical connections and electrophysiological properties of toral and dorsal tegmental neurons in the terrestrial urodele *Salamandra salamandra*. J Hirnforsch 31:65–76.

Margoliash D, Fortune ES (1992) Temporal and harmonic sombination-sensitive neurons in the zebra finch's HVc. J Neurosci 12:4309–4326.

Masino T, Knudsen EI (1992) Anatomical pathways from the optic tectum to the spinal cord subserving orienting movements in the barn owl. Exp Brain Res 92:194–208.

Masterton RB, Diamond IT (1967) Medial superior olive and sound localization. Science 155:1696–1697.

May BJ (2000) Role of the dorsal cochlear nucleus in the sound localization behavior of cats. Hear Res 148:74–87.

McAlpine D, Grothe B (2003) Sound localization and delay lines—do mammals fit the model? Trends Neurosci 26:347–350.

McCormick CA (1999) Anatomy of the central auditory pathways of fish and amphibians. In: Fay RR, Popper AN, Fay RR, Popper AN (eds) Comparative Hearing: Fish and Amphibians. New York: Springer-Verlag, pp. 155–217.

McCormick CA (2002) Variations on a vertebrate theme: central anatomy of the auditory system in fish. Bioacoust 12:134–137.

McKibben JR, Bass AH (1999) Peripheral encoding of behaviorally relevant acoustic signals in a vocal fish: single tones. J Comp Physiol [A] 184:563–576.

Meininger V, Pol D, Derer P (1986) The inferior colliculus of the mouse. A Nissl and Golgi study. Neurosci 17:1159–1179.

Mello CV, Vates GE, Okuhata S, Nottebohm F (1998) Descending auditory pathways in the adult male ebra finch (Taeniopygia guttata). J Comp Neurol 395:137–160.

Merzenich MM, Schreiner CE (1992) Mammalian auditory cortex: some comparative observations. In: Webster DB, Day RR, Popper AN (eds) The Evolutionary Biology of Hearing. New York: Springer-Verlag, pp. 673–690.

Metzner W (1996) Anatomical basis for audio-vocal integration in echolocating horseshoe bats. J Comp Neurol 368:252–269.

Middlebrooks JC, Dykes RW, Merzenich MM (1980) Binaural response-specific bands in primary auditory cortex (AI) of the cat: topographical organization orthogonal to isofrequency contours. Brain Res 181:31–48.

Middlebrooks JC, Xu L, Furukawa S, Macpherson EA (2002) Cortical neurons that localize sounds. Neuroscientist 8:73–83.

Miller MR (1975) The cochlear nuclei of lizards. J Comp Neurol 159:375–405.

Miller MR (1980) The cochlear nuclei of snakes. J Comp Neurol 192:717–736.

Miller MR, Kasahara M (1979) The cochlear nuclei of some turtles. J Comp Neurol 185:221–235.

Mittmann DH, Wenstrup JJ (1995) Combination-sensitive neurons in the inferior colliculus. Hear Res 90:185–191.

Moiseff A (1989) Bi-coordinate sound localization by the barn owl. J Comp Physiol [A] 164:637–644.

Monsivais P, Rubel EW (2001) Accommodation enhances depolarizing inhibition in central neurons. J Neurosci 21:7823–7830.

Monsivais P, Yang L, Rubel EW (2000) GABAergic inhibition in nucleus magnocellularis: implications for phase locking in the avian auditory brainstem. J Neurosci 20:2954–2963.

Morest DK (1968) The growth of synaptic endings in the mammalian brain: a study of the calyces of the trapezoid body. Z Anat Entwicklungsgesch 127:201–220.

Moss CF, Schnitzler H-U (1995) Behavioral studies of auditory information processing. In: Fay RR, Popper AN (eds) Springer Handbook of Auditory Research: Hearing by Bats. Berline, Heidelberg, New York: Springer-Verlag, pp. 87–145.

Narins PM, Capranica RR (1980) Neural adaptations for processing the two-note call of the Puerto Rican treefrog, *Eleutherodactylus coqui*. Brain Behav Evol 17:48–66.

Norberg RA (1978) Occurrence and independent evolution of bilateral ear asymmetry in owls and implications on owl taxonomy. Philos Trans R Soc Lond [B] Biol Sci 280:375–408.

Northcutt RG (1981) Evolution of the telencephalon in nonmammals. Annu Rev Neurosci 4:301–350.

Northcutt RG (1990) Ontogeny and phylogeny: a re-evaluation of conceptual relationships and some applications. Brain Behav Evol 36:116–140.

Northcutt RG (1996) The Agnathan ark: the origin of craniate brains. Brain Behav Evol 48:237–247.

Northcutt RG, Kaas JH (1995) The emergence and evolution of mammalian neocortex. Trends Neurosci 18:373–379.

Oertel D, Bal R, Gardner SM, Smith PH, Joris PX (2000) Detection of synchrony in the activity of auditory nerve fibers by octopus cells of the mammalian cochlear nucleus. Proc Natl Acad Sci USA 97:11773–11779.

Oliver DL, Huerta MF (1992) Inferior and superior colliculi. In: Webster DB, Popper AN, Fay RR (eds) The Mammalian Auditory Pathway: Neuroanatomy. New York: Springer-Verlag, pp. 168–221.

Overholt E, Rubel EW, Hyson RL (1992) A circuit for coding interaural time differences in the chick brainstem. J Neurosci 12:1698–1708.

Owen R (1848) On the archetype and homologies of vertebrate skeleton. Assoc Adv Sci 16:169–340.

Parks TN, Rubel EW (1975) Organization and development of brain stem auditory nuclei of the chicken: organization of projections from n. magnocellularis to n. laminaris. J Comp Neurol 164:435–448.

Pauley S, Wright TJ, Pirvola U, Ornitz D, Beisel K, Fritzsch B (2003) Expression and function of FGF10 in mammalian inner ear development. Dev Dyn 227:203–215.

Pena JL, Konishi M (2001) Auditory spatial receptive fields created by multiplication. Science 292:249–252.

Pena JL, Viete S, Albeck Y, Konishi M (1996) Tolerance to sound intensity of binaural coincidence detection in the nucleus laminaris of the owl. J Neurosci 16:7046–7054.

Pichaud F, Desplan C (2002) Pax genes and eye organogenesis. Curr Opin Genet Dev 12:430–434.

Pollak GD, Burger RM, Klug A (2003) Dissecting the circuitry of the auditory system. Trends Neurosci 26:33–39.

Popper AN, Fay RR (1999) The acoustic periphery in fishes. In: Fay RR, Popper AN (eds) Comparative Hearing: Fish and Amphibians. New York: Spriner-Verlag, pp. 43–100.

Portfors CV, Wenstrup JJ (1999) Delay-tuned neurons in the inferior colliculus of the mustached bat: implications for analyses of target distance. J Neurophysiol 82:1326–1338.

Potter DH (1965) Mesencephalic auditory region of the bullfrog. J Neurophysiol 28:1132–1154.

Pritz MB (1974) Ascending connections of a thalamic auditory area in a crocodile, Caiman crocodilus. J Comp Neurol 153:199–214.

Pritz MB, Stritzel ME (1989) Reptilian somatosensory midbrain: identification based on input from the spinal cord and dorsal column nucleus. Brain Behav Evol 33:1–14.

Pritz MB, Stritzel ME (1994) Anatomical identification of a telencephalic somatosensory area in a reptile, Caiman crocodilus. Brain Behav Evol 43:107–127.

Puelles L, Kuwana E, Puelles E, Rubenstein JL (1999) Comparison of the mammalian and avian telencephalon from the perspective of gene expression data. Eur J Morphol 37:139–150.

Qian Y, Fritzsch B, Shirasawa S, Chen CL, Choi Y, Ma Q (2001) Formation of brainstem (nor)adrenergic centers and first-order relay visceral sensory neurons is dependent on homeodomain protein Rnx/Tlx3. Genes Dev 15:2533–2545.

Ramon y Cajal S (1907) Histologie du Systeme Nerveux de l'Homme et des Vertebrates. Paris: Malonie.

Rauschecker JP, Tian B (2000) Mechanisms and streams for processing of "what" and "where" in auditory cortex. Proc Natl Acad Sci USA 97:11800–11806.

Rayleigh, Lord Strutt JW 3rd, Baron Rayleigh (1907) On our perception of sound direction. Philos Mag 13:214–232.

Reichert C (1837) Ueber die Visceralbogen der Wierbelthiere im Allgemeinen und deren Metamorphose bei den Saeugethieren und Voegeln. Arch Anat Phys Med 120–222.

Reyes AD, Rubel EW, Spain WJ (1996) In vitro analysis of optimal stimuli for phase-locking and time-delayed modulation of firing in avian nucleus laminaris neurons. J Neurosci 16:993–1007.

Rheinländer J, Gerhardt HC, Yager DD, Capranica RR (1979) Accuracy of phono taxis by the green tree frog (Hyla cinerea). J Comp Physiol 133:247–256.

Rheinländer J, Walkowiak W, Gerhardt HC (1981) Directional hearing in the green tree frog: a variable mechanism? Naturwissenschaften 68:430–431.

Rhode WS, Greenberg S (1992) Physiology of the cochlear nuclei. In: Popper AN, Fay RR (eds) The Mammalian Auditory Pathway: Neurophysiology. New York: Springer-Verlag, pp. 94–152.

Roberts BL, Meredith GE (1992). The efferent innervation of the ear: variations on an enigma. In: Webster DB, Popper AN, Fay RR (eds) The Evolutionary Biology of Hearing. New York: Springer-Verlag, pp. 182–210.

Rosowski JJ, Saunders JC (1980) Sound transmission through the avian interaural pathways. J Comp Physiol [A] 136:183–190.

Rubel EW, Fritzsch B (2002) Auditory system development: primary auditory neurons and their targets. Annu Rev Neurosci 25:51–101.

Rubel EW, Parks TN (1975) Organization and development of brain stem auditory nuclei of the chicken: tonotopic organization of N. magnocellularis and N. laminaris. J Comp Neurol 164:411–433.

Ryugo DK, Parks TN (2003) Primary innervation of the avian and mammalian cochlear nucleus. Brain Res Bull 60:435–456.

Ryugo DK, Rouiller EM (1988) Central projections of intracellularly labeled auditory nerve fibers in cats: morphometric correlations with physiological properties. J Comp Neurol 271:130–142.

Sachs MB, Sinnott JM (1978) Responses to tones of single cells in Nucleus magnocellularis and nucleus angularis of the redwing blackbird (*Agelaius phoeniceus*). J Comp Physiol 126:347–361.

Scheich H (1990) Representational geometries of telencephalic auditory maps in birds and mammals. In: Scheich H (ed) The Neocortex—Ontogeny and Phylogeny. New York: Plenum, pp. 119–135.

Scheich H, Langner G, Koch R (1977) Coding of narrow-band and wide-band vocalizations in the auditory midbrain nucleus (MLD) of the guinea fowl (*Numida meleagris*). J Comp Physiol 117:245–265.

Scheich H, Langner G, Bonke D (1979) Responsiveness of units in the auditory neostriatum of the guinea fowl (numida meleagris) to species specific calls and synthetic stimuli. II. Discrimination of iambus-like calls. J Comp Physiol 132:257–276.

Schwartz IR (1992) The superior olivary complex and lateral lemniscal nuclei. In: Webster DB, Popper AN, Fay RR (eds) The Mammalian Auditory Pathway: Neuroanatomy. New York: Springer-Verlag, pp. 117–167.

Shofner WP, Dye RH Jr (1989) Statistical and receiver operating characteristic analysis of empirical spike-count distributions: quantifying the ability of cochlear nucleus units to signal intensity changes. J Acoustl Soc Am 86:2172–2184.

Smith PH, Spirou GA (2002) From the cochlea to the cortex and back. In: Oertel D, Fay RR, Popper AN (eds) Integrative Functions in the Mammalian Auditory Pathway. New York: Springer-Verlag, pp. 6–71.

Smith PH, Joris PX, Yin TC (1993) Projections of physiologically characterized spherical bushy cell axons from the cochlear nucleus of the cat: evidence for delay lines to the medial superior olive. J Comp Neurol 331:245–260.

Smith PH, Joris PX, Yin TC (1998) Anatomy and physiology of principal cells of the medial nucleus of the trapezoid body (MNTB) of the cat. J Neurophysiol 79:3127–3142.

Soares D (2002) Diversity and common themes in the archosaur auditory brainstem. Ph.D. thesis, Dept. of Biology, University of Maryland, College Park, Maryland.

Soares D, Carr CE (2001) The cytoarchitecture of the nucleus angularis of the barn owl (*Tyto alba*). J Comp Neurol 429:192–205.

Soares D, Chitwood RA, Hyson RL, Carr CE (2002) The intrinsic neuronal properties of the chick nucleus angularis. J Neurophysiol 88:152–162.

Spirou GA, Berrebi AS (1996) Organization of ventrolateral periolivary cells of the cat superior olive as revealed by pep-19 immunocytochemistry and nissl stain. J Comp Neurol 368:100–120.

Striedter GF (1991) Auditory, electrosensory, and mechanosensory lateral line pathways through the forebrain in channel catfishes. J Comp Neurol 312:311–331.

Striedter GF (1994) The vocal control pathways in budgerigars differ from those in songbirds. J Comp Neurol 343:35–56.

Striedter GF (1997) The telencephalon of tetrapods in evolution. Brain Behav Evol 49: 179–213.

Striedter GF, Keefer BP (2000) Cell migration and aggregation in the developing telencephalon: pulse-labeling chick embryos with bromodeoxyuridine. J Neurosci 20: 8021–8030.

Suga N (1988) Parallel-heirarchical processing of biosonar information in the mustached bat. In: Nachtigall P, Moore P (eds) Animal Sonar. New York: Plenum, pp. 149–159.

Suga N, O'Neill WE, Manabe T (1978) Cortical neurons sensitive to combinations of information-bearing elements of biosonar signals in the mustache bat. Science 200: 778–781.

Suga N, Gao E, Zhang Y, Ma X, Olsen JF (2000) The corticofugal system for hearing: recent progress. Proc Natl Acad Sci USA 97:11807–11814.

Sullivan WE (1985) Classification of response patterns in cochlear nucleus of barn owl: correlation with functional response properties. J Neurophysiol 53:201–216.

Szpir MR, Sento S, Ryugo DR (1990) Central projections of cochlear nerve fibers in the alligator lizard. J Comp Neurol 295:530–547.

Szpir MR, Wright DD, Ryugo DK (1995) Neuronal organization of the cochlear nuclei in alligator lizards: a light and electron microscopic investigation. J Comp Neurol 357:217–241.

Takahashi TT, Keller CH (1992) Commissural connections mediate inhibition for the computation of interaural level differences in the barn owl. J Comp Physiol [A] 170: 161–169.

Takahashi TT, Konishi M (1988a) Projections of the cochlear nuclei and nucleus laminaris to the inferior colliculus of the barn owl. J Comp Neurol 274:190–211.

Takahashi TT, Konishi M (1988b) Projections of nucleus angularis and nucleus laminaris to the lateral lemniscal nuclear complex of the barn owl. J Comp Neurol 274:212–238.

Takahashi Y, Konishi M (2002) Manipulation of inhibition in the owl's nucleus laminaris and its effects on optic tectum neurons. Neuroscience 111:373–378.

Takahashi TT, Barberini CL, Keller CH (1995) An anatomical substrate for the inhibitory gradient in the VLVp of the owl. J Comp Neurol 358:294–304.

Ten Donkelaar HJ (1998) Anurans. In: Nieuwenhuys R, Donkelaar HJT, Nicholson C (eds) The Central Nervous System of Vertebrates. New York: Springer-Verlag, pp. 1151–1314.

Thompson SP (1882) On the function of the two ears in the perception of space. Philos Mag 13:406–416.

Trussell LO (2002) Cellular mechanisms for information coding in auditory brainstem nuclei. In: Oertel D, Fay RR, Popper AN (eds) Integrative Functions in the Mammalian Auditory Pathway. New York: Springer-Verlag, pp. 72–98.

Venter JC, Adams MD, Myers EW, Li PW, Mural RJ, Sutton GG, Smith HO, Yandell M, Evans CA, Holt RA, Gocayne JD, Amanatides P, Ballew RM, Huson DH, Wortman JR, Zhang Q, Kodira CD, Zheng XH, Chen L, Skupski M, Subramanian G, Thomas PD, Zhang J, Gabor Miklos GL, Nelson C, Broder S, Clark AG, Nadeau J, McKusick VA, Zinder N, Levine AJ, Roberts RJ, Simon M, Slayman C, Hunkapiller M, Bolanos R, Delcher A, Dew I, Fasulo D, Flanigan M, Florea L, Halpern A, Hannenhalli S, Kravitz S, Levy S, Mobarry C, Reinert K, Remington K, Abu-Threideh J, Beasley E, Biddick K, Bonazzi V, Brandon R, Cargill M, Chandramouliswaran I, Charlab R, Chaturvedi K, Deng Z, Di Francesco V, Dunn P, Eilbeck K, Evangelista C, Gabrielian AE, Gan W, Ge W, Gong F, Gu Z, Guan P, Heiman TJ, Higgins ME, Ji RR, Ke Z, Ketchum KA, Lai Z, Lei Y, Li Z, Li J, Liang Y, Lin X, Lu F, Merkulov GV, Milshina N, Moore HM, Naik AK, Narayan VA, Neelam B, Nusskern D, Rusch DB, Salzberg S, Shao W, Shue B, Sun J, Wang Z, Wang A, Wang X, Wang J, Wei M, Wides R, Xiao C, Yan C et al. (2001) The sequence of the human genome. Science 291:1304–1351.

Viemeister NF, Plack CJ (1993) Time analysis. In: Yost WA, Popper AN, Fay RR (eds) Human Psychophysics. New York: Springer-Verlag, pp. 116–154.

Volman SF, Konishi M (1990) Comparative physiology of sound localization in 4 species of owls. Brain Behav Evol 36:196–215.

von Gersdorff H, Borst JG (2002) Short-term plasticity at the calyx of held. Nat Rev Neurosci 3:53–64.

von Melchner L, Pallas SL, Sur M (2000) Visual behaviour mediated by retinal projections directed to the auditory pathway. Nature 404:871–876.

Wang VY, Hassan BA, Bellen HJ, Zoghbi HY (2002) Drosophila atonal fully rescues the phenotype of Math1 null mice: new functions evolve in new cellular contexts. Curr Biol 12:1611–1616.

Wang X, Kadia SC (2001) Differential representation of species-specific primate vocalizations in the auditory cortices of marmoset and cat. J Neurophysiol 86:2616–2620.

Warchol ME, Dallos P (1990) Neural coding in the chick cochlear nucleus. J Comp Physiol [A] 166:721–734.

Warr WB (1966) Fiber degeneration following lesions in the anterior ventral cochlear nucleus of the cat. Exp Neurol 14:453–474.

Wenthold RJ, Huie D, Altschuler RA, Reeks KA (1987) Glycine immunoreactivity localized in the cochlear nucleus and superior olivary complex. Neurosci 22:897–912.

Wilczynski W (1984) Central neural systems subserving a homoplasous periphery. Am Zool 24:755–763.

Wilczynski W (1988) Brainstem auditory pathways in anurans. In: Fritzsch B, Wilczynski W, Ryan MJ, Hetherington TE, Walkowiak W (eds) The Evolution of the Amphibian Auditory System. New York: Wiley, pp. 209–231.

Wilczynski W, Allison JD, Marler CA (1993) Sensory pathways linking social and environmental cues to endocrine control regions of amphibian forebrains. Brain Behav Evol 42:252–264.

Wild JM (1993) Descending projections of the songbird nucleus robustus archistriatalis. J Comp Neurol 338:225–241.
Wild JM (1994) The auditory-vocal-respiratory axis in birds. Behav Evol Brain 44:192–209.
Wild JM (1997) Neural pathways for the control of birdsong production. J Neurobiol 33:653–670.
Wild JM, Karten HJ, Frost BJ (1993) Connections of the auditory forebrain in the pigeon (Columba livia). J Comp Neurol 337:32–62.
Winer JA (1992) The functional architecture of the medial geniculate body and the primary auditory cortex. In: Popper AN, Fay RR, Webster DB (eds) The Mammalian Auditory Pathway: Neuroanatomy. New York: Springer-Verlag, pp. 222–409.
Winter P, Schwartzkopff J (1961) Form und Zellzahl der akustischen Nervenzentren in der Medulla oblongata von Eulen (Striges). Experientia 17:515–518.
Woodworth RS (1962) Experimental Psychology. New York: Rinehart and Winston.
Woolley SMN, Casseday JH (2001a) Tuning properties of auditory midbrain neurons in adult male zebra finches. Soc Neurosci Abstr 17:724.
Woolley SMN, Casseday JH (2001b) Acoustic tuning to songs and calls in songbird auditory midbrain neurons. Int Soc Neuroethol abstr 144.
Wright TJ, Mansour SL (2003) Fgf3 and Fgf10 are required for mouse otic placode induction. Development 130:3379–3390.
Wu SH (1999) Synaptic excitation in the dorsal nucleus of the lateral lemniscus. Prog Neurobiol 57:357–375.
Xiang M, Maklad A, Pirvola U, Fritzsch B (2003) Brn3c null mutant mice show long-term, incomplete retention of some afferent inner ear innervation. BMC Neurosci 4:2.
Yang L, Monsivais P, Rubel EW (1999) The superior olivary nucleus and its influence on nucleus laminaris: a source of inhibitory feedback for coincidence detection in the avian auditory brainstem. J Neurosci 19:2313–2325.
Yin TC (2002) Neural mechanisms of encoding binaural localization cues. In: Oertel D, Fay RR, Popper AN (eds) Integrative Functions in the Mammalian Auditory Pathway. New York: Springer-Verlag, pp. 99–159.
Young ED, Davis KA (2002) Circuitry and function of the dorsal cochlear nucleus. In: Oertel D, Fay RR, Popper AN (eds) Integrative Functions in the Mammalian Auditory Pathway. New York: Springer-Verlag, pp. 160–206.
Young ED, Rubel EW (1983) Frequency-specific projections of individual neurons in chick brainstem auditory nuclei. J Neurosci 3:1373–1378.
Zhang S, Oertel D (1993) Giant cells of the dorsal cochlear nucleus of mice: intracellular recordings in slices. J Neurophysiol 69:1398–1408.
Zhang S, Oertel D (1994) Neuronal circuits associated with the output of the dorsal cochlear nucleus through fusiform cells. J Neurophysiol 71:914–930.
Zook JM, Winer JA, Pollak GD, Bodenhamer RD (1985) Topology of the central nucleus of the mustache bat's inferior colliculus: correlation of single unit properties and neuronal architecture. J Comp Neurol 231:530–546.

11
Advances and Perspectives in the Study of the Evolution of the Vertebrate Auditory System

GEOFFREY A. MANLEY

1. Introduction

The study of the evolution of the ear is only possible using two approaches. The first approach, looking at evolutionary changes in the structure of the ear in fossil organisms, is the obvious and most direct. In fact, this turns out to be remarkably difficult, since the ear region is not always well preserved in fossils, and soft tissue components of the ear almost never fossilize. Moreover, since (based on comparisons to modern animals) the bony components of the inner ear provide only a very rough idea of the full structure of the ear, this approach is obviously of limited usefulness. That is why most of the material of this book does not concern fossil organisms. Where useful information is to be had, for example concerning the structure of the middle ear (see Clack and Allin, Chapter 5) or of the dimensions of the basilar membrane in therian mammals (which alone among amniotes is supported by bone, see Vater et al., Chapter 9), it is discussed in this book. Of course, fossil organisms provide the foundation for our general—and in some cases rather detailed—understanding of the evolutionary relationships of the vertebrates (see Manley and Clack, Chapter 1), without which the second approach could not be used.

The second, and most useful, approach is a comparison of the structure and function of the ears of modern organisms (Webster et al. 1992). This approach is possible only in the context of the known phylogeny of the different groups of vertebrates, which allows for interpretation of the observed differences and similarities. This comparative approach can even be extended to comparative studies of the physiology of the ear and associated brain centers, although only with greater uncertainty regarding the conclusions reached. In the absence of alternatives, this latter approach dominates the themes discussed in this book.

This chapter summarizes the main conclusions reached in this book regarding the evolution of the ear and provides a perspective for future studies.

2. The Importance of Evolutionary Studies

It is a sad fact that mainly due to the separation of scientific disciplines, there are still too many scholars of hearing research who are unaware of potentially highly revealing studies that are carried out using groups of vertebrates that the investigators themselves do not deal with. Research reports thus tend to deal with material only within the narrow context of one animal group or even one species. The authors thus neglect the fact that the object they are studying is linked genetically to other organisms and neglect the potentially useful—perhaps even critical—information available from the historical (i.e., evolutionary) perspective. For obvious reasons, this problem is most acute in the area of mammalian and human auditory research. The significance of this omission varies a great deal, but is perhaps most acute either when more global explanations for phenomena are sought or where complex phenomena are explained in the context of a detailed model. It is often the case that, although many phenomena in auditory physiology are almost universal (Fay and Popper 2000; Manley 2000), explanations for the substrate of these phenomena in mammals only take, for example, mammalian cochlear structure into account. This is one area where an improvement is relatively easily possible and would result in a much better foundation for more comprehensive explanations of auditory phenomena.

It is hoped that this book will encourage such a development, especially the use of a bottom-up approach to cochlear modeling (working from the simplest ears to more complex), rather than a top-down approach (first trying to explain all phenomena in terms of mammalian complexities). Particular interest should be given to cases in which there are exceptions to otherwise universal phenomena. These exceptions can provide valuable clues to the underlying mechanisms and the reason(s) for the exceptions. Thus the variety of structures and of physiological characteristics in the ears of vertebrates belonging to different groups should be viewed as natural experiments that can and should be intelligently exploited by researchers.

3. The Establishment of Receptors for Sensing Mechanical Disturbances

One of the characteristics of animals is that of movement. Even the earliest of one-celled organisms were capable of movement, since movement of some kind—even just creating water currents to filter food—is essential for obtaining and ingesting food. It is safe to assume that the accompanying mechanical disturbances of the medium (first in water, later in air) unavoidably created signals of potential use to the organisms itself (sensing of own movements) and to other organisms. Sensory cells that respond to various aspects of mechanical energy (displacement, velocity, or acceleration of the medium) are so universal in animal species that it can be safely assumed that such receptor cells were developed very early in evolution. Thus common ancestry presumably explains why there are important similarities between mechanoreceptors of different animal groups at the structural and molecular-biological and developmental levels (see Coffin et al., Chapter 3). Initially, such cells were placed superficially, probably localized to specific areas of the body surface for detecting water motion. Later, many groups internalized those receptors. Some became associated with dense materials that were specialized for gravity detection (linear acceleration) and some became sensitive to body or head rotation (radial acceleration). In this respect, there is a remarkable parallel evolution of what we term *vestibular systems* in vertebrates and cephalopod mollusks (octopuses, squid, and cuttlefish).

We can define the beginning of hearing as the ability to detect oscillating mechanical pressure and pressure gradient disturbances in the medium, whose energy penetrated the body directly from the water (see also Webster et al. 1992). Initially, the sounds detected would have been incidental noises associated with the animal's own movement, feeding, and other interactions with the environment but also stimuli originating from other sources. As noted by Lewis and Fay, Chapter 2, such acoustic waves were propagated over large distances when compared to animal body dimensions, and the waves contained information about the location and the nature of the source. The increasing importance of the sense of hearing during the evolution of vertebrates reflects the usefulness of this information to the receiver. The detectors of "auditory" signals developed specializations that made it possible to separate different sources from each other (source segregation), to distinguish them from environmental or self-generated noise, and to extract information about the nature of the sender (species, size, etc.) (see also Lewis and Fay, Chapter 2). A deeper understanding of these environmental constraints and the ecological context of evolutionary developments would further our understanding of the sense of hearing.

In vertebrate animals, the mechanoreceptors responsible for sensing disturbances in water, for detecting head rotation, gravity, and hearing are very similar to each other in structure and basic physiology. These so-called hair cells are secondary sensory cells, that is, derivatives of epithelial cells that make direct contact to the axons of nearby ganglion cells that feed the information to brain centers. These hair cells may not be unique, since secondary hair cells also occur in cephalopod molluscs, but the whole question of homology among mechanoreceptors is still open, and much more work needs to be done before the ancestry or phylogeny of hair cells can be established (see Coffin et al., Chapter 3). This is one of the areas that should and will receive increased attention in the near future. A recent study of tunicate chordates, the ascidians or sea squirts, revealed that they possess a sensory organ containing hair cells with obvious similarities to those of the vertebrates (Burighel et al., 2003), and the study of these and other nonvertebrate chordates offers an exciting and very promising new approach to understanding the early evolution of these important receptors.

The basic structure of vertebrate hair cells was thus established very early, and almost certainly before vertebrates arose. The further evolution of hair cells into specialized types is still only poorly understood, especially the earlier steps that presumably occurred in the various groups of early fishes well before land vertebrates arose. There are hair cells to be found in organs that differ from one another to a greater or lesser degree (superficial neuromast cells, canal neuromast cells, crista hair cells associated with the semicircular canals, hair cells within the various maculae—saccule, utricle, etc.). Although there are numerous studies of these different hair cell types, especially of their structure, a comprehensive description of the correlation between specific stimulus parameters being detected and the structural and physiological specializations associated with them is still lacking. It was only relatively recently that it was discovered that in the saccule, lagena, and utricle of fish, only molecular markers permit the identification of two types of hair cells (e.g., Saidel et al. 1990; see also Ladich and Popper, Chapter 4) that may be related to specific structurally distinguishable hair cell types of land vertebrates. Many more studies in this area are necessary before we understand the rules that governed early hair cell specialization and evolution.

4. Strategies for Increasing Sensitivity

Water is a rather dense medium and it offers considerable resistance to objects moved within it. For very small objects such as the stereovillar bundles of hair cells, there is a thin layer of water molecules essentially bound to the surface that move with the bundle, effectively increasing its mass and dimensions. Moving such an object in water, which in one form or another surrounds all hair cells, requires substantial quantities of energy. Early in hair cell evolution, this energy would have been exclusively provided either by Brownian motion, which would, if at all, only have moved the bundle randomly and over very small distances, or by the mechanical stimuli present in the environment. Thus fluid viscosity and the drag on the bundle would have strongly limited the hair cells' ability to detect small signals. Apart from becoming frequency-selective and therefore shutting out interfering signals from other parts of the spectrum (see Lewis and Fay, Chapter 2), there are two additional ways to increase sensitivity in such systems that were realized, to a good degree independently, by various groups.

The first possibility is to increase the magnitude of the signal that arrives at the hair cells. This can be done by using an impedance transition that can be created at an interface between media of differing density. Such a density interface is presented by the otoliths of many sensory maculae, but potentially also by the interface between gas chambers such as swim bladders and body fluids and tissues in a number of bony fishes (see Ladich and Popper, Chapter 4). The transition in densities from high to low leads to larger amplitudes of displacement at such interfaces, and in some cases these interface tissues are connected by specially modified bones to the inner ear (as in the well-known Weberian ossicles, see Ladich and Popper, Chapter 4).

The impedance-matching system of land vertebrate middle ears is designed to prevent the large losses of amplitudes that would normally be incurred at the transition from a low-density medium (air) to a high-density medium (the water of body fluids and tissues). Structural evolution leading to new openings in the inner ear to act as pressure releasers and to the development of impedance-matching (tympanic) middle ears was remarkably complex, with often parallel developments in different taxonomic groups (see Clack and Allin, Chapter 5). Any future fossil findings that clearly show the structures of and near the inner ear would be very welcome, since at present a good deal of speculation is necessary to piece together the story of the evolution of the tympanic ears of the different groups, their times of origin, and their relationships. It is, however, a very exciting finding that the major vertebrate groups apparently all developed their middle-ear structures—the tympanic ear—independently of one another and roughly in the same geological period (the Triassic; see Clack and Allin, Chapter 5). Unfortunately, we are still largely ignorant of the trigger(s) that set off this unprecedented set of evolutionary developments, and there is a great need for further work in this area of paleontology. It would be very enlightening to have much more information on animals at the transition to mammals, to better understand the selection pressures that drove the evolution of the unique mammalian middle ear that is capable of transmitting much higher frequencies than the middle ear of nonmammals. If, as comparative evidence suggests, the ancestral mammal-like reptiles did not have an inner ear capable of high-frequency hearing, then it would appear to be a remarkable accident of evolution that just at the time that mammals were developing a tympanic middle ear, a secondary jaw joint evolved and gave up some little bones to the middle ear. This may have been one of the most fortuitous accidents of evolution, empowering later mammals to develop inner-ear epithelia capable of responding to the higher frequencies that were then able to traverse their middle ear.

The second possibility for increasing sensitivity is for the sensory cell itself to amplify the signal it has received. The curious and almost counter-intuitive phenomenon of active amplification by hair cells was a late discovery in auditory research and became possible only after technical advances in equipment in the late 1980s. Since then, evidence from many fields of study confirms the notion that not only are hair cells in a position to add acoustic energy to an existing signal, but that this energy is often produced in the absence of external signals and results in the production of spontaneous emissions, low-level acoustic energy measurable in the external ear canal (see Coffin et al., Chapter 3). Theoretical considerations and modeling work indicate that such internal sources of energy really can amplify sound stimuli. As discussed by Coffin et al., Chapter 3; Manley, Chapter 7; and Gleich et al., Chapter 8, the source of the energy for this amplification in nonmammalian systems (both vestibular and auditory) is active motion of the hair cell bundle (Hudspeth 1997; Manley 2001). It appears to be driven by premature closure of transduction channels induced by the binding of calcium ions that have entered through the channel during its opening phase. The beautiful thing about this model is that it can potentially work as fast as the transduction channels themselves (which is extremely fast) and that the phase of energy injection relative to bundle movement and thus to the phase of imposed signals is stable and integral to the system (Hudspeth 1997; Manley 2000). A recent article has now provided the first evidence that such a bundle motor is also operative in mammals (Kennedy et al. 2003), strengthening the notion that the bundle motor is plesiomorphic and has been retained in all land-vertebrate groups studied. In the mammalian cochlea, much evidence speaks for the existence of an additional, newly evolved mechanism using preexisting protein complexes (prestin; see Vater et al., Chapter 9) in the lateral cell membrane to change the length of outer hair cells during stimulation. This length change then feeds back into the motion of the organ of Corti.

The advantages of concentrating amplification in one set of sensory hair cells led to the independent evolution of a division of labor between hair cells located at different positions across the hearing organ in birds (see Gleich et al., Chapter 8) and in mammals (Vater et al., Chapter 9). In both cases, the evidence suggests that the outer group of hair cells is the more mechanically active and that it amplifies the sensory signal, whereas the inner group of hair cells detects the net result of the motion of the organ. More study needs to be given to the question of the phase relationships in the mammalian system, since there, the amplification provided by prestin in the cell membrane is only indirectly coupled to the opening state of the transduction channels. It is possible that mammals have exploited phase delays to influence the net response differently in different frequency ranges. The mammalian amplification system apparently works as well as, but probably no better than, the nonmammalian system. Certainly, in terms of hearing sensitivity and frequency selectivity, many nonmammals easily match or exceed the mammalian performance (Manley and Köppl 1998; Manley 2000). New work is needed to find out how these two motors are integrated in the mammalian cochlea and what advantage the mammals gained through their newly evolved cell-membrane motor.

5. The Independent Development and Specialization of the Ears of Major Groups of Land Vertebrates

The descriptions of the evolution of hearing in the various vertebrate lineages in Chapters 4 to 9 suggest that there were two major phases in the evolution of vertebrate hearing systems. The first phase was during the long period of the evolution of the different groups of fishes (at least 150 million years up to the origin of terrestrial tetrapods, but even up to the present day, see Manley and Clack, Chapter 1) and was a time of adaptation to the use and detection of acoustic signals in water. The evolution of the fishes was highly complex and even among the groups of modern fishes, a number of independent lineages can be discerned. Thus the sensory organ responsible for hearing in different fish groups is by no means constant (see Ladich and Popper, Chapter 4), and the presence of a gas (swim) bladder in bony fishes offered the opportunity for additional specializations for greater sensitivity, with the devices connecting the gas bladder to the ear often independently evolved in different fish groups.

The second major phase of ear evolution in vertebrates was triggered by the independent evolution of tympanic ears in land vertebrate groups. This relatively late development (about 100 million years after the origin of terrestrial tetrapods) occurred at a time when the different lineages of terrestrial vertebrates had already diverged and the preceding, presumably relatively weak, selection pressures had already established lineage-typical inner-ear epithelial patterns. From this basis, the patterns typical for each group were the subject of increased selection pressure due to the amplification provided by the tympanic ear and the accompanying ability to detect a wider range of signals, both in terms of their amplitude and their frequency range. The remarkably different structural patterns we see today in amphibians, lepidosaurs, archosaurs, and mammals are the result of independent selection processes in the different lineages.

In spite of the independence of the groups, the underlying principles of sensory processing, however, dictated that parallel developments would be inevitable (Fay and Popper 2000; see also Lewis and Fay, Chapter 2). The most obvious parallel development was the origin of two groups of hair cells across the hearing epithelium in archosaurs and in mammals (see above), with a similarly specialized innervation pattern (Köppl et al. 2000; see also Gleich et al., Chapter 8). Now that a large number of molecular-biological tools are becoming available, it is possible to intensify the comparison of hair-cell types at the level of the molecules they utilize. Thus the near future should bring new insights not only into the subtleties of hair-cell specialization and their functional consequences, but also the interrelationships of hair-cell types and hearing organs, so that we can better understand their evolution.

6. The Need for More Work on Neglected Groups

Not all groups of vertebrates have received the same degree of research attention. For evolutionary studies, it would be very useful if some of the neglected groups would receive more attention. Even within the mammals, which understandably have been most intensively studied, the hearing of marsupials and monotremes are very poorly known and understood, mainly due to their poor accessibility and in some cases their protected status. Although some recent studies have begun to rectify this situation, more work is needed, for example on the innervation patterns of the hair cells and their physiological response patterns.

It is encouraging to see that the birds are not only studied for their own sake but increasingly also as highly vocal animals as models for acoustic communication mechanisms. Amazing parallels are seen in the neural pathways for vocal production and processing in primates and songbirds, despite their independent evolution. This continues to be one of the prime examples illustrating the value of basic, comparative research for questions also relating to human performance. Even within the birds, complex vocalization patterns seem to have developed independently in the songbirds and parrot groups (see Gleich et al., Chapter 8). The birds continue to offer the opportunity—in parallel to mammalian studies—to explore the fundamental mechanisms of the analysis of complex communication signals. In addition, among the birds the barn owl, with the extreme specialization of its auditory system including the asymmetry of its external ears, can be directly compared in its performance and the mechanisms used with mammals that also localize sounds well. It should then be possible to observe the parallels developed in response to similar selection pressures and determine the forces that drove these evolutionary specializations.

7. Brain Pathway Evolution and Neural Processing Patterns—The Consequences of Parallel Evolution

A consequence of the major formative evolution of the ear in the separate lineages of amniotes was the independent specialization of brain centers of their respective auditory pathways. The question of the homology of these neural processing centers is largely unresolved for many parts of the brain stem auditory system. In the past, a confusion of naming arose since it was often assumed that if a nucleus was in more or less the same place and carried out a similar function, the nuclei were probably homologous and received the same name. This imprecise and often incorrect procedure that often totally ignored the long phases of independent evolution of the different groups of vertebrates has led to considerable confusion. Now that it has become clear that the ear and accompanying brain centers evolved largely independently, this problem is finally receiving more attention and a serious attempt is being made to determine to what extent homology can be determined (see Grothe et al., Chapter 10). Comparative studies of the adult conditions (structure and function) are of limited value here, because it appears that there are many examples of both convergent evolution and highly modified plesiomorphic structures (see Grothe et al., Chapter 10). Significant advances, however, may come from future studies of development, particularly the early development of the brain and the molecular cues guiding it.

The coding of interaural time differences in the brain stem has also recently become a topic of intensified debate. New data from small mammals have led researchers to question the generality of the classical textbook version involving axonal delay lines and postsynaptic coincidence detectors, which had been developed largely on the basis of bird data (see Grothe et al., Chapter 10). Data from a greater variety of species are needed to resolve this issue and to distinguish between a phylogenetic difference due to the clearly independent evolution of the avian and mammalian system and differences driven by the character and dimensions of the physical parameters, such as head size.

8. Summary

The remarkable diversity of the peripheral hearing organs of vertebrates reflects the long periods of independent and parallel evolution of the different groups both of fishes and of land vertebrates. In spite of this variety, many common functional principles were rooted very early and these are reflected in the greater similarity of the input to the brain than is seen in the peripheral structure. It is therefore not surprising that the evolution of brain centers utilized common principles of processing that result in many similarities in the auditory pathway of the brains of different vertebrates. Most of these similarities probably do not, however, derive from a true homology of the nuclei. Although much progress has been made and the contours have become clearer, the elucidation of the evolutionary-historical stages of the making of vertebrate ears and brain pathways will continue to provide fascinating science for years to come.

References

Burighel P, Lane NJ, Fabio G, Stefano T, Zaniolo G, Carnevali MDC, Manni L (2003) Novel, secondary sensory cell organ in ascidians: In search of the ancestor of the vertebrate lateral line. J Comp Neurol 461:236–249.

Fay RR, Popper AN (2000) Evolution of hearing in vertebrates: the inner ears and processing. Hear Res 149:1–10.

Hudspeth AJ (1997) Mechanical amplification of stimuli by hair cells. Curr Opin Neurobiol 7:480–486.

Kennedy HJ, Evans MG, Crawford AC, Fettiplace RR (2003) Fast adaptation of mechanoelectrical transduction channels in mammalian cochlear hair cells. Nat Neurosci 6:832–836.

Köppl C, Manley GA, Konishi M (2000) Auditory processing in birds. Curr Opinion Neurobiol, 10:474–481.

Manley GA (2000) Cochlear mechanisms from a phylogenetic viewpoint. Proc Natl Acad Sci USA 97:11736–11743.

Manley GA (2001) Evidence for an active process and a cochlear amplifier in nonmammals. J Neurophysiol 86:541–549.

Manley GA, Köppl C (1998) Phylogenetic development of the cochlea and its innervation. Curr Opinion Neurobiol 8:468–474.

Saidel WM, Presson JC, Chang JS (1990) S100 immunoreactivity identifies a subset of hair cells in the utricle and saccule of a fish. Hear Res 47:139–146.

Webster DB, Fay RR, Popper AN (eds) (1992). Evolutionary Biology of Hearing. New York: Springer-Verlag.

Appendix: Useful Concepts from Circuit Theory

EDWIN R. LEWIS

1. Introduction

For understanding the physics of sound and the biophysics of acoustic sensors in animals, the concepts of impedance and admittance, impedance matching, transducer, transformer, passive and active, bidirectional coupling, and resonance are widely used. They all arise from the circuit-theory metamodel, which has been applied extensively in acoustics. As the name suggests, the circuit-theory metamodel provides recipes for constructing models. The example applications of the metamodel in this appendix involve a few elementary concepts from classical physics: the adiabatic gas law, Newton's laws of motion and of viscosity, and the definitions of work and of Gibbs free energy. They also involve a few elements of calculus and of the arithmetic of complex numbers. All of these should be familiar to modern biologists and clinicians. For further examples of the application of circuit concepts to acoustical theory and acoustical design, see Baranek (1954), Olsen (1957), or Morse (1981).

1.1 The Recipe

Applied to an elementary physical process, such as sound conduction in a uniform medium, the appropriate recipe comprises the following steps:

1. Identify an entity (other than Gibbs free energy, see below) that is taken to be conserved and to move from place to place during the process.
2. Imagine the space in which the process occurs as being divided into such places (i.e., into places in which the conserved entity can accumulate).
3. Construct a map of the space (a circuit graph) by drawing a node (e.g., a small dot) for each place.
4. Designate one place as the reference or ground place for the process.
5. Assign the label "0" to the node representing the ground place.
6. Assign a unique, nonzero real integer (i) to each of the other nodes in the graph.
7. Define the potential difference between each place (i) and the ground place (0) to be F_{i0}, the corresponding Gibbs potential for the conserved entity

(change in Gibbs free energy of the system per unit of conserved entity moved from place 0 to place i).
8. From basic physical principles (or by measurement), determine the relationship between the amount, Q_i, of conserved entity accumulated at place i and the potential F_{i0}.
9. From basic physical principles (or by measurement), determine the relationship between potential difference F_{ij} between each pair of neighboring places and the flow J_{ij} of conserved entity between those places.

The Gibbs free energy of a system is defined as the total work available from the system less that portion of the work that must be done against the atmosphere. For one-dimensional motion from point a to point b, the work (W_{ab}) done by the system is defined by the expression $W_{ab} = \int_{x=a}^{b} f(x)dx$ where $f(x)$ is the force applied by the system to the object on which the work is being done. The force is taken to be positive if it is applied in the direction in which the values of x increase; otherwise it is taken to be negative. For tables of Gibbs potentials for various conserved entities, see Lewis (1996).

1.2 Example: An Acoustic Circuit Model

As an example application of the recipe, consider the conduction of sound, axially, through an air-filled cylindrical tube with rigid walls. In that case, one might identify the conserved entity as the air molecules that move from place to place along the tube. As air molecules accumulate at a given place, the pressure at that place increases. Assuming that the air pressure outside the tube is uniform and equal to atmospheric pressure, it is convenient to take that place (the outside of the tube) to be the reference place. The SI unit for number of air molecules is 1.0 mol. Therefore, the SI unit for the flow of air molecules will be 1.0 mol/s; and the SI unit for the Gibbs potential will be 1.0 joule/mol. If one takes p_i to be the absolute pressure at place i along the inside of the tube (SI unit 1.0 nt/m² = 1.0 Pa) and p_0 to be the absolute atmospheric pressure, then the sound pressure (Δp_i) at place i will be the difference between those two pressures:

$$\Delta p_i = p_i - p_0 \qquad (1)$$

Acousticians usually assume that the processes involved in sound propagation are adiabatic and that sound pressures are exceedingly small in comparison to atmospheric pressure:

$$\frac{\Delta p_i}{p_0} << 1.0 \qquad (2)$$

The Gibbs potential difference, F_{i0}, between place i and the reference place is given by:

$$F_{i0} = \frac{\Delta p_i}{c_0} \qquad (3)$$

Appendix: Useful Concepts from Circuit Theory 371

where c_0 is the resting concentration of air molecules (SI unit = 1.0 mol/m³).

$$c_0 = \frac{n_0}{V_0} \qquad (4)$$

where n_0 is the resting number of air molecules at place i, and V_0 is the volume of place i (assumed to be constant). The resting absolute pressure at place i is taken to be p_0.

Equation 3 for the Gibbs potential can be derived easily from the equation for work. Invoking the adiabatic gas law, one can derive the relationship between the Gibbs potential at place i and the excess number of air molecules at place i, Δn_i, where

$$\Delta n_i = n_i - n_0 \qquad (5)$$

n_i being the total number of air molecules at place i. For $\Delta p_i/p_0$ exceedingly small, the adiabatic gas law can be stated as follows:

$$\frac{\Delta p_i}{p_0} = \gamma \frac{\Delta n_i}{n_0} \qquad (6)$$

where γ is 1.4, the ratio of specific heat of air at constant pressure to that at constant volume. From this,

$$F_{i0} = \frac{\Delta p_i}{c_0} = \frac{\gamma p_0}{c_0 n_0} \Delta n_i \qquad (7)$$

In the circuit-theory metamodel, this is a capacitive relationship. The general constitutive relationship for a linear, shunt capacity at place i in any circuit is

$$F_{i0} = \frac{Q_i}{C_i} \qquad (8)$$

where the parameter C_i often is called the *capacitance* or *compliance*. In this case, Q_i, the accumulation of conserved entity at place i, is taken to be to be Δn_i,

$$C_i = \frac{c_0 n_0}{\gamma p_0} \qquad (9)$$

Dividing the cylindrical tube axially into a large number of segments of equal volume, V_0, brings the first eight steps of the metamodel recipe to completion. For the ninth step one might assume that the acoustic frequencies are sufficiently high to preclude significant propagation of drag effects from the wall of the cylinder into the mainstream of axial air flow. In other words, one might ignore the effects of viscosity (see subsection 8.2). Absent viscosity, the axial flow of air molecules would be limited only by inertia. Lumping the total mass of air molecules at a single place, and invoking Newton's second law of motion, one has the following relationship for the flow, J_{ij}, of molecules from place i to its neighboring place, j:

$$\frac{dJ_{ij}}{dt} = \frac{A_0^2 c_0^2}{V_0 \rho_0} F_{ij} \tag{10}$$

where t is time, ρ_0 is the resting density of the air (SI unit equal 1.0 kg/m³), and A_0 is the luminal area of the cylindrical tube. In the circuit-theory metamodel, this is an inertial relationship. The general constitutive relationship for a linear, series inertia between place i and j in any circuit is

$$F_{ij} = I_{ij} \frac{dJ_{ij}}{dt} \tag{11}$$

where the parameter I_{ij} often is called the *inertia* or *inertance*. In this case,

$$I_{ij} = \frac{V_0 \rho_0}{A_0^2 c_0^2} \tag{12}$$

2. Derivation of an Acoustic Wave Equation from the Circuit Model

Having thus completed the circuit model, one now can extract from it an equation describing sound conduction through the tube. This normally is accomplished by writing explicit expressions for the conservation of (conserved) entity at node i and the conservation of energy in the vicinity of node i, and then incorporating the constitutive relations obtained in steps 8 and 9 of the metamodel recipe. Let node i have two neighbors, node h on its left and node j on its right. Conservation of air molecules requires that

$$J_{hi} - J_{ij} = \frac{dQ_i}{dt} \tag{13}$$

From Equation 8 one notes that

$$\frac{dQ_i}{dt} = C_i \frac{dF_{i0}}{dt} \tag{14}$$

Therefore,

$$J_{hi} - J_{ij} = C_i \frac{dF_{i0}}{dt} \tag{15}$$

Conservation of energy requires that

$$F_{ij} = F_{i0} - F_{j0} \tag{16}$$

Therefore,

$$F_{i0} - F_{j0} = I_{ij} \frac{dJ_{ij}}{dt} \tag{17}$$

2.1 Translation to a Transmission-Line Model

In the manner of Lord Rayleigh, one can simplify further analysis by taking the limit of each side of Equations 15 and 17 as one allows the number of places represented in the model to become infinite and the extent of each place infinitesimal. Each place represented in the circuit model has length L_0 given by

$$L_0 = \frac{V_0}{A_0} \tag{18}$$

Dividing C_i and I_i by L_0, one has the capacitance and inertance per unit length of the uniform, air-filled tube:

$$\hat{C} = \frac{C_i}{L_0} = \frac{A_0 c_0^2}{\gamma p_0}$$

$$\hat{I} = \frac{I_i}{L_0} = \frac{\rho_0}{A_0 c_0^2} \tag{19}$$

Now, let x be the distance along the tube and let the length of each place be Δx. One can rewrite Equations 15 and 17 as follows:

$$-\Delta J = J_{hi} - J_{ij} = \hat{C} \, \Delta x \, \frac{dF}{dt}$$

$$-\Delta F = F_{i0} - F_{j0} = \hat{I} \, \Delta x \frac{dJ}{dt} \tag{20}$$

Dividing each side of each equation by Δx and taking the limit as Δx approaches zero, one has

$$-\frac{d}{dx}J(x,t) = \lim_{\Delta x \to 0}\left[-\frac{\Delta J}{\Delta x}\right] = \hat{C}\frac{d}{dt}F(x,t)$$

$$-\frac{d}{dx}F(x,t) = \lim_{\Delta x \to 0}\left[-\frac{\Delta F}{\Delta x}\right] = \hat{I}\frac{d}{dt}J(x,t) \tag{21}$$

These are the classic transmission-line equations, variations of which have been used widely in neurobiology. Traditionally, they are written in a more general form:

$$\frac{dF}{dx} = -zJ$$

$$\frac{dJ}{dx} = -yF \tag{22}$$

where z is the series impedance per unit length of the transmission line, and y is the shunt admittance per unit length; and F and J are linear transforms of F and J. Although the notions of impedance and admittance are treated here in a general sense, when they involve time derivatives (as they do here), circuit theorists normally apply them to one of two carefully restricted situations: (1)

sinusoidal steady state (a pure tone has been applied for an indefinitely long time); or (2) zero state (a temporal waveform is applied; at the instant before it begins, all Fs and Js in the circuit model are equal to zero). For sinusoidal steady state, circuit theorists traditionally use the phasor transform; for zero state they traditionally use the Laplace transform. Yielding results interpretable directly in terms of frequency, the former is widely used in classical acoustics texts.

2.2 Derivation of the Phasor Transform

The definition of the phasor transform is based on the presumption that exponentiation of an imaginary number, such as ix (where i is the square root of -1), obeys the same rules as exponentiation of a real number (such as x). Specifically, it is presumed that e^{ix} can be evaluated by substituting ix for x in the Maclaurin series for e^x. In that case,

$$e^{ix} = \cos x + i \sin x \qquad (23)$$

and

$$\cos x = \operatorname{Re}\{e^{ix}\} \qquad (24)$$

where $\operatorname{Re}\{z\}$ is the real part of the complex number, z:

$$\begin{aligned} z &= a + ib \\ i &= \sqrt{-1} \\ \operatorname{Re}\{z\} &= a \end{aligned} \qquad (25)$$

The phasor transform applies to sinusoidal steady state, with the radial frequency of the sinusoid being ω (SI unit = 1.0 rad/s).

$$\omega = 2\pi f \qquad (26)$$

where f is the conventional frequency (unit = 1.0 Hz). One can represent the phasor transformation as follows:

$$\begin{aligned} H(\omega) &= \Phi\{H(t)\} \\ H(t) &= \Phi^{-1}H(\omega) \end{aligned} \qquad (27)$$

where the function of frequency, $H(\omega)$, is the phasor transform of the time function $H(t)$. The transform is defined by its inverse

$$H(t) = \Phi^{-1}\{H(\omega)\} \triangleq \operatorname{Re}[H(\omega)e^{i\omega t}] \qquad (28)$$

Combining this with Equations 23–25, one finds the following basic transform pairs:

$$\begin{aligned} \Phi\{\cos \omega t\} &= 1 \\ \Phi^{-1}\{1\} &= \cos \omega t \\ \Phi\{A \cos(\omega t + \alpha)\} &= Ae^{i\alpha} = A \cos \alpha + iA \sin \alpha \\ \Phi^{-1}\{Ae^{i\alpha}\} &= \Phi^{-1}\{A \cos \alpha + iA \sin \alpha\} = A \cos(\omega t + \alpha) \end{aligned} \qquad (29)$$

Notice that
$$\Phi^{-1}\{Ae^{i\alpha}Be^{i\beta}\} = AB\cos(\omega t + \alpha + \beta) \tag{30}$$

Thus, multiplying the phasor transform of a sinusoidal function of time (of any amplitude and phase) by the factor $Ae^{i\alpha}$ corresponds to two mathematical operations on the function itself: (1) multiplying the amplitude of the function by the factor A, and (2) adding α radians to the phase of the function. Taking the first derivative of any sinusoidal function of time can be described in terms of the same two operations: (1) the amplitude of the function is multiplied by ω, and (2) $\pi/2$ radians (90 degrees) are added to the phase of the function. Thus,

$$\Phi\left\{\frac{dH(t)}{dt}\right\} = \omega e^{i\frac{\pi}{2}}H(\omega) \tag{31}$$

Applying Equation 23, one has
$$e^{i\frac{\pi}{2}} = i$$
$$\Phi\left(\frac{dH(t)}{dt}\right) = i\omega H(\omega) \tag{32}$$

2.3 Phasor Transform of the Transmission-Line Model

Based on Equation 32, one can write the transformed version of Equation 21 as follows:

$$\frac{d}{dx}F(x,\omega) = -i\omega\hat{I}J(x,\omega)$$
$$\frac{d}{dx}J(x,\omega) = -i\omega\hat{C}F(x,\omega) \tag{33}$$

Thus, for the transmission-line model (Eq. 22), under sinusoidal steady state at radial frequency ω, z and y would be written as follows:

$$z = i\omega\hat{I}$$
$$y = i\omega\hat{C} \tag{34}$$

To simplify notation in transformed equations, ω often is omitted in the arguments of the dependent variables J and F. For z and y both independent of x, as they are taken to be in this case, the spatial solutions to these equations usually are written in one of two forms:

Form 1

$$J(x) = F_f(x) + F_r(x) \qquad J(x) = J_f(x) - J_r(x)$$

$$J_f(x) = Y_0 F_f(x) \qquad J_r(x) = Y_0 F_r(x)$$

$$F_f(x) = Z_0 J_f(x) \qquad F_r(x) = Z_0 J_r(x)$$

$$Z_0 = \frac{1}{Y_0} = \sqrt{\frac{z}{y}}$$

$$F_f(x) = F_f(0)e^{-\sqrt{zy}x} \qquad F_r(x) = F_r(0)e^{\sqrt{zy}x} \tag{35}$$

Form 2

$$F(x) = F(0)\cosh(\sqrt{zy}\,x) - Z_0 J(0)\sinh(\sqrt{zy}\,x)$$
$$J(x) = J(0)\cosh(\sqrt{zy}\,x) - Y_0 F(0)\sinh(\sqrt{zy}\,x)$$

$$Z_0 = \frac{1}{Y_0} = \sqrt{\frac{z}{y}} \tag{36}$$

2.4 Forward and Reverse Waves

Define the forward direction for the transmission line as the direction of increasing values of x. In the first form, $F_f(x)$ and $J_f(x)$ are interpreted as being the potential and flow components of a wave traveling in the forward direction; $F_r(x)$ and $J_r(x)$ are interpreted as the components of a wave traveling in the reverse direction. At any instant, the product $F_f(x)J_f(x)$ of the corresponding time functions is the rate at which the forward wave carries Gibbs free energy (in the forward direction) past location x, and the product $F_r(x)J_r(x)$ is the rate at which the reverse wave carries Gibbs free energy past x (in the reverse direction).

2.4.1 Speed of Sound and Characteristic Impedance

For the simple transmission-line model under consideration here, z is purely inertial and y is purely capacitive. The wave components in form 1 of the solution become

$$F_f(x) = F_f(0)e^{-i\sqrt{\hat{l}\hat{C}}\omega x}$$
$$F_r(x) = F_r(0)e^{+i\sqrt{\hat{l}\hat{C}}\omega x} \tag{37}$$

Both components are sinusoids, radial frequency ω. For the forward component, the phase at distance x from the origin lags that at the origin by an amount that is directly proportional to x. That is the behavior of a wave traveling at constant speed in the direction of increasing x (the forward direction). For the reverse component, the phase at distance x leads that at the origin by an amount proportional to x. That is the behavior of a wave traveling at constant speed in the direction of decreasing x (the reverse direction). For both components, the amplitude is independent of x. The speed, ϕ, of the wave is the same in both cases. The solution is valid for all frequencies, so one can generalize it to waves of arbitrary shape that travel at constant speed without changing shape:

$$F_f(x,t) = F_f\left(0, t - \frac{x}{\phi}\right) \qquad F_r(x,t) = F_r\left(0, t + \frac{x}{\phi}\right)$$
$$J_f(x,t) = Y_0 F_f(x,t) \qquad J_r(x,t) = Y_0 F_r(x,t) \tag{38}$$

$$Y_0 = \frac{1}{Z_0} = \sqrt{\frac{\hat{C}}{\hat{l}}} = \frac{A_0 c_0^2}{\sqrt{\rho_0 \gamma p_0}} \tag{39}$$

Appendix: Useful Concepts from Circuit Theory 377

$$\phi = \frac{1}{\sqrt{\hat{I}\hat{C}}} = \sqrt{\frac{\gamma p_0}{\rho_0}} \qquad (40)$$

The parameter Z_0, usually called the characteristic impedance, is a positive real number in this case. For sinusoidal steady state this implies that at each place (each value of x) along the acoustic path, F_f and J_f are perfectly in phase with one another, as are F_r and J_r. The relationship between Z_0 and the physical parameters of the system and its SI unit both depend on the choice one made at step 1 of the metamodel recipe. If one carries out the model construction correctly, however, the wave speed will be independent of that choice.

3. Alternative Formulations of the Circuit Model

Instead of taking the number of air molecules as the conserved entity, one might have taken their mass. Assuming that the density changes induced in the air by the propagating sound wave are negligible (the analysis in section 2 was based on the same assumption), acousticians often use fluid volume as the conserved entity (as a surrogate for mass). Sometimes mass itself is used. All of these choices lead to equivalent formulations and moving from one to another requires only the application of appropriate multiplicative factors. The SI unit of impedance for each choice of conserved entity is easily determined. If the selected entity were apples, with an SI unit of 1.0 apple, then the SI unit of flow would be 1.0 apple/s, and the SI unit of Gibbs potential would be 1.0 joule/apple. The SI unit of impedance would be that of Gibbs potential divided by flow—1.0 joule s per apple². For the acousticians' choice of volume, the SI unit of conserved entity is 1.0 m³. The SI unit of impedance becomes 1.0 joule s per m⁶. For the choice made at the beginning of section 1.2 (SI unit of conserved entity = 1.0 mol), the SI unit of impedance is 1.0 joule s/mol².

3.1 Conventional Acousticians' Formulation

To obtain the acousticians' J (SI unit 1.0 m³/s), one divides the J of section 2 (SI unit 1.0 mol/s) by c_0 (SI unit 1.0 mol/m³). To obtain the acousticians' F (SI unit 1.0 joule/m³), one multiplies the F of section 2 (SI unit 1.0 joule/mol) by c_0. When these changes are made, the inertance and compliance per unit length become.

$$\hat{C} = \frac{A_0}{\gamma p_0}$$

$$\hat{I} = \frac{\rho_0}{A_0} \qquad (41)$$

and the characteristic impedance, Z_0, becomes

$$Z_0 = \frac{\sqrt{\gamma p_0 \rho_0}}{A_0} \tag{42}$$

Recognizing γp_0 as the adiabatic bulk modulus of an ideal gas at pressure p_0, one can translate Equations 40 and 42 into general expressions for compressional (longitudinal) acoustic waves in fluids:

$$\hat{C} = \frac{A_0}{E_B}$$

$$\phi = \sqrt{\frac{E_B}{\rho}} \tag{43}$$

$$Z_0 = \frac{\sqrt{E_B \rho}}{A_0}$$

$$E_B \triangleq -V_0 \frac{\Delta p}{\Delta V} \tag{44}$$

where ϕ is the speed of sound in the fluid; Z_0 is the characteristic acoustic impedance of the fluid; E_B is the fluid's adiabatic bulk modulus; ρ is its density, and A_0 is the cross-sectional area of the acoustic path (the luminal area of the cylindrical tube in the system being modeled here). Once again, these expressions were based on the presumption that changes in the fluid density, ρ, produced by the waves are exceedingly small.

3.2 Modeling Perspective

Following a path already well established by physicists, with the circuit-theory metamodel as a guide, the development to this point has combined that presumption (exceedingly small changes in fluid density) with two conservation principles (energy is conserved, matter is conserved) and with two of the empirical linear laws of physics (the adiabatic form of the ideal gas law—translated to a form of Hooke's law, and Newton's second law of motion) to reconstruct a theory of sound propagation. With the theory constructed, mathematics was used to determine where that combination of presumptions, principles, and empirical laws would lead. What the analysis showed was that the local elastic and inertial properties of fluids (gases and liquids) should combine to allow spatially extensive sound propagation. It also showed that if that were the case, then the speed of propagation should be related to the elastic and inertial parameters (adiabatic bulk modulus and density) in the specific manner described in Equation 43. In fact, the derived relationship (Eq. 43) predicts extremely well the actual speeds of sound in fluids (gases and liquids). This should bolster one's faith not only in the specific theory, but also in the modeling process itself. It follows the pattern established by Isaac Newton and practiced successfully by celebrated physical scientists over and over again since his time. Newton combined a presumption about gravity with a conservation law (momentum is conserved) and with his second law of motion to construct a theory of planetary motion. He used mathematics to see where that theory would lead. Where it led was directly to Kepler's laws, which are empirical descriptions of the actual motions of planets. Recently, writing about Newton's *Principia*, where this theory and its analysis first were published, a noted physicist said that the astounding conclusion one would draw from the work is the fact that nature obeys the laws of mathematics. I would put it differently. Physical nature appears to behave in an astoundingly consistent manner. In mathematics, mankind has invented tools that allow concise and precise statements of apparently consistent local behaviors (e.g., empirically derived laws), tools (such as the circuit-theory metamodel) that allow those statements to be combined to construct theories of spatially (in its general sense) more extensive behaviors, and tools to derive from those theories their predictions. The fact that the predicted behaviors so often match, extremely well, observed behaviors, is attributable to physical nature's consistency.

Returning to the metamodel and to the first step in the recipe, one can see that momentum (Newton's choice) is a particularly interesting alternative to be taken as the conserved entity. Newton's second law equates the force two particles exert on one another to the flow of momentum between them, and (for nonrelativistic systems) the Gibbs potential difference for that momentum flow is the difference between the particles' velocities. When one chooses fluid mass, fluid volume, or number of fluid molecules as the conserved entity, J is proportional to velocity and F is proportional to force. With momentum as the choice, F is proportional to velocity, J is proportional to force. This is an example of the celebrated duality that arises in the circuit-theory metamodel.

4. Generalization to Plane Waves

As long as the consequences of viscosity were negligible, the presence of the rigid cylindrical walls posited in the previous subsections had only one effect on the propagation of sound waves through the fluid inside the cylinder. It forced that propagation to be axial. The transmission-line equations derived for the fluid in the cylinder would apply perfectly well to propagation of a plane wave through any cross-sectional area A_0, normal to the direction of propagation, in an unbounded fluid medium (e.g., air or water). The propagation speed derived for the waves confined to the fluid in the cylinder is the same as it would have been for unconfined plane waves in the same fluid. The expression for characteristic impedance (Eq. 44) would apply to an interface, normal to the direction of propagation, with cross-sectional area A_0. Viewed locally, propagating sound waves usually approximate plane waves, making Equation 44 and its antecedents very general. In an unbounded fluid, however, the propagation of sound waves is not confined to a single coordinate. A local volume in such a fluid could have sound waves impinging upon it from all directions at once.

The instantaneous local sound-pressure component of each of these waves is a scalar, and the pressures from all of the waves together would sum to a single value. The local volume flow component of each wave will now be a vector (having an amplitude and a direction), and the flows from all of the waves would sum vectorially, at any given instant (t), to a single amplitude and direction. For a sound wave from a given direction, the local flow of Gibbs free energy through a cross-sectional area A_0 would be given by the product of the local flow vector ($J(t)$) for that wave and the corresponding local Gibbs potential ($F(t)$) for that wave. Notice that the product itself now is a vector, with amplitude and direction. The direction of the vector is the direction of energy flow. Notice that it is the same as the direction of $J(t)$ whenever $F(t)$ is positive. The cross-sectional area, A_0, would be taken in a plane normal to the direction of energy flow.

4.1 Particle Velocity

Noting that $J(t)$ for a uniform plane wave would be directly proportional to A_0, one can normalize it to form another variable—flow density ($\vartheta(t)$).

$$\vartheta(t) = \frac{J(t)}{A_0} \qquad (45)$$

When J is volume flow (the acousticians' choice, SI unit = 1.0 m³/s), ϑ is the sound-induced component of the velocity of the fluid (SI unit = 1.0 m/s). In that case, it traditionally is labeled the *particle velocity*. F, in that case (volume taken to be the conserved entity), is the local sound pressure. Whatever the choice of conserved entity, the instantaneous product $F\vartheta$ will be a vector whose amplitude is the instantaneous flow of Gibbs free energy per unit area (SI unit = 1.0 W/m²), and whose direction is the direction of that energy flow.

It is interesting to contemplate the vector sum of particle velocities of several unrelated sound waves impinging on a small local volume from different directions in an extended fluid. The local sound pressure would vary in time much as it might for a plane wave. The particle-velocity vector, on the other hand, would vary not only in amplitude, but also in direction. Try to imagine the computational task one would face trying to reconstruct the original sound waves from the combination of that scalar function, $F(t)$, and that vector function, $\vartheta(t)$.

4.2 Particle Displacements

For an acoustical sine wave, the local particle velocity is a sinusoidal function of time, as is the corresponding local particle displacement, $\delta(x,t)$. One can compute the amplitude, $D(x)$, of the latter directly from the amplitude, $B(x)$, of the former:

$$\vartheta(x,t) = B(x)\cos(\omega t)$$
$$\delta(x,t) = D(x)\sin(\omega t) = \int\vartheta(x,t)dt = \frac{B(x)\sin(\omega t)}{\omega}$$
$$D(x) = \frac{B(x)}{\omega} \qquad (46)$$

This, in turn, can be related to the local peak sound pressure, $\Delta p(x)$, through the characteristic impedance:

$$F(x,t) = \Delta p(x)\cos(\omega t)$$
$$\vartheta(x,t) = \frac{J(x,t)}{A_0} = \frac{F(x,t)}{A_0 Z_0} = \frac{F(x,t)}{\sqrt{E_B \rho}}$$
$$B(x) = \frac{\Delta p(x)}{\sqrt{E_B \rho}}$$
$$D(x) = \frac{\Delta p(x)}{\omega \sqrt{E_B \rho}}$$
$$\sqrt{E_B \rho} = \rho\phi = \frac{E_B}{\phi} \qquad (47)$$

In water, ρ is approximately 10^3 kg/m^3 and ϕ is approximately 1.5×10^3 m/s, making $\rho\phi$ approximatelly 1.5×10^6 kg/m^{2}s. Standard pressure of 1.0 atm is defined to be 101,325 Pa. The velocity of sound in dry air at 1.0 atm and 20°C is approximately 340 m/s. Therefore, for dry air at 1.0 atm and 20°C, E_B/ϕ (which is $1.4p_0/\phi$) is approximately 400 kg/m^{2}s. For a given sound pressure, this means that the peak local particle displacement in air will be almost 4000 times as great as it is in water.

For a 100-Hz tone with a peak sound pressure of 1.0 Pa (the upper end of the usual sound intensity range), for example, the peak local particle displacement in water would be approximately 10 nm. In air, it would be 40 µm. For 1-kHz tones, those particle displacements would be 1 nm and 4 µm, respectively, and for 10 kHz they would be 0.1 nm and 0.4 µm. At the nominal threshold level for human hearing (0.00003 Pa) at 1.0 kHz, they would be 0.03 pm and 0.12 nm. If an evolving animal were able to couple a strain-sensitive device (neural transducer) directly to a flexible structure (e.g., a cuticular sensillum, an antenna, hair) projecting into the air, then a sense of hearing could evolve based on direct coupling of particle-velocity to that structure. One would guess that such a sense would be limited to low-frequency, high-intensity sounds. For a strain-sensitive device (e.g., a bundle of stereovilli) projected into water, on the other hand, an evolving sense of hearing based on direct coupling of particle velocity to that device would be much more limited (4,000 times less effective at every frequency, every intensity). For such a device, one might expect evolution to have incorporated auxiliary structures to increase the displacement of the strain-sensor well beyond the particle displacement of the acoustic wave in the medium. For detection of airborne sound, such auxiliary structures also would have to include a mechanism for transferring acoustic energy from the air to the strain sensor. In the language of the circuit-theory metamodel, the latter would be a *transducer*, the former a *transformer* or an *amplifier*.

5. Passive Transducers

In the basic circuit model, one tracks the flow of a single entity from place to place. By taking F to be the Gibbs potential for that entity, one is able to track the flow of Gibbs free energy as well. The conserved entity will be confined to a particular realm. Air molecules, for example, are confined to the volume containing the air, which one might label the *pneumatic realm*. The Gibbs free energy (acoustic energy) that they carry, on the other hand, can be transferred in and out of that realm. In the circuit-theory metamodel, a device that transfers energy from one physical realm to another is a transducer. A piston can be modeled as such a device. It can translate sound pressure, F_1, and flow of air molecules, J_1, in a pneumatic realm to a Gibbs potential, F_2, and flow, J_2, appropriate to a translational rigid-body mechanical realm, and vice versa. In doing so, it carries Gibbs free energy back and forth between those two realms. For a rigid-body mechanical realm, there are two common choices for conserved entity: (1) the shapes of the rigid elements, or (2) momentum. In rigid-body mechanics, it is convenient to treat each of the orthogonal directions of translational and rotational motion as a separate realm. For the piston, the motion of interest is axial (translational motion perpendicular to the face of the piston). Following standard engineering practice, one would select axial displacement as the conserved entity (as a surrogate for shape, displacement is conserved as it passes through an element whose shape does not change). The SI unit of displacement is 1.0 m. The flow, J_2, becomes axial velocity (SI unit = 1.0 m/s); and the Gibbs potential, F_2, becomes force (SI unit = 1.0 joule/m). The directions assigned in Fig. A.1 are the conventional associated reference directions for the circuit-theory construction known as a two-port element. When F_1 is positive, it tends to push the piston to the right, the direction assigned to positive values of J_1. When F_2 is positive, it tends to push the piston to the left, the direction assigned to positive values of J_2.

Constitutive relationships for the piston can be expressed conveniently in the following form:

$$F_1 = h_{11}J_1 + h_{12}F_2$$
$$J_2 = h_{21}J_1 + h_{22}F_2 \tag{48}$$

where the parameters h_{12} and h_{21} are transfer relationships; h_{11} is an impedance (ratio of a Gibbs potential to the corresponding flow); and h_{22} is an admittance (ratio of a flow to the corresponding Gibbs potential). One estimates these parameters individually (by measurement or by invoking simple physics) from the following equations:

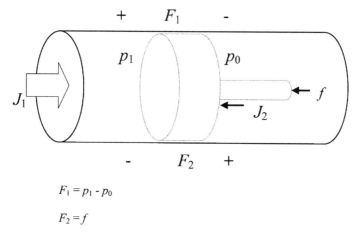

$F_1 = p_1 - p_0$

$F_2 = f$

FIGURE A.1. Depiction of a piston inside a cylinder. From the acoustic (fluid) realm, the piston is subjected to a pressure difference, $p_1 - p_0$; and translational motion of the piston corresponds to volume flow, J_1, of the fluid. From the rigid-body mechanical realm, the piston is subjected to a force, f, and translational motion of the piston corresponds to translational velocity, J_2.

$$h_{11} = \left.\frac{F_1}{J_1}\right|_{F_2=0} \qquad h_{12} = \left.\frac{F_1}{F_2}\right|_{J_1=0}$$
$$h_{21} = \left.\frac{J_2}{J_1}\right|_{F_2=0} \qquad h_{22} = \left.\frac{J_2}{F_2}\right|_{J_1=0} \qquad (49)$$

If, for now, one assumes that the piston is perfectly rigid, that its mass is negligible, and that it operates with no resistance, then h_{11} and h_{22} can be taken to be zero. That leaves one with the equations for an ideal transducer:

$$F_1 = h_{12}F_2 \qquad J_2 = h_{21}J_1 \qquad (50)$$

5.1 Reciprocity

Onsager (1931) demonstrated that, ideal or not, as long as it does not involve spinning elements (e.g., a gyroscope or spinning charge), a linear, passive transduction process such as this must be reciprocal. For the selected parameters (applied to the associated reference directions of Fig. A.1), that means

$$h_{12} = -h_{21} \qquad (51)$$

As a specific test of the Onsager reciprocity theorem, one can invoke the acousticians' choice of conserved entity (volume) in the pneumatic realm, so that F_1 is the sound pressure on the left side of the piston ($F_1 = \Delta p_1 = p_1 - p_0$) and J_1 is the volume flow on the left. From elementary physics one deduces that J_1

will be zero when F_1 (the net pressure pushing from the left) times the luminal area, A_0, is equal to the force pushing from the right. Therefore,

$$h_{12} = \left.\frac{F_1}{F_2}\right|_{J_1=0} = \frac{1}{A_0} \qquad (52)$$

When there is no force from the right (F_2 is zero), so that the piston could not be deformed even if it were not perfectly rigid, then from elementary geometry one knows that the (rightward) volume flow into the piston, J_1, must equal the luminal area, A_0, times the rightward velocity of the piston. Therefore,

$$h_{21} = \left.\frac{J_2}{J_1}\right|_{F_2=0} = -\frac{1}{A_0} \qquad (53)$$

and $h_{12} = -h_{21}$, as the Onsager theorem says it must.

5.2 Bidirectionality and Its Implications: Driving-Point Impedance

Reciprocity implies bidirectionality (Gibbs free energy can be transferred in both directions across the transducer), which in turn implies that what happens in either of the physical realms connected to a transducer will affect what happens in the other physical realm. This implication becomes explicit in the expressions for driving-point impedance or admittance. Imagine an acoustic source on the left side of the piston, and mechanical impedance (e.g., a combination of masses, springs, and dashpots) on the right, as in Fig. A.2. Here, the mechanical impedance is designated simply Z_M. For the general linear situation, one would have

$$F_1 = h_{11}J_1 + h_{12}F_2$$
$$J_2 = h_{21}J_1 + h_{22}F_2 = -\frac{F_2}{Z_M} \qquad (54)$$

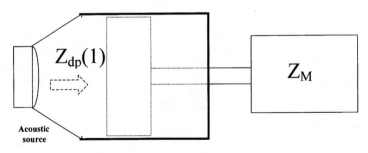

FIGURE A.2. Depiction of an acoustic source connected to a mechanical load (impedance Z_M) through a piston. The resulting pneumatic impedance, $Z_{dp}(1)$, faced by the source depends on both Z_M and the properties of the piston (see text).

from which

$$Z_{dp}(1) = \frac{F_1}{J_1} = h_{11} - \frac{h_{12}h_{21}}{h_{22} + 1/Z_M} \quad (55)$$

where $Z_{dp}(1)$ is the (driving-point) impedance that the acoustic source faces as it drives the transducer from the pneumatic realm. It should not be surprising that $Z_{dp}(1)$ depends strongly on what is connected to the transducer in the other realm (i.e., on Z_M). For this passive transducer (which has no spinning elements), with h_{11} and h_{22} both equal zero,

$$Z_{dp}(1) = \frac{Z_M}{A_0^2} \quad (56)$$

Note that with the current choices of conserved entities (volume in the pneumatic realm, axial displacement in the rigid-body mechanical realm) the SI unit of $Z_{dp}(1)$ is 1.0 joule s per m^6 and that of Z_M is 1.0 joule s per m^2. The factor $1/A_0^2$ translates the rigid-body mechanical impedance, Z_M, into the pneumatic impedance, $Z_{dp}(1)$.

6. Terminal Impedances and Reflected Waves

Imagine the piston with its very short cylinder (luminal area A_0) connected to the end ($x = L$) of an air-filled tube (also with luminal area A_0). $Z_{dp}(1)$ now forms what is known as a *terminal impedance* for the acoustic path through the air in the tube. Applying form 1 of the transmission-line equation (Eq. 35), one has

$$F_1 = F_f(L) + F_r(L) = Z_{dp}(1)J_1 \qquad J_1 = J_f(L) - J_r(L)$$
$$F_f(L) = Z_0 J_f(L) \qquad F_r(L) = Z_0 J_r(L) \quad (57)$$

where F_1 and J_1 are the sound pressure and volume flow applied to the piston by the air in the tube; Z_0 is the characteristic impedance of the acoustic path through the air in the tube. From Equation 57 one can derive the reflection coefficient $F_r(L)/F_f(L)$ of the air-transducer interface:

$$\frac{F_r}{F_f} = \frac{Z_{dp}(1) - Z_0}{Z_{dp}(1) + Z_0} \quad (58)$$

If the goal is to maximize the transfer of Gibbs free energy from the pneumatic realm (acoustic energy) to the rigid-body mechanical realm (mechanical energy), then one wants to minimize the amplitude of the reflected wave, F_r. It is clear from Equation 58 that F_r will be reduced to zero if $Z_{dp}(1) = Z_0$. In other words, there will be no reflection if the acoustic path is terminated with an impedance equal to the characteristic impedance of the path. If that is not so, one often can make it so by inserting an appropriate passive transformer on one side or the other of the transducer (the piston).

6.1 Transformers and Impedance Matching

The simplest passive transformers are constructed as a cascade of two passive transducers connected back to back. Consider the system depicted in Figure A.3. Here are two rigid-walled cylindrical tubes containing different fluids. Between the two fluids is a cascade of two pistons connected back to back in a translational rigid-body mechanical realm. Being reversed, the second piston is represented by a two-port model in which F_1 is force, J_1 is rightward translational (axial) velocity, F_2 is pressure, and J_2 is leftward volume flow. In that case, if one assumes as before that the piston is perfectly rigid, that its mass is negligible, and that it operates with no resistance,

$$h_{11} = 0 \quad h_{12} = A_2 \quad h_{12} = -A_2 \quad h_{22} = 0 \tag{59}$$

Applying Equation 55 to this situation (Z_M becomes Z_{02}), one finds that the mechanical driving-point impedance at the left side of the second piston is

$$Z_{dp}(p2) = A_2^2 Z_{02} \tag{60}$$

Applying the same equation to the first piston (the piston is not reversed, Z_M becomes $Z_{dp}(p2)$), one finds that the driving-point impedance at the left side of the complete transformer is

$$Z_{dp}(p1) = \frac{A_2^2}{A_1^2} Z_{02} = \tau^2 Z_{02} \tag{61}$$

To avoid reflections of sound waves as they pass from fluid 1 to fluid 2 through the transformer formed by the pistons, one requires

$$\frac{A_2^2}{A_1^2} Z_{02} = Z_{01} \tag{62}$$

Substituting the expression for Z_0 from Equation 44, one can restate this requirement in terms of the densities and adiabatic bulk moduli of the two fluids:

$$A_2 \sqrt{\rho_2 E_{B2}} = A_1 \sqrt{\rho_1 E_{B1}} \tag{63}$$

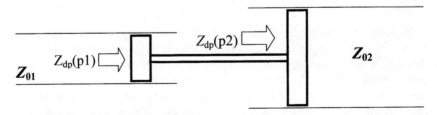

FIGURE A.3. Depiction of an acoustic transformer comprising back-to-back pistons rigidly connected in the rigid-body mechanical realm. The transformer couples an acoustic medium with characteristic impedance Z_{01} to one with characteristic impedance Z_{02}.

The symmetry of this relationship implies that when it is true, sound waves pass through the transformer in either direction without reflection.

Equation 61 is a form of the classic transformer equation, τ being the *transformer ratio*. By adjusting τ (the ratio of piston areas), one could, in principle, transform any real-valued impedance (Z_{02}) to match any other real-valued impedance (e.g., Z_{01}). A mechanical transformer in which τ is less than 1.0 is said to provide *mechanical advantage*. The passive device that does so comprises either a single pair of back-to-back passive transducers (a simple machine), or multiple pairs of such transducers (a compound machine). In addition to pistons and their relatives (including flexible diaphragms), the repertoire of passive mechanical transducers includes the wheel (transfers free energy between translational and rotational realms), which is the basis of gear systems and pulley systems, and the semilever—a rigid bar between a fulcrum and a point of applied force (also transfers free energy between translational and rotational realms), which is the basis of lever systems. For gear or pulley systems, τ is the ratio of wheel (e.g., pulley or gear) radii; for lever systems it is the ratio of semilever lengths.

7. Driving-Point Impedance of a Terminated Acoustic Path

In the previous subsection, the piston was modeled as a circuit element with two distinct ports: port 1 facing the pneumatic realm, and port 2 facing the rigid-body mechanical realm. The same sort of model can be useful as well in a single realm. Suppose the length of a fluid-filled, rigid-walled cylindrical tube were L. If one were concerned only with the acoustic variables (Fs and Js) at the ends of the tube, one could translate the transmission-line model (Eq. 36) to a two-port model, one port representing one end of the tube, the other port representing the other end. For that purpose, it is useful to use the two-port impedance (z) parameters instead of the two-port hybrid (h) parameters:

$$F_1 = z_{11}J_1 + z_{12}J_2 \qquad F_2 = z_{21}J_1 + z_{22}J_2 \qquad (64)$$

where F_1 and J_1 in Equation 64 correspond to $F(0)$ and $J(0)$, respectively, in Equation 36; F_2 and J_2 in Equation 64 correspond to $F(x)$ and $J(x)$ in Equation 36 when x is set equal to L. The corresponding z parameters are derived as follows:

$$z_{11} = \left.\frac{F_1}{J_1}\right|_{J_2=0} = \left.\frac{F(0)}{J(0)}\right|_{J(L)=0}$$

$$J_2 = J(L) = 0 = J(0)\cosh(\sqrt{zy}\,L) - Y_0 F(0)\sinh(\sqrt{zy}\,L)$$

$$z_{11} = Z_0 \frac{\cosh(\sqrt{zy}\,L)}{\sinh(\sqrt{zy}\,L)}$$

$$z_{21} = \left.\frac{F_1}{J_2}\right|_{J_1=0} = \left.\frac{F(0)}{J(L)}\right|_{J(0)=0} = \frac{Z_0}{\sinh(\sqrt{zy}\,L)} \qquad (65)$$

and, by symmetry, $z_{22} = z_{11}$, $z_{21} = z_{12}$. The equality of z_{12} and z_{21} (or, equivalently, $h_{12} = -h_{21}$) also is stated in a general reciprocity theorem that applies to all passive, linear circuit models in which rotation (sometimes called gyration) is not represented. Proof of the theorem can be found in standard circuit-theory texts. For the z parameter set, Equation 55 becomes

$$Z_{dp}(1) = \frac{F_1}{J_1} = z_{11} - \frac{z_{12}z_{21}}{z_{22} + Z_L} \tag{66}$$

where Z_L is the (*load*) impedance connected to port 2. In this case, Z_L is the external impedance that the sound wave encounters as it reaches the far end of the tube (where $x = L$). Combining Equations 65 and 66, one finds the general expression for the driving-point impedance that an acoustic wave would face as it enters the tube (at $x = 0$).

$$Z_{dp} = Z_0 \frac{Z_L \cosh(\sqrt{zy}\, L) + Z_0 \sinh(\sqrt{zy}\, L)}{Z_0 \cosh(\sqrt{zy}\, L) + Z_L \sinh(\sqrt{zy}\, L)} \tag{67}$$

Notice that if Z_L equals Z_0, so that there is no reflected wave, then Z_{dp} also equals Z_0. What Equation 67 describes is the impact of the reflected wave on the driving point impedance when Z_L does not equal Z_0. For the fluid-filled tube, with a steady tone (frequency $\omega_0 = 2\pi f_0$) applied to the near end ($x = 0$), the driving-point impedance becomes

$$Z_{dp} = Z_0 \frac{Z_L \cos(2\pi \frac{L}{\lambda}) + iZ_0 \sin(2\pi \frac{L}{\lambda})}{Z_0 \cos(2\pi \frac{L}{\lambda}) + iZ_L \sin(2\pi \frac{L}{\lambda})} \tag{68}$$

where λ is the wavelength of the tone in the fluid (e.g., in air or in water):

$$\lambda = \frac{\phi}{f_0} = \frac{2\pi \phi}{\omega_0} \tag{69}$$

As the frequency of the tone increases and L/λ shifts from 0 to 0.25, Z_{dp} will shift from Z_L to Z_0^2/Z_L. For intermediate values of L/λ, the values of Z_{dp} will be complex numbers, implying phase differences between F_1 and J_1. The size of the phase difference will depend strongly on the frequency of the tone.

7.1 The Significance of Phase Difference: Passivity Constraint

To understand the significance of a phase difference between F_1 and J_1, one can consider the flow, $P_1(t)$, of energy into the tube at port 1, instant by instant:

$$P_1(t) = F_1(t) J_1(t) \tag{70}$$

Appendix: Useful Concepts from Circuit Theory 389

From this, one computes the total energy, E_1, entering the tube over each full cycle of the sinusoidal waveform. Let

$$J_1(t) = A \cos \omega t$$
$$Z_{dp} = K e^{i\beta}$$
$$F_1(t) = KA \cos(\omega t + \beta) \qquad (71)$$

In that case,

$$E_1 = \int_0^{1/f_0} KA^2 \cos(2\pi f_0 t) \cos(2\pi f_0 t + \beta) dt = \frac{A^2}{2f_0} K \cos \beta = \frac{A^2}{2f_0} \text{Re}\{Z_{dp}\} \qquad (72)$$

E_1 is maximum when $\beta = 0$; it is positive when $0 < \beta < \pi/2$ or $0 > \beta > -\pi/2$; it is zero when $\beta = \pi/2$ or $\beta = -\pi/2$; and it is negative when $\pi/2 < \beta < \pi$ or $-\pi/2 > \beta > -\pi$. The combination of the tube and Z_L, as posited here, is passive; it cannot deliver more acoustic energy output than that which has been put into it. Therefore,

$$E_1 \geq 0$$

Passivity constraint

$$-\frac{\pi}{2} \leq \beta \leq \frac{\pi}{2} \qquad (73)$$

Which requires simply that

$$\text{Re}\{Z_{dp}\} \geq 0 \qquad (74)$$

If Z_{dp} were purely inertial or purely capacitive, then $\text{Re}\{Z_{dp}\}$ would be zero. In that case, the energy delivered into the tube during one half cycle will be delivered back out again during the other half cycle.

7.2 The Evolving Ear's Options for Maximizing Free-Energy Transfer

If, under sinusoidal steady-state conditions, one wishes to transfer Gibbs free energy from one acoustic path to another, and if the second path is finite in length (terminated), and if one wishes to use a transformer to translate the driving-point impedance of the second path to the characteristic impedance of the first, so as to maximize free-energy transfer over a wide range of frequencies, then Equations 68 and 72 establish the constraints that are imposed by the physics of the situation. The driving-point impedance of the second path must be a positive real number that is independent of frequency. Assuming Z_0 is such a number, there seem to be two ways to make that happen: (1) terminate the second path with an impedance equal to the characteristic impedance of that path, or (2) make Z_L real and make the second path so short that L/λ remains very small for all frequencies of interest. These would have been the possibilities open to evolution as it sculpted the paths and transformers for conveying acoustic energy into the ear.

8. Wall Effects

8.1 Effects of Wall Compliance

If the wall of the cylinder in the original scheme were compliant, if the diameter of the cylinder were small in comparison to the wavelengths of the sound, and if the pressure outside the wall remained fixed at p_0 (e.g., atmospheric pressure), then the expression for capacitance per unit length (Eq. 43) would become

$$\hat{C} = \frac{A_0}{E_B} + \hat{C}_{\text{wall}} \tag{75}$$

In other words, the compliance per unit length of the wall would add to that of the fluid. In the original formulation, the mass-related conserved entity (mass, particle-population, or volume) was accumulated in local compression of the fluid alone. Now it also is accumulated in local expansion of the cylinder wall. Thus, for a given incremental change in pressure, more of the conserved entity would be stored per unit length. This would make both the speed of the sound and the characteristic impedance of the sound path smaller than they would be in an extended volume of the fluid (see Eqs. 39 and 40). In the mammalian cochlea, the local characteristic impedance and speed of the sound wave evidently are determined almost entirely by the local compliance of the wall (the basilar membrane in that case), the contributions of the endolymph and perilymph compliance (i.e., the compliance of water) being negligible in comparison to that of the wall. Recall that the wavelength of a tone equals the speed of sound divided by the frequency of the tone. If the local speed of sound in the cochlea were that of the inner-ear fluids (i.e., in the neighborhood of 1500 m/s), then the wavelength of a 10-kHz tone, for example, would be in the neighborhood of 15 cm—several times the length of the cochlea. In modern cochlear models, the actual wavelength of a 10-kHz tone is a small fraction of the total cochlear length, implying that

$$C_{\text{wall}} \gg \frac{A_0}{E_B} \tag{76}$$

and that the bulk modulus of water contributes almost nothing to the characteristic impedance or the speed of the wave. Among other things, this means that the local particle displacements for waves propagating in the cochlear fluids will be much greater than for similar waves propagating in unbounded aqueous media (see subsection 4.2).

8.2 Effects of Viscosity

For a sinusoidal wave propagating axially in the forward direction through the fluid in the cylinder, the local volume velocity vector, $J(x,t)$, would reverse periodically (with the same period as the sine wave). The flow would be in the forward direction for one half cycle (when the local sound pressure, F, was positive), and in the opposite direction for the next half cycle (when the local sound pressure was negative). In developing the transmission-line model (see Eqs. 10 and 17) one assumed that the frequencies of the sign waves were sufficiently high to avoid the effects of viscosity. Under this assumption, the local axial particle velocity, $\vartheta(x,t)$, is independent of distance from the wall of the cylinder. Strictly speaking, this implies the existence of an infinite radial gradient of velocity at the interface between the wall and the fluid. Fluid dynamicists generally assume that the particle velocity of the fluid immediately adjacent to the wall must be zero (i.e., that the fluid sticks to the wall), which would move the infinite gradient slightly into the fluid itself. Newton's law of viscosity states that shearing tension, $\tau_{\theta z}$ (tangential force per unit area), between fluid layers is directly proportional to the particle velocity gradient taken normal to those layers. For the fluid in the cylinder, this translates to

$$\tau_{\theta z} = \eta \frac{d\vartheta}{dr} \tag{77}$$

r is the where radial coordinate, with its origin taken to be the central axis of the cylinder, and η is the viscosity of the fluid. What one has assumed, then, are extremely large (near-infinite) shearing forces in the fluid layers immediately adjacent to the wall. These would impose a drag on the more central layers, slowing them down and thus reducing the shear gradient, $d\vartheta/dr$. In this manner, the drag effects would move radially inward, toward the central axis of the cylinder. If the duration of one half cycle of the propagating sine wave were sufficiently short, the drag effects would not have time to spread far from the wall. At the end of the half cycle, the particle velocity would reverse and the spreading process would start over again, with shearing forces in the opposite direction. Throughout the full period of the sine wave, the local axial particle velocity would be nearly independent of r over most of the cross-sectional area (A_0) of the cylinder. In the development of Equations 10 to 12, it had been assumed to be independent of r over the entire area.

If the period of the propagating sine wave were sufficiently long, on the other hand, the drag effects would have time to reach the central axis of the cylinder and approach very close to a steady-state condition known as fully developed flow. Because its effects take time to propagate to the central axis, however, the viscous drag on the wall does not become fully effective until the fluid has moved a sufficient distance from the entrance of the cylindrical tube. For small-amplitude acoustic waves, that distance is comparable to or less than the diameter of the tube (e.g., see Talbot 1996). When the flow through the cylinder is fully developed, the particle velocity is a parabolic function of r (the radial velocity profile has the shape of a parabola centered on the central axis of the cylinder). The relationship between F (pressure) and J (axial volume flow) in that case will be given by Poiseuille's equation for flow in a cylinder,

$$\frac{dF}{dx} = -z_R J$$

$$z_R = \frac{8\pi\eta}{A_0^2} \tag{78}$$

This is a resistive relationship. Thus, for low-frequency waves, the series impedance, z, in Equation 22 will be resistive rather than inertial. For high-frequency waves it will be inertial, with inertance per unit length given by Equation 41 (see also Eq. 34).

$$\frac{dF}{dx} = -z_I J$$

$$z_I = i2\pi f \frac{\rho_0}{A_0} \tag{79}$$

where f is the frequency of the sine wave. For fixed total volume flow, as the drag effects spread toward the center of the tube, the volume velocity there increases (making up for the reduction in volume velocity near the walls). This increases the effective inertance of the fluid slightly. For frequencies in the neighborhood of f_c (see Eq. 80) acousticians use $\hat{I} = \frac{4\rho}{3A_0}$.

8.2.1 A Cutoff Frequency for Acoustic Waves in Tubes or Ducts

When

$$f = f_c = \frac{4\eta}{A_0 \rho} \tag{80}$$

the magnitudes of the inertial and resistive components of impedance per unit length will be approximately equal. For long tubes, therefore, the assumption that viscous effects can be ignored will be applicable if $f >> f_c$. If f is comparable to or less than f_c, the effects of viscosity will be severe, and the energy of sound waves in the tube will be dissipated (converted to heat) very quickly. For pure water at 20°C (Table A.1),

TABLE A.1. Properties of pure water.

Temperature (°C)	Density (kg/m³)	Viscosity (kg/m s)
3.98	1000.0	0.1567
10	999.7	0.1307
15	999.1	0.1139
20	998.2	0.1002
25	997.0	0.0890
30	995.7	0.0798
35	994.1	0.0719

$$f_c \approx \frac{0.0004}{A_0} \tag{81}$$

where the cross-sectional area (A_0) of the tube is given in square meters, and f_c is given in Hz. For a periotic duct whose cross-sectional area is n square millimeters, for example, the cutoff frequency would be approximately $400/n$ Hz. With Equation 81 in hand, it is interesting to contemplate the cross-sectional areas of various ducts and canals in both the auditory and the vestibular sides of the vertebrate inner ear, as well as the canals associated with lateral-line structures and those in the bases of the antennules of crustaceans (e.g., see Lewis 1984; Lewis and Narins 1998).

9. Resonance

9.1 Helmholtz Resonators

When a lumped inertance is connected in series with a lumped capacitance, the resulting impedance is the sum of the impedances of the two elements:

$$Z_s(\omega) = i\omega I + \frac{1}{i\omega C} = \frac{1 - \omega^2 IC}{i\omega C} \tag{82}$$

Similarly, when a lumped inertance is connected in parallel with a lumped capacitance, the resulting admittance is the sum of the admittances of the two elements:

$$Y_p(\omega) = i\omega C + \frac{1}{i\omega I} = \frac{1 - \omega^2 IC}{i\omega I} \tag{83}$$

Z_s and Y_p approach 0 as ω approaches $1/\sqrt{IC}$. This means that their reciprocals, Y_s and Z_p, approach infinity under the same circumstances. This critical value of ω is the resonant frequency, ω_r,

$$\omega_r = \frac{1}{\sqrt{IC}} \tag{84}$$

A commonly cited acoustic example is a volume of air enclosed in a rigid-walled container, connected to the rest of the acoustic circuit through a short, small-diameter, rigid-walled air-filled tube. This forms a series IC circuit, with the values of I and C given by Equations 9 and 12, respectively:

$$\omega_r = \sqrt{\frac{V_{01}\rho_0}{A_{01}^2 c_0^2} \frac{c_0 n_{02}}{\gamma P_0}} = \sqrt{\frac{L_{01} V_{02}}{A_{01}} \frac{\rho_0}{\gamma p_0}} \tag{85}$$

where the subscript 01 refers to the tube, and the subscript 02 refers to the enclosed volume of air. For short tubes, acousticians sometimes apply a correction to account for the momentum carried beyond the end of the tube, replacing the tube length, L_{01}, with $L_{01} + 1.7R_{01}$, where R_{01} is the tube radius. This configuration, known as a *Helmoholtz resonance*, is used in sound production by animals (Ewing 1989) and some musical instruments (whistle, ocarina, body of violin; see Olson 1967).

Imagine that, beginning at some initial time ($t = 0$) a constant-amplitude, sinusoidal sound pressure waveform at frequency ω_r were applied (from a source of some sort) at the mouth of the tube (the end of the tube opposite the enclosed volume). The corresponding volume flow through the tube, into the enclosed volume, would approximate a ramp-modulated sine wave (e.g., $t\sin\omega t$). Cycle by cycle, the amplitude of the sinusoidal volume flow of air through the tube would grow. Correspondingly, the amplitude of the sinusoidal pressure variations in the enclosed volume would grow. Those pressure variations would be precisely 180 degrees out of phase with the pressure variations at the source (at the other end of the tube). Thus the two pressure variations would reinforce one another in the creation of a net pressure difference from one end of the tube to the other. It is that net pressure difference that drives the flow through the tube. The growing flow through the tube and pressure in the enclosure represent a growing amount of acoustic energy stored in the resonant system. In the idealized example presented here, with no mechanism for dissipation of acoustic energy, it would continue to accumulate forever; sinusoidal steady state would never occur. In a real situation, at least two things will happen to the acoustic energy: (1) some fraction of it will be converted to heat by viscous resistance in the system, and (2) some fraction will be radiated from the mouth of the tube and the walls of the enclosure. When the total amount of acoustic energy gained from the source during each cycle is balanced precisely by the acoustic energy lost during each cycle, then sinusoidal steady state will have been achieved.

9.2 Standing-Wave Resonators

In the Helmholtz resonator, one presumes that the wavelength of sound at the resonant frequency is very long in comparison to the dimensions of the enclosed volume and the tube. When that is not so, another form of resonance—a *standing-wave resonance*—can occur. Consider, for example, a rigid-walled tube filled with air. The acoustical driving-point impedance of the air in the tube, viewed from one end of the tube, is given by Equation 68. If the other end of the tube were sealed, making Z_L exceedingly large, then the driving-point impedance would be approximated very well by the following expression:

$$Z_{dp} = Z_0 \frac{\cos\left(2\pi\frac{L}{\lambda}\right)}{i \sin\left(2\pi\frac{L}{\lambda}\right)} \qquad (86)$$

which is zero for all values of wavelength (λ for which the tube length (L) divided by wavelength is an odd multiple of one quarter (e.g., $L/\lambda = 1/4, 3/4, 5/4$ ...). The inertial effects on the flow of air in the tube extend slightly beyond the open end of the tube. Acousticians correct for this by adding (for each open end) $0.82R$ to L in these expressions, where R is the tube radius (Rayleigh 1926; see Olson 1957). At each one of the corresponding frequencies (see Eq. 69), the acoustic path through the air in the pipe exhibits a resonance. As in the Helmholtz resonator, imagine that, beginning at some initial time ($t = 0$) a constant-amplitude, sinusoidal sound pressure waveform at one of the resonant frequencies were applied at the mouth of the tube (the open end of the tube). The time course of the amplitude of the corresponding sinusoidal flow would be essentially the same as that of the Helmholtz resonator, growing linearly with time in the idealized tube. Within the tube would be a standing wave whose amplitude also increases linearly with time. Sinusoidal steady-state conditions would never be achieved. In a real tube, the energy of the standing wave would continue to grow until the energy lost to viscous resistance and acoustic radiation during each cycle of the sound waveform was precisely equal to the energy provided during each cycle by the sound pressure source driving the tube. Sinusoidal steady state would be achieved when that energy balance was achieved.

Standing-wave resonances occur in a wide range of mechanical systems, including flexible strings under tension, rigid bars, stretched membranes, and rigid circular plates (e.g., see Olson 1957). The resonances in strings and bars are easily identified by appropriate application of the transmission-line model.

10. Active Transducers and Amplifiers

A circuit model can be used to represent the flow of free energy from physical realm to physical realm and from place to place within any given physical realm. For a model of the auditory periphery, one form of free energy that would be of interest would be that which begins as the free energy being carried by an acoustical input signal (e.g., a tone), representing a stimulus applied to the external ear or its piscine equivalent. Any element of the model that merely stores that energy, dissipates that energy, or transfers that energy from one place to another or from one physical realm to another is, by definition, *passive*. Circuit models also may include elements in which free energy that originated in the input signal is used to gate the flow of free energy into the model from another source (e.g., a battery of some kind). These often are represented as two-port elements. The free energy that originated in the input signal enters port 1. The gated energy from the battery flows out of port 2. If ports 1 and 2 are represented in the model as being in the same physical realm (e.g., both in the pneumatic realm), then the element is an (*active*) *amplifier*. If the two ports are depicted as being in different physical realms, then the element is an *active transducer*.

10.1 Passivity Implies Bidirectionality

The active nature of the two-port element is reflected in its parameters. Among other things, a passive, linear process must be reciprocal or antireciprocal. Lack of reciprocity or antireciprocity in a linear two-port element implies that the element is active. As a corollary of the Onsager reciprocity theorem, for example, one can state the following: if a transduction process (1) is not based on spinning elements (e.g., a spinning top or rotating charge), (2) operates linearly (e.g., over a small range about the origin), and (3) is not bidirectional, then it *must be active*. Linear, passive transduction processes based on spinning elements are antireciprocal ($h_{12} = h_{21}$, $z_{12} = -z_{21}$) (see Lewis 1996). To be passive, not only must a linear transduction process be bidirectional, it must be either reciprocal or antireciprocal (equally effective in both directions).

10.2 Gated Channels Are Active

On the other hand, gated channels are abundant in nervous systems. In circuit models, the elements representing gated channels would be active. The mechanical input to a stereovillar bundle in an inner-ear auditory sensor carries free energy that originated with the acoustical input to the external ear. This input modulates the opening and closing of the strain-gated channels, through which flows electric current driven by the endolymphatic voltage. The source of the free energy carried by the resulting electrical signal (hair cell receptor potential) is the endolymph battery, not the acoustical input to the ear. Acoustically derived energy modulates the flow of electrical energy from a battery—a quintessential active transduction process. In some instances the hair cell receptor potential, in turn, is applied to a transducer that produces force and motion in the stereovillar bundle (see Manley and Clack, Chapter 1, section 5.1). It is not clear, yet, whether this process is passive, involving merely the transfer of free energy from the receptor potential to the mechanical motion of the bundle, or active, involving the gating of energy from some other source, such as an adenosine triphosphate (ATP) battery, to the motion of the bundle.

10.3 Positive Feedback Can Undamp a Resonance for Sharper Tuning

Whether or not the reverse transduction process (from receptor potential to stereovillar motion) is active, the forward transduction process (from stereovillar motion to receptor potential) definitely is. Thus there is a loop in which there is the possibility of power gain. It has been proposed that such gain could compensate, in part, for the power loss that must be imposed by the viscosity of the water surrounding the stereovillar bundle (see Manley and Clack, Chapter 1, section 5). Assuming that tuning of the hair cell is accomplished by mechanical resonance associated with the stereovillar bundle, for example, one can depict the essential ingredients of this proposition with the circuit model of

Figure A.4. Here the mechanical side is depicted on the left, the electrical side on the right. F_{in} is the force applied to the apex of the bundle by the acoustic input signal; Z_b is the mechanical impedance of the bundle. The upper two-port element represents an idealized active forward transducer [zero driving-point impedance into port 1 (left), zero driving-point admittance into port 2 (right), zero coupling from port 2 to port 1]. The lower two-port element represents an idealized active reverse transducer [zero driving-point admittance into port 2 (right), zero driving-point impedance into port 1 (left), zero coupling from port 1 to port 2]. The hair cell electrical properties are represented by an admittance, Y_m, which relates the receptor potential, F_m, to the receptor current, J_m. The translational motion (velocity J_b) of the apex of the stereovillar bundle is applied through the mechanical sides of both the forward and reverse transducers. For that reason, the corresponding ports are represented as being in series. This in turn means that the force (F_{fb}) developed by the reverse transducer is depicted as being subtracted from the input force:

$$J_b = \frac{F_{in} - F_{fb}}{Z_b}$$

$$J_m = -k_b(i\omega)J_b$$

$$F_m = \frac{J_m}{Y_m} = \frac{-k_b(i\omega)J_b}{Y_m}$$

$$F_{fb} = k_m(i\omega)F_m = \frac{-k_m(i\omega)k_b(i\omega)J_b}{Y_m}$$

$$F_{in} = Z_b J_b + F_{fb} = \left(Z_b - \frac{k_m(i\omega)k_b(i\omega)}{Y_m}\right)J_b \qquad (87)$$

If it were a simple resonance, the impedance of the bundle (to very small translational motion of its apex) could be written as follows:

$$Z_b = i\omega l_b + \frac{1}{i\omega C_b} + R_b \qquad (88)$$

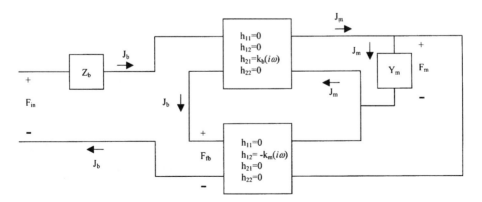

FIGURE A.4. Circuit model depicting forward and reverse transduction in a hair cell.

where I_b is the inertance of the bundle, C_b is its elastic compliance, and R_b is the resistance of the bundle (owing at least in part to the viscosity of water). With the transducers in place, the relationship between bundle motion (velocity) and input force becomes

$$J_b = \frac{F_b}{Z_b - \dfrac{k_m(i\omega)k_b(i\omega)}{Y_m}} = \frac{F_b}{i\omega I_b + \dfrac{1}{i\omega C_b} + R_b - \dfrac{k_m(i\omega)k_b(i\omega)}{Y_m}} \qquad (89)$$

The net resistance, which damps the resonance, is equal to

$$R_b - \mathrm{Re}\left\{\frac{k_m k_b}{Y_m}\right\} \qquad (90)$$

To be effective for undamping the resonance, the quantity $\mathrm{Re}\left\{\dfrac{k_m k_b}{Y_m}\right\}$ should be greater than zero and less than R_b. If it is greater than R_b, the resonance will oscillate spontaneously. If it remains in the proper range, the pair of active transducers will operate as a stable amplifier (Manley and Clack, Chapter 1, section 5). The effect of this on tuning is depicted in Figure A.5, which shows the phase and amplitude tuning curves of a damped resonance. The bottom panel in Figure A.5 shows the temporal response (e.g., velocity) of the resonance as its residual energy decays after the input has ceased. Notice that as the resonance becomes less damped, its tuning peak becomes sharper but it requires more time to rid itself of residual energy. The latter reflects both the advantage and the disadvantage of resonance tuning. It allows the response to an ongoing input signal to accumulate over time. In a quiet world, where the threshold of hearing is limited by the internal noise of the ear, this should enhance the ability of the hearer to detect the presence of the signal. At the same time, it would reduce the temporal resolution with which the hearer could follow amplitude changes in the signal, once the signal has been detected.

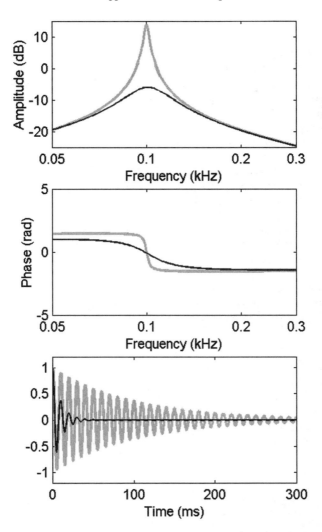

FIGURE A.5. Tuning of a 100-Hz resonance with a high degree of damping (solid black line) and a reduced degree of damping (thick gray line). The bottom panel shows the time course of decay for residual excitation in the resonance.

References

Baranek LL (1954) Acoustics. New York: McGraw-Hill.
Ewing AW (1989) Arthropod Bioacoustics. Ithaca, NY: Cornell University Press.
Lewis ER (1984) Inertial motion sensors. In: Bolis L, Keynes RD, Maddrell SHP (eds) Comparative Physiology of Sensory Systems. Cambridge: Cambridge University Press, pp. 587–610.
Lewis ER (1996) A brief introduction to network theory. In: Berger SA, Goldsmith W, Lewis ER (eds) Introduction to Bioengineering. Oxford: Oxford University Press, pp. 261–338.
Lewis ER, Narins PM (1998) The acoustic periphery of amphibians: anatomy and physiology. In: Fay RR, Popper AN (eds) Comparative Hearing: Fish and Amphibians. New York: Springer-Verlag, pp. 101–154.
Morse PM (1981) Vibration and Sound. New York: Acoustical Society of America.
Olson HF (1957) Acoustical Engineering. Princeton, NJ: D. van Nostrand.
Olson HF (1967) Music, Physics and Engineering. New York: Dover.
Onsager L (1931) Reciprocal relations in irreversible processes: I, II. Phys Rev 37:405–426; 38:2265–2279.
Rayleigh (1926) Theory of Sound. London: Arnold.
Talbot L (1996) Fundamentals of fluid mechanics. In: Berger SA, Goldsmith W, Lewis ER (eds) Introduction to Bioengineering. Oxford: Oxford University Press, pp. 101–132.

Index

Absorption, sound in air, 36–37
Acanthostega, evolution, 165
   inner ear, 133–134
   skull, 131, 132–133
Acetylcholine, hair cell efferent neurotransmitter, 59
Acoustic physics, 369ff
Acoustic scene, hearing evolution, 100
Acoustic wave equation, derivation, 372ff
Acoustic waves, properties, 41ff
Acoustics, resonance, 393–395
   review, 369ff
Actinopterygia (ray-finned fishes), 6
Active amplification, hair cell, 18–19, 364–365
Active process, lizards, 18–19, 207
   reptiles, 69
   reptiles and amphibians, 211
Afferent terminal number, birds, 246ff
Afferents, S1 and S2 in saccular nerve, 62
African clawed frog, *see Xenopus laevis*
Agamid lizards, hair cell types, 69
Agnatha, evolution of ear structures, 100ff
   earliest ears, 97
   inner ear, 75–76
Allosaurus fragilis, vestibular apparatus, 234ff
Alosa sapidissima (American shad), ultrasonic hearing, 117
Ambystoma tigrinum (tiger salamander), amphibian papilla, 185
   torus semicircularis, 322
American shad, *see Alosa sapidissima*
Amia calva (bowfin), labyrinth, 106–107

Amniote, basilar papilla origin, 8ff
   evolution of stapes, 139, 140ff
   hair cell tuning, 15ff
   hearing principles, 15ff
   inner ear, 8
   middle ear evolution, 137ff
   phylogeny, 10
   skull evolution, 137ff
Amos gene, mechanoreceptors, 84
Amphibian papilla, 7–8, 66, 169ff
   Ambystoma tigrinum, 186
   electrical tuning of hair cells, 187, 188
   epithelia structure, 185ff
   hair cell orientation patterns, 188–189
   pressure reception, 182ff
   relationship to macula neglecta, 183–184
   tadpole, 177
Amphibians, saccule, 169
   amphibian papilla, 7–8, 66, 169ff, 187–189
   auditory brainstem, 300
   auditory midbrain organization, 327
   basilar papilla, 7–8, 172–173
   development of inner ear, 189–190
   diversity of ear structure, 168ff
   ear anatomy, 166ff
   ear evolution, 135–137, 164ff
   evolution of middle ear, 173ff
   hair cells, 64ff
   hearing, 166–168
   impedance match, 173
   middle ear, 173ff
   origin and evolution, 2–4, 165–166
   origin of basilar papilla, 184–185
   papilla neglecta, 169–170, 183–184

Amphibians, saccule *(continued)*
  periodic canal and hearing, 179–181
  saccule and seismic detection, 178, 181–182
  sound pathways, 166–168
  stapes, 166–168, 173
  time pattern discrimination, 331
  tympanic ear, 135–137, 173
  underwater hearing, 164
  see also Frog, Salamander, Tadpole
Amphioxus (lancelet), hearing, 97
Amphisbaenids, basilar papilla, 214
Amplitude modulation, ITD processing, 319–320
  rate representation in midbrain, 331
Anabantids, hearing mechanisms, 110, 113
  sound production, 116
Analogy, defined, 290
Anas platyrhynchos (mallard duck), audiogram, 232–233
  hair cells, 71
Ankle links, hair cells, 59
Anterior forebrain pathway, birdsong, 342
Anterior octaval nucleus, fish hearing, 299–300
Antrozous fuscus, duration coding in midbrain, 333
Antrozous pallidus (pallid bat), duration coding in midbrain, 333
Anurans, *see* Amphibians
Archaeopteryx (primitive bird), otic region, 143
  archosaur evolution, 226
Archistriatum, and arcopallium in birds, 337
Archistriatum, birdsong, 329, 342–343
Archosaur, basilar papilla compared to primitive mammals, 249–250
  hair cells, 11–13, 69ff
  hearing organ evolution and specialization, 13–14, 224ff
  nucleus magnocellularis, 301
Archosauromorphs, middle ear, 142–143
Arcopallium, and archistriatum in birds, 337
Arius felis (marine catfish), labyrinth, 108
  low-frequency hearing, 108
Ascaphus truei, hair cells, 64–65

Astronotus ocellatus (oscar), gentamycin effects, 62
  hair cells, 61–62
  lagena and gentamycin, 62
Atonal gene, 290
Atonal transcription factor, hair cells, 83–84
Audiograms, various fish species, 112ff
Auditory CNS, basic vertebrate plan, 297
  evolution, 289ff
  see also CNS
Auditory forebrain, 335ff
  common features, 336ff, 345
  dorsal and ventral pallium, 335ff
  primates, 339–340
  "what" and "where" functions, 339–340
Auditory midbrain, 298
  common features of processing, 335
  core and belt organization, 326–327
  descending connections, 329
  fishes and amphibians, 323ff
  inputs, 327–328
  and motor systems, 328–329
  nucleus centralis, 323ff
  projections to forebrain, 337
  response property transformations, 328
  temporal processing, 329ff
  various species, 325
Auditory space, population codes in forebrain, 340
Aves, *see* Birds
Aythya fuligula (tufted duck), basilar papilla dimensions, 237–238

Bandwidth of hearing, early mammals, 279–281
  evolution in fishes, 99
Barn owl, *see Tyto alba*
Basilar membrane, high-frequency hearing, 272
Basilar papilla, amphibian, 7–8, 66, 172–173
  archosaurs, 69ff
  evolution, 8ff, 184–185, 294
  evolution in reptiles, 213ff
  evolution of elongation, 19–20
  Gekko, 207
  hair cell orientation patterns, 188–189

hair cell resonances, 187–188
homology between amphibian and amniote structures, 184–185
innervation in amphibians, 185
Latimeria, 7
lizards, 200ff
mechanics, 211
pressure reception, 182ff
relationship to lagena, 184–185
reptiles, 11ff, 66ff
sauropsids, 300
stereovilli height in birds, 239–240
subpapillae in lizards, 220
tadpole, 177
turtle, 10–11
Basilar papilla frequency map, various bird species, 228ff
Basilar papilla innervation, implications in birds, 249ff
Basilar papilla length, frequency range of hearing in birds, 232–233
Basilar papilla width, various bird species, 238ff
Bats, cortical mechanisms, 344–345
  IC frequency maps, 327
bHLH gene, auditory CNS, 295
Bidirectional-type hair cells, hair cell groups in reptiles, 69, 209ff
Binaural cues, processing in birds and mammals, 310ff
Binaural processing, auditory forebrain, 339–340
  delay lines, 313
Biologically significant sounds, forebrain pathways, 338
Birchir, *see Polypterus*
Bird, ear, 11ff
  cochlear amplification, 271
  cochlear nuclei, 304ff
  hair cell innervation and specialization, 21, 23
  hair cells, 69ff
  hearing organ evolution and specialization, 224ff
  middle ear in early species, 143
  nucleus angularis, 301
  relationship to dinosaurs, 6, 235
  stereovilli height, 239–240
Bird basilar papilla, widths, 238ff

Bird body size, basilar papilla length, 234ff
Bird body weight, hair cell number, 236–237
Bird common ancestor, basilar papilla features, 237
Birdsong, and human speech, 340ff
BK channels, turtle hair cells, 67
Bloch's catfish, *see Pimelodus blochi*
Bobtail skink, *see Tiliqua regosa*
Bony fish, *see* Osteichthyes
Borylloides violaceus (sea squirt), coronal organ, 78–79
Botryllus schlosseri (sea squirt), coronal organ, 78–79
Bowfin, *see Amia calva*
Brachiosaurus brancai (sauropod dinosaur), inner ear, 234ff
Brainstem, special adaptations, 299ff
Branchiostoma floridae (lancelet), hair cells, 76
Brass instruments, sound sources, 34–35
Brevoortia patronis (menhaden), ultrasonic hearing, 117
Brienomyrus niger, lacking sound production, 116
Broca's area, speech, 341–342
Brown bullhead catfish, *see Ictalurus nebulosus*
Budgerigar, *see Melopsittacus undulatus*
Bullfrog, *see Rana catesbeiana*
Bullhead catfish, *see Ictalurus nebulosus*
Bushy cells, ITD processing, 310

Caecilians, hair cells, 64
Caenorhabditis elegans, mechanoreceptor channel, 80–81
Caiman crocodilus (caiman), basilar papilla, 248
  frequency map, 228ff
  hearing organ evolution and specialization, 224ff
  papillar length and body size, 236–237
  tall and short hair cells, 241
Calcium binding protein (S-100), hair cells, 61
Calcium/calmodulin kinase II, outer hair cells, 73
Calyx, Type I hair cells, 60

Calyx of Held, MNTB, 317
Calyx terminals, birds, 306
Carassius auratus (goldfish), audiogram, 112
 saccular hair cells, 62, 63
 sound discrimination, 99
 Weberian ossicles, 111
Catfish, *see also Ictalurus* sp., *Silurus* sp.
 hearing specializations, 109
 labyrinth, 107–108
 sound production, 116
Cave swiftlet, *see Collocalia linch*
Central auditory pathways, evolution, 289ff
 see also CNS
Central auditory system, *see* CNS
Cephalochordates, hair cells, 75ff
 inner ear, 76–77
Cephalopods, ciliated mechanoreceptors, 79
Ceratodus forsteri, labyrinth, 106–107
Channel catfish, *see Ictalurus punctatus*
Characids, sound production, 116
Chelonia, *see* Turtles
Chicken, *see Gallus gallus domesticus*
Chiclids, hearing mechanisms, 110
Chimaera monstrosa (ratfish), lateral line hair cells, 63
 labyrinth, 103
Chondrichthyes, labyrinth, 102ff
Chopper cells, in various taxa, 308
Chordates, non-vertebrate hair cells, 75ff
Chordotonal organs, *Drosophila*, 80
Cilial motility, *Drosophila*, 80
Circuit theory metamodel, 30–31, 369ff
 physics of sound, 369ff
Cladistic analysis, 1–2
Cladogram, fish hearing organs, 96
 lizard families, 212
 vertebrate, 3
Clupeids, ultrasonic hearing, 117
CNS, basic design, 296ff
 evolution, 289ff
 evolution of lateral line system, 292–293
 origins, 289ff
Cochlea, and lagena, 278–279
 coiling, 275, 278–279
 comparative anatomy, 265ff

 extinct mammals, 272ff
 function, 263–265
 marsupials, 265–267
 Mesozoic mammals and monotremes, 267ff, 273ff
 Morganucodon, 154
 therian mammals, 275–278
 tuning, 263–265
 see also Inner ear
Cochlear amplifier, birds and reptiles, 271
 evolution, 263–265, 270–272
 monotremes, 270–271
 otoacoustic emissions, 271–272
Cochlear function, *Monodelphis*, 264
 monotremes and marsupials, 264
Cochlear nucleus, and airborne sound, 304
 mammalian cell types, 306ff
 parallel evolution of cell types, 304
 special adaptations, 299ff
 species diversity, 297–298
 subdivisions, 301–302
 terrestrial vertebrates, 298
Cod, *see Gadus morhua*
Coelacanth, fossil ears, 130–132
 see also Latimeria chalumnae
Collocalia linch (cave swiftlet), vocalization, 224–225
Columba livia (pigeon), hair cells, 71
 infrasound, 225–226
Columbiformes, evolution, 227
Columella, origin in early tetrapods, 133
Combination-sensitive neurons, bat CNS, 333ff
Communication, archosaurs, 224ff
 in fishes, 115–116
 later evolution of, 98
Communication sounds, early evolution, 98
 forebrain processing, 335, 337
 HVc response in birds, 338
 midbrain processing, 331–332
 transmission effects in fishes, 99
Cortex, and dorsal pallium, 338
Cortical maps, bats, 344–345
Corydoras paleatus (lumpsucker), hearing ability, 113
Craniates, hair cells, 75ff
Croaking gourami, *see Trichopsis* sp.

Crocodilia, ear, 11ff
  hair cells, 69ff
  hearing organ evolution and specialization, 224ff
  middle ear, 143
Cross-correlation, nucleus laminaris, 311ff
Crossopterygia, 2
C-start, fishes, 117
Cuticular plate, archosaurs, 71
Cyclopterus lumpus (lumpfish), labyrinth, 107
Cyprinids, sound communication, 116

Damselfish, *see Eupomacentrus*
Delay lines, and auditory midbrain, 327–328
  binaural circuits, 313
Delay tuning, bat midbrain, 333–334
Descending auditory system, 299
Descending octaval nucleus, fish hearing, 299–300
Diapsids, ear, 11ff
Didelphis virginiana (Virginia opossum), hearing range, 262
Dinosaurs, *see also Brachiosaurus* sp.
  middle ear, 142–143
  relationship to birds, 6
Distalless gene, 290
Distortion Product Otoacoustic Emission, *see* DPOAE
DNA hybridization, evolution of birds, 227–228
Doppler shift, 42–43
Dorsal ventricular ridge (DVR), auditory forebrain, 337–338
Dorsolateral nucleus (DLN), anurans, 300ff
DPOAE, marsupial, 264–265
  Monodelphis, 264–265
Dromaius novaehollandiae (emu), auditory nerve frequency response, 228
  basilar papilla dimensions, 237–238
  hair cells, 70–71
Drosophila (fruit fly), ciliated mechanoreceptors, 75, 79ff
Duck-billed platypus, *see Onnithorhynchus anatinus*
Duration encoding, midbrain, 331, 333

Ear, *Acanthostega*, 132–133
  amphibian structure and diversity, 166ff
  approaches to study of evolution, 360
  bird, 11ff
  crocodilian, 11ff
  diapsids, 11ff
  early tetrapods, 132ff
  evolution in amphibians, 135–137
  lissamphibians, 135–137
  mammals, 15, 256ff
  salamander, 137, 174
  shape in different mammalian taxa, 258
  snakes and lizards, 13–14
  synapsids, 15
  tadpole, 177–178
  Temnospondyls, 135–137
  Xenopus laevis, 173–174
Ear evolution, and other sensory pathways, 292ff
Ear structure, evolution in fishes, 100ff
Eavesdropping, early hearing function, 98
Echnidna, *see Tachyglossus aculeatus*
Echolocation, midbrain processing, 333ff
Efferent innervation, bird basilar papilla, 245ff
  Carassius saccular hair cells, 62
Efferent synapses, archosaurs, 69
  hair cells, 59–60
Efferents, evolution, 291–292
  outer hair cell motility, 73
Elasmobranchs (cartilaginous fish), vestibular hair cells, 63
Electrical tuning, amphibian papilla hair cells, 187, 188
  amphibian saccular hair cells, 187, 188
  hair cell, 16–17, 46
  turtle hair cells, 203ff
Electromotility, outer hair cells, 73
Electroreception, relation to CNS evolution, 293
Elephant-nose fish, *see Gnathonemus petersii*
Eleutherodactylus coqui (coqui frog), sound duration processing, 331
Emu, *see Dromaius novaehollandiae*
Entosphenus japonicus (Japanese lamprey), hair cells, 63
Environment, combined effects on sound, 35ff, 39–41

406 Index

Eptesicus fuscus (little brown bat), duration coding in midbrain, 333
Eptesicus fuscus, IC frequency map, 327
Eupomacentrus dorsopunicans (dusky damselfish), audiogram, 112
Eupomacentrus partitus (bicolor damselfish), hearing ability, 109
  sound production, 116
Eurasian ruffe, *see Gymnocephalus cernuus*
European catfish, *see Silurus glanis*
Eusthenopteron, skull, 131, 132–133
Eutherians, *see* Mammals
Evolution, *Acanthostega*, 165
  active amplification by hair cells, 364–365
  amniote middle ear, 137ff
  amphibian ear, 164ff
  amphibian middle ear, 173ff, 176
  amphibian papilla, 169ff
  amphibians, 165–166
  auditory CNS, 368
  basilar papilla, 19–20
  cochlear amplifier, 270–272
  columella, 133
  ear in fishes, 365–366
  ear in tetrapods, 365–366
  ear ontogeny, 189–190
  hair cell, 6–8
  hair cell specializations, 20–21
  hair cell tuning, 15ff
  hair cells of different types, 366
  high-frequency hearing, 17–18
  impedance matching, 365–366
  lizard ear, 14
  mammals, 257ff
  mechanoreceptor cells, 361–363
  middle ear, 128ff, 145ff
  middle ear in archosauromorphs, 142–143
  middle ear in neodiapsids, 140ff
  middle ear in spenodontids, 141
  middle ear in synapsids, 145ff
  optimization of free energy transfer in ear, 389–390
  pressure receiver, 176ff
  selective pressures on hearing, 27ff
  sense of hearing, 362
  stapes, 133–135, 137
  stream segregation, 28–29
  tetrapod hearing, 128ff
  therapsid middle ear, 148–149
  tympanic ear, 146ff, 366
Extinct mammals, cochlea, 272ff
Extinct monotremes, cochlea, 273–275

Far field, 98
Fence lizard, *see Sceloporus occidentalis*
Fibroblast growth factor (FGF), 290
Field L, bird forebrain, 337, 343
  sharp tuning, 338
Fire salamander, *see Salamandra salamandra*
Fish, definition of term, 2
  ears and auditory CNS, 299–300
  evolution of ear, 6, 365–366
  hearing organs and parallel evolution, 95ff
  hearing sensitivity, 176, 363
  inner ear hair cells, 61ff
  temporal processing in midbrain, 330
Forebrain, *see also* Auditory forebrain
  uniqueness among mammals, 343ff
Fossil ears, coelacanth, 130–132
Fossils, early tetrapods, 132ff
Frequency map in birds, compression at low frequencies, 229ff
  slope, 231–232
Frequency range of hearing, bird basilar papilla, 232ff
Frog, *see also* Amphibians
  auditory forebrain, 337
  ear, 137
  interaural cues, 310
  sound production mechanisms, 136
Fundus, basilar papilla in geckos, 211

GABA, and MSO inhibition, 316
  hair cell efferent neurotransmitter, 59
  lateral lemniscus, 320–321
  NA and NL inputs in birds, 313ff
Gadus morhua (Atlantic cod), audiogram, 112
  hearing abilities, 109ff
  ultrasonic hearing, 117
Galliformes, evolution, 227
Gallus gallus domesticus (chicken), hair cell currents and channels, 71

hair cells, 71
  inner ear, 224–225
  nucleus laminaris, 312
Gastrulation homeobox gene (gbx), 290–291
GATA3 gene, auditory nerve, 295
Gecko, *see Gekko gecko*
Gekko gecko (Tokay gecko), basilar papilla, 210–211
  basilar papilla tonotopy, 207
  hair cell innervation, 216ff
Geological time scale, 5
Gerrhonotus multicarinatus (alligator lizard), frequency selectivity, 218ff
Globular bushy cells, in various taxa, 309
Glutamate, hair cell neurotransmitter, 59–60
Glycine, and MSO inhibition, 316–317
  hair cells, 60
Gnathonemus petersii (elephant-nose fish), audiogram, 112
Goldfish, *see Carassius auratus*
Gonorhynchiforms (milkfish), relationship to Otophysi, 128
Granite spiny lizard, *see Sceloporous orcutti*
Grass frog, *see Rana esculenta*
Guinea fowl, *see Numida meliagris*
Gymnocephalus cernuus (Eurasian ruffe), audiogram, 112
Gymnophiona, middle ear, 174–175

Habitat, hearing mechanisms in fishes, 114–115
Hagfish, inner ear, 75–76
  see also Myxine glutinosa
Hair cell, 55ff
  active amplification, 18–19
  archosaur, 11–13
  basilar papilla, 187–188
  channels and currents, 203ff
  currents in goldfish saccule, 62–63
  electrical resonance, 16–17, 46
  electrical tuning in amphibian papilla, 187, 188
  evolution, 6–8, 55ff, 81ff, 361–363
  evolution of active amplification, 364–365
  evolution of different types, 366

functional gradient, 21–23
general morphology, 56ff
and invertebrate ciliated mechanoreceptors, 81ff
lateral line, 7
mammalian cochleae, 265–267
marsupial cochlea, 265–267
Math1 gene, 290–291
mechanical tuning, 17–18
monotreme cochlea, 267ff
number of stereovilli in birds, 239–240
and phylogeny, 56ff
specialization of function, 20–21
spikes in *Carassius* saccule, 62
tuning, 15ff
tuning mechanisms in reptiles, 204–205
types, 7
Hair cell conductances, amphibians, 65
Hair cell development, genes, 270
Hair cell dimensions, basilar papilla of birds, 241ff, 245ff
Hair cell innervation, birds, 23
  lizard basilar papilla, 217
  mammals, 22
Hair cell motility, monotremes, 271
Hair cell number, papillar length in birds, 234ff
Hair cell orientation patterns, amphibian papilla, 188–189
  basilar papilla, 188–189
  lizards, 67ff
Hair cell specializations, reptiles, 206–207
Hair cell tuning, evolution, 15ff
  temperature dependence, 205
Hair cell types, innervation patterns in birds, 244ff
Hearing specializations, amphibians, 166–168
  definition, 31
  Didelphis virginiana, 262
  early mammals, 279–281
  evolution, 27ff
  fishes, 99, 108ff, 176
  high frequency in mammals, 279–281
  high-frequency specializations, 272
  increasing sensitivity, 363–364
  Morganucodon, 154, 279–280

Hearing specializations, amphibians (*continued*)
  origins, 362
  osteolepiforms, 132
  range in mammals, 260ff
  range in marsupials, 261, 262
  range in *Monodelphis*, 261, 262
  range in monotremes, 261, 262–263
  range in *Ornithorhynchus anatinus*, 262–263
  role of periotic canal in amphibians, 179–181
  salamander, 174
  tadpoles, 177–178
  tetrapod ancestors, 130
  underwater by *Xenopus*, 174
Herrings, ultrasonic hearing, 117
Hes1 gene, hair cell development, 270
Higher vocal center (HVc), bird forebrain, 338, 342
High-frequency hearing, advantages, 279–281
  specializations, 272
Homology, defined, 290
  lacking between NL and MSO, 318
Horseshoe bat, *see Rhinolophus rouxi*
Human beings, stream segregation, 29, 50–51
Hyomandibular, conversion to columella, 133
Hypothalamus, and auditory pathway, 336

IC, *see* Inferior colliculus, Auditory midbrain
Ictalurus nebulosus (brown bullhead catfish), hearing sensitivity, 112
Ictalurus punctatus (channel catfish), hearing ability, 113
Impedance, driving point, 387ff
  terminal, 385–387
Impedance matching, amphibians, 173
  auditory system, 48–49
  coming on to land, 364
  evolution, 365–366
  fish ears, 96
Impedance mismatch, 173
Implosions, as sound source, 32
Inferior colliculus, *see also* Auditory midbrain
  evolution and homology, 322ff

IID processing, 320–321
  various species, 325
Infrasound, fish behaviors, 118
Infrasound detection, archosaurs, 225–226, 233–234
Inner ear, *see also* Cochlea, Ear
  Acanthostega, 133–134
  amniote, 8
  development in amphibians, 189–190
  early tetrapods, 133–134
Inner hair cells, 72ff
  marsupials, 265–267
  monotremes, 267ff
Interaural cues, processing in various taxa, 310ff
Interaural intensity differences, processing in birds and mammals, 320ff
Interaural level differences (IID), 310ff
Interaural time differences (ITD), 310ff
  avian vs. mammalian processing, 317
Interaural time processing, archosaurs, 311ff
Interaural time processing, circuits in mammals and birds, 313ff
Isocortex, origins, 338

Japanese lamprey, *see Entosphenus japonicus*
Jawless fishes, hair cells, 63–64
Jeffress, space map, 313
Johnston's organ, *Drosophila*, 80

Kinocilium, 58
Kinocilium bulbs, amphibians, 65

Lagena, dinosaur fossils, 234ff
  hearing in fishes, 108
  origins of cochlea, 278–279
  relationship to basilar papilla, 184–185
Lambdopsalis bulla (a multituberculate mammal), inner ear, 274
Laminar nucleus, torus semicircularis, 326
Lampetra fluviatilis (river lamprey), hair cells, 55, 61, 63–64
  labyrinth, 101–102
Lamprey, 63
  see also Branchiostoma floridae, Entosphenus, Lampetra, Petromyzon

Lateral lemniscus, IID processing, 320–321
  species diversity, 297–298
Lateral line neuromasts, gentamycin, 63
Lateral line system, 55
  hair cells, 7, 61ff
Lateral superior olive, IID processing, 321ff
Latimeria chalumnae (coelacanth), basilar papilla, 7
  ear, 7
  hearing, 130
  phylogeny, 2
Learning, birdsong and human speech, 340ff
Leopard frog, *see Rana pipiens*
Lepidosaurs, lizard evolution, 201–202
Lissamphibia, *see also* Amphibians
  ear, 135–137
Little brown bat, *see Eptesicus fuscus*
Lizards, afferent and efferent synapses, 69
  ear, 13–14
  evolution of basilar papilla, 14, 67ff, 200ff, 214ff
  hair cells, 67ff
  nucleus magnocellularis, 301
  papillar lengths, 217–218
Lobe-fins, *see* Sarcopterygia
Low-frequency hearing, importance for birds, 250
Lumpfish, *see Cyclopterus lumpus*
Lumpsucker, *see Corydoras* sp.
Lungfish, hearing, 130
  see also Ceratodus forsteri

Macromechanics, lizard basilar papilla, 211
Macropodus opercularis (paradise fish), hearing mechanisms, 113–114
Macula communis, *Myxine*, 101–102
Macula neglecta, amphibians, 183–184
  relationship to amphibian papilla, 183–184
  sharks, 104
  see also Papilla neglecta
Mallard duck, *see Anas platyrhynchos*
Mammals, cochlea in extinct species, 272ff
  cochlear function, 263–265
  ear, 15, 256ff
  ear shape in different taxa, 258
  evolution, 257ff
  fossil record, 259–260
  hair cell innervation, 22
  hair cell specialization, 21, 22
  hair cells, 72ff
  hearing in early species, 259, 279–281
  hearing range, 260ff
  middle ear evolution, 153ff
  systematics and phylogeny, 257ff
  unique forebrain mechanisms, 343ff
Mandible, *see* Middle ear
Marine catfish, *see Arius felis*
Marmoset, twitter calls and cortex, 344
Marsupials, cochlear anatomy, 265–267
  cochlear function, 264
  DPOAE, 264–265
  hearing range, 261
  systematics, 257, 259
Masking, fishes, 100
Math1 gene, 290, 295
  hair cell development, 270
Mauthner cell, 117, 296
Mechanical tuning, hair cell, 17–18
Mechanoreceptor cells, evolution, 361–363
Mechanotransduction, defined, 59
Medial geniculate complex, mammals, 299, 338
Medial olivary efferent system, 73
Medial superior olive, and frequency range of hearing, 318–319
  evolution, 317ff
  interaural time processing, 315ff
  and nucleus laminaris, 315ff
  space map, 317
Melopsittacus undulatus (parakeet), critical masking ratio, 225–226
Mesozoic mammals, cochlea, 273ff
Mesozoic therian mammals, inner ear, 275–278
Micromechanical tuning, reptiles, 204ff, 207
Midbrain, evolution and homology, 322ff
  see also Auditory brainstem, Brainstem, Inferior colliculus, Torus semicircularis
Middle ear binaural coupling, 310–311
  amphibians, 173ff
  archosauromorphs, 142–143
  crocodilians, 143

Middle ear binaural coupling *(continued)*
  dinosaur, 142–143
  early turtle, 143–145
  evolution, 128ff
  evolution in amniotes, 137ff
  evolution in amphibians, 176
  evolution in mammals, 155–157
  evolution in neodiapsids, 140ff
  evolution in pelycosaurs, 151
  evolution in stem mammals, 153–155
  evolution in therapsids, 152
  evolution in theriodonts, 152–153
  evolution of three ossicles, 146ff
  frequency range of hearing in birds, 233
  Gymnophiona, 174–175
  lizards, 201
  monotreme, 155, 156
  Morganucodon, 150, 153–154
  periotic canal in amphibians, 179–181
  pressure reception, 178–179
  salamander, 174
  sound production in *Rana catesbeiana*, 174–175
  spenodontids evolution, 141
  squamate evolution, 141–142
  synapsids, 145ff
  theories of evolution, 145ff
  turtle, 10
Midshipman fish, *see Porichthys notatus*
Monodelphis domestica (short-tailed opossum), cochlear function, 264
  cochlea in Mesozoic species, 273–275
  cochlear amplifier, 270–271
  cochlear anatomy, 267ff
  cochlear function, 264
  hair cells, 72
  hearing range, 261, 262
  middle ear, 155, 156
  outer hair cells, 271
  pillar cells, 269–270
  stereovilli, 272
  systematics, 257, 259
Morganucodon (primitive mammal), binaural processing, 319
  cochlea, 154, 250
  hearing capabilities, 154, 279–280
  inner ear, 273
  middle ear, 150, 153–154

Mormyrids, hearing mechanisms, 113, 115
Motility, outer hair cells, 73
  stereovilli, 69
Motor systems, and auditory midbrain, 328–329
Multituberculates, cochlea, 272–273
Musical instruments, as sound sources, 34–35
Mustached bat, *see Pteronotus parnellii*
Myripristus kuntee (squirrelfish), audiogram, 112, 114
  hearing mechanisms, 110
Myxine glutinosa (hagfish), evolution of ear structures, 100–101
  hair cells, 55
  inner ear, 75–76, 101–102
Myxiniformes (hagfishes), inner ear, 75–76
Myxiniformes, lateral line system, 76

Nautilus, ciliated mechanoreceptors, 79
Near field, sound detection evolution, 98
Neodiapsids, middle ear evolution, 140ff
Neurogenin, auditory CNS, 295
Neutral evolution, lizard basilar papilla, 200ff
NMDA receptors, Type I hair cells, 60
NompC, *Drosophila* mechanosensory channel, 80–81
Norhern tegu lizard, *see Tupinambis nigropuncatus*
Notch pathway, 84
Nucleus angularis, 301
  chopper cells, 308
  IID processing, 320–321
  sauropsids, 300
  type IV response type, 306
Nucleus centralis, auditory midbrain, 323ff
Nucleus laminaris, and superior olive, 311ff
  archosaurs, 298
  evolution, 317ff
Nucleus magnocellularis, sauropsids, 300
Nucleus mesencephalicus lateralis pars dorsalis (MLD), auditory midbrain, 322ff

Numida meliagris (Guinea fowl), midbrain vocalization processing, 332–333

Octaval nerve, 296
Octopus cells, in various taxa, 309
Octopus vulgaris, ciliated mechanoreceptors, 79
Oil bird, *see Steatornis caripensus*
Onset responses, in various taxa, 308–309
Ontogeny, evolution of ear, 189–190
Opossum, *see Didelphis, Monodelphis*
Opsanus beta (toadfish), torus semicircularis, 324
Opsanus tau (oyster toadfish), audiogram, 112
  hair cell physiology, 62–63
Organ of Corti, *see Cochlea*
Ornithorhynchus anatinus (duck-billed platypus), hearing range, 262–263
Orthodenticle (otx) transcription factor, 290–291
Oscar, *see Astronotus ocellatus*
Ostariophysi, 6
  evolution of Weberian apparatus, 128, 130
  hearing mechanisms, 114–115
  labyrinth, 107–108
  relationship to Otophysi, 128
Osteichthyes, cladistics, 2
  labyrinth, 104ff
Osteolepiforms, hearing, 132
Ostracoderms, vestibular labyrinth, 97
Otoacoustic emissions, cochlear amplifier, 20–21, 271–272
  reptiles, 211
  see also DPOAE
Otolith organs, fishes, 61ff
Otoliths, and otoconia, 105–106
  coelacanth, 130
  fish hearing, 104ff
Otophysi, relationships in Ostariophysi, 128
Outer hair cells, 72ff
  currents and channels, 73
  marsupials, 265–267
  monotremes, 267ff, 271
  tuning, 263–265

Paleozoic, tetrapod ears, 132ff
Papilla neglecta, amphibians, 169–170
  see also Macula neglecta
Paradise fish, *see Macropodus opercularis*
Parakeet, *see Melopsittacus undulatus*
Parallel evolution, fish hearing organs, 95ff
Parallel evolution, sound source localization circuits, 309ff
Parareptiles, *see* Turtles
Pars inferior, labyrinth, 170
Pars superior, labyrinth, 169–170
Particle displacements, 380–382
Particle velocity, 380
Passeriformes, evolution, 227
Pax6 gene, 290
Pelycosaurs (basal synapsids), middle ear evolution, 151
Periotic canal, comparative, 179–181
Periotic space, fish and amphibians, 176
Petromyzon fluviatilis (river lamprey), labyrinth, 101–102
Petromyzon marinus (sea lamprey), hair cells, 63
Phylogeny, amniotes, 10
  lizards, 201–202
  mammals, 257ff
  vertebrate, 1ff
Physics of sound, 369ff
Pigeon, *see Columba livia*
Pike perch, *see Stizostedion lucioperca*
Pillar cells, monotremes, 269–270
Pimelodus blochi (Bloch's catfish), muscles on Weberian ossicles, 119
Pimelodus pictus, hearing ability, 113
Platydoras costatus (striped Rahael catfish), audiogram, 112–113
Platypus, *see Ornithorhynchus anatinus*
Podarcis (European wall lizard), tonotopy, 208
Pollimyrus adspersus (weakly electric fish), torus semicircularis, 323–324
Pollimyrus isidori, sound production, 116
Polypterus (bichir), labyrinth, 106–107
Polypterus, sound production, 98
Population codes, auditory space, 340
Porichthys notatus (midshipman), vocal signal processing, 330–331
  torus semicircularis, 324

Posterior forebrain pathway, vocal production, 342
Potassium currents, voltage-activated, 60
Pressure detection by tetrapods, evolution, 176ff, 294ff
Pressure receiver, middle ear, 178–179
Pressure-gradient receiver, anuran ears, 310
Prestin, hair cell motility, 73
Primary-like cells, in various taxa, 309
Primitive mammal cochlea, compared to archosaurs, 249–250
Pseudemys scripta elegans (red-eared slider turtle), basilar papilla, 67
  torus semicircularis, 324–325
Pteronotus parnellii (mustached bat), cortex areas, 344–345
  delay tuning in midbrain, 333–334
  IC frequency map, 327
Pterosaurs, otic region, 143
Pygocentrus nattereri (red-bellied piranha), acoustic behaviors, 117
  audiogram, 112

Rana catesbeiana (bullfrog), hair cells, 64–65
  saccular response, 178
  sound duration processing, 331
  tympanic membrane, 173, 174–175
Rana esculenta (grass frog), hair cell conductances, 65–66
Rana pipiens (leopard frog), DLN cell types, 302ff
  hair cell electrical tuning, 66, 188
  saccular hair cells, 66
  torus semicircularis, 324
Rana sp., periotic canal, 179–181
Ratfish, *see Chimaera monstrosa*
Rays, vestibular hair cells, 63
Red-bellied piranha, *see Pygocentrus nattereri*
Red-eared slider turtle, *see Pseudemys scripta elegans*
Reflected waves, 385–387
Reflection, effects on sound, 35ff
Reptile, auditory midbrain, 324
  basilar papilla, 11ff, 200ff, 206
  cochlear amplification, 271
  evolution of basilar papilla, 213ff
  hair cells, 66ff
  origins, 6
Resonance, acoustics, 393–395
  goldfish saccular hair cells, 62
  hair cells in reptiles, 69
  turtle hair cells, 67, 203ff
Resonances, as sound sources, 33
Rhinolophus rouxi (horseshoe bat), efferent system, 73
Rhipidistia, 2
Ring dove, *see Streptopelia risoria*
River lamprey, *see Lampetra fluviatilis*
Rutilus rutilus (roach), hearing ability, 113
Rynchocephalians (sphenodons), basilar papilla, 66–67

Saccular nerve, goldfish S1 and S2 afferents, 62–63
Saccular otolith, Ostariophysi, 107–108
Saccule, amphibian, 169, 178
  anatomy in amphibians, 182
  electrical tuning in hair cells, 187
  response in *Rana catesbeiana*, 178
  seismic detection in amphibians, 181–182
Salamander, ear, 137, 174
  seismic stimulation of ear, 174
  see also Ambystoma tigrinum, Salamandra salamandra
Salamandra salamandra (fire salamander), torus semicircularis, 322
Sallet, lizard basilar papilla, 213
Salmo salar (Atlantic salmon), audiogram, 112
  infrasound, 118
Sarcopterygia (lobe-fins), cladistics, 2–3
Scaphirhynchus (shovel-nosed sturgeon), labyrinth, 106
Sceloporous occidentalis (fence lizard), hair cells, 68
Sceloporous orcutti (granite spiny lizard), papillar shortening, 217–218
  tonotopy, 208
Scyliorhinus canicula (small-spotted catshark), labyrinth, 103
Sea squirt, *see Borylloides, Botryllus*

Segregation, sound sources, 49–51, 100
Seismic detection, amphibian saccule, 178, 181–182
Seismic stimulation of ear, salamander, 174
Selective pressures, on hearing, 27ff
Sensitivity of hearing, evolution in fishes, 99
Sensory hair cell, see Hair cell
Sharks, labyrinth, 102ff
　sound attraction, 103, 118
　sound transmission to ear, 104
　vestibular hair cells, 63
Short bare-tailed opossum, see Monodelphis domestica
Short hair cells, archosaurs, 69ff, 249
　lacking afferent innervation, 71
Short-beaked echidna, see Tachyglossus aculeatus
Shovel-nose sturgeon, see Scaphirhynchus
Silurus glanis (wels catfish), labyrinth, 105–106
Skinks, hair cell types, 69
Skull, Acanthostega, 131, 132–133
　Eusthenopteron, 131, 132–133
　evolution in amniotes, 137ff
Small-spotted catshark, see Scyliorhinus canicula
Snakes, ear, 13–14
　evolution, 212–213
　hair cells, 67ff
　nucleus angularis, 301
　stapes, 142
Song systems, independent evolution in birds, 343
Songbirds, evolution, 227
　vocalization processing in midbrain, 332–333
Sound, definition, 31
　environmental effects on, 35ff
　pathways in amphibians, 166–168
Sound, physical sources, 32ff
　properties of acoustic waves, 41ff
Sound attraction, fishing, 117–118
Sound pressure receptor, origins and evolution, 182ff
Sound production, fishes, 115–116
　Rana catesbeiana, 175–176
　various fish species, 115

Sound source localization, parallel evolution of circuits, 309ff
Sound source segregation, mechanisms, 44
　see also Stream segregation
Sound sources, liquids, 32ff
Sound sources, solids, 32ff
Sounds of the environment, fishes, 98
Speech, and birdsong, 340ff
Sphenodon (Tuatara "lizard"), 201–202
　basilar papilla, 66–67, 206
　reptile evolution, 226
Sphenodontids, middle ear evolution, 141
Spherical bushy cells, and archosaur NM, 304
Spontaneous activity, preferred intervals in reptiles, 211
Spontaneous otoacoustic emissions, birds, 71
　lizards, 211
Squamates, basilar papilla, 201ff
　hair cells, 66ff
　middle ear evolution, 141–142
Stapes, amphibians, 166–168, 173
　archosauromorphs, 142
　early amniotes, 139, 140ff
　evolution in neodiapsids, 140ff
　origins, 133–135, 137
　snakes, 142
　synapsids, 145ff
　therapsids, 152
Statoliths, lungfish, 130
Steatornis caripensis (oilbird), vocalization, 224–225
Stem mammals, evolution of middle ear, 153–155
Stem reptiles, basilar papilla, 202ff
Stereocilia, see Stereovilli
Stereovilli, 58
　monotremes, 272
Stizostedion lucioperca (pike perch), labyrinth, 105–106
Stream segregation, 28–29, 49–51
　fishes, 100
　human beings, 29
Streptopelia risoria (ring dove), midbrain sensory-motor connections, 329
Strigiformes, evolution, 227
Striola, vestibular epithelia, 61–62

Striped Rahael catfish, *see Platydoras costatus*
Superior colliculus, 298–299, 328–329
Superior olive, binaural interactions, 298
species diversity, 297–298
Swim bladder, 363
hearing in fishes, 109ff
Synapses, bird basilar papilla, 246ff
Synapsids (mammals), ear, 15
ear evolution, 145ff
stapes, 145ff

Tachyglossus aculeatus (short-beaked echidna), kinocilia, 72
Tadpoles, ear and hearing, 177–178
Tall hair cells, archosaurs, 69ff
Tectorial membrane, basilar papilla size, 218
Tectorial structures, reptile basilar papilla, 213
Telencephalon, auditory areas, 299
Temnospondyls, ear, 135–137
Temporal patterns, discrimination by fishes, 99
Temporal processing, midbrain, 329ff
Temporal resolution, in fishes, 99
Tetrapod ancestors, hearing, 130
Tetrapodomorpha, 2
Tetrapods, ear evolution, 132ff, 365–366
ears in Paleozoic, 132ff
evolution of hearing, 128ff
impedance matching, 364
origin of columella, 133
Thalamus, projections from auditory midbrain, 326, 337
Therapsids, middle ear evolution, 152
origin, 148–149
stapes, 152
Therian mammals, Mesozoic inner ear, 275–278
Therians, *see* Mammals
Theriodonts, middle ear, 152–153
Three-spot gourami, *see Trichogaster* sp.
Tiger salamander, *see Aurbystoma tigrinum*
Tiliqua rugosa (bobtail skink), active process, 211
frequency selectivity, 218ff
tonotopy, 208

Time pattern processing, midbrain, 329ff
Tip links, 58–59
Toadfish, *see Opsanus tau*, *Porichthys notatus*, *Opsanus beta*
Tokay gecko, *see Gekko gecko*
Tonotopy, *Carassius* saccule, 62
cochlear nucleus, 301
field-L, 337–338
forebrain, 338–339
lizard basilar papilla, 208, 220
reptile basilar papilla, 213
turtle basilar papilla, 205
Torus semicircularis, auditory midbrain and IC, 298, 322ff
core and belt organization, 326–327
nucleus centralis, 323ff
see also Auditory midbrain
Transducers, passive, 382ff
Transducers and amplifiers, acoustics, 395ff
Transduction channels, stereovilli, 59
Trichogaster trichopterus (three-spot gourami), audiogram, 112ff
Trichopsis vittata (croaking gourami), sound production, 116
Tuatara "lizard," *see Sphenodon*
Tufted duck, *see Aythya fuligula*
Tunicates, *see* Urochordates
Tupinambis nigropuncatus (northern tegu lizard), tonotopy in IC, 327
Turbulence, as sound source, 32
Turtle
basilar papilla, 10–11, 202ff
hair cell currents and channels, 63ff, 203
middle ear, 10, 143–145
nucleus angularis, 301
nucleus laminaris, 318
nucleus magnocellularis, 300–301
reptile evolution, 205, 226
Turtle hair cells, active processes, 211
Tympanic ear, evolution, 205ff, 366
Tympanic membrane, amphibians, 173
Rana catesbeiana, 173, 174–175
Tympanum, fishes, 118–119
sound production in *Rana catesbeiana*, 174–175
Xenopus, 173–174

Type I, hair cells, 59–60
Type II, hair cells, 59
Type IV cell type, birds and mammals, 308
Tyto alba (barn owl), auditory midbrain motor systems, 328–329
  basilar papilla specializations, 70, 226, 232–233, 238ff
  frequency map, 228ff
  IID processing, 320
  ITD processing, 313ff
  nucleus laminaris, 312
  octaval column cell types, 304ff
  papillar length and hair cell number, 236ff
  space-tuned neurons in forebrain, 338

Ultrasound, hearing in herrings, 99
Underwater hearing, amphibians, 164, 174
Unidirectional-type hair cells, reptiles, 169
Urochordates, ciliated mechanoreceptors, 77ff
  cupular organs, 77–78
Urodeles, hair cells, 65ff

Vertebrate, cladogram, 3
  on geological time scale, 5
  phylogeny, 1ff
  taxa hierarchy, 4
Vertebrate hair cells, summary, 73–74
Vestibular epithelia, 59ff
Vestibular labyrinth, evolution of hearing, 96ff
Violin, sound production, 34
Virginia opossum, *see Didelphis virginiana*
Vocalization, and evolution of hearing, 328–329
  archosaurs, 224ff
  pathways in birds and mammals, 340–341
  sound processing in midbrain, 331–332

Weberian ossicles, 107ff, 128, 130
  extirpation, 113

Xenopus laevis (African clawed frog), ear, 173ff
  auditory midbrain organization, 326
  periotic canal, 179–181
  underwater hearing, 174

SPRINGER HANDBOOK OF AUDITORY RESEARCH *(continued from page ii)*

Volume 22: Evolution of the Vertebrate Auditory System
Edited by Geoffrey A. Manley, Arthur N. Popper and Richard R. Fay

For more information about the series, please visit www.springer-ny.com/shar.

DATE DUE	
~~MAY 1 1 2007~~	
~~RET'D NOV 1~~	~~6 2017~~

DEMCO INC 38-2971